Biointegration of Med Materials

Woodhead Publishing Series in Biomaterials

Biointegration of Medical Implant Materials

Second Edition

Edited by
Chandra P. Sharma

WOODHEAD PUBLISHING
ELSEVIER An imprint of Elsevier

Woodhead Publishing is an imprint of Elsevier
The Officers' Mess Business Centre, Royston Road, Duxford, CB22 4QH, United Kingdom
50 Hampshire Street, 5th Floor, Cambridge, MA 02139, United States
The Boulevard, Langford Lane, Kidlington, OX5 1GB, United Kingdom

Copyright © 2020 Elsevier Ltd. All rights reserved.

No part of this publication may be reproduced or transmitted in any form or by any means, electronic or mechanical, including photocopying, recording, or any information storage and retrieval system, without permission in writing from the publisher. Details on how to seek permission, further information about the Publisher's permissions policies and our arrangements with organizations such as the Copyright Clearance Center and the Copyright Licensing Agency, can be found at our website: www.elsevier.com/permissions.

This book and the individual contributions contained in it are protected under copyright by the Publisher (other than as may be noted herein).

Notices
Knowledge and best practice in this field are constantly changing. As new research and experience broaden our understanding, changes in research methods, professional practices, or medical treatment may become necessary.

Practitioners and researchers must always rely on their own experience and knowledge in evaluating and using any information, methods, compounds, or experiments described herein. In using such information or methods they should be mindful of their own safety and the safety of others, including parties for whom they have a professional responsibility.

To the fullest extent of the law, neither the Publisher nor the authors, contributors, or editors, assume any liability for any injury and/or damage to persons or property as a matter of products liability, negligence or otherwise, or from any use or operation of any methods, products, instructions, or ideas contained in the material herein.

Library of Congress Cataloging-in-Publication Data
A catalog record for this book is available from the Library of Congress

British Library Cataloguing-in-Publication Data
A catalogue record for this book is available from the British Library

ISBN: 978-0-08-102680-9

For information on all Woodhead Publishing publications visit our website at https://www.elsevier.com/books-and-journals

Publisher: Matthew Deans
Acquisition Editor: Sabrina Webber
Editorial Project Manager: Emma Hayes
Production Project Manager: Joy Christel Neumarin Honest Thangiah
Cover Designer: Christian J. Bilbow

Typeset by TNQ Technologies

Contents

Contributors		xi
Preface		xv

1　Biointegration: an introduction　　1
Sunita Prem Victor, C.K.S. Pillai and Chandra P. Sharma

1.1	Introduction	1
1.2	Biointegration of biomaterials for orthopedics	2
1.3	Biointegration of biomaterials for dental applications	7
1.4	AlphaCor artificial corneal experience	8
1.5	Biointegration and functionality of tissue engineering devices	9
1.6	Percutaneous devices	10
1.7	Future trends	11
	References	12

Part One　Soft tissue biointegration　　17

2　Interface biology of stem cell–driven tissue engineering: concepts, concerns, and approaches　　19
Soumya K. Chandrasekhar, Finosh G. Thankam,
Devendra K. Agrawal and Joshi C. Ouseph

2.1	Introduction	19
2.2	Stem cells for tissue engineering	21
2.3	Mesenchymal stem cells in a nutshell	23
2.4	Mesenchymal stem cell action in wound healing	23
2.5	Biomaterials in stem cell–based soft tissue engineering	26
2.6	Influence of scaffold patterns in stem cell behavior	29
	References	37

3	Replacement materials for facial reconstruction at the soft tissue—bone interface	45

E. Wentrup-Byrne, Lisbeth Grøndahl and A. Chandler-Temple

3.1	Introduction	45
3.2	Facial reconstruction	48
3.3	Materials used in traditional interfacial repair	53
3.4	Surface modification of facial membranes for optimal biointegration	64
3.5	Future trends	70
	Acknowledgments	70
	References	70

4	Tissue engineering of small-diameter vascular grafts	79

Kiran R. Adhikari, Bernabe S. Tucker and Vinoy Thomas

4.1	Background	79
4.2	Clinical significance	80
4.3	Tissue engineering approach	81
4.4	Surface modification approach	85
4.5	Cellular interactions	88
4.6	Emerging perspectives	94
4.7	Conclusion	96
	References	96

5	Clinical applications of mesenchymal stem cells	101

Rani James, Namitha Haridas and Kaushik D. Deb

5.1	Introduction	102
5.2	Mesenchymal stem cells	104
5.3	Sources of mesenchymal stem cells	104
5.4	Properties of mesenchymal stem cells	106
5.5	Clinical applications of mesenchymal stem cells	107
5.6	Stem cell banking	111
	References	112
	Further reading	116

Part Two   Tissue regeneration   117

6	Cardiac regeneration	119

Raghav Murthy and Aditya Sengupta

6.1	Introduction	120
6.2	The controversy	120
6.3	Mechanisms	121
6.4	Stem cell therapies	125
6.5	Barriers in stem cell therapy	127

6.6	Tissue engineering	**130**
6.7	Cellular reprogramming	**132**
6.8	Stem cell−derived exosomes and small vesicles	**132**
6.9	Hydrogels	**132**
6.10	Cardiac regeneration in children	**133**
6.11	Valves	**133**
6.12	Biointegration	**136**
6.13	Conclusion	**138**
	References	**138**

7 Tissue-based products — **145**
Umashankar P.R. and Priyanka Kumari

7.1	Introduction	**145**
7.2	Acellular tissue products	**147**
7.3	Chemically cross-linked tissue products	**158**
7.4	Tissue-derived products	**159**
7.5	Host response to tissue products	**171**
7.6	Sterilization of tissue-based/tissue-derived products	**174**
7.7	Risk management of tissue-based products	**176**
7.8	Conclusion	**177**
	References	**177**
	Further reading	**185**

8 Tendon Regeneration — **187**
Jeffery D. St. Jeor, Donald E. Pfeifer and Krishna S. Vyas

8.1	Tendon cells and composition	**188**
8.2	Internal architecture	**189**
8.3	Importance of the complex three-dimensional structure	**190**
8.4	Tendon to bone insertion	**191**
8.5	Pure dense fibrous connective tissue	**192**
8.6	Uncalcified fibrocartilage	**192**
8.7	Tidemark	**192**
8.8	Calcified fibrocartilage	**193**
8.9	Bone	**193**
8.10	Supporting structures	**193**
8.11	Blood supply	**194**
8.12	Biomechanical properties	**195**
8.13	Impacting factors	**196**
8.14	Effects of aging	**197**
8.15	Effects of exercise	**198**
8.16	Effects of immobilization	**199**
8.17	Tendon injury	**199**
8.18	Types of injury	**200**

	8.19	Tendon healing	200
	8.20	Mechanisms of healing	201
	8.21	Surgical intervention	202
	8.22	Tendon regeneration	202
	8.23	Utilization of growth factors in tendon healing	203
	8.24	Stem cell−based approaches to tendon healing	207
	8.25	The role of biologic and synthetic scaffolds in tendon healing	210
	8.26	The role of gene transfer in tendon healing	213
	8.27	Future of tendon regeneration	214
		References	214
9	**Integration of dental implants: molecular interplay and microbial transit at tissue−material interface**		221
	Smitha Chenicheri and Remya Komeri		
	9.1	Evolution of the concept of biointegration of dental implants	222
	9.2	Mechanisms of biointegration of dental implants	223
	9.3	Establishing biological gingival seal	223
	9.4	Early inflammatory phase	224
	9.5	Neovascularization at peri-implant zone	225
	9.6	Osteoconduction	225
	9.7	Soft tissue healing and biointegration	227
	9.8	Cell signaling and integration of dental implants	228
	9.9	Genetic networks in osseointegration	229
	9.10	Microbial interplay in osseointegration of dental implants	232
	9.11	Interface biofilms: a unique pulpit for microbial homing	234
	9.12	Implant failure and enhancement of biointegration	236
	9.13	ECM disorganization	237
	9.14	Microbial versus host cell signaling at the interface	237
	9.15	Conclusions	239
		References	240
10	**Biointegration of bone graft substiutes from osteointegration to osteotranduction**		245
	F.B. Fernandez, Suresh S. Babu, Manoj Komath and Harikrishna Varma		
	10.1	Introduction	245
	10.2	Bone, the hard tissue	246
	10.3	Bone grafts	246
	10.4	Synthetic bone graft substitutes	247
	10.5	Biointegration of synthetic bone graft substitutes	251
	10.6	Conclusion	257
		References	257

11	**Stem cell–based therapeutic approaches toward corneal regeneration**	263
	Balu Venugopal, Bernadette K. Madathil and Anil Kumar P.R.	
	11.1 Introduction	264
	11.2 Corneal blindness and current therapies	268
	11.3 Other cell-based approaches—nonlimbal sources	273
	11.4 Biomaterials in corneal reconstruction	278
	11.5 Translational and clinical perspective	286
	References	286

Part Three   Drug delivery   295

12	**Biocompatibility of materials and its relevance to drug delivery and tissue engineering**	297
	Thomas Chandy	
	12.1 Biocompatibility of materials and medical applications	298
	12.2 Biomaterials for controlled drug delivery	304
	12.3 Biomaterials for tissue engineering and regenerative medicine	310
	12.4 Role of scaffold and the loaded drug/growth factor in the integration of extracellular matrix and cells at the interface	315
	12.5 Future outlook on combination devices with drug delivery and tissue engineering	321
	References	322
13	**Inorganic nanoparticles for targeted drug delivery**	333
	Willi Paul and Chandra P. Sharma	
	13.1 Introduction	333
	13.2 Calcium phosphate nanoparticles	337
	13.3 Gold nanoparticles	349
	13.4 Iron oxide nanoparticles	355
	13.5 Conclusion	362
	13.6 Biointegration concept and future perspective	363
	Acknowledgments	363
	References	364
14	**Applications of alginate biopolymer in drug delivery**	375
	Lisbeth Grøndahl, Gwendolyn Lawrie, A. Anitha and Aparna Shejwalkar	
	14.1 Introduction	375
	14.2 Alginate biopolymer	376
	14.3 Drug delivery using alginate matrices	384
	14.4 Concluding remarks and future directions	395
	Acknowledgments	395
	References	396

Part Four Design considerations   405

15 Failure mechanisms of medical implants and their effects on outcomes   407
A. Kashi and S. Saha
- 15.1 Introduction   407
- 15.2 Manufacturing deficiencies   410
- 15.3 Mechanical factors (e.g., fatigue, overloading, and off-axis loading)   410
- 15.4 Wear   413
- 15.5 Corrosion   416
- 15.6 Clinical factors for implant success and failure   417
- 15.7 Failure mechanisms of non–load-bearing implants   418
- 15.8 Failure analysis of medical implants   420
- 15.9 Multivariate analysis   422
- 15.10 Ethical issues   423
- 15.11 Conclusion   424
- References   426

16 Biointegration of three-dimensional–printed biomaterials and biomedical devices   433
Vamsi Krishna Balla, Subhadip Bodhak, Pradyot Datta, Biswanath Kundu, Mitun Das, Amit Bandyopadhyay and Susmita Bose
- 16.1 Introduction   434
- 16.2 Metallic implants via three-dimensional printing   438
- 16.3 Bioceramic scaffolds using three-dimensional printing   446
- 16.4 Bioprinting   455
- 16.5 Current challenges and future directions   467
- 16.6 Summary   467
- References   468

Index   483

Contributors

Kiran R. Adhikari Polymers & Healthcare Materials/Devices, Department of Materials Science & Engineering, University of Alabama at Birmingham, Birmingham, AL, United States

Devendra K. Agrawal Clinical & Translational Sciences, Creighton University, Omaha, NE, United States

A. Anitha School of Chemistry and Molecular Biosciences, The University of Queensland, Brisbane, QLD, Australia

Suresh S. Babu Division of Bioceramics, Department of Biomaterial Sciences and Technology, Biomedical Technology Wing, Sree Chitra Tirunal Institute for Medical Sciences and Technology, Trivandrum, Kerala, India

Vamsi Krishna Balla Bioceramics & Coating Division, CSIR-Central Glass & Ceramic Research Institute, Kolkata, West Bengal, India; Department of Mechanical Engineering, University of Louisville, Louisville, KY, United States

Amit Bandyopadhyay School of Mechanical and Materials Engineering, Washington State University, Pullman, WA, United States

Subhadip Bodhak Bioceramics & Coating Division, CSIR-Central Glass & Ceramic Research Institute, Kolkata, West Bengal, India

Susmita Bose School of Mechanical and Materials Engineering, Washington State University, Pullman, WA, United States

A. Chandler-Temple The University of Queensland, St Lucia, QLD, Australia

Soumya K. Chandrasekhar Department of Zoology, KKTM Govt. College, Pullut, Calicut University, Kerala, India; Department of Zoology, Christ College, Calicut University, Iringalikkuda, Kerala, India

Thomas Chandy Philips Medisize LLC, Hudson, WI, United States

Smitha Chenicheri Microbiology Division, Biogenix Research Center for Molecular Biology and Applied Sciences, Thiruvananthapuram, Kerala, India; Department of Microbiology, PMS College of Dental Science and Research, Thiruvananthapuram, Kerala, India

Mitun Das Bioceramics & Coating Division, CSIR-Central Glass & Ceramic Research Institute, Kolkata, West Bengal, India

Pradyot Datta Bioceramics & Coating Division, CSIR-Central Glass & Ceramic Research Institute, Kolkata, West Bengal, India

Kaushik D. Deb DiponEd Biointelligence, Bangalore, India

F.B. Fernandez Division of Bioceramics, Department of Biomaterial Sciences and Technology, Biomedical Technology Wing, Sree Chitra Tirunal Institute for Medical Sciences and Technology, Trivandrum, Kerala, India

Lisbeth Grøndahl School of Chemistry and Molecular Biosciences, The University of Queensland, Brisbane, QLD, Australia

Namitha Haridas DiponEd Biointelligence, Bangalore, India

Rani James DiponEd Biointelligence, Bangalore, India

A. Kashi Private Practice, Rochester, NY, United States

Manoj Komath Division of Bioceramics, Department of Biomaterial Sciences and Technology, Biomedical Technology Wing, Sree Chitra Tirunal Institute for Medical Sciences and Technology, Trivandrum, Kerala, India

Remya Komeri Regional Cancer Center, Division of Biopharmaceuticals and Nanomedicine, Thiruvananthapuram, Kerala, India

Priyanka Kumari Bioemedical Technology Wing, Sree Chitra Tirunal Institute for Medical Sciences and Technology, Trivandrum, Kerala, India

Biswanath Kundu Bioceramics & Coating Division, CSIR-Central Glass & Ceramic Research Institute, Kolkata, West Bengal, India

Gwendolyn Lawrie School of Chemistry and Molecular Biosciences, The University of Queensland, Brisbane, QLD, Australia

Bernadette K. Madathil Division of Tissue Culture, Department of Applied Biology, Biomedical Technology Wing, Sree Chitra Tirunal Institute for Medical Sciences and Technology, Thiruvananthapuram, Kerala, India

Raghav Murthy Assistant Professor, Assistant Director of Pediatric Cardiac Transplantation, Department of Cardiovascular Surgery, Mount Sinai Hospital, Icahn School of Medicine at Mount Sinai, New York, NY, United States

Joshi C. Ouseph Department of Zoology, Christ College, Calicut University, Iringalikkuda, Kerala, India

Umashankar P.R. Bioemedical Technology Wing, Sree Chitra Tirunal Institute for Medical Sciences and Technology, Trivandrum, Kerala, India

Anil Kumar P.R. Division of Tissue Culture, Department of Applied Biology, Biomedical Technology Wing, Sree Chitra Tirunal Institute for Medical Sciences and Technology, Thiruvananthapuram, Kerala, India

Contributors

Willi Paul Sree Chitra Tirunal Institute for Medical Sciences and Technology, Thiruvananthapuram, Kerala, India

Donald E. Pfeifer Mayo Clinic School of Medicine, Rochester, MN, United States

C.K.S. Pillai Sree Chitra Tirunal Institute for Medical Sciences and Technology, Thiruvananthapuram, Kerala, India

S. Saha University of Washington, WA, Seattle, United States

Aditya Sengupta Cardiothoracic Surgery Resident, Department of Cardiovascular Surgery, Mount Sinai Hospital, Icahn School of Medicine at Mount Sinai, New York, NY, United States

Chandra P. Sharma Sree Chitra Tirunal Institute for Medical Sciences and Technology, Thiruvananthapuram, Kerala, India

Aparna Shejwalkar School of Chemistry and Molecular Biosciences, The University of Queensland, Brisbane, QLD, Australia

Jeffery D. St. Jeor Mayo Clinic School of Medicine, Rochester, MN, United States

Finosh G. Thankam Clinical & Translational Sciences, Creighton University, Omaha, NE, United States

Vinoy Thomas Polymers & Healthcare Materials/Devices, Department of Materials Science & Engineering, University of Alabama at Birmingham, Birmingham, AL, United States

Bernabe S. Tucker Polymers & Healthcare Materials/Devices, Department of Materials Science & Engineering, University of Alabama at Birmingham, Birmingham, AL, United States

Harikrishna Varma Division of Bioceramics, Department of Biomaterial Sciences and Technology, Biomedical Technology Wing, Sree Chitra Tirunal Institute for Medical Sciences and Technology, Trivandrum, Kerala, India; Head, Biomedical Technology Wing, Sree Chitra Tirunal Institute for Medical Sciences and Technology, Trivandrum, Kerala, India

Balu Venugopal Division of Tissue Culture, Department of Applied Biology, Biomedical Technology Wing, Sree Chitra Tirunal Institute for Medical Sciences and Technology, Thiruvananthapuram, Kerala, India

Sunita Prem Victor Sree Chitra Tirunal Institute for Medical Sciences and Technology, Thiruvananthapuram, Kerala, India

Krishna S. Vyas Department of Plastic Surgery, Mayo Clinic, Rochester, MN, United States

E. Wentrup-Byrne Queensland University of Technology, Brisbane, Australia

Preface

The first edition of "Biointegration of Medical Implant Materials - Science and Design" has been very much accepted by the scientific community; therefore, the second edition of the book is developed with the same theme of advancing our understanding of the interfacial interaction and integration of implant materials with hard/soft tissue. This edition contains four parts—Part 1: Soft tissue biointegration, Part 2: Tissue regeneration, Part 3: Drug delivery, and Part 4: Design considerations.

Each part contains several chapters by experts with the latest advanced information in the respective specialization. This is expected to be an excellent reference book for graduate students, researchers, biomedical engineers, and their industrial partners who are interested in translational and advanced research in their area of specialty.

I thank all the authors for their excellent contribution and Ms. Emma Hayes for her very effective coordination of this project.

I also very much appreciate and thank my wife Aruna Sharma for her sustained support during the course of this project.

Chandra P. Sharma

Biointegration: an introduction

Sunita Prem Victor, C.K.S. Pillai, Chandra P. Sharma
Sree Chitra Tirunal Institute for Medical Sciences and Technology, Thiruvananthapuram, Kerala, India

Chapter outline

1.1 Introduction 1
1.2 Biointegration of biomaterials for orthopedics 2
1.3 Biointegration of biomaterials for dental applications 7
1.4 AlphaCor artificial corneal experience 8
1.5 Biointegration and functionality of tissue engineering devices 9
1.6 Percutaneous devices 10
1.7 Future trends 11
References 12

1.1 Introduction

Biointegration has been defined as an interconnection between a biomaterial and the recipient tissue at the microscopic level. This implant-to-tissue interface could be a distinct boundary or a zone of interaction between the corresponding biomaterial and tissue. So when biomaterials are designed, a set of properties are built in such a way as to ensure that, after implantation, they will assist the body to heal itself. Thus, it is of critical importance that these materials are integrated into organ-specific repair mechanisms such as the physiologic process required for the biologization of implants (Amling et al., 2006). It should further involve a direct structural and functionally stable connection between the living part and the surface of an implant. Although various materials have been developed in recent years with enhanced physical, surface, and mechanical properties, the use of these materials in certain biological applications is often limited by minimal tissue integration. The rate of implant failures and economic impacts are slowly being documented (Carr et al., 2012; Zhang et al., 2011). This takes us back to the query on how biomaterials could be converted to "living tissues" after implantation. Biomaterials are defined as "materials intended to interface with biological systems to evaluate, treat, augment, or replace any tissue, organ, or function of the body" (Williams, 1999). To cite an example, the bonding of hydroxyapatite (HA) to bone, which is considered as a true case of biointegration, is thought to involve a direct biochemical bond of the bone to the surface of an implant at the electron microscopic level and is independent of any mechanical interlocking mechanism (Cochran, 1996; Meffert et al., 1987). To this end, several groups working on various aspects of the design, development, and application of improved devices are concerned with how the physical, chemical, and biological properties of materials can be tuned to integrate with soft and hard tissues in the human body (Kim et al., 2013; Yu et al., 2013; Patel et al., 2010).

1.2 Biointegration of biomaterials for orthopedics

The biological interface between an orthopedic implant and the host tissue results from a direct, structural, and functional connection between ordered, living bone and the surface of a load-carrying implant (Branemark et al., 1985). Desired effects include osteoinduction, vascularization, and osseointegration, leading to augmented mechanical stability (Svensson et al., 2013; Jain and Kapoor, 2015). Orthopedic research is developing and advancing at a rapid pace as latest techniques are being applied in the repair of musculoskeletal tissues (Jain and Kapoor, 2015). The discovery of biological solutions to important problems, such as fracture-healing, soft tissue repair, osteoporosis, and osteoarthritis, continues to be an important research focus. In orthopedic applications, there is a significant need and demand for the development of bone substitutes that are bioactive and exhibit material properties (mechanical and surface) comparable with those of natural, healthy bone. Conventional biomaterials including ceramics, polymers, metals, and composites have limitations involving suitable cellular responses resulting in diminished osteointegration (Balasundaram and Webster, 2006; Barrere et al., 2008). Numerous strategies are now being envisaged to prevent potential drawbacks such as inflammation, infection, and fibrous encapsulation resulting in implant loosening with deleterious clinical effects (Hacking et al., 2000). For instance, the surfaces of orthopedic implants have dramatically evolved from solid to porous structures in an effort to increase vascularization and promote optimal and unparalleled biological fixation (www.azom, 2004; http://academic.uprm.edu).

Several metallic materials based on titanium, tantalum, cobalt, stainless steel, and magnesium have been widely used for orthopedic applications (Dabrowski et al., 2010; Jafari et al., 2010; Babis and Mavrogenis, 2014). Of these, titanium-based materials in particular are considered for load-bearing applications, which include pin structure, and the fabrication of plates and femoral stems. Titanium's ability to integrate with bone has been known for quarter of a century, and it forms the basis of orthopedic research and implant technology today (Meneghini et al., 2010). These titanium implant surfaces are subjected to various surface treatments and are designed to exhibit varying roughness to promote osseointegration between the implant and host tissue (Barbas et al., 2012). Roughness was evaluated by measuring root mean square (RMS) values and RMS/average height ratio, in different dimensional ranges, varying from 100 microns square to a few hundreds of nanometers. The results showed that titanium presented a lower roughness than the other materials analyzed, frequently reaching statistical significance (Covani et al., 2007). In particular, this study focuses

Biointegration: an introduction

attention on AP40 and especially RKKP, which proved to have a significant higher roughness at low dimensional ranges. This determines a large increase in surface area, which is strongly connected with osteoblast adhesion and growth in addition to protein adsorption. In addition, roughened titanium surfaces augment focal contacts for cellular adherence and in turn enhance cytoskeletal assembly and membrane receptor organization (Wall et al., 2009; Borsari et al., 2005; Stevens and George, 2005).

One should mention here the famous osseointegration concept evolved by Per Ingvar Brånemark, closely coupled with the design of a cylindrical titanium screw (Albrektsson et al., 1981; Branemark et al., 1985) having a specific surface treatment to enhance its bioacceptance (Adell et al., 1990). The titanium screw underwent many animal and, subsequently, human clinical trials to test the concept, design, and success rate associated with the implant. A fixture is believed to have osseointegrated if it provides a stable and apparently immobile support of prosthesis under functional loads, without pain, inflammation, or loosening. Direct bonding of orthopedic biomaterials with collagen is rarely considered; however, several noncollagenic proteins have been shown to adhere to biomaterial surfaces (Rey, 1981). According to Yazdani et al. (2018), osseointegration involves close proximity between implant and bone with no collagen or fibrous tissue, and biointegration represents a continuity of implant to bone without intervening space (Fig. 1.1).

Osseointegration of an implant is thus defined as a direct structural and functionally stable connection between living bone and the surface of an implant that is exposed to mechanical load (http://www.dental-oracle.org). This integration of the implant to bone essentially involves two processes: interlocking with bone tissue and chemical interactions with the surrounding bone constituents. The two processes in turn depend on the chemical composition and the morphology of the surface of the implant. Irradiation with laser light (Nd:YAG [$\lambda = 1064$ nm, $\tau = 100$ ns]) is used for the surface

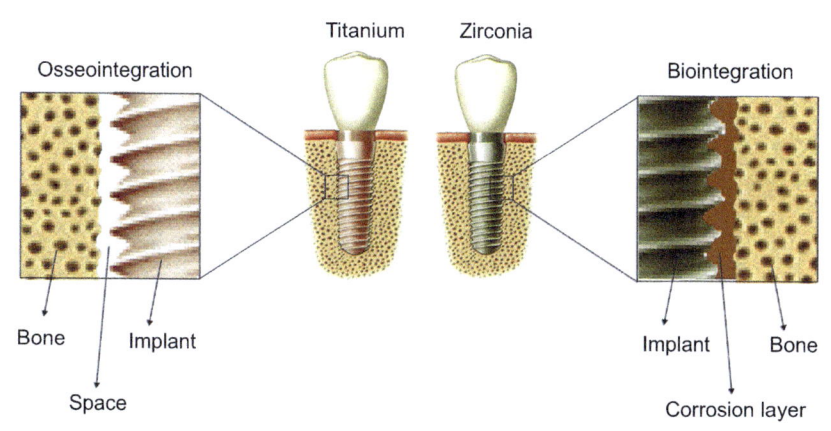

Figure 1.1 Osseointegration versus biointegration.
Reproduced from Yazdani, J., Ahmadian, E., Sharifi, S., Shahi, S., Maleki Dizaj, S., 2018. A short view on nanohydroxyapatite as coating of dental implants. Biomed. Pharmacother. 105, 553–557 with permission from Elsevier Paris.

modification of Ti-6Al-4V—which is widely used in implantation to enhance biointegration (Mirhosseini et al., 2007). Conventional sputtering techniques have shown some advantages over the commercially available plasma spraying method for generating HA films on metallic substrates to promote osteoconduction and oesteintegration; however, the as-sputtered films are usually amorphous, which can cause some serious adhesion problems when postdeposition heat treatment is necessitated. Nearly stoichiometric, highly crystalline HA films strongly bound to the substrate were obtained by an opposing radio frequency magnetron sputtering approach. HA films have been widely recognized for their biocompatibility and utility in promoting biointegration of implants in both osseous and soft tissue (Hong et al., 2007). Oudadesse et al. (Oudadesse et al., 2007) studied the in vitro behavior of com pounds in contact with simulated body fluid (SBF) and in vivo experiments in a rabbit's thigh bones. The inductively coupled plasma optical emission spectroscopy method was used to study the eventual release of Al from composites to SBF and evaluate the chemical stability of composites characterized by the succession of SiO_4 and AlO_4 tetrahedra. The results obtained show the chemical stability of composites. At the bone—implant interface, the intimate links revealed the high quality of the biointegration and the bioconsolidation between composites and bony matrix. Histological studies confirmed good bony bonding and highlighted the total absence of inflammation or fibrous tissues, indicating good biointegration (Oudadesse et al., 2007). Zanotti and Verlicchi (2006) proposed a bioglass alumina spacer that could perform an excellent arthrodesis by a mechanical stabilization (primary one) and a biological biomimetic stabilization (biointeraction, biointegration, and biostimulation). Other advantages are easy use, even in the low somatic interspaces (C7-D1 type), reduction of convalescence, and reduced costs of this type of device. That makes it interesting even in societies with low economic well-being (Zanotti and Verlicchi, 2006). Porous alumina has also been used as a bone spacer to replace large sections of bone that have been removed due to diseased conditions (http://academic.uprm.edu).

In addition to promoting interlocking with bone tissue, understanding pore parameters such as porosity, pore size, and pore interconnectivity are of utmost importance when designing porous metal orthopedic implants. The percentage of porosity and construct geometry controls the surface area available for cell adhesion and vascularization (Jones et al., 2009). Pore sizes influence and determine the types of cells that are likely to penetrate the implant. As examples, osteoid tissue grows into pores of size 75−100 μm; mineralized bone tissue penetrates pores of ∼ 100 μm; and fibrous tissue grows into pore sizes of 10−75 μm. To obtain optimal and efficient bone infiltration, a pore size between 150 and 500 μm is recommended. Also the potential for effective osseointegration is controlled by optimal perfusion of oxygen and nutrients through the newly grown tissue, which is augmented by open pore interconnectivity (Raggatt and Partridge, 2010).

The process by which these materials are integrated into the organ-specific repair cascade is termed "bone remodeling," which is the concerted interplay of two cellular activities: osteoclastic bone resorption and osteoblastic bone formation. The latter physiologic process not only maintains bone mass, skeletal integrity, and skeletal function but also is the cellular process that controls structural and functional integration of bone substitutes. A molecular understanding of this process is therefore of paramount importance for almost all aspects of the body's reaction to biomaterials and might help us to understand, at least in part, the success or the failure of various materials used as bone substitutes. Recent genetic studies have demonstrated that there is no tight cross-control of bone formation and bone resorption in vivo and that there is also a central axis controlling bone remodeling, radically enhancing our understanding of this process. Amling et al. (2006) demonstrated how an understanding of the physiologic process in bone remodeling could provide a platform for the design, development, and application of improved biomaterials.

Many studies have been reported on the biointegration of orthopedic devices. Typically, with regard to orthopedic devices, the primary concerns are wear, infection, and failure of biointegration (Harris and Richards, 2006; Viceconti et al., 2004). The survival rate of an implant under optimal conditions is at least 96% after 5 years. Zim reported the development of a highly porous expanded polytetrafluoroethylene that has been fabricated to provide a softer feel with less shrinkage and migration because of better biointegration and cellular ingrowth (Zim, 2004). Long-term results with porous polyethylene (PE) have demonstrated superior biocompatibility and minimal complications. In addition, HA films have been widely recognized for their biocompatibility and utility in promoting biointegration of implants in both osseous and soft tissue. In this regard, it was observed that a HA coating employed for surface improvement resulted in rapid osseointegration and biointegration in 4 weeks with 90% implant−bone contact observed at 10 months in contrast to titanium alone, which required 10 weeks to be osseointegrated with 50% implant−bone contact at 10 months. Similarly in a study on HA-coated (by electroplating) commercially pure titanium (cpTi) implants, Badr and El Hadary (2007) showed the formation of recognizable osseointegration of bone regeneration with denser bone trabeculae and concluded that electroplating provided a thin and uniform pure crystalline HA coating, which assisted in more bone formation. The characterization of the precipitated HA film is promising for clinically successful long-term bone fixation. On parallel lines, an AFM analysis of roughness on seven materials widely used in bone reconstruction carried out by Covani et al. (2007) showed that the biointegration properties of bioactive glasses can also give an answer in terms of surface structures on how chemical composition can directly (e.g., with chemical exchanges and development of specific surface electrical charge) and indirectly, via the properties induced on tribological behavior that expresses itself during the smoothing of the surfaces, influence the biological system (Covani et al., 2007). Another material, osteopontin, an extracellular glycosylated bone phosphoprotein with a polypeptide back bone, has also been noticed to make dead metal "come alive." Surrounding cells "do not see an inert piece of metal, they see a protein and it's a protein they know." Other novel innovations include the development of nanostructure materials and diagnostic techniques for both in vitro and in vivo applications (Navamar, 2003). HA cement has been associated with an immunoguided delayed inflammatory reaction that leads to thinning of the overlying skin and exposure of the implant. Applications of distraction osteogenesis are rapidly expanding and include deformities of the mandible, midface, and cranium. There has been a trend toward the use of internal hardware, and internal devices are being developed to deliver a greater degree of vector control. Biodegradable devices have been developed to eliminate the second surgical procedure necessary for hardware removal. In the future, successful tissue engineering could eliminate many of the drawbacks associated with implants and osteotomies. Further, the ability to stimulate stem cells to generate autogenous bone has been demonstrated. Computer technology has been successfully used to integrate laser surface scanning and digitizing with computer-aided design and manufacturing to produce facial prostheses. Technological advances in biomaterials, computer modeling, and tissue engineering continue to supply the surgeon's repertoire with improved methods to augment and restore the craniomaxillofacial skeleton. A recent study by Noushin et al. (Nasiri et al., 2018) demonstrated that extracellular nano- and microstructural hierarchy could provide microenvironmental cues to promote adhesion, survival, proliferation, and scaffold ingrowth.

Several chemical strategies have also been considered toward the central goal of improving osseointegration and assisting biological fixation (Amin Yavari et al., 2014). Extracellular ligands, bone morphogenetic proteins, insulin-like and transforming growth factors stimulate osteogenesis. For example (Clark et al., 2008), Clark et al. controlled the release of transforming growth factor-beta from metallic porous implants and demonstrated augmented bone osseointegration and bone-to-implant contact in a rabbit model. Additional agents including ascorbic acid, platelet-derived growth factors, and interleukins enhance osteointegration potential (Galante et al., 1971; Guo et al., 2007). In recent years, infusion of growth factors, small molecules, and cells onto device surfaces coupled with electromechanical stimulation of the implant interface has been successfully investigated. These breakthroughs could revolutionize the biology of implant osseointegration.

Whenever metallic devices are implanted in vivo, successful biointegration requires that host cells colonize the highly reactive implant surface (Schmidt and Swiontkowski, 2000). Physicochemical surface properties of the biomaterial including chemical activity, electric potential, surface energy, hydrophobicity, and surface topography coupled with bacteria cell surface have been shown to play a major role in this adhesion phenomenon (An and Friedman, 1998). Bacteria such as *Staphylococci* can also become adherent to metallic or polymeric implants and will compete with host cells for colonization of the implant surface. It has been demonstrated in animal models that contaminated fractures without internal fixation develop clinical infection more commonly than similar fractures treated with internal fixation at the time of colonization. Besides bacterial infections, fungal mycoses incidents are increasing as a result of latest practices that include immunosuppressive therapies, central venous access devices, broad-spectrum antibiotics, and stem cell transplantation (Hospenthal and Rinaldi, 2015). For example, in vascular catheter—related infections, the most commonly isolated microbial species are Gram-positive *Staphylococcus epidermidis* and *Staphylococcus aureus*, along with the fungus *Candida albicans* (Hampton and Sherertz, 1988). Typically, biomaterials possess a potential risk of biofilm formation and infection, but the properties of biofilm and its virulent degree are highly specific in nature. Because of the potential for infection, appropriate prophylactic antibiotic coverage for *Staphylococci* and Gram-negative organisms should be provided. Another option to minimize infection is doping biomaterials with ions such as silver that have an oligodynamic effect and are capable of reducing biofilm formation (Secinti et al., 2011).

1.3 Biointegration of biomaterials for dental applications

The biomaterials community is rapidly producing improved implant materials and techniques to meet the growing demands for biomaterials in dental applications. Dental implant systems should possess a firm attachment of the implant to the bone for long-term stability and should not illicit an adverse reaction when placed into service (www.azom.com). Dental implants face innumerable mechanical challenges that include mastication and grinding along with biological challenges in the oral environment that cause micromotions on the tissue interface (Nogueira et al., 2014). However, a variety of materials—metallic (pins for anchoring tooth implants and as parts of orthodontic devices), ceramics (tooth implants including alumina and dental porcelains), and polymeric (orthopedic devices such as plates and dentures)—with excellent biointegration have been developed and have been placed into service.

Starting with the "Integral Biointegrated Dental Implant System," consisting of a titanium implant cylinder coated with calcitite that permitted the bone to actually bond with the implant surface in a jaw restoration (Roling, 1989), dental surgery has undergone a revolution in both implant techniques and material technology. It was shown early on that the surface oxide of titanium appears to be central to the ability of titanium implants to achieve osseointegration, and ceramic coatings appear to improve the in-growth of bone and promote chemical integration of the implant with the bone (Wataha, 1996). Badr and El Hadary (Badr and El Hadary, 2007) have reported development of osseointegration of the regenerated bone when HA was coated onto the surface of cpTi implants using an electroplating technique. Pelsoczi et al. (2004) obtained more effective osseointegration on surface modifications of titanium implants with an excimer laser. In this case, it was easy to achieve the desired morphology (microstructure) and physicochemical properties that control the biointegration process. X-ray photoelectron spectroscopy studies show that laser treatment, in addition to microstructural and morphological modification, results in a decrease of surface contamination and thickening of the oxide layer. Thin film deposition of ceramic oxides onto titanium by excimer lasers and pulsed lasers has been successfully employed by other groups to improve the surface characteristics for facilitating biointegration, e.g., pulsed laser deposition of bioceramic thin films from human teeth (Smausz et al., 2004) and surface modifications induced by ns and sub-ps excimer laser pulses on titanium implant material (Bereznai et al., 2003). In recent times, highly porous scaffold that have eminent osteoinductive and osteogenic potential promote higher vascularization and support firm bone attachment and growth pivotal for dental applications (Holzapfel et al., 2013).

1.4 AlphaCor artificial corneal experience

AlphaCor is a biocompatible, flexible, one-piece artificial cornea (keratoprosthesis) designed to replace a scarred or diseased native cornea. It is a one-piece convex disc consisting of a central transparent optic and an outer skirt that is entirely manufactured from poly(2-hydroxyethyl methacrylate) or PHEMA. AlphaCor's material and patented features are designed to promote retention and optimize patient outcome. The outer skirt is an opaque, high water content, PHEMA sponge. The porosity of the sponge encourages biointegration with host tissue and thus promotes retention of the implanted device. The central optic core is a transparent PHEMA gel, providing a refractive power similar to that of the human cornea. The optic core is designed to allow the patients' visual potential to be achieved. The junctional zone between the skirt and central optic is the interpenetrating polymer network or IPN. This is a permanent bond formed at the molecular level and is designed to prevent the downgrowth of cells around the optic, which can lead to the formation of retroprosthetic membranes, one of the major complications historically associated with artificial corneas (http://www.medcompare.com; http://www.pricevisiongroup.com).

The WHO reports that corneal blindness affects more than 10 million people worldwide; however, only 100,000 people received corneal transplants each year. This shortfall is due to a combination of inadequate supply of donor corneas and the unsuitability of some patients to receive a corneal graft. AlphaCor is designed for use in patients who have had multiple failed corneal transplants or in those patients in whom a donor graft is likely to fail. Its patented design features are aimed to promote retention, minimize postoperative complications, and restore vision in patients who cannot receive, or are unlikely to have, a beneficial outcome from a human donor graft.

AlphaCor is available in two versions, to suit those with a natural lens (phakic) or artificial lens (pseudophakic) and for those without a lens (aphakic). Keratoprosthesis for artificial cornea surgery is a procedure for restoring the sight of patients suffering from a severely damaged anterior segment due to trauma, chemical burns, infections, etc. The ideal keratoprosthesis would be inert and not be rejected by the patient's immune system, be inexpensive, and maintain long-term clarity. In addition, it would be quick to implant, easy to examine, and allow an excellent view of the retina.

Coassin et al. (2007) reported histopathologic and immunologic characteristics of late artificial corneal failure in a small series of patients who underwent AlphaCor implantation, but light microscopic examination of the specimens disclosed adequate biointegration with no foreign body response. Immunofluorescence studies of the skirt exhibited expression of inflammatory cytokines such as interleukin-1β and tumor necrosis factor α, and some interferon γ. The keratocytes stained positively for Thy-1 and smooth muscle actin but negatively for CD34.

The AlphaCor implant (Fig. 1.2) is a viable method of treatment for multiple failed PKPs, but it may be associated with unique complications, including corneal stromal melting, focal calcification, and retroprosthetic membrane formation (Crawford, 2016). Infectious keratitis may be a risk factor for corneal stromal melting and needs to be managed aggressively. Explantation of the implant is essential if the skirt is exposed (Chow et al., 2007). Hicks et al. (Hicks et al., 2005) showed that histologic findings of the AlphaCor skirt in humans are consistent with earlier animal studies.

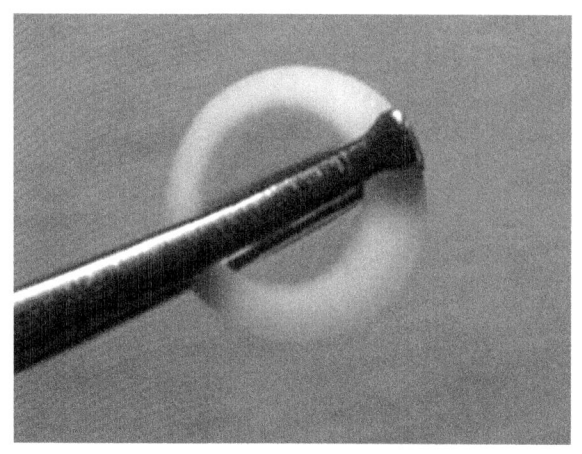

Figure 1.2 AlphaCor artificial cornea is a two-part single-piece design.
Reproduced with permission from Crawford, G.J., 2016. The development and results of an artificial cornea: AlphaCor. In: Chirila, T.V., Harkin, D.G. (Eds.), Biomaterials and Regenerative Medicine in Ophthalmology, Second ed. Woodhead Publishing, Duxford.

Their study confirmed that biointegration by host fibroblastic cells, with collagen deposition, occurred after AlphaCor implantation in humans. In cases in which stromal melting had occurred, biointegration was seen to be reduced. On correlating preoperative clinical factors with biointegration observed histologically, preoperative vascularization appears not to be required for AlphaCor biointegration (Hicks et al., 2005). Another study demonstrated that systemic factors affected the risk of retroprosthetic membrane formation with AlphaCor (Hicks and Hamilton, 2005). Hicks et al. (2005) further showed that device biointegration gets reduced in the cases of patients with a history of ocular herpes simplex virus (HSV) because the extensive lamellar corneal surgery involved in AlphaCor implantation may precipitate reactivation of latent HSV, such that reactivation and resultant inflammation could facilitate melting of corneal stromal tissue anterior to the device (Hicks et al., 2002).

Chirila (2006), in a recent publication, claims that the first keratoprosthesis based on polyurethane was made in 1985 by Lawrence Hirst, an Australian ophthalmologist then working in St Louis, USA. This keratoprosthesis, which also had a porous skirt, was inserted intralamellarly in a monkey cornea and was followed up clinically for about 3 months. There were no significant postoperative complications, and the histology of the explant indicated proper biointegration of the prosthetic skirt within the host stromal tissue (Chirila, 2006; Chirila et al., 1998). Hydrogel lenses may even make their way deeper into the eye, as replacements for inner eye lenses were damaged by cataracts.

1.5 Biointegration and functionality of tissue engineering devices

Experiments on animals have underlined the importance of vascularization for biointegration and functionality of any given tissue engineering device. Polykandriotis et al. recently showed that the presence of a vascular bed before cell transplantation might protect against hypoxia-induced cellular death, especially at central portions of the matrix, and therefore ensure physiological function of the device. The generation of vascularized bioartificial tissue substitutes might offer new modalities of surgical reconstruction for use in reparative medicine (Polykandriotis et al., 2006).

1.6 Percutaneous devices

Percutaneous devices among surgical implants have become ubiquitous in modern medicine. These biointerfaces between tissue and device being prone to infection are not conducive to efficient integration leading to implant failures. Various approaches are underway including repetitive local delivery of cells, drugs, and the implementation of regenerative strategies to hasten the biointegration process. Another approach to circumvent infection would be to heal the wound around the devices by promoting skin cell attachment (Fukano et al., 2006).

Biointegration through human fibronectin (FN) plays a key role in the biointegration of implants, as the success depends on adsorption of proteins such as FN. Indeed, FN can be an intermediary between the biomaterial surface and cells (Sousa et al., 2005). Isenhath et al. (2007) developed an in vivo model that permits examination of the implant/skin interface and that will be useful for future studies designed to facilitate skin cell attachment where percutaneous devices penetrate the skin. The space created between the skin and the device becomes a haven for bacterial invasion and biofilm formation, and this results in infection. Sealing this space via integration of the skin into the device is expected to create a barrier against bacterial invasion. Porous PHEMA rods were implanted for 7 days in the dorsal skin of C57 BL/6 mice. The porous PHEMA rods were surface-modified with carbonyldiimidazole (CDI) or CDI plus laminin 5, with unmodified rods serving as the control. Implant sites were sealed with 2-octyl cyanoacrylate; corn pads and adhesive dressings were tested for stabilization of implants. All rods remained intact for the duration of the study. There was histological evidence of both epidermal and dermal integration into all PHEMA rods, regardless of treatment.

The effects of hyperbaric oxygen (HBO) therapy on biointegration of porous PE implanted beneath dorsal burn scar and normal skin of Sprague–Dawley rats were microscopically examined, and the ratio of fibrovascular ingrowth was determined for each rat (Dinar et al., 2008). The results showed that HBO therapy enhanced biointegration of porous PE in hypoxic burn scar areas via improving collagen synthesis and neovascularization; otherwise, it apparently delayed tissue ingrowth into a porous structure implanted in normal healthy tissues. Three in vitro model systems have been presented as a proof-of-concept pilot study to deliver materials at the skin percutaneous device interface (Peramo et al., 2009, 2010; Peramo, 2010) enabling a close monitoring of the interface between the skin and respective implant. In addition, the possibility of delivering autologous keratinocytes through the implant has been established (Peramo, 2010). The cells delivered exhibited differentiation toward keratinocyte phenotype and did not reveal any apoptosis suggesting that this technique could be useful in promoting better integration in percutaneous implants.

1.7 Future trends

The future of biointegration and the future of implants are considered to be bright, as advancements in frontier biomaterials are advancing rapidly to unravel the physiologic process required for the biologization of the implants and developing materials that become intrinsically integrated into the organ-specific repair mechanisms. One example is the "designer implant," which could carry different types of proteins, one set to spur soft tissue healing and another to encourage hard tissue growth on another front. According to Rush, future development of osteobiologic materials will no doubt replace materials currently being used (Rush, 2005). Techniques to improve biointegration and manipulation of the healing environment will be developed such that future graft substitutes may exceed even autogenous bone in their reliability. An understanding of the cascade of events that occurs with bone healing and graft incorporation will enhance the chances to augment or manipulate the grafting process. The biomaterials community is producing new and improved implant materials and techniques to meet this demand. A counterforce to this technological push is the increasing level of regulation and the threat of litigation. To meet these conflicting needs, it is necessary to have reliable methods of characterization of the material and material/host tissue interactions (www.azom.com). In addition, progress on the road to regenerating major body parts, salamander-style, could transform the treatment of amputations and major wounds (Muneoka et al., 2008).

At the same time, the newly emerging nanobiotechnology is revolutionizing capability to resolve biological and medical problems by developing subtle biomimetic techniques. The use of nanoscale materials is expected to increase dramatically in many applications of medicine and surgery. It is interesting to note that the size of these nanomaterials is comparable to many biological systems, and so there is large scope for biointegration. Additionally, nanomaterials exhibit fundamentally different properties from their properties in bulk, such that they can be tailor-made to provide properties to suit a specific application (Navamar, 2003; Christenson et al., 2007). Significant advancements are expected to achieve the desired goals and their clinical use, especially in areas such as nanostructured coatings, nanostructured porous scaffoldings, and other nanobiomaterials. The current trends in nanobiotechnology, thus, offer a bright future through the use of nanobiomaterials in achieving better biointegration properties. A recent review by Ye et al. (Ye and Peramo, 2014) addresses in depth the current concepts available in achieving permanent integration between the implant and the tissues surrounding them and describes the various clinical trials where novel techniques are being investigated. We can envisage that in the future the areas of medical implants, regenerative medicine, and tissue engineering will converge and apply translational solutions to address implant failure. This would enable the biomaterial community to establish techniques to develop advanced biomaterials and to promote a deeper and differentiated understanding of the concepts in biointegration (Holzapfel et al., 2013). Moreover, current regenerative technologies including bioprinting have recently ushered in an era of patient-specific organ development and are revolutionizing the future of medicine (Chin et al., 2018). Eventually, appropriate infrastructure with hospitals coupled with the development and implementation of new guidelines and a collaborative spirit with scientists would ultimately improve our health-care system and usher us into a new era in the service of humanity.

References

Adell, R., Eriksson, B., Lekholm, U., Branemark, P.I., Jemt, T., 1990. Long-term follow-up study of osseointegrated implants in the treatment of totally edentulous jaws. Int. J. Oral Maxillofac. Implant. 5, 347—359.

Albrektsson, T., Branemark, P.I., Hansson, H.A., Lindstrom, J., 1981. Osseointegrated titanium implants. Requirements for ensuring a long-lasting, direct bone-to-implant anchorage in man. Acta Orthop. Scand. 52, 155—170.

Amin Yavari, S., Van Der Stok, J., Chai, Y.C., Wauthle, R., Tahmasebi Birgani, Z., Habibovic, P., Mulier, M., Schrooten, J., Weinans, H., Zadpoor, A.A., 2014. Bone regeneration performance of surface-treated porous titanium. Biomaterials 35, 6172—6181.

Amling, M., Schilling, A.F., Pogoda, P., Priemel, M., Rueger, J.M., 2006. Biomaterials and bone remodeling: the physiologic process required for biologization of bone substitutes. Eur. J. Trauma 32, 5.

An, Y.H., Friedman, R.J., 1998. Concise review of mechanisms of bacterial adhesion to biomaterial surfaces. J. Biomed. Mater. Res. 43, 338—348.

Babis, G.C., Mavrogenis, A.F., 2014. Cobalt—chrome porous-coated implant-bone interface in total joint arthroplasty. In: Karachalios, T. (Ed.), Bone-Implant Interface in Orthopedic Surgery. Springer, London.

Badr, N.A., EL Hadary, A.A., 2007. Hydroxyapatite-electroplated cp-titanium implant and its bone integration potentiality: an in vivo study. Implant Dent. 16, 297—308.

Balasundaram, G., Webster, T.J., 2006. Nanotechnology and biomaterials for orthopedic medical applications. Nanomedicine 1, 169—176.

Barbas, A., Bonnet, A.S., Lipinski, P., Pesci, R., Dubois, G., 2012. Development and mechanical characterization of porous titanium bone substitutes. J. Mech. Behav. Biomed. Mater. 9, 34—44.

Barrere, F., Mahmood, T.A., De Groot, K., Van Blitterswijk, C.A., 2008. Advanced biomaterials for skeletal tissue regeneration: instructive and smart functions. Mater. Sci. Eng. R Rep. 59, 38—71.

Bereznai, M., Pelsoczi, I., Toth, Z., Turzo, K., Radnai, M., Bor, Z., Fazekas, A., 2003. Surface modifications induced by ns and sub-ps excimer laser pulses on titanium implant material. Biomaterials 24, 4197—4203.

Borsari, V., Giavaresi, G., Fini, M., Torricelli, P., Tschon, M., Chiesa, R., Chiusoli, L., Salito, A., Volpert, A., Giardino, R., 2005. Comparative in vitro study on a ultra-high roughness and dense titanium coating. Biomaterials 26, 4948—4955.

Branemark, P.I., Zarb, G.A., Albrektsson, T., 1985. Tissue-Integrated Prostheses: Osseointegration in Clinical Dentistry. Quintessence Publishing Company, Inc, Chicago, Ill.

Carr, A.J., Robertsson, O., Graves, S., Price, A.J., Arden, N.K., Judge, A., Beard, D.J., 2012. Knee replacement. Lancet 379, 1331—1340.

Chin, S.O., Lucy, N., Kingsfield, O., Aravind, K., Chen, Y.H., Takuma, F., Narutoshi, H., 2018. 3D and 4D bioprinting of the myocardium: current approaches, challenges, and future prospects. BioMed Res. Int. 6497242.

Chirila, T.V., 2006. First development of a polyurethane keratoprosthesis and its Australian connection: an unbeknown episode in the history of artificial cornea. Clin. Exp. Ophthalmol. 34, 485−488.
Chirila, T.V., Hicks, C.R., Dalton, P.D., Vijayasekaran, S., Lou, X., Hong, Y., Clayton, A.B., Ziegelaar, B.W., Fitton, J.H., Platten, S., Crawford, G.J., Constable, I.J., 1998. Artificial cornea. Prog. Polym. Sci. 23, 447−473.
Chow, C.C., Kulkarni, A.D., Albert, D.M., Darlington, J.K., Hardten, D.R., 2007. Clinicopathologic correlation of explanted AlphaCor artificial cornea after exposure of implant. Cornea 26, 1004−1007.
Christenson, E.M., Anseth, K.S., Van Den Beucken, J.J., Chan, C.K., Ercan, B., Jansen, J.A., Laurencin, C.T., Li, W.J., Murugan, R., Nair, L.S., Ramakrishna, S., Tuan, R.S., Webster, T.J., Mikos, A.G., 2007. Nanobiomaterial applications in orthopedics. J. Orthop. Res. 25, 11−22.
Clark, P.A., Moioli, E.K., Sumner, D.R., Mao, J.J., 2008. Porous implants as drug delivery vehicles to augment host tissue integration. FASEB J. 22, 1684−1693.
Coassin, M., Zhang, C., Green, W.R., Aquavella, J.V., Akpek, E.K., 2007. Histopathologic and immunologic aspects of AlphaCor artificial corneal failure. Am. J. Ophthalmol. 144, 699−704.
Cochran, D., 1996. Implant therapy I. Ann. Periodontol. 1, 707−791.
Covani, U., Giacomelli, L., Krajewski, A., Ravaglioli, A., Spotorno, L., Loria, P., Das, S., Nicolini, C., 2007. Biomaterials for orthopedics: a roughness analysis by atomic force microscopy. J. Biomed. Mater. Res. A 82, 723−730.
Crawford, G.J., 2016. The development and results of an artificial cornea: AlphaCor. In: Chirila, T.V., Harkin, D.G. (Eds.), Biomaterials and Regenerative Medicine in Ophthalmology, Second ed. Woodhead Publishing, Duxford.
Dabrowski, B., Swieszkowski, W., Godlinski, D., Kurzydlowski, K.J., 2010. Highly porous titanium scaffolds for orthopaedic applications. J. Biomed. Mater. Res. B Appl. Biomater. 95B, 53−61.
Dinar, S., Agir, H., Sen, C., Yazir, Y., Dalcik, H., Unal, C., 2008. Effects of hyperbaric oxygen therapy on fibrovascular ingrowth in porous polyethylene blocks implanted under burn scar tissue: an experimental study. Burns 34, 467−473.
Fukano, Y., Knowles, N.G., Usui, M.L., Underwood, R.A., Hauch, K.D., Marshall, A.J., Ratner, B.D., Giachelli, C., Carter, W.G., Fleckman, P., Olerud, J.E., 2006. Characterization of an in vitro model for evaluating the interface between skin and percutaneous biomaterials. Wound Repair Regen. 14, 484−491.
Galante, J., Rostoker, W., Lueck, R., Ray, R.D., 1971. Sintered fiber metal composites as a basis for attachment of implants to bone. J. Bone Joint Surg. Am. 53, 101−114.
Guo, J., Padilla, R.J., Ambrose, W., De Kok, I.J., Cooper, L.F., 2007. The effect of hydrofluoric acid treatment of TiO_2 grit blasted titanium implants on adherent osteoblast gene expression in vitro and in vivo. Biomaterials 28, 5418−5425.
Hacking, S.A., Bobyn, J.D., Toh, K.K., Tanzer, M., Krygier, J.J., 2000. Fibrous tissue ingrowth and attachment to porous tantalum. J. Biomed. Mater. Res. 52, 631−638.
Hampton, A.A., Sherertz, R.J., 1988. Vascular-access infections in hospitalized patients. Surg. Clin. 68, 57−71.
Harris, L.G., Richards, R.G., 2006. Staphylococci and implant surfaces: a review. Injury 37 (Suppl. 2), S3−14.
Hicks, C.R., Crawford, G.J., Tan, D.T., Snibson, G.R., Sutton, G.L., Gondhowiardjo, T.D., Lam, D.S., Downie, N., 2002. Outcomes of implantation of an artificial cornea, AlphaCor: effects of prior ocular herpes simplex infection. Cornea 21, 685−690.

Hicks, C.R., Hamilton, S., 2005. Retroprosthetic membranes in AlphaCor patients: risk factors and prevention. Cornea 24, 692−698.
Hicks, C.R., Werner, L., Vijayasekaran, S., Mamalis, N., Apple, D.J., 2005. Histology of AlphaCor skirts: evaluation of biointegration. Cornea 24, 933−940.
Holzapfel, B.M., Reichert, J.C., Schantz, J.T., Gbureck, U., Rackwitz, L., Noth, U., Jakob, F., Rudert, M., Groll, J., Hutmacher, D.W., 2013. How smart do biomaterials need to be? A translational science and clinical point of view. Adv. Drug Deliv. Rev. 65, 581−603.
Hong, Z.D., Luan, L., Paik, S.B., Deng, B., Ellis, D.E., Ketterson, J.B., Mello, A., Eon, J.G., Terra, J., Rossi, A., 2007. Crystalline hydroxyapatite thin films produced at room temperature − an opposing radio frequency magnetron sputtering approach. Thin Solid Films 515, 6773−6780.
Hospenthal, D.R., Rinaldi, M.G., 2015. Diagnosis and Treatment of Fungal Infections. Springer, London.
Isenhath, S.N., Fukano, Y., Usui, M.L., Underwood, R.A., Irvin, C.A., Marshall, A.J., Hauch, K.D., Ratner, B.D., Fleckman, P., Olerud, J.E., 2007. A mouse model to evaluate the interface between skin and a percutaneous device. J. Biomed. Mater. Res. A 83A, 915−922.
Jafari, S.M., Bender, B., Coyle, C., Parvizi, J., Sharkey, P.F., Hozack, W.J., 2010. Do tantalum and titanium cups show similar results in revision hip arthroplasty? Clin. Orthop. Relat. Res. 468, 459−465.
Jain, R., Kapoor, D., 2015. The dynamic interface: a review. J. Int. Soc. Prev. Community Dent. 5, 354−358.
Jones, A.C., Arns, C.H., Hutmacher, D.W., Milthorpe, B.K., Sheppard, A.P., Knackstedt, M.A., 2009. The correlation of pore morphology, interconnectivity and physical properties of 3D ceramic scaffolds with bone ingrowth. Biomater. 30, 1440−1451.
Kim, D.Y., Kim, Y.Y., Lee, H.B., Moon, S.Y., Ku, S.Y., Kim, M.S., 2013. In vivo osteogenic differentiation of human embryoid bodies in an injectable in situ-forming hydrogel. Materials 6, 2978−2988.
Meffert, R.M., Block, M.S., Kent, J.N., 1987. What is osseointegration? Int. J. Periodontics Restor. Dent. 7, 9−11.
Meneghini, R.M., Ford, K.S., Mccollough, C.H., Hanssen, A.D., Lewallen, D.G., 2010. Bone remodeling around porous metal cementless acetabular components. J. Arthroplast. 25, 741−747.
Mirhosseini, N., Crouse, P.L., Schmidth, M.J.J., Li, L., Garrod, D., 2007. Laser surface microtexturing of Ti-6Al-4V substrates for improved cell integration. Appl. Surf. Sci. 253, 7738−7743.
Muneoka, K., Han, M., Gardiner, D.M., 2008. Regrowing human limbs. Sci. Am. 298, 56−63.
Nasiri, N., Mukherjee, S., Panneerselvan, A., Nisbet, D.R., Tricoli, A., 2018. Optimally hierarchical nanostructured hydroxyapatite coatings for superior prosthesis biointegration. ACS Appl. Mater. Interfaces 10, 24840−24849.
Navamar, F., 2003. Applications of nantechnology for alternative bearing surfaces in orthopaedics. In: Ravaglioli, A., Krajewski, A. (Eds.), 8th Ceramics, Cells and Tissues Meeting − Seminar, March 2003. ISTEC-CNR, Faenza, Italy.
Nogueira, A.V., Nokhbehsaim, M., Eick, S., Bourauel, C., Jager, A., Jepsen, S., Rossa Jr., C., Deschner, J., Cirelli, J.A., 2014. Biomechanical loading modulates proinflammatory and bone resorptive mediators in bacterial-stimulated PDL cells. Mediat. Inflamm. 2014, 425421.
Oudadesse, H., Derrien, A.C., Mami, M., Martin, S., Cathelineau, G., Yahia, L., 2007. Aluminosilicates and biphasic HA-TCP composites: studies of properties for bony filling. Biomed. Mater. 2, S59−64.

Patel, R.B., Solorio, L., Wu, H.P., Krupka, T., Exner, A.A., 2010. Effect of injection site on in situ implant formation and drug release in vivo. J. Control. Release 147, 350—358.
Pelsoczi, K.I., Bereznai, M., Toth, Z., Turzo, K., Radnai, M., Bor, Z., Fazekas, A., 2004. Surface modifications of titanium implant material with excimer laser for more effective osseointegration. Fogorv. Szle. 97, 231—237.
Peramo, A., 2010. Autologous cell delivery to the skin-implant interface via the lumen of percutaneous devices in vitro. J. Funct. Biomater. 1, 14—21.
Peramo, A., Marcelo, C.L., Goldstein, S.A., Martin, D.C., 2009. Novel organotypic cultures of human skin explants with an implant-tissue biomaterial interface. Ann. Biomed. Eng. 37, 401—409.
Peramo, A., Marcelo, C.L., Goldstein, S.A., Martin, D.C., 2010. Continuous delivery of biomaterials to the skin-percutaneous device interface using a fluid pump. Artif. Organs 34, E27—33.
Polykandriotis, E., Arkudas, A., Euler, S., Beier, J.P., Horch, R.E., Kneser, U., 2006. Prevascularisation strategies in tissue engineering. Handchir. Mikrochir. Plast. Chir. 38, 217—223.
Raggatt, L.J., Partridge, N.C., 2010. Cellular and molecular mechanisms of bone remodeling. J. Biol. Chem. 285, 25103—25108.
Rey, C., 1981. Orthopedic biomaterials, bioactivity, biodegradation; a physical-chemical approach. In: 11th Conference of the ESB. Toulouse, France: Journal of Biomechanics.
Roling, T., 1989. Biointegration revolutionizes dental surgery. Sulzer Tech. Rev. 71, 7—10.
Rush, S.M., 2005. Bone graft substitutes: osteobiologics. Clin. Podiatr. Med. Surg. 22, 619—630 viii.
Schmidt, A.H., Swiontkowski, M.F., 2000. Pathophysiology of infections after internal fixation of fractures. J. Am. Acad. Orthop. Surg. 8, 285—291.
Secinti, K.D., Ozalp, H., Attar, A., Sargon, M.F., 2011. Nanoparticle silver ion coatings inhibit biofilm formation on titanium implants. J. Clin. Neurosci. 18, 391—395.
Smausz, T., Hopp, B., Huszar, H., Toth, Z., Kecskemeti, G., 2004. Pulsed laser deposition of bioceramic thin films from human tooth. Appl. Phys. Mater. Sci. Process 79, 1101—1103.
Sousa, S.R., Moradas-Ferreira, P., Barbosa, M.A., 2005. TiO_2 type influences fibronectin adsorption. J. Mater. Sci. Mater. Med. 16, 1173—1178.
Stevens, M.M., George, J.H., 2005. Exploring and engineering the cell surface interface. Science 310, 1135—1138.
Svensson, S., Suska, F., Emanuelsson, L., Palmquist, A., Norlindh, B., Trobos, M., Backros, H., Persson, L., Rydja, G., Ohrlander, M., Lyven, B., Lausmaa, J., Thomsen, P., 2013. Osseointegration of titanium with an antimicrobial nanostructured noble metal coating. Nanomed. Nanotechnol. Biol. Med. 9, 1048—1056.
Viceconti, M., Davinelli, M., Taddei, F., Cappello, A., 2004. Automatic generation of accurate subject-specific bone finite element models to be used in clinical studies. J. Biomech. 37, 1597—1605.
Wall, I., Donos, N., Carlqvist, K., Jones, F., Brett, P., 2009. Modified titanium surfaces promote accelerated osteogenic differentiation of mesenchymal stromal cells in vitro. Bone 45, 17—26.
Wataha, J.C., 1996. Materials for endosseous dental implants. J. Oral Rehabil. 23, 79—90.
Williams, D.F., 1999. The Williams Dictionary of Biomaterials. Liverpool University Press, Liverpool.
Yazdani, J., Ahmadian, E., Sharifi, S., Shahi, S., Maleki Dizaj, S., 2018. A short view on nanohydroxyapatite as coating of dental implants. Biomed. Pharmacother. 105, 553—557.

Ye, D., Peramo, A., 2014. Implementing tissue engineering and regenerative medicine solutions in medical implants. Br. Med. Bull. 109, 3–18.

Yu, L., Li, Y., Zhao, K., Tang, Y.F., Cheng, Z., Chen, J., Zang, Y., Wu, J.W., Kong, L., Liu, S.A., Lei, W., Wu, Z.X., 2013. A novel injectable calcium phosphate cement-bioactive glass composite for bone regeneration. PLoS One 8.

Zanotti, B., Verlicchi, A., 2006. Is one cervical prosthesis equal to another?'. Riv. Med. 12, 349–356.

Zhang, L., Gowardman, J., Rickard, C.M., 2011. Impact of microbial attachment on intravascular catheter-related infections. Int. J. Antimicrob. Agents 38, 9–15.

Zim, S., 2004. Skeletal volume enhancement: implants and osteotomies. Curr. Opin. Otolaryngol. Head Neck Surg. 12, 349–356.

Part One

Soft tissue biointegration

Interface biology of stem cell−driven tissue engineering: concepts, concerns, and approaches

Soumya K. Chandrasekhar[1,2], Finosh G. Thankam[3], Devendra K. Agrawal[3], Joshi C. Ouseph[2]
[1]Department of Zoology, KKTM Govt. College, Pullut, Calicut University, Kerala, India;
[2]Department of Zoology, Christ College, Calicut University, Iringalikkuda, Kerala, India;
[3]Clinical & Translational Sciences, Creighton University, Omaha, NE, United States

Chapter outline

2.1 Introduction 19
2.2 Stem cells for tissue engineering 21
2.3 Mesenchymal stem cells in a nutshell 23
2.4 Mesenchymal stem cell action in wound healing 23
2.5 Biomaterials in stem cell−based soft tissue engineering 26
2.6 Influence of scaffold patterns in stem cell behavior 29
    2.6.1 Scaffold decoration with stem cells 31
        2.6.1.1 *Chemokine-decorated scaffolds as stem cell recruiter* 34
    2.6.2 Summary and future directions 36
References 37

2.1 Introduction

End-stage organ failure resulting from disease and/or trauma relies on various therapeutic approaches to activate repair response, replacement, and/or regeneration (O'Brien, 2011). Major strategy adopted for most of the end-stage clinical conditions is the transplantation of organ or tissue from healthy donors. However, the shortage of organ donors and lack of proper techniques for organ storage hurdle the life expectancy of the millions of sufferers across the globe. According to a recent report by World Health Organization, more than 90% of the global organ transplantation remains unaddressed. In United States alone, roughly a million deaths occur because of end-stage organ failure; however, approximately 5% of them are being added to transplant waiting list, every year. The availability of organs for transplantation in the United States could double the life expectancy, thus reducing more than 30% mortality (Giwa et al., 2017). Moreover, these statistics hikes significantly when the global population is considered. According to Global Observatory of Donation and Transplantation statistics, 126,670 organs are transplanted annually with approximately 14.5 transplants every hour (Survey, 2016).

The increased demand of organs for transplantation has been hurdled by lack of availability that has accelerated the international attempts for encouraging organ donation. Various approaches including live donation, split organ donation, and paired donor exchange have been adopted for transplantation. Despite the whole organ transplantation, treatment strategies have also been focused on autografting and allografting of tissues. Still, these efforts and approaches are insufficient to meet the growing demand for organs, and the proportionate increase in the mortality rate of patients in the waiting list continues (Heidary Rouchi and Mahdavi-Mazdeh, 2015). In spite of the lifesaving potential, these treatment methods are limited by several constrains including cost, trauma, anatomical limitations, associated donor site morbidity, and chances of immunological rejection.

The increasing demand and scarcity of organs/tissues has led the modern medicine to seek for alternatives to meet the increasing organ demands. The concept of tissue engineering (TE) and regenerative medicine has opened new opportunities to address the limited availability of organs for transplantation and can form an ideal alternative for organ transplantation. Recent advances in TE are encouraging and are expected to meet the ever-increasing demand for organs by serving as an inexhaustible and immunocompatible source of tissue substitutes (Orlando et al., 2013). Although the term "TE" was addressed in discussions very earlier, a precise definition remains obscure. TE incorporates the principles of engineering and basic sciences to construct 3D templates that facilitate a suitable microenvironment for tissue regeneration. These scaffolds when seeded with cells and/or other biological cues such as growth factors can serve as biological substitutes to restore, maintain, or improve tissue or organ functions (Howard et al., 2008).

Recent decade witnessed a dramatic evolution in the field of TE, and several efforts in TE have found clinical applications especially for skin and cartilage repair. In addition, TE has created several tissue substitutes such as blood vessels, liver, nerve conduits, heart, bones, and teeth (Berthiaume et al., 2011). The emergence of stem cell—based therapy as a promising approach in TE is one of the most exciting advances in modern medicine and life science (Wei et al., 2017). A lion's share of studies employing stem cells has revealed encouraging results implying their potential in maintaining homeostasis in healthy tissues and eliciting reparative response to pathological insults (Kränkel et al., 2011).

Stem cells are explained as a class of undifferentiated pluripotent cells that have the unique ability to differentiate into same or other cell types on activation by specific signals. Two types of stem cells are identified based on their origin: adult stem cells (ASCs) and embryonic stem cells (ESCs). ESCs are derived from embryonic tissue, specifically from the blastocyst stage, whereas ASCs are derived from tissues of different organs from adult individuals (Ghieh et al., 2015). Common sources of

Table 2.1 Markers for adult stem cells (Beane and Darling, 2012; Ha et al., 2015; Yang et al., 2017).

Stem cells	Positive markers
HPSC	CD14, CD45, CD11a, CD34
ASC	CD10, CD13, CD29, CD34, CD44, CD49d, CD54, CD71, CD73, CD90, CD105, CD166
SDSC	CD10, CD13, CD44, CD49a, CD73, CD90, CD105, CD166
MDSC	CD10, CD13, CD 56, CD90, CD105
PBMSC	CD44, CD54, CD90, CD106, CD117, CD133, CD166
DPSCs	CD105, CD13, CD73
DFSCs	CD44, CD90, CD150, STRO-1

ASCs for TE include bone marrow, adipose tissue, umbilical cord, lung, and muscle (Pacelli et al., 2017). The major types of stem cells used in TE approaches and their characteristic positive markers are given in Table. 2.1.

Studies have demonstrated that specific signals arising from a postischemic or posttraumatic site in the body trigger the recruitment of stem cells that undergo differentiation into specific lineages and initiate healing response by secreting chemokines and growth factors (Kränkel et al., 2011; Wei et al., 2017). However, this system collapses when the causality is too extensive. It is possible that the administration or recruitment of sufficient stem cell population to the injury zone would enhance the repair responses leading to the restoration of tissue homeostasis. The regenerative strategies based on stem cells relies either on the direct administration of in vitro cultured stem cells by growing them with natural or synthetic scaffolds or by combining scaffolds with bioactive molecules that attracts stem cells to the site of injury.

2.2 Stem cells for tissue engineering

The efficacy of stem cell—based TE in the treatment of various clinical conditions has undergone extensive evaluation, globally. Although the success rate of such efforts is variable, studies are still ongoing to optimize the several crucial elements for the successful translation of TE strategies to clinical arena (Park et al., 2017). The principal factor that influences the success of any such application is the type of stem cells used. In vitro engineering of tissues requires stem cells that proliferate onto the interstices of a scaffold to form a matrix that resembles native extracellular matrix (ECM). Therefore, the source of stem cells has a profound impact on the success of this strategy. ESCs, extraembryonic stem cells (EESCs), induced pluripotent stem cells (iPSCs), and ASCs are being preferred to design TE strategies for in vivo implantation. The scope of ESC in TE unraveled with the isolation of human

ESC by Thompson in 1998 from the inner cell mass (ICM) of blastocyst stage of embryo. Being pluripotent, they can differentiate into more than 200 cell types including beta cells of islets of Langerhans, cardiomyocytes, and hepatocyte (Mahla, 2016) (Khan et al., 2018), suggesting their immense application in the treatment of several casualties. ESCs are characterized by several markers such as SSEA3, SSEA4, TRA-1-60, TRA-1-8, OCT4, SOX2, and NANOG (Beane and Darling, 2012). Specifically, the homeodomain transcription factor OCT4, the high mobility group—box transcription factor SOX2, and the variant homeodomain transcription factor NANOG are responsible for the maintenance of a network of key transcription factors within the cell that controls pluripotency (Johnson et al., 2008).

Fresh and/or frozen embryos serve as preferred sources of ESCs. First stage in generation of ESCs involves isolation of ICM by removing trophectoderm, which is the outer layer of the blastocyst. Several techniques, such as microdissection, mechanical dissection, laser dissection, and immunosurgery, are being employed for removing trophectoderm. Micro, mechanical, and laser dissection techniques are superior as they provide xeno-free cell lines. Immunosurgery on the other hand employs culture media containing guinea pig serum, which is not suitable for generating clinical grade cell lines. After isolation, ICM is grown to form ESC using feeder layers, extracellular matrices, proteins, peptides, or synthetic polymers, which enhances the proliferation of stem cells and prevents their differentiation (Khan et al., 2018). Added to pluripotency, the easy maintenance in the culture for providing a nearly infinite source of cells for tissues is another advantage of ESC. However, the ethical issues of harvesting ESCs from ICM of the embryo limit their use for regenerative therapy. iPSCs are produced by the reprogramming of adult somatic cells by the specific introduction of four genes viz *Oct3/4*, *Sox2*, *Klf4*, and *c-Myc*, referred to as Yamanaka factors in recognition of Shinya Yamanaka, the inventor of this method. The cells thus formed are pluripotent and resembles ESCs with respect to genomics, cell biology, and phenotypic characteristics (Han and Yoon, 2011). However, there are high chances that the techniques used for genetic manipulations would induce undesirable genetic modifications leading to teratoma formation, which raises serious concerns in the clinical application of iPSCs.

EESCs are isolated from the discarded tissues following the child birth. Being in a more developmentally primitive state, EESC exhibits significantly higher proliferation and multilineage differentiation potential. However, for the autologous use of these stem cells, cryopreservation is recommended, which is not cost effective. Also, prolonged storage of such cells can lead to deleterious effects. The major sources for EESCs include Wharton's jelly, umbilical cord blood, amniotic fluid, and placenta (Beane and Darling, 2012).

ASCs are resident stem cells of adult tissues that have self-renewal potential and function in the repair of damaged tissues. The multipotent lineage, limited chances for tumor formation, and easy isolation are the added benefits of using ASCs in regenerative medicine. However, these cells have limited self-renewal capacity when compared with other stem cell types mentioned earlier. Limited proliferation rates and decreased differentiation potential are other drawbacks associated with these cell types. Furthermore, the invasive procedures for stem cell harvest offer challenges to this strategy (Howard et al., 2008). However, the ASCs isolated from various sources such as bone marrow, adipose tissue, hair follicle, and foreskin have excellent regenerative potential.

Mesenchymal stem cells (MSCs) isolated form versatile sources are the most explored stem cells in the literature. The mechanism of integration to TE scaffolds and regenerative functions of MSCs share similarities with the stem cells from other sources. Therefore, the present chapter mainly focuses on MSCs, where the same principles are applicable to other stem cell phenotypes as well.

2.3 Mesenchymal stem cells in a nutshell

Discovered by Alexander Friedenstein in 1974, MSCs are the most investigated stem cell for TE applications. These are the subset of ASCs originating from the mesoderm and localized in almost all tissues including bone marrow and adipose tissue. Although mesodermal in origin, MSCs can differentiate into cells of ectodermal, endodermal, and mesodermal lineages, which upgrade MSCs to be the most preferred candidate in stem cell—based therapies. Moreover, the ease to harvest, maintain, and manipulate MSCs onto scaffold materials are added advantage for its TE applications (Finosh and Jayabalan, 2012; Wei et al., 2013; Griffin et al., 2015). MSCs can be harvested by bone marrow aspiration, density gradient separation, and fluorescence- or magnetic-activated cell sorting (Grisendi et al., 2010; Gronthos et al., 2003; Miltenyi et al., 1990). In addition, the MSCs can address ethical and legal concerns as well as teratoma formation. Although specific biomarkers are unavailable, several features such as adherence to plastic, ability to self-renew, ability to differentiate into osteoblast, chondrocytes, and adipocytes have been exploited to characterize the MSCs (Graham and Qian, 2018). As MSCs are highly diverse and exhibit considerable plasticity, it is difficult to define them using a unique subset of biomarkers. Yet, some of the markers exhibited by MSCs include CD105, CD73, CD90, STRO-1, CD146, SSEA-4, CD271, MSCA-1, and PDGFR-α (Samsonraj et al., 2017). A diagrammatic representation of different biomarkers is depicted in Fig. 2.1.

2.4 Mesenchymal stem cell action in wound healing

The inherent repair/regenerative capacity of tissues and organs depend on the ability to recruit and home stem cells including MSCs. Cell homing refers to the mobilization of resident MSCs from their anatomical niches to the site of an injury for initiating the normal healing response. Earlier in the process of homing, the adhesion of MSCs to the matrix proteins is disrupted by proteolytic enzymes such as matrix metallo

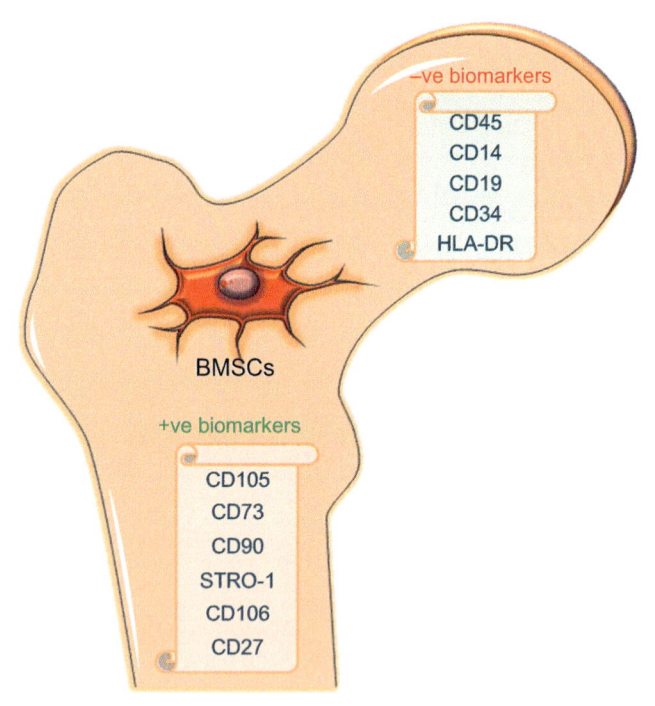

Figure 2.1 Positive and negative biomarkers of bone marrow mesenchymal stem cells (BMSCs).

proteases and cathepsins released from the site of injury. Later, chemokine receptors expressed on MSC membrane mediate their recruitment to the site of injury under the influence of posttraumatic inflammatory chemokines. C-X-C motif chemokine receptor 4/stromal cell—derived factor-1 (CXCR4/SDF1) axis plays a major role in the mobilization of MSCs from their niches to the peripheral blood circulation and to the damaged tissue. SDF1 is a chemokine protein expressed by several tissues including bone marrow, endothelium, and other stromal cells, while CXCR4 is its receptor found on MSCs. Under physiological conditions, the interaction between CXCR4 and SDF1 retains stem cells in their niches. However, hypoxia due to reduced tissue perfusion during trauma causes the production of nitric oxide and hypoxia-inducible factor 1 (HIF-1) resulting in the secretion of SDF1 to the injury site. Subsequently, a concentration gradient builds up in the injury site, which is several folds higher than the native stem cell niches which facilitates the migration of MSCs from bone marrow to site of injury (Fong et al., 2011; Thurairajah et al., 2017). At the target site, the migratory movements of the cells slow down; however, contact with endothelial wall is established by the process of tethering and rolling. Interplay between various adhesion molecules such as E- and P-selectin, VCAM-1, and integrins expressed by the stem cells and endothelial cell mediate this homing process.

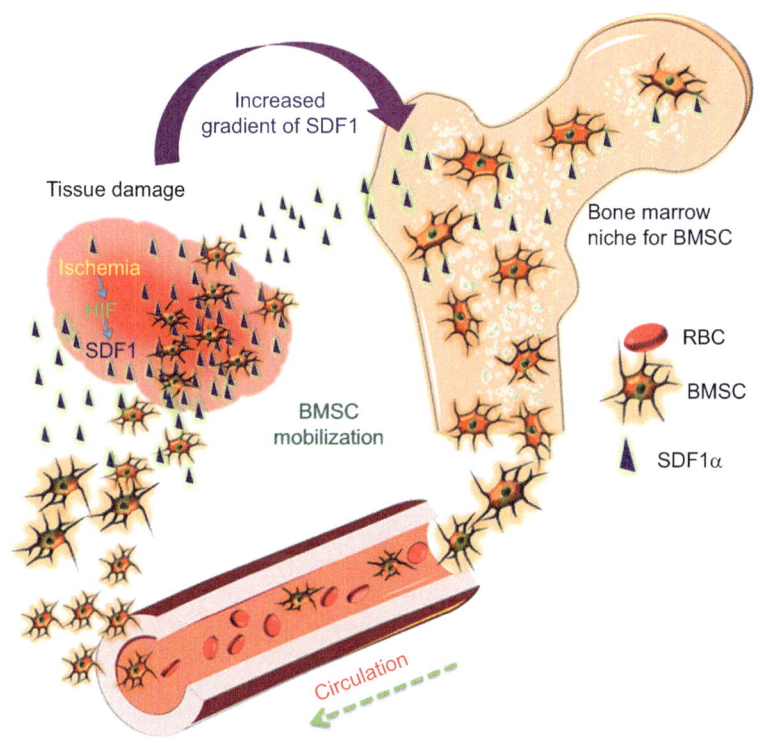

Figure 2.2 Mobilization and homing of stem cells to the site of injury. Chemokines released from the site of injury are a strong inducer for the movement of mesenchymal stem cells (MSCs) from their niche to injury site for reparative responses. Hypoxia-inducible factor (HIF) produced at injury site causes a concentration gradient of stromal cell−derived factor-1 (SDF1) between bone marrow and injury site attracting bone marrow mesenchymal stem cells (BMSCs) to injury site.

This is followed by the transmigration through the endothelium into the tissues. This general mechanism of mobilization and subsequent homing of MSCs under the influence of chemokines is represented in Fig. 2.2.

At the injury zone, the MSCs cross talk with local stimuli such as inflammatory cytokines, toll-like receptors, and hypoxia to secrete a multitude of growth factors. Earlier, it was believed that the MSCs mediate the regeneration by engraftment into the tissues. However, the mounting evidences suggest that engraftment and differentiation of MSCs at wound site is not mandatory for eliciting the repair responses. Rather, the MSCs mediate healing response by secreting bioactive factors that promote immunomodulation of inflammatory cells that participate in tissue repair. These growth factors include interleukin-10, indoleamine 2,3-dioxygenase, vascular endothelial growth factor (VEGF), chemokine (C-C motif) ligand 5, prostaglandin E2, interleukin-6, transforming growth factor, monocyte chemoattractant protein-1(MCP-1), and hepatocyte growth factor (HGF). In addition, the extracellular vesicles containing proteins of diverse functions mediate several processes involved in tissue regeneration including prevention of apoptosis, cell survival, collagen synthesis, and neoangiogenesis.

2.5 Biomaterials in stem cell–based soft tissue engineering

The success of TE strategies largely depends on the advancement of our understanding regarding the potential of biomaterials to be effectively served as scaffolds/templates for tissue regeneration (Chen and Liu, 2016). Any material that functions as a substitute for an organ or organ part and/or performs or assists the function of the body can be referred as a "biomaterial." Biomaterials for stem cell–based TE applications can be either synthetic or natural. Natural biomaterials are subdivided as protein- and polysaccharide-based materials, whereas polymeric- and ceramic-based materials constitute synthetic biomaterials (Sakiyama-Elbert, 2008). The suitability of natural materials for developing engineered scaffold is rendered to its ability to better interact and integrat with the host tissues. However, the clinical trials with such scaffolds remain far from reality owing to some inherent demerits such as low mechanical strength and instability in the in vivo system. However, synthetic polymeric biomaterials superior in mechanical, chemical, and structural properties required for a biomaterial. However, uncontrolled degradation, toxicity of degradation products, lack of biological cues, and immune reactivity limit the application of synthetic polymers in regenerative medicine (Nooeaid et al., 2012; Thankam et al., 2013). Interestingly, hybrid scaffolds developed by combining synthetic and natural polymers have been gaining significance owing to their capability to overcome the limitations of their individual applications (O'Brien, 2011; Thankam and Muthu, 2014a). A novel approach employing the ECM of decellularized organs or tissues as TE scaffold has been benefitted by excellent biocompatibility and bioactivity (Stratton et al., 2016).

An optimal scaffold for TE would be able to incorporate the mechanical properties of synthetic materials and the biological characteristics of natural materials (Thankam and Muthu, 2014b). Biodegradability and biocompatibility is the first and foremost feature required for a TE scaffold, where the scaffold as a whole and/or the degradation products should be nontoxic and immunocompatible. The mechanical characteristics of the biomaterial need to be at par with that of the host tissue that is essential to maintain cell mechanoregulation and structural integrity. Controlled degradation is another essential attribute of a scaffold material, in which the rate of degradation of the biomaterial is proportionate with neotissue formation. Moreover, the 3D architecture of the scaffold determines the cell viability, and the pore size and density influence nutrient traffic, angiogenesis, cell migration, proliferation, and signaling. The ease to manipulate the scaffold material into different dimensions for the purpose of implantation is another desirable feature (Turnbull et al., 2018).

As mentioned above, natural polymers investigated for TE include protein- and carbohydrate-based polymers such as collagen, fibrin, silk, hyaluronin, chitosan, and alginate. Several studies have proven the efficiency of natural biomaterials to support the differentiation of versatile stem cell types. Collagen is the most preferred biomaterial for TE applications owing to its biocompatibility and biodegradability. Amino acid domains such as Gly—Pro—hydroxyproline, Gly—Phe—hydroxyproline, and Arg-Gly-Asp (RGD) present in collagen molecule interact with specific membrane receptors such as glycoprotein-IV, discoidin domain receptor-1(DDR1), and discoidin domain receptor-2 (DDR2), facilitating cell adhesion (Parenteau-Bareil et al., 2010). For example, the collagen I and hyaluronic acid (HA) scaffolds supported the chondrogenic differentiation of bone marrow mesenchymal stem cells (BMSC) when cultured under hypoxic conditions (Bornes et al., 2015). Assi et al. reported that activated MSCs delivered in a biomimetic collagen scaffold enhanced wound healing in a translationally relevant diabetic mouse model (Assi et al., 2016). In another recent study, the behavior of human induced pluripotent stem cell—derived neural progenitors on 3D laminin-coated collagen scaffolds was demonstrated to be an efficient candidate for neuronal TE applications (Khayyatan et al., 2014). Also, the collagen-based scaffolds support the differentiation and development of human ESCs into blood vessels (Gerecht-Nir et al., 2003). Similarly, 2D and 3D collagen scaffolds were used for the differentiation of human ESCs into hepatocytes, which expressed several features of hepatocyte including urea production (Baharvand et al., 2006). A similar study showed the potency of primary preadipocytes to differentiate into mature adipocytes on both 2D and 3D collagen scaffolds. Moreover, adipocytes cultured in 3D collagen scaffolds remain confined within the matrix and remain intact during biochemical analysis when compared with 2D system (Daya et al., 2007).

Another natural protein widely used for preparing TE scaffold is fibrin. Fibrin, the final product of the blood coagulation cascade, is formed by the polymerization of the precursor molecule fibrinogen. Several properties such as the ease of polymerization and content of biological cues for cell differentiation promote fibrin to be a favorite candidate of tissue engineers (Brown and Barker, 2014). To cite, mouse ESCs were induced to differentiate to neural progenitor cells in 2D and 3D fibrin scaffolds. After 14 days, it was observed that the stem cells differentiated into neurons and astrocytes, where the 3D fibrin scaffold cultures were superior to 2D culture in terms of cell proliferation and differentiation potential (Willerth et al., 2006). Liu et al. evaluated the efficiency of fibrin gel and PEGylated fibrin gel scaffolds for promoting vasculature formation from mouse ESCs. The findings revealed that ESC proliferation was higher in fibrin and PEGylated fibrin scaffolds when compared with conventional 2D suspension culture and methylcellulose scaffolds (Liu et al., 2006). Also, hMSCs were viable on fibrin scaffolds in which the concentration of fibrinogen and thrombin solution influenced the proliferation of the cells and maintenance of 3D morphology (Ho et al., 2006). In another study, myoblast cells suspended in fibrin gel was injected into the ischemic myocardium of rat model, which increased the survival rate of transplanted cells, reduced the infarct size, and reversed the blood flow to the infarct zone (Christman et al., 2004). Encapsulation of human umbilical cord MSCs in fibrin hydrogel microbeads greatly enhanced cell viability and successful myogenic differentiation for muscle TE (Liu et al., 2013).

Silk fibroin is another superior natural material for TE in regard to biocompatibility and mechanical properties. Advanced technologies in structural and chemical fabrications enhance the dependency of silk fibroin—based material for various TE applications (Correia et al., 2012). Silk fibroin is ideal for ligament TE and extensive research have been conducted in this direction in conjunction with MSCs (Sakiyama-Elbert, 2008). Human BMSCs seeded in macroporous 3D aqueous-derived silk fibroin scaffolds showed higher proliferation rate and exhibited appreciable outcomes for orthopedic TE (Kim et al., 2005). Meinel et al. demonstrated the suitability of silk-based biomaterials for the culture of BMSCs (Meinel et al., 2005). The silk fibroin harvested from *Antheraea mylitta* displayed superior characteristics for cardiac TE. The presence of RGD domain and fibronectin-like properties in silk fibroin upgrades it to be an ideal candidate for cardiac TE scaffolds (Patra et al., 2012). A recent study showed that human skeletal muscle myoblasts adhered and deposited matrixes onto fibroin scaffolds to create a permissive environment for neomuscle formation (Chaturvedi et al., 2017).

Agarose represents an efficient polymeric system with respect to safety, effectiveness, cost, and availability. With water, agarose forms a rigid and transparent 3D network with a highly porous reticulum. A handful of studies demonstrated that agarose gel has unique physical and chemical features for stem cell promoting the growth and proliferation of cell. Agarose hydrogels support chondrogenic differentiation of human adipose-derived adult stem cells revealing its potential application in cartilage TE (Awad et al., 2004). Also, the bovine MSCs seeded in agarose undergone chondrogenesis; however, the mechanical properties of the matrix formed were inferior to that of cartilage tissue (Mauck et al., 2006). Dopamine-releasing cells isolated from monkey ESCs revealed colonization in agarose microcapsules suggesting the potential of agarose gels to promote stem cell homing and differentiation (Ando et al., 2007).

Owing to the superior viscoelastic nature and intrinsic hydrophilicity, the marine polysaccharide, alginate, has gained much attention as a natural biomaterial for TE. As a scaffold, alginate facilitates the uniform distribution of cells mimicking the native ECM (Gnanaprakasam Thankam et al., 2013). Human MSCs cultured on calcium alginate scaffolds remain viable for several weeks and secrete healing factors such as VEGF and basic fibroblast growth factor, suggesting the application of alginate-based scaffolds onto surgical sites for local maintenance of growth factors for effective regeneration of the injured tissue (Schmitt et al., 2015). Adipose tissue—derived stem cells (ADMSCs) loaded in alginate microspheres were transplanted into the liver in mouse model, where the transplanted ADMSCs underwent hepatogenic differentiation (Chen et al., 2015). Another study demonstrated that the MSCs isolated from orofacial tissue encapsulated in RGD-modified alginate scaffold promoted craniofacial bone regeneration (Moshaverinia et al., 2014). An injectable 3D RGD-coupled alginate scaffold with multiple growth factor delivery capacity encapsulated with gingival-derived MSCs was found to be a promising candidate for muscle TE (Ansari et al., 2016).

The native ECM protein HA/hyaluronin has also been used in manipulating TE scaffolds due to its biocompatibility, increased water solubility, and low immunogenicity. The scope of HA modification into versatile physical forms by chemical treatments increases the demand of HA as an ideal scaffold material (Burdick and

Prestwich, 2011). Complete hyaline cartilage regeneration using composites of human umbilical cord blood-derived mesenchymal stem cells and HA hydrogel has been demonstrated in mini pig model (Ha et al., 2015). In another study, HA gel was used to culture human ADMSCs to create soft tissue filler that promoted the growth of de novo adipose tissue. The cells adhered and proliferated on the scaffold and in vivo grafts showed the formation of well-organized adipose tissue (Huang et al., 2015). Injection of spheroids were prepared by culturing human ADMSCs in non—cross-linked HA gel, which promoted angiogenesis and tissue regeneration following the injury in mice models (Mineda et al., 2015).

Debanath et al. investigated the efficiency of chitosan hydrogel for supporting the proliferation and differentiation of human ADMSCs and observed appreciable cell proliferation without compromising the cell viability (Debnath et al., 2015). The suitability of chitosan for promoting the differentiation of MSCs to chondrogenic lineage suggested that medium molecular weight chitosan was more efficient for inducing the differentiation than lower and high molecular weight HA (Lu et al., 2017). However, the effect of the molecular weight of chitosan on MSC differentiation is largely unknown, which warrants further research.

A few synthetic polymers widely used in regenerative medicine include poly(caprolactone) (PCL) and, in particular, poly(α-hydroxy esters):poly(lactic acid) (PLA), poly(glycol alcohol) (PGA), the copolymer of PLA and PGA, poly(vinyl alcohol), and poly(ethylene glycol) (PEG) (Martin et al., 2007; O'Shea and Miao, 2008). Recent findings suggest that PCL nanofibres blended with PLA supported the osteogenic differentiation of human MSCs. In vitro and in vivo examinations showed that the blend scaffold possessed appreciable mechanical strength and bioactivity to maintain cell viability and differentiation (Yao et al., 2017). Also, the murine iPSCs seeded within 3D PCL nanofibrous scaffold differentiated into functional cardiomyocytes more efficiently when compared with the cells grown in tissue culture plates (Chen et al., 2015). In a similar study, biodegradable poly(L-lactide)/PEG scaffolds seeded with human amniotic mesenchymal cells implanted in rabbit model of urethral defect revealed faster healing outcomes (Lv et al., 2016).

The abovementioned reports are only a few among the myriads of trials conducted around the globe to explore the suitability of biomaterials as carriers of stems cells for TE and regenerative medicine. The biomaterial- and stem cell—mediated TE requires more volumes to explain. However, it is noteworthy that the molecular interplay at the stem cell—biomaterial interface determines the sustainability and success of TE approaches. The following sections throw light to the interface biology and the strategies adopted to improve TE-based regenerative therapies.

2.6 Influence of scaffold patterns in stem cell behavior

In native niches, the stem cells encounter various anatomical topographies such as bones, ligaments, cells, and several macromolecules, which influence their behavior and fate. A vital structure that the stem cells interact with is the basement membrane that contains numerous nanoscale pores, ridges, and fibers. With the advent of micro- and

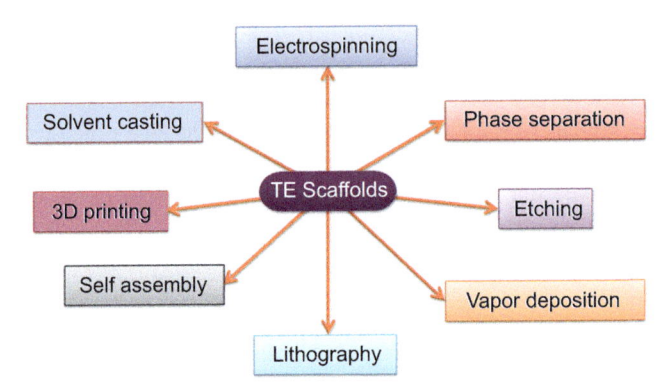

Figure 2.3 Several techniques employed to manipulate nanoscale topographies such as porosity roughness, etc., on the scaffold surface. This is very essential for the successful adherence, migration, and proliferation of cells on scaffold surface. These nanoscale topographies are found even to dictate the fate of stem cells.

nanofabrication techniques, extensive studies have been carried out to investigate the response of stem cells to the topographical features of the biomaterial scaffolds. The biophysical signals present in the substratum such as nanotopography, mechanical forces, stiffness of the matrix, and roughness of the biomaterial determine the adhesion, migration, proliferation, and differentiation of cells. Several techniques are employed to introduce nanoscale features in scaffold materials (Fig. 2.3).

Managing pore size, porosity, and permeability is a key factor for the fruitful operation of the stem cell–seeded scaffolds. It is suggested that the pore should be large enough to ensure nutrient trafficking while not too large to hinder cell attachment and migration. Pore size and porosity are entirely different entities, where pore size refers to the geometry of the pore and porosity refers to the density of pores per unit area. The pores can be nanometric, micrometric, or macrometric. Nanopore sizes are found to enhance cell attachment improving further cell functions such as proliferation, differentiation, and migration. Anchorage-dependent cell–cell interactions require micropores, whereas cell migration and direct cell–cell contact demand still larger pores. The effect of pore size and cell behavior is depicted in Table 2.2. Elasticity of the scaffolds also impacts stem cell differentiation. Matrix stiffness varies for different cells and thus it is reasonable to assume that stiffness exhibits profound influence on lineage differentiation. Various studies have established that MSCs differentiate differently in scaffolds of varying stiffness. For example, the MSCs differentiate into neuronal lineage in soft scaffolds, while into osteogenic lineage in harder substrates, whereas the medium stiff scaffolds resulted in myogenic lineage (Ghasemi-Mobarakeh et al., 2015). Moreover, the scaffold stiffness affects the differentiation of MSCs to vascular lineage, which was evident that MSCs cultured on soft scaffolds expressed Flk-1 endothelial markers, while those cultured on a much harder scaffold expressed α-actin markers of smooth muscle cells (Wingate et al., 2012). Integrin-mediated adhesions are the major players, which facilitate stem cells to sense and respond to the biophysical cues. Several integrin-mediated signaling mechanisms

Table 2.2 Effect of Pore size on stem cells (Bružauskaitė et al., 2016).

Type of scaffold	Pore size	Effect
2D Scaffold Membrane	Less than 1 μm	Better cell attachment
2D Scaffold Membrane	1–3 μm	Anchorage-dependent cell–cell interaction
2D Scaffold Membrane	3–12 μm	Direct cell–cell contact and migration
3D Scaffold	1–3 μm with porous internal structure	Indirect cell–cell or cell–ECM interaction
3D Scaffold	100–800 μm	Cell migration in and out of the scaffold

are proposed to explain how topographic features that influence the stem cell fate; however, a complete picture remains elusive. The balance of force generated during the contractility of endogenous cytoskeleton and external mechanical forces that are transmitted across cell ECM adhesions regulate such signaling. A few pathways include Ras/MAPK, PI3K/Akt, RhoA/ROCK, and Wnt/βcatenin, and these determine the fate of stem cells onto the interstices of scaffolds.

2.6.1 Scaffold decoration with stem cells

The adhesion of stem cells on to the surface of the scaffold material is very crucial for the biological performance of TE implants. Various strategies have been adopted to improve the adhesion of stem cells on to the surface of the scaffold materials. A major approach is modulating the surface properties of the scaffold material by mosaicking biomolecules that interact with specific membrane receptors of the stem cells (Fig. 2.4). Native ECM proteins such as collagen subtypes, fibronectin, and vitronectin, which bind with transmembrane receptors, are the primary molecules targeted for this strategy. These proteins contain several amino acid domains including RGD and Pro-His-Ser-Arg-Asn, which interact with the integrin molecules of the cells. However, chances of immunological insults, conformational changes arising because of tethering, and pathogen transfer hurdle the use of these proteins as a whole. Such drawbacks led to the use of amino acid domains in the form of short adhesive peptides to bridge the interface between cells and biomaterial surface (Bellis, 2011; Mao and Schwarzbauer, 2006; Pacelli et al., 2017).

Adhesive peptides can overcome the abovementioned demerits of using whole proteins. Added to this, the ease of bulk production of these peptides in short time span enhances their chances in clinical applications. Furthermore, the peptide–biomaterial surface interaction can be improved with several physical and chemical strategies. Physical interactions including the hydrogen bonding, electrostatic forces, and hydrophobic interactions have been exploited for the better entanglement of these

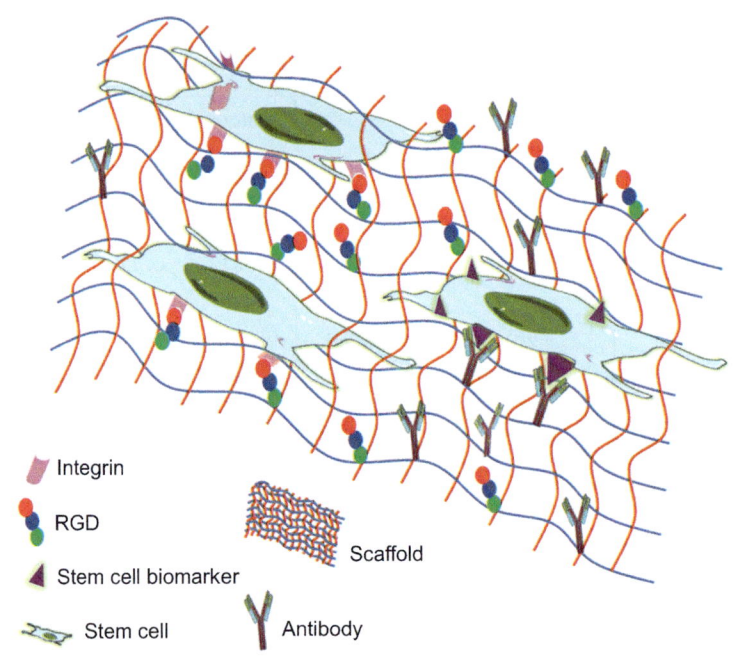

Figure 2.4 Depiction of a biomaterial scaffold containing biomolecules such as peptides and antibodies that immobilizes stem cells onto scaffold surface. Several chemical and physical interactions between biomolecules and reactive groups present on the scaffold are exploited to fix these biomolecules on to the scaffolds.

adhesive cues within the scaffolds. Physical bonds have the advantage of application to biomaterial irrespective of their natural or synthetic origin without harsh chemical treatments. However, the instability of binding due to fluctuations in the environmental conditions such as pH, ionic strength, fluidity, and temperature offers challenge. Chemical conjugation has been found superior to physical method in this context. In addition, the unreactive scaffold surfaces can be chemically modified to introduce binding moieties, which improve stem cell binding. There are also instances where physical and chemical strategies are being used together to facilitate better adhesion of reactive biomolecules on to the surface of the scaffolds (Pacelli et al., 2017).

Among the peptides, RGD sequence is the most investigated, which is found in most native ECM proteins that specifically bind with integrin moieties of the cells. The peptide conformation and the degree of spacing between clusters of RGD are the main factors that influence the extent of stem cell adhesion (Frith et al., 2012). Also, the cyclic form of RGD facilitates stable interactions than the linear form (Hsiong et al., 2009). In addition, RGDs coupled with other sequences such as Tyr—Ile—Gly—Ser—Arg and triple helical Gly-Phe-hydroxyproline-Gly-Glu-Arg provide extra binding loci with other proteins such as laminin receptor and collagens. Recently, phage display technology has been introduced to engineer synthetic peptides that bind to specific stem cell population (Wu et al., 2016).

Surface coating of scaffold material with antibodies is another strategy adopted to immobilize stem cells. The antibodies can be attached to the scaffold surfaces by chemical and physical interactions with the reactive functional groups present on the surface of the biomaterial. Antigens expressed on the surface of stem cells bind specifically with the immobilized antibodies in the scaffolds (Sin et al., 2005). In the case of inert scaffolds that are devoid of reactive moieties, biomolecules including hydrophobins, collagens, albumin, and heparin are introduced to the surface, which facilitates the attachment of the antibodies. In addition, the coating antibodies attract purified colonies of specific subpopulations of stem cells. For example, CD34 antibody immobilized on HA hydrogels selectively recruits EPCs (Camci-Unal et al., 2010). In a recent study, a biomimetic chitosan membrane was developed onto which specific antibodies were covalently immobilized using bis(sulfosuccinimidyl) suberate. These antibodies maintained the bioactivity and attracted specific subpopulation of cells from a heterogeneous population (Custódio et al., 2012). Also, by immobilizing stem cell−specific antibodies onto collagen scaffolds and their implantation into the wound site has been proven to promote stem cell enrichment for cardiac regeneration. To achieve this, functional collagen scaffolds were generated by coating antibodies of Sca-1 antigen expressed on cardiac stem cells. The Sca-1^{+} cells adhered to the scaffolds due to antigen−antibody interaction, both in vitro and in vivo, and the subsequent transplantation to the defective heart in the mouse model resulted in appreciable regeneration of cardiomyocytes (Shi et al., 2011). In another approach, titanium coated with anti-CD34 antibody was implanted into canine femoral artery and was found to augment EPC attachment leading to a rapid and complete endothelialization of the lumenal surface (Li et al., 2010).

Aptamers represent another set of valid biomolecule that act at interface between stem cells and scaffold materials (Pacelli et al., 2017). Aptamers are single-stranded oligonucleotides either DNA or RNA, which fold into 3D structures to bind with a variety of targets such as proteins, peptides, enzymes, antibodies, and various cell surface receptors. Aptamers can be of 100 bases long, which are generated by a method called systematic evolution of ligands by exponential enrichment (SELEX) (Guo et al., 2006). In this method, the target molecule is incubated with the random nucleic acid pool, and later, the more strongly bound nucleic acid target complex is eluted after several levels of selection processes. The nucleic acid thus obtained is amplified to obtain the single aptamer that is specific for the selected target. And, the aptamers exhibit specificity for targets similar to antibodies (Yoon et al., 2015) and therefore exhibit the advantages of antibodies including low immunogenicity, efficient entry into biological compartments due to smaller size, contamination-free production, stability in storage, easy and rapid production, and flexibility of modifications of functional groups during synthesis (Keefe et al., 2010). In addition, aptamers possess certain unique features such as thermal stability, low cost, and versatility in applications (Song et al., 2012). Currently, aptamers are being used for the detection of cancerous cells; however, it is suggested that aptamers with high specificity can be used to isolate and recruit specific stem cell populations (Wiraja et al., 2014). In an attempt, adult MSCs were isolated from bone marrow using aptamers. As specific surface markers of MSCs were lacking, SELEX technique was used to obtain aptamers with high binding affinity, in which the combination of multiple markers are used for stem cell isolation (Table 2.1) (Guo et al., 2006). It is hypothesized that aptamers that recognize the specific markers expressed on the surface of stem cells can be used to immobilize them on the scaffold surface. However, till now no reports on such trials appear in the literature.

2.6.1.1 Chemokine-decorated scaffolds as stem cell recruiter

Although stem cell—based treatment is a promising strategy in regenerative medicine, certain challenges restrict the in vivo delivery of ex vivo expanded stem cells. As mentioned in the previous section, the limited availability of stem cell sources, the excessive cost of commercialization, the anticipated difficulties of clinical translation, and regulatory approval are a few to name (Chen et al., 2011). Also, the ASCs used in regenerative treatments are restricted to certain lineages and their regenerative capacity is declined by age (Lemcke et al., 2018). Moreover, most of the clinical and translational results have failed to fulfill the expectations laid. These challenges can be addressed, to a greater extent, by employing chemokines that promote stem cell homing. Versatile chemical attractants released as a result of the wound mediate this physiological response. The discovery of the role of such chemoattractants paved the way for an alternative treatment strategy, wherein exogenously administered chemokines are used to trigger the mobilization of stem cell from their native niches to the site of damage. Moreover, more precise delivery of such signals to the site of injury can be achieved by incorporating such cues to biocompatible biomaterials and transplanting to the damaged tissues, which triggers the natural healing mechanisms. Also, it is extremely important that the implanted biomaterial should provide the microenvironment required for homing and subsequent regeneration of the tissues (Ko et al., 2013). Among the mediators employed for enhancing the stem cell mobilization, a few are used in combination with biomaterials including SDF1, substance P, granulocyte-macrophage colony-stimulating factor (G-CSF), monocyte chemotactic protein-3 (MCP-3), and others (Ko et al., 2013; Li et al., 2017).

A novel collagen scaffold designed to contain radially oriented channels mosaicked with SDF1 promoted osteochondral repair by enhancing the migration of BMSCs in rabbit models (Chen et al., 2015). In a similar study, comparable results were obtained, where the titanium implants were employed to deliver SDF-1α to enhance bone formation. Moreover, it has been suggested that the controlled delivery of SDF1 enhances the local recruitment of progenitor cells promoting the osseointegration (Karlsson et al., 2016). In another study, biocompatible chitosan (Ch)/poly(γ-glutamic acid) scaffold was assembled layer by layer incorporated with SDF-1α that enhanced the stem cells migration. Therefore, the controlled release of SDF1 possesses greater therapeutic potential in regard to MSC homing to injured tissues (Goncalves et al., 2012).

A bioactive knitted silk-collagen sponge scaffold incorporated with exogenous SDF-1α enhanced the recruitment of fibroblast-like cells resulting in improved efficacy for tendon regeneration. In vitro studies showed that CXCR4 gene expression, migration of bone mesenchymal stromal cells, and hypodermal fibroblasts were more sensitive to exogenous SDF-1α (Shen et al., 2010). Recently, a recombinant

SDF-1α with collagen-binding domain was developed for facilitating the sustained release to an ischemic heart, which binds to collagen gel, and facilitating controlled release both in vitro and in vivo, resulting in the mobilization of endogenous stem cells to the ischemic heart that improved the cardiac function (Sun et al., 2016).

Substance P has been targeted for the recruitment of endogenous stem cell, which is a short peptide neurotransmitter secreted by peripheral sensory neurons in response to injury that acts through the receptor neurokinin 1 (NK1). It has been postulated that creating a substance P gradient between blood and bone marrow mobilizes NK1 expressing cells from bone marrow central niche to peripheral blood circulation (Li et al., 2017). Recently, several targeted biomaterial implants utilized substance P as a homing factor. For instance, substance P bound to small diameter PLCL grafts recruited endogenous progenitor cells for vascular regeneration (Shafiq et al., 2015). In another study, the controlled administration of substance P from electron spun membranes of vascular grafts generated with PLCL enhanced the recruitment of human bone marrow MSCs for vascular regeneration (Shafiq et al., 2016). Moreover, the substance P bound to self-assembled peptide matrices inhibits the progression of osteoarthritis by recruiting MSCs (Kim et al., 2016). PCL/collagen grafts with substance P were found to mobilize significantly higher population of MSCs when implanted in rat abdominal aorta. Substance P containing grafts also promoted smooth muscle cell regeneration, endogenous stem cell recruitment, and blood vessel formation, suggesting the significance of substance P in situ vascular regeneration (Shafiq et al., 2018).

G-CSF, an autocrine signaling molecule, stimulates the mobilization of HSCs and BMMSCs from their niche into the blood stream. The intramuscular administration of G-CSF encapsulated in PEG diacrylate−poly(ethylene imine) hydrogel scaffold resulted in enhanced mononuclear cells and EPC mobilization into the blood stream. Similarly, stem cell factor (SCF), another endogenous growth factor, functions to mobilize HSCs by binding to tyrosine kinase receptor c-kit, which is expressed on HSCs (Li et al., 2017). Moreover, recombinant SCFs have been shown to act in synergy with G-CSF in mobilization of bone marrow−derived HSCs (Ko et al., 2013).

MCP-3 modulates the migration and homing of stem cells. MCP-3 belonging to CC chemokine family is found to induce myocardial stem cell homing. There are evidences that release of MCP-3 from synthetic grafts caused the homing of cells from circulation effecting the regeneration of small diameter blood vessels in rat models. The inflammatory cytokine TNF-α triggers MSC migration and upregulates the intercellular adhesion molecule-1 on MSCs, prompting the cells to be more receptive to chemoattractants. However, in vivo studies are limited in this direction (Fu et al., 2009). The neuropeptide, galanin, promote bone marrow−derived MSC migration through activation of galanin receptor (Louridas et al., 2009).

In addition to the mentioned mediators, several factors such as platelet-derived growth factor (PDGF)-BB, PDGF-AB, epidermal growth factor (EGF), heparin-binding epidermal growth factor (HB-EGF), insulin growth factor, HGF, fibroblast growth factor 2 (FGF-2), and thrombin were also found to enhance migration of MSCs at appropriate concentrations (Ozaki et al., 2007). These mediators act by versatile signaling mechanisms depending on the nature of biomaterials employed and type of injury. Apart from the choice of biological signals, the methodology adopted for the loading of these mediators onto the scaffolds and the release kinetics greatly influence the stem cell homing. The present chapter is limited to a few models in regard to the interface biology of soft TE. The rapid advancements in the field of biomaterial science throw light to the emergence of novel strategies to improve the interface biology of stem cells, which offers pleasant hope to millions suffering from end-stage organ failure across the globe.

2.6.2 Summary and future directions

The understanding of interface biology is necessary to mend the interface of dissimilar tissues; hence, it warrants the complete restoration of tissue function followed by the implantation. The transition of physiochemical, biochemical, mechanical, and functional properties at the interface requires compatibility for the better performance of the seeded stem cells for soft tissue regeneration. The design of stem cell−based strategies to bridge the interface for creating homogeneous tissues warrants further investigation. The currently available techniques have limited ability to examine the interface owing to the smaller dimension of the interface (ranging from 100 μm to 1 mm in length). The mechanism of stem cell−mediated regeneration of soft tissues, especially at the interface, is largely unknown, and how the distinct boundaries between scaffolds and soft tissues are reestablished postinjury is also known. Also, the structure−function relationship of cell moieties with the scaffold interface and the coordinated interactions of multiple cell types represent complex challenges. This recommends a stratified scaffold design to incorporate the stem cells that can be achieved by 3D organ printing approaches, which may ensure the growth and differentiation of stem cells to multiple lineages and drive heterotypic and homotypic cellular interactions. In addition, the genetic and epigenetic mechanisms need to be considered to examine the gene expression of interface proteins. It is likely that the homeostasis at the interface is the cumulative effect of heterotypic cellular interactions and adhesion/regenerative signals triggered by the surviving and/or recruited cells.

Although significant progress has been made in stem cell−based regenerative medicine, this field of medicine is still in its infancy. Newer biocompatible materials, natural synthetic or hybrid, that can better accommodate and proliferate stem cells that are superior to the currently available material should be sought for. Detailing into the interactions of various stem cells with native ECM of different tissues may aid in developing novel scaffolds that can shelter stem cells in a better way. Strategies are needed to enhance the attachment of stem cells on to the surface of the scaffold material, and intelligent materials that can manipulate the stem cells depending on the pathological signals are wanting. Modification of scaffolds using proteins of native ECM that attach to stem cells in vivo needs to be investigated for stem cell recruitment. Also, the surface markers of different stem cells demand further attention. Along with this, minimal invasive methodologies for harnessing stem cells without losing their vigor are a need of the hour. Harvesting and subsequent in vitro expansion of stem cells lead to a decrease in their differentiation potential. The different cellular events and signaling pathways initiated by stem cells in regenerative process warrants further investigation for the better understanding of the underlying mechanism of wound healing and tissue regeneration. Stem cell programming, which promotes their homing in biomaterials, also needs to be investigated. Exosomal proteins from stem cells are found to mediate several signaling pathways involved in tissue regeneration. An insight into the role of these exosomal proteins in mediating the different events of regeneration can be beneficial for the future use of such proteins in combination with biomaterials for initiating wound healing events. Nonetheless, the interface biology of stem cells versus scaffolds has been proven successful in addressing several limiting factors of TE, which significantly improved our understanding on regenerative medicine. A handful of technologies on this aspect have been approved by FDA and a good number of them are in clinical trials. The rapid advancements in the field of biomaterial science throw light to the emergence of novel strategies to improve the interface biology of stem cells, which offers pleasant hope to millions suffering from end-stage organ failure across the globe.

References

Ando, T., Yamazoe, H., Moriyasu, K., Ueda, Y., Iwata, H., 2007. Induction of dopamine-releasing cells from primate embryonic stem cells enclosed in agarose microcapsules. Tissue Eng 13, 2539–2547. https://doi.org/10.1089/ten.2007.0045.

Ansari, S., Chen, C., Xu, X., Annabi, N., Zadeh, H.H., Wu, B.M., Khademhosseini, A., Shi, S., Moshaverinia, A., 2016. Muscle tissue engineering using gingival mesenchymal stem cells encapsulated in alginate hydrogels containing multiple growth factors. Ann. Biomed. Eng. 44, 1908–1920. https://doi.org/10.1007/s10439-016-1594-6.

Assi, R., Foster, T.R., He, H., Stamati, K., Bai, H., Huang, Y., Hyder, F., Rothman, D., Shu, C., Homer-Vanniasinkam, S., Cheema, U., Dardik, A., 2016. Delivery of mesenchymal stem cells in biomimetic engineered scaffolds promotes healing of diabetic ulcers. Regen. Med. 11, 245–260. https://doi.org/10.2217/rme-2015-0045.

Awad, H.A., Wickham, M.Q., Leddy, H.A., Gimble, J.M., Guilak, F., 2004. Chondrogenic differentiation of adipose-derived adult stem cells in agarose, alginate, and gelatin scaffolds. Biomater 25, 3211–3222. https://doi.org/10.1016/j.biomaterials.2003.10.045.

Baharvand, H., Hashemi, S.M., Kazemi Ashtiani, S., Farrokhi, A., 2006. Differentiation of human embryonic stem cells into hepatocytes in 2D and 3D culture systems in vitro. Int. J. Dev. Biol. 50, 645–652. https://doi.org/10.1387/ijdb.052072hb.

Beane, O.S., Darling, E.M., 2012. Isolation, characterization, and differentiation of stem cells for cartilage regeneration. Ann. Biomed. Eng. 40, 2079–2097. https://doi.org/10.1007/s10439-012-0639-8.

Bellis, S.L., 2011. Advantages of RGD peptides for directing cell association with biomaterials. Biomaterials 32, 4205–4210. https://doi.org/10.1016/j.biomaterials.2011.02.029.

Berthiaume, F., Maguire, T.J., Yarmush, M.L., 2011. Tissue engineering and regenerative medicine: history, progress, and challenges. Annu. Rev. Chem. Biomol. Eng. 2, 403–430. https://doi.org/10.1146/annurev-chembioeng-061010-114257.

Bornes, T.D., Jomha, N.M., Mulet-Sierra, A., Adesida, A.B., 2015. Hypoxic culture of bone marrow-derived mesenchymal stromal stem cells differentially enhances in vitro chondrogenesis within cell-seeded collagen and hyaluronic acid porous scaffolds. Stem Cell Res. Ther. 6, 84. https://doi.org/10.1186/s13287-015-0075-4.

Brown, A.C., Barker, T.H., 2014. Fibrin-based biomaterials: modulation of macroscopic properties through rational design at the molecular level. Acta Biomater. 10, 1502–1514. https://doi.org/10.1016/j.actbio.2013.09.008.

Bružauskaitė, I., Bironaitė, D., Bagdonas, E., Bernotienė, E., 2016. Scaffolds and cells for tissue regeneration: different scaffold pore sizes-different cell effects. Cytotechnology 68, 355–369. https://doi.org/10.1007/s10616-015-9895-4.

Burdick, J.A., Prestwich, G.D., 2011. Hyaluronic acid hydrogels for biomedical applications. Adv. Mater. Deerfield Beach Fla 23, H41–56. https://doi.org/10.1002/adma.201003963.

Camci-Unal, G., Aubin, H., Ahari, A.F., Bae, H., Nichol, J.W., Khademhosseini, A., 2010. Surface-modified hyaluronic acid hydrogels to capture endothelial progenitor cells. Soft Matter 6, 5120–5126. https://doi.org/10.1039/c0sm00508h.

Chaturvedi, V., Naskar, D., Kinnear, B.F., Grenik, E., Dye, D.E., Grounds, M.D., Kundu, S.C., Coombe, D.R., 2017. Silk fibroin scaffolds with muscle-like elasticity support in vitro differentiation of human skeletal muscle cells. J. Tissue Eng. Regenerat. Med. 11, 3178–3192. https://doi.org/10.1002/term.2227.

Chen, F.-M., Liu, X., 2016. Advancing biomaterials of human origin for tissue engineering. Prog. Polym. Sci. 53, 86–168. https://doi.org/10.1016/j.progpolymsci.2015.02.004.

Chen, F.-M., Wu, L.-A., Zhang, M., Zhang, R., Sun, H.-H., 2011. Homing of endogenous stem/progenitor cells for in situ tissue regeneration: promises, strategies, and translational perspectives. Biomaterials 32, 3189–3209. https://doi.org/10.1016/j.biomaterials.2010.12.032.

Chen, M.-J., Lu, Y., Simpson, N.E., Beveridge, M.J., Elshikha, A.S., Akbar, M.A., Tsai, H.-Y., Hinske, S., Qin, J., Grunwitz, C.R., Chen, T., Brantly, M.L., Song, S., 2015. In situ transplantation of alginate bioencapsulated adipose tissues derived stem cells (ADSCs) via hepatic injection in a mouse model. PLoS One 10, e0138184. https://doi.org/10.1371/journal.pone.0138184.

Christman, K.L., Vardanian, A.J., Fang, Q., Sievers, R.E., Fok, H.H., Lee, R.J., 2004. Injectable fibrin scaffold improves cell transplant survival, reduces infarct expansion, and induces neovasculature formation in ischemic myocardium. J. Am. Coll. Cardiol. 44, 654–660. https://doi.org/10.1016/j.jacc.2004.04.040.

Correia, C., Bhumiratana, S., Yan, L.-P., Oliveira, A.L., Gimble, J.M., Rockwood, D., Kaplan, D.L., Sousa, R.A., Reis, R.L., Vunjak-Novakovic, G., 2012. Development of silk-based scaffolds for tissue engineering of bone from human adipose-derived stem cells. Acta Biomater. 8, 2483–2492. https://doi.org/10.1016/j.actbio.2012.03.019.

Custódio, C.A., Frias, A.M., del Campo, A., Reis, R.L., Mano, J.F., 2012. Selective cell recruitment and spatially controlled cell attachment on instructive chitosan surfaces functionalized with antibodies. Biointerphases 7, 65. https://doi.org/10.1007/s13758-012-0065-3.

Daya, S., Loughlin, A.J., Macqueen, H.A., 2007. Culture and differentiation of preadipocytes in two-dimensional and three-dimensional in vitro systems. Differ. Res. Biol. Divers. 75, 360–370. https://doi.org/10.1111/j.1432-0436.2006.00146.x.

Debnath, T., Ghosh, S., Potlapuvu, U.S., Kona, L., Kamaraju, S.R., Sarkar, S., Gaddam, S., Chelluri, L.K., 2015. Proliferation and Differentiation Potential of Human Adipose-Derived Stem Cells Grown on Chitosan Hydrogel. PLoS ONE 10. https://doi.org/10.1371/journal.pone.0120803.

Finosh, G.T., Jayabalan, M., 2012. Regenerative therapy and tissue engineering for the treatment of end-stage cardiac failure: new developments and challenges. Biomatter 2, 1–14. https://doi.org/10.4161/biom.19429.

Frith, J.E., Mills, R.J., Cooper-White, J.J., 2012. Lateral spacing of adhesion peptides influences human mesenchymal stem cell behaviour. J. Cell Sci. 125, 317–327. https://doi.org/10.1242/jcs.087916.

Fong, E.L.S., Chan, C.K., Goodman, S.B., 2011. Stem cell homing in musculoskeletal injury. Biomaterials 32, 395−409. https://doi.org/10.1016/j.biomaterials.2010.08.101.

Fu, X., Han, B., Cai, S., Lei, Y., Sun, T., Sheng, Z., 2009. Migration of bone marrow-derived mesenchymal stem cells induced by tumor necrosis factor-α and its possible role in wound healing. Wound Repair Regen. 17, 185−191. https://doi.org/10.1111/j.1524-475X.2009.00454.x.

Gerecht-Nir, S., Ziskind, A., Cohen, S., Itskovitz-Eldor, J., 2003. Human embryonic stem cells as an in vitro model for human vascular development and the induction of vascular differentiation. Lab. Investig. J. Tech. Methods Pathol. 83, 1811−1820.

Ghasemi-Mobarakeh, L., Prabhakaran, M.P., Tian, L., Shamirzaei-Jeshvaghani, E., Dehghani, L., Ramakrishna, S., 2015. Structural properties of scaffolds: crucial parameters towards stem cells differentiation. World J. Stem Cell. 7, 728−744. https://doi.org/10.4252/wjsc.v7.i4.728.

Ghieh, F., Jurjus, R., Ibrahim, A., Geagea, A.G., Daouk, H., El Baba, B., Chams, S., Matar, M., Zein, W., Jurjus, A., 2015. The Use of Stem Cells in Burn Wound Healing: A Review. BioMed Res. Int. 2015 684084. https://doi.org/10.1155/2015/684084.

Giwa, S., Lewis, J.K., Alvarez, L., Langer, R., Roth, A.E., Church, G.M., Markmann, J.F., Sachs, D.H., Chandraker, A., Wertheim, J.A., Rothblatt, M., Boyden, E.S., Eidbo, E., Lee, W.P.A., Pomahac, B., Brandacher, G., Weinstock, D.M., Elliott, G., Nelson, D., Acker, J.P., Uygun, K., Schmalz, B., Weegman, B.P., Tocchio, A., Fahy, G.M., Storey, K.B., Rubinsky, B., Bischof, J., Elliott, J.A.W., Woodruff, T.K., Morris, G.J., Demirci, U., Brockbank, K.G.M., Woods, E.J., Ben, R.N., Baust, J.G., Gao, D., Fuller, B., Rabin, Y., Kravitz, D.C., Taylor, M.J., Toner, M., 2017. The promise of organ and tissue preservation to transform medicine. Nat. Biotechnol. 35, 530−542. https://doi.org/10.1038/nbt.3889.

Gnanaprakasam Thankam, F., Muthu, J., Sankar, V., Kozhiparambil Gopal, R., 2013. Growth and survival of cells in biosynthetic poly vinyl alcohol−alginate IPN hydrogels for cardiac applications. Colloids Surf. B Biointerfaces 107, 137−145. https://doi.org/10.1016/j.colsurfb.2013.01.069.

Goncalves, R.M., Antunes, J.C., Barbosa, M.A., 2012. Mesenchymal stem cell recruitment by stromal derived factor-1-delivery systemsbased on chitosan/poly(gamma-glutamic acid) polyelectrolyte complexes. Eur. Cells Mater. 23, 249−260.

Graham, N., Qian, B.-Z., 2018. Mesenchymal stromal cells: emerging roles in bone metastasis. Int. J. Mol. Sci. 19. https://doi.org/10.3390/ijms19041121.

Griffin, M.F., Butler, P.E., Seifalian, A.M., Kalaskar, D.M., 2015. Control of stem cell fate by engineering their micro and nanoenvironment. World J. Stem Cells 7, 37−50. https://doi.org/10.4252/wjsc.v7.i1.37.

Grisendi, G., Annerén, C., Cafarelli, L., Sternieri, R., Veronesi, E., Cervo, G.L., Luminari, S., Maur, M., Frassoldati, A., Palazzi, G., Otsuru, S., Bambi, F., Paolucci, P., Pierfranco, C., Horwitz, E., Dominici, M., 2010. GMP-manufactured density gradient media for optimized mesenchymal stromal/stem cell isolation and expansion. Cytotherapy 12, 466−477. https://doi.org/10.3109/14653241003649510.

Gronthos, S., Zannettino, A.C.W., Hay, S.J., Shi, S., Graves, S.E., Kortesidis, A., Simmons, P.J., 2003. Molecular and cellular characterisation of highly purified stromal stem cells derived from human bone marrow. J. Cell Sci. 116, 1827−1835.

Guo, K.-T., SchAfer, R., Paul, A., Gerber, A., Ziemer, G., Wendel, H.P., 2006. A new technique for the isolation and surface immobilization of mesenchymal stem cells from whole bone marrow using high-specific DNA aptamers. Stem Cells Dayt. Ohio 24, 2220−2231. https://doi.org/10.1634/stemcells.2006-0015.

Ha, C.-W., Park, Y.-B., Chung, J.-Y., Park, Y.-G., 2015. Cartilage repair using composites of human umbilical cord blood-derived mesenchymal stem cells and hyaluronic acid hydrogel in a minipig model. Stem Cells Transl. Med. 4, 1044−1051. https://doi.org/10.5966/sctm.2014-0264.

Han, J.W., Yoon, Y.-S., 2011. Induced pluripotent stem cells: emerging techniques for nuclear reprogramming. Antioxid. Redox Signal. 15, 1799−1820. https://doi.org/10.1089/ars.2010.3814.

Heidary Rouchi, A., Mahdavi-Mazdeh, M., 2015. Regenerative Medicine in Organ and Tissue Transplantation: Shortly and Practically Achievable? Int. J. Organ Transplant. Med. 6, 93−98.

Kränkel, N., Spinetti, G., Amadesi, S., Madeddu, P., 2011. Targeting stem cell niches and trafficking for cardiovascular therapy. Pharmacol. Ther. 129, 62−81. https://doi.org/10.1016/j.pharmthera.2010.10.002.

Ho, W., Tawil, B., Dunn, J.C.Y., Wu, B.M., 2006. The behavior of human mesenchymal stem cells in 3D fibrin clots: dependence on fibrinogen concentration and clot structure. Tissue Eng. 12, 1587−1595. https://doi.org/10.1089/ten.2006.12.1587.

Howard, D., Buttery, L.D., Shakesheff, K.M., Roberts, S.J., 2008. Tissue engineering: strategies, stem cells and scaffolds. J. Anat. 213, 66−72. https://doi.org/10.1111/j.1469-7580.2008.00878.x.

Hsiong, S.X., Boontheekul, T., Huebsch, N., Mooney, D.J., 2009. Cyclic arginine-glycine-aspartate peptides enhance three-dimensional stem cell osteogenic differentiation. Tissue Eng. A 15, 263−272. https://doi.org/10.1089/ten.tea.2007.0411.

Huang, S.-H., Lin, Y.-N., Lee, S.-S., Chai, C.-Y., Chang, H.-W., Lin, T.-M., Lai, C.-S., Lin, S.-D., 2015. New adipose tissue formation by human adipose-derived stem cells with hyaluronic acid gel in immunodeficient mice. Int. J. Med. Sci. 12, 154−162. https://doi.org/10.7150/ijms.9964.

Johnson, B.V., Shindo, N., Rathjen, P.D., Rathjen, J., Keough, R.A., 2008. Understanding pluripotency—how embryonic stem cells keep their options open. Mol. Hum. Reprod. 14, 513−520. https://doi.org/10.1093/molehr/gan048.

Karlsson, J., Harmankaya, N., Palmquist, A., Atefyekta, S., Omar, O., Tengvall, P., Andersson, M., 2016. Stem cell homing using local delivery of plerixafor and stromal derived growth factor-1alpha for improved bone regeneration around Ti-implants. J. Biomed. Mater. Res. A 104, 2466−2475. https://doi.org/10.1002/jbm.a.35786.

Keefe, A.D., Pai, S., Ellington, A., 2010. Aptamers as therapeutics. Nat. Rev. Drug Discov. 9, 537−550. https://doi.org/10.1038/nrd3141.

Khan, F.A., Almohazey, D., Alomari, M., Almofty, S.A., 2018. Isolation, culture, and functional characterization of human embryonic stem cells: current trends and challenges. Stem Cell. Int. 2018, 1429351. https://doi.org/10.1155/2018/1429351.

Khayyatan, F., Nemati, S., Kiani, S., Hojjati Emami, S., Baharvand, H., 2014. Behaviour of human induced pluripotent stem cell-derived neural progenitors on collagen scaffolds varied in freezing temperature and laminin concentration. Cell J 16, 53−62.

Kim, H.J., Kim, U.-J., Vunjak-Novakovic, G., Min, B.-H., Kaplan, D.L., 2005. Influence of macroporous protein scaffolds on bone tissue engineering from bone marrow stem cells. Biomater. 26, 4442−4452. https://doi.org/10.1016/j.biomaterials.2004.11.013.

Kim, S.J., Kim, J.E., Kim, S.H., Kim, S.J., Jeon, S.J., Kim, S.H., Jung, Y., 2016. Therapeutic effects of neuropeptide substance P coupled with self-assembled peptide nanofibers on the progression of osteoarthritis in a rat model. Biomater. 74, 119−130. https://doi.org/10.1016/j.biomaterials.2015.09.040.

Ko, I.K., Lee, S.J., Atala, A., Yoo, J.J., 2013. In situ tissue regeneration through host stem cell recruitment. Exp. Mol. Med. 45, e57. https://doi.org/10.1038/emm.2013.118.

Lemcke, H., Voronina, N., Steinhoff, G., David, R., 2018. Recent progress in stem cell modification for cardiac regeneration. Stem Cell. Int. 2018, 1−22. https://doi.org/10.1155/2018/1909346.

Li, Q.-L., Huang, N., Chen, C., Chen, J.-L., Xiong, K.-Q., Chen, J.-Y., You, T.-X., Jin, J., Liang, X., 2010. Oriented immobilization of anti-CD_{34} antibody on titanium surface for self-endothelialization induction. J. Biomed. Mater. Res. A 94, 1283−1293. https://doi.org/10.1002/jbm.a.32812.

Li, X., He, X.-T., Yin, Y., Wu, R.-X., Tian, B.-M., Chen, F.-M., 2017. Administration of signalling molecules dictates stem cell homing for in situ regeneration. J. Cell Mol. Med. 21, 3162−3177. https://doi.org/10.1111/jcmm.13286.

Liu, H., Collins, S.F., Suggs, L.J., 2006. Three-dimensional culture for expansion and differentiation of mouse embryonic stem cells. Biomaterials 27, 6004−6014. https://doi.org/10.1016/j.biomaterials.2006.06.016.

Liu, J., Xu, H.H.K., Zhou, H., Weir, M.D., Chen, Q., Trotman, C.A., 2013. Human umbilical cord stem cell encapsulation in novel macroporous and injectable fibrin for muscle tissue engineering. Acta Biomater. 9, 4688−4697. https://doi.org/10.1016/j.actbio.2012.08.009.

Louridas, M., Letourneau, S., Lautatzis, M.-E., Vrontakis, M., 2009. Galanin is highly expressed in bone marrow mesenchymal stem cells and facilitates migration of cells both in vitro and in vivo. Biochem. Biophys. Res. Commun. 390, 867−871. https://doi.org/10.1016/j.bbrc.2009.10.064.

Lu, T.-J., Chiu, F.-Y., Chiu, H.-Y., Chang, M.-C., Hung, S.-C., 2017. Chondrogenic Differentiation of Mesenchymal Stem Cells in Three-Dimensional Chitosan Film Culture. Cell Transplant 26, 417−427. https://doi.org/10.3727/096368916X693464.

Lv, X., Guo, Q., Han, F., Chen, C., Ling, C., Chen, W., Li, B., 2016. Electrospun Poly(l-lactide)/Poly(ethylene glycol) Scaffolds Seeded with Human Amniotic Mesenchymal Stem Cells for Urethral Epithelium Repair. Int. J. Mol. Sci. 17 https://doi.org/10.3390/ijms17081262.

Mahla, R.S., 2016. Stem cells applications in regenerative medicine and disease therapeutics. Int. J. Cell Biol. 2016, 6940283. https://doi.org/10.1155/2016/6940283.

Mao, Y., Schwarzbauer, J.E., 2006. Accessibility to the fibronectin synergy site in a 3D matrix regulates engagement of alpha5beta1 versus alphavbeta3 integrin receptors. Cell Commun. Adhes. 13, 267−277. https://doi.org/10.1080/15419060601072215.

Martin, I., Miot, S., Barbero, A., Jakob, M., Wendt, D., 2007. Osteochondral tissue engineering. J. Biomech. 40, 750−765. https://doi.org/10.1016/j.jbiomech.2006.03.008.

Mauck, R.L., Yuan, X., Tuan, R.S., 2006. Chondrogenic differentiation and functional maturation of bovine mesenchymal stem cells in long-term agarose culture. Osteoarthritis Cartilage 14, 179−189. https://doi.org/10.1016/j.joca.2005.09.002.

Meinel, L., Fajardo, R., Hofmann, S., Langer, R., Chen, J., Snyder, B., Vunjak-Novakovic, G., Kaplan, D., 2005. Silk implants for the healing of critical size bone defects. Bone 37, 688−698. https://doi.org/10.1016/j.bone.2005.06.010.

Miltenyi, S., Müller, W., Weichel, W., Radbruch, A., 1990. High gradient magnetic cell separation with MACS. Cytometry 11, 231−238. https://doi.org/10.1002/cyto.990110203.

Mineda, K., Feng, J., Ishimine, H., Takada, H., Doi, K., Kuno, S., Kinoshita, K., Kanayama, K., Kato, H., Mashiko, T., Hashimoto, I., Nakanishi, H., Kurisaki, A., Yoshimura, K., 2015. Therapeutic potential of human adipose-derived stem/stromal cell microspheroids prepared by three-dimensional culture in non-cross-linked hyaluronic acid gel. Stem Cells Transl. Med. 4, 1511−1522. https://doi.org/10.5966/sctm.2015-0037.

Moshaverinia, A., Chen, C., Xu, X., Akiyama, K., Ansari, S., Zadeh, H.H., Shi, S., 2014. Bone regeneration potential of stem cells derived from periodontal ligament or gingival tissue sources encapsulated in RGD-modified alginate scaffold. Tissue Eng. Part A 20, 611−621. https://doi.org/10.1089/ten.TEA.2013.0229.

Nooeaid, P., Salih, V., Beier, J.P., Boccaccini, A.R., 2012. Osteochondral tissue engineering: scaffolds, stem cells and applications. J. Cell Mol. Med. 16, 2247−2270. https://doi.org/10.1111/j.1582-4934.2012.01571.x.

O'Brien, F.J., 2011. Biomaterials & scaffolds for tissue engineering. Mater. Today 14, 88−95. https://doi.org/10.1016/S1369-7021(11)70058-X.

Orlando, G., Soker, S., Stratta, R.J., Atala, A., 2013. Will regenerative medicine replace transplantation? Cold Spring Harb. Perspect. Med. 3 https://doi.org/10.1101/cshperspect.a015693.

Ozaki, Y., Nishimura, M., Sekiya, K., Suehiro, F., Kanawa, M., Nikawa, H., Hamada, T., Kato, Y., 2007. Comprehensive analysis of chemotactic factors for bone marrow mesenchymal stem cells. Stem Cell. Dev. 16, 119−129. https://doi.org/10.1089/scd.2006.0032.

O'Shea, T.M., Miao, X., 2008. Bilayered scaffolds for osteochondral tissue engineering. Tissue Eng. B Rev. 14, 447−464. https://doi.org/10.1089/ten.teb.2008.0327.

Pacelli, S., Basu, S., Whitlow, J., Chakravarti, A., Acosta, F., Varshney, A., Modaresi, S., Berkland, C., Paul, A., 2017. Strategies to develop endogenous stem cell-recruiting bioactive materials for tissue repair and regeneration. Adv. Drug Deliv. Rev. 120, 50−70. https://doi.org/10.1016/j.addr.2017.07.011.

Parenteau-Bareil, R., Gauvin, R., Berthod, F., 2010. Collagen-based biomaterials for tissue engineering applications. Materials 3, 1863−1887. https://doi.org/10.3390/ma3031863.

Park, S.S., Moisseiev, E., Bauer, G., Anderson, J.D., Grant, M.B., Zam, A., Zawadzki, R.J., Werner, J.S., Nolta, J.A., 2017. Advances in bone marrow stem cell therapy for retinal dysfunction. Prog. Retin. Eye Res. 56, 148−165. https://doi.org/10.1016/j.preteyeres.2016.10.002.

Patra, C., Talukdar, S., Novoyatleva, T., Velagala, S.R., Mühlfeld, C., Kundu, B., Kundu, S.C., Engel, F.B., 2012. Silk protein fibroin from *Antheraea mylitta* for cardiac tissue engineering. Biomaterials 33, 2673−2680. https://doi.org/10.1016/j.biomaterials.2011.12.036.

Sakiyama-Elbert, 2008. Combining stem cells and biomaterial scaffolds for constructing tissues and cell delivery. StemBook. https://doi.org/10.3824/stembook.1.1.1.

Samsonraj, R.M., Raghunath, M., Nurcombe, V., Hui, J.H., van Wijnen, A.J., Cool, S.M., 2017. Concise review: multifaceted characterization of human mesenchymal stem cells for use in regenerative medicine. Stem Cells Transl. Med. 6, 2173−2185. https://doi.org/10.1002/sctm.17-0129.

Schmitt, A., Rödel, P., Anamur, C., Seeliger, C., Imhoff, A.B., Herbst, E., Vogt, S., van Griensven, M., Winter, G., Engert, J., 2015. Calcium alginate gels as stem cell matrix-making paracrine stem cell activity available for enhanced healing after surgery. PLoS One 10, e0118937. https://doi.org/10.1371/journal.pone.0118937.

Shafiq, M., Jung, Y., Kim, S.H., 2015. In situ vascular regeneration using substance P-immobilised poly(L-lactide-co-ε-caprolactone) scaffolds: stem cell recruitment, angiogenesis, and tissue regeneration. Eur. Cell. Mater. 30, 282−302.

Shafiq, M., Jung, Y., Kim, S.H., 2016. Covalent immobilization of stem cell inducing/recruiting factor and heparin on cell-free small-diameter vascular graft for accelerated in situ tissue regeneration. J. Biomed. Mater. Res. A 104, 1352−1371. https://doi.org/10.1002/jbm.a.35666.

Shafiq, M., Zhang, Q., Zhi, D., Wang, K., Kong, D., Kim, D.-H., Kim, S.H., 2018. In situ blood vessel regeneration using SP (substance P) and SDF (stromal cell-derived factor)-1α peptide eluting vascular grafts. Arterioscler. Thromb. Vasc. Biol. 38, e117−e134. https://doi.org/10.1161/ATVBAHA.118.310934.

Shen, W., Chen, X., Chen, J., Yin, Z., Heng, B.C., Chen, W., Ouyang, H.-W., 2010. The effect of incorporation of exogenous stromal cell-derived factor-1 alpha within a knitted silk-collagen sponge scaffold on tendon regeneration. Biomater. 31, 7239−7249. https://doi.org/10.1016/j.biomaterials.2010.05.040.

Shi, C., Li, Q., Zhao, Y., Chen, W., Chen, B., Xiao, Z., Lin, H., Nie, L., Wang, D., Dai, J., 2011. Stem-cell-capturing collagen scaffold promotes cardiac tissue regeneration. Biomater. 32, 2508−2515. https://doi.org/10.1016/j.biomaterials.2010.12.026.

Sin, A., Murthy, S.K., Revzin, A., Tompkins, R.G., Toner, M., 2005. Enrichment using antibody-coated microfluidic chambers in shear flow: model mixtures of human lymphocytes. Biotechnol. Bioeng. 91, 816−826. https://doi.org/10.1002/bit.20556.

Song, K.-M., Lee, S., Ban, C., 2012. Aptamers and their biological applications. Sensors 12, 612−631. https://doi.org/10.3390/s120100612.

Stratton, S., Shelke, N.B., Hoshino, K., Rudraiah, S., Kumbar, S.G., 2016. Bioactive polymeric scaffolds for tissue engineering. Bioact. Mater. 1, 93−108. https://doi.org/10.1016/j.bioactmat.2016.11.001.

Sun, J., Zhao, Y., Li, Q., Chen, B., Hou, X., Xiao, Z., Dai, J., 2016. Controlled release of collagen-binding SDF-1α improves cardiac function after myocardial infarction by recruiting endogenous stem cells. Sci. Rep. 6, 26683. https://doi.org/10.1038/srep26683.

Survey, G., 2016. Global Observatory on Donation and Transplantation.

Thankam, F.G., Muthu, J., 2014a. Biosynthetic hydrogels-Studies on chemical and physical characteristics on long-term cellular response for tissue engineering: biosynthetic Hydrogels. J. Biomed. Mater. Res. A 102, 2238−2247. https://doi.org/10.1002/jbm.a.34895.

Thankam, F.G., Muthu, J., 2014b. Influence of physical and mechanical properties of amphiphilic biosynthetic hydrogels on long-term cell viability. J. Mech. Behav. Biomed. Mater. 35, 111−122. https://doi.org/10.1016/j.jmbbm.2014.03.010.

Thankam Finosh, G., Jayabalan, M., 2013. Reactive oxygen species—control and management using amphiphilic biosynthetic hydrogels for cardiac applications. Adv. Biosci. Biotechnol. 04, 1134−1146. https://doi.org/10.4236/abb.2013.412150.

Thurairajah, K., Broadhead, M., Balogh, Z., 2017. Trauma and Stem Cells: Biology and Potential Therapeutic Implications. Int. J. Mol. Sci. 18, 577. https://doi.org/10.3390/ijms18030577.

Wei, L., Wei, Z.Z., Jiang, M.Q., Mohamad, O., Yu, S.P., 2017. Stem cell transplantation therapy for multifaceted therapeutic benefits after stroke. Prog. Neurobiol. 157, 49−78. https://doi.org/10.1016/j.pneurobio.2017.03.003.

Turnbull, G., Clarke, J., Picard, F., Riches, P., Jia, L., Han, F., Li, B., Shu, W., 2018. 3D bioactive composite scaffolds for bone tissue engineering. Bioact. Mater. 3, 278−314. https://doi.org/10.1016/j.bioactmat.2017.10.001.

Wei, X., Yang, X., Han, Z., Qu, F., Shao, L., Shi, Y., 2013. Mesenchymal stem cells: a new trend for cell therapy. Acta Pharmacol. Sin. 34, 747−754. https://doi.org/10.1038/aps.2013.50.

Willerth, S.M., Arendas, K.J., Gottlieb, D.I., Sakiyama-Elbert, S.E., 2006. Optimization of fibrin scaffolds for differentiation of murine embryonic stem cells into neural lineage cells. Biomater. 27, 5990−6003. https://doi.org/10.1016/j.biomaterials.2006.07.036.

Wingate, K., Bonani, W., Tan, Y., Bryant, S.J., Tan, W., 2012. Compressive elasticity of three-dimensional nanofiber matrix directs mesenchymal stem cell differentiation to vascular cells with endothelial or smooth muscle cell markers. Acta Biomater. 8, 1440–1449. https://doi.org/10.1016/j.actbio.2011.12.032.

Wiraja, C., Yeo, D., Lio, D., Labanieh, L., Lu, M., Zhao, W., Xu, C., 2014. Aptamer technology for tracking cells' status & function. Mol. Cell. Ther. 2, 33. https://doi.org/10.1186/2052-8426-2-33.

Wu, C.-H., Liu, I.-J., Lu, R.-M., Wu, H.-C., 2016. Advancement and applications of peptide phage display technology in biomedical science. J. Biomed. Sci. 23, 8. https://doi.org/10.1186/s12929-016-0223-x.

Yang, B., Qiu, Y., Zhou, N., Ouyang, H., Ding, J., Cheng, B., Sun, J., 2017. Application of stem cells in oral disease therapy: progresses and perspectives. Front. Physiol. 8, 197. https://doi.org/10.3389/fphys.2017.00197.

Yao, Q., Cosme, J.G.L., Xu, T., Miszuk, J.M., Picciani, P.H.S., Fong, H., Sun, H., 2017. Three dimensional electrospun PCL/PLA blend nanofibrous scaffolds with significantly improved stem cells osteogenic differentiation and cranial bone formation. Biomaterials 115, 115–127. https://doi.org/10.1016/j.biomaterials.2016.11.018.

Yoon, J.W., Jang, I.H., Heo, S.C., Kwon, Y.W., Choi, E.J., Bae, K.-H., Suh, D.-S., Kim, S.-C., Han, S., Haam, S., Jung, J., Kim, K., Ryu, S.H., Kim, J.H., 2015. Isolation of foreign material-free endothelial progenitor cells using CD_{31} aptamer and therapeutic application for ischemic injury. PLoS One 10, e0131785. https://doi.org/10.1371/journal.pone.0131785.

Replacement materials for facial reconstruction at the soft tissue—bone interface

E. Wentrup-Byrne[1], Lisbeth Grøndahl[2], A. Chandler-Temple[3]
[1]Queensland University of Technology, Brisbane, Australia; [2]School of Chemistry and Molecular Biosciences, The University of Queensland, Brisbane, QLD, Australia; [3]The University of Queensland, St Lucia, QLD, Australia

Chapter outline

3.1 Introduction 45
3.2 Facial reconstruction 48
    3.2.1 Tissues at the bone interface 49
    3.2.2 Organs of special senses: eye, nose, and ear 51
3.3 Materials used in traditional interfacial repair 53
    3.3.1 Naturally derived materials 56
    3.3.2 Bioresorbable and nonbiodegradable materials 57
    3.3.3 The polytetrafluoroethylenes 59
3.4 Surface modification of facial membranes for optimal biointegration 64
    3.4.1 Calcium phosphate coatings 64
    3.4.2 Plasma treatment and ion implantation 66
    3.4.3 Gamma irradiation—induced grafting 67
3.5 Future trends 70
Acknowledgments 70
References 70

3.1 Introduction

In spite of the beauty and complexity of the magnificently engineered structure that constitutes the human face, it appears that humankind has never been "satisfied" with their faces and how they "look." In his fascinating book, Landau states that documentation dating back to the Minoan Bronze Age around 3500 BC confirms that altering the human head and face occurred in several civilizations, and he speculates that more recent evidence dates the practice even further back in the mists of time. The human face is intrinsically linked to a person's identity. Our fascination with the human face ranges from the artist who continually paints self-portraits—none more famous than Rembrandt who painted over 90 (Osmond, 2000)—to one of the most recognized faces of all, "The Mona Lisa," and the speculation about her identity (Rising, 2008). Scientists such as the renowned Erik Erikson spent a lifetime exploring the meaning of identity; in fact, the term "identity crisis" is attributed to him. One aspect of his research involved the "loss" of identity in World War II soldiers as a result of facial injury (Landau, 1989).

The psychological repercussions and effects on a person's life of severe deformities resulting from either trauma or genetic malformations cannot be underestimated. As evidenced by the recent global publicity afforded to the first reports of facial or "near-total" facial transplants, interest in restoring the esthetic appearance is seen as just as important as restoration of function (BBC, 2009; Gonzalez, 2008; AFP, 2008).

Fig. 3.1 shows some of the principal bones of the craniofacial skeleton, which forms the foundations for the esthetic features of the human face. To quote from a presentation by Fialkov et al. (2000) at the Bone Engineering Workshop, held in Toronto in 1999, "The morphology of the entire facial skeleton, in particular the upper region, has significant cultural, sexual, and social implications. Concepts of beauty, youth and intelligence are associated with particular facial morphologies, and vary with ethnicity. Much of this morphology is based on the underlying bony structure of the facial skeleton." The main functions of the human craniofacial skeleton are threefold. First, the multilayered bone framework provides protection for vital structures such as the ocular and aural systems, the central nervous system, and the upper

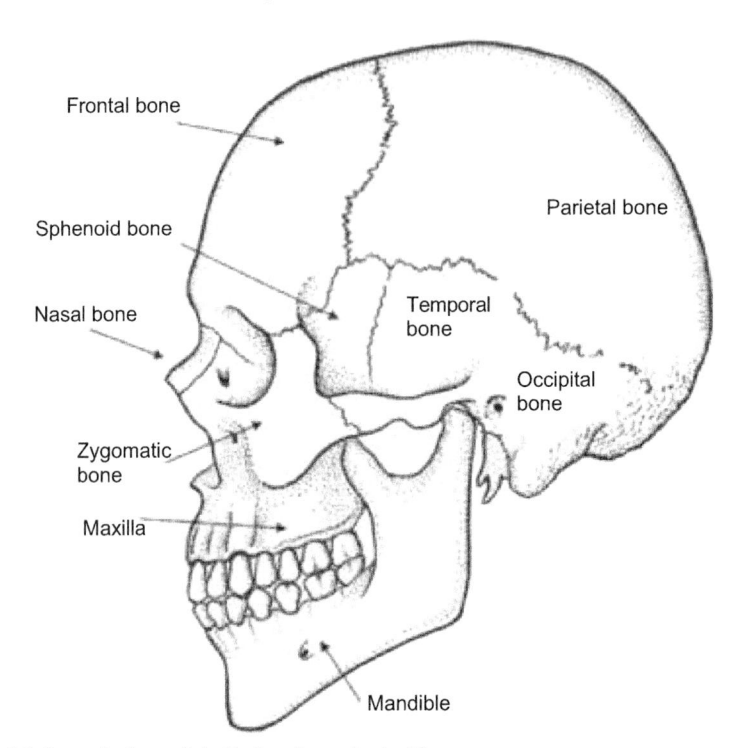

Figure 3.1 Lateral view of skull showing principal bones.

aerodigestive tract. Second, it also forms the foundations for the esthetic features of the human face. The third and final function is that, through providing the structural framework, lever, and fulcrum, it renders mastication possible.

Another important component of one's facial features is the soft adipose (subcutaneous fat) tissue that forms the interface between the bones and the skin. Fig. 3.2 shows an anterior cut away view of the face showing the soft tissue as well as a radiograph showing the underlying bone structure. The anatomical interrelationship between these "structures" is a complex one, and this, of course, means that to fulfill their functions, their repair, and regeneration are also interdependent.

Plastic surgery constitutes an enormous and sometimes much criticized industry, but it also includes the extremely important challenge of trauma repair. One example where facial surgery may be required and which is particularly relevant to this chapter is the case of human immunodeficiency virus (HIV)−related lipoatrophy (LA). As far as can be ascertained, Carr et al. (1998) were the first to describe this relatively newly discovered condition (HIV-LA). It involves loss of facial soft tissues, leading to serious changes (other areas of the body such as buttocks and feet can be also affected) associated with generalized LA, apparently triggered by the antiretroviral therapy. Many patients perceive it as a highly stigmatizing manifestation of their HIV infection as loss of facial fatty tissue leads to a gaunt appearance and may lead to issues in work, social, and personal relationships. Another serious issue is that it has been reported that patients reduce their adherence to their antiretroviral medication regime to avoid this highly visible wasting effect, thus jeopardizing their treatment (Collins et al., 2000).

Whether by choice or as a result of trauma, it is clear that the challenges faced by any repair and regeneration (RR) process are many and complex. Although it is, of

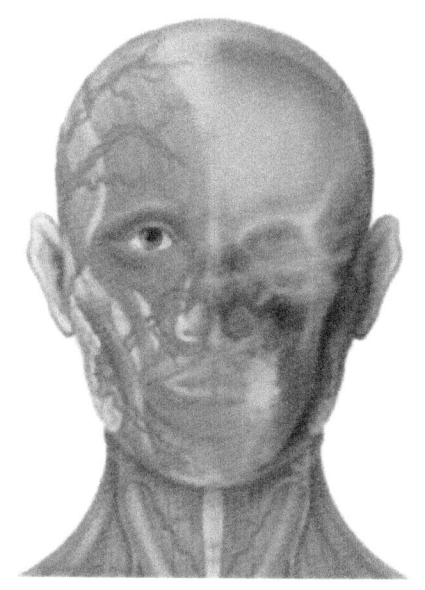

Figure 3.2 Anterior cut away view of the face, showing the soft tissue (left), and a radiograph showing the underlying bone structure (right).

course, impossible to identify any one anatomical region as being the most demanding in this respect, the craniofacial region must surely be a strong contender. The long healing and repair process starts with the triage team before finally reaching the reconstruction surgical team. Even before the RR process can properly begin, there is a long and difficult road for the trauma patient. For nonmedical specialists, even a quick perusal of books such as "*Head, Face and Neck Trauma*" edited by Stewart (2005) or "*Evaluation and Treatment of Orbital Fractures: A Multidisciplinary Approach*" edited by Holck and Ng (2006) brings home the enormity of the task, not to mention the costs involved. Bone and soft tissue injuries are not always of immediate priority when other potentially vision- or life-threatening injuries are involved. However, once the immediate medical aspects of the trauma have been satisfied, then the multidisciplinary team of reconstructive surgeons becomes involved. A multidisciplinary approach and multiple skills are required in the ultimate effort to restore both function and appearance requiring access to a range of accredited, well-defined, and tested repair materials to be used in the reconstruction process.

To develop clinically useable products, the reconstructive surgeons and indirectly the materials scientists and engineers must become conversant not only with their own fields of expertise as well as with the anatomy and specific functions of the repair sites involved but also with the healing process. In addition to fractures of the facial and orbital bones, injuries to eyes, nose, mandibular, and lachrymal systems and soft tissues such as cartilage, muscles, tendons, ligaments, and nerves may all have to be considered. A comprehensive coverage of all of these is beyond the scope of this chapter. Hence, we shall focus on one aspect of facial reconstruction that has been less well addressed in the literature, namely an overview of some of the materials used to repair and regenerate soft tissue both in terms of fillers and in terms of materials used at the hard—soft tissue interface. We will focus on the current range of commercially available repair materials, some relevant current research aimed at improving these materials, and finally the next generation of repair materials and new RR strategies.

3.2 Facial reconstruction

There is a plethora of useful books describing the relevant tissues as well as the complexity of their interrelationships. As a starting point an almost historic, compact, and very readable text is "*Anatomy of the Head, Neck, Face and Jaws*" by Fried. It provides the anatomical language necessary to form a deep understanding of all the tissue structures from bone and cartilage to nerves and muscles (Fried, 1980).

From a clinical perspective, the craniofacial skeleton is divided into upper and lower regions. The main function of the lower facial skeleton is occlusal and includes the masticatory structures of the mandible and maxilla, whereas the upper facial skeleton serves as a protective device for housing vital organs. In their chapter on "Strategies for Bone Substitution in Craniofacial Surgery," Fialkov et al. (2000), in addition to describing the interacting forces involved between the muscles and facial bones, also specify the importance of the overlying soft tissues. The morphology of the face depends not only on the underlying bony structure but also on the subcutaneous fat or adipose layer that interfaces between the bones and skin. Hence, in any major trauma or malformation, the repair and regeneration of more than one tissue will be required. Soft tissue repair requirements are to some extent dependent on bone repair. Some of the important and relevant conclusions these authors arrive at are: in order to withstand the overlying soft tissue compression, bone substitutes, resorbable scaffolds and osteogenic carriers all require a degree of rigidity in order to be effective in the reconstruction and augmentation of the upper facial skeleton and secondly, the requirements of bone substitutes in the upper facial skeleton are different from other parts of the body (Fialkov, 2000). As will be discussed in the following sections, the same can be said for the requirements of many soft tissue substitutes that form the interface with the hard tissue. One exception is in cases where the soft tissue is repaired using subcutaneous injections of fillers such as autologous adipose tissue (AT).

One significant problem encountered in facial repair is the fact that autologous bone resorption, as well as the short- to medium-term lifetimes of many soft tissue replacement materials, leads to volume changes that are highly visible in the face. Hence, where the repair and regeneration of both soft and hard facial tissues is concerned, the search for the "ideal" implant or material that will maintain both volume and contour has lead to the widespread use of alloplastic (see Section 3.3.2) materials.

From the available literature, it is clear that there are still aspects of many craniofacial repair procedures and the materials used that would benefit from further in-depth fundamental research. Before embarking on soft tissue repair, a brief description of the tissues involved is pertinent.

3.2.1 Tissues at the bone interface

Bone is a vascularized tissue consisting of cells and a mineralized extracellular matrix (ECM) scaffold. The principal mineral component of the scaffold is carbonated hydroxyapatite (HAP) that, although closely related to HAP ($Ca_{10}(PO_4)_6(OH)_2$), is unique. The most abundant of the collagens present is collagen Type I, which is also the second highest component. It provides flexibility and structure to the tissue. Bone also contains collagen XI, V, and III, as well as proteoglycans (PGs) such as chondroitin, versican and syndecan, serum proteins, phosphoproteins, and gamma carboxy glutamate (Siebel et al., 2006). The craniofacial skeleton is made up of two distinct bone types: membranous bone and endochondral bone. In membranous bone formation, which is responsible for the majority of the bones, ossification through direct mineral deposition into the extracellular produced matrix is followed by transformation of the mesenchymal cells into osteoblasts. Ultimately, this leads to the formation of the frontal, parietal, nasal, zygoma, maxilla, and mandibular bones (Fig. 3.1). Endochondral bone formation, on the other hand, is responsible for the occipital bone, nasal septum, and cranial base. Here, the initial cartilaginous template is mineralized. Osteoclasts invade and are replaced by osteoblasts that ultimately lead to bone formation. As a result of extensive remodeling, at maturity, there is minimal original skeletal bone, and this has important implications for the development of the materials used in RR strategies.

Cartilage is an avascular tissue and one of four mineralized tissues found in the body. It consists of an ECM comprised mainly of a three-dimensional hydrated network of collagen fibers in which chondrocytes and PGs are embedded. Depending on the type of cartilage, it may or may not be mineralized. For example, articular cartilage contains about 60% collagen Type II and 40% Types I, VI, IX, X, and XI (Walsh, 2006). The ECM contains 4%−7% (wet weight) PGs, with aggrecan being the most abundant. From a material and chemical point of view, cartilage is a hydrogel due to its high water content, and hence, it is capable of resisting pressure. The PGs interact with water that swells them, thus stabilizing the tissue and imparting compressive stiffness. This contributes to the viscoelastic properties of cartilage. Because it is anaerobic, it also has low oxygen consumption. Bone and cartilage are very different materials, and the interface between them is an indistinct "calcified zone." Although collagen Type II is the most abundant cartilage collagen, calcifying cartilage is enriched with collagen Type X and contains ECM vesicles. Currently, there are as many as 28 different collagen types described in the literature. During the bone mineralization process, which is regulated by growth factors, cytokines, and hormones, apatite deposition occurs onto the collagen Type I matrix (this being the most abundant bone collagen). According to one source, the distinction between bone and cartilage was already recognized by Aristotle, who separated fish into either cartilaginous or bony categories (Hall, 2005). However, it was not until the 1690s and the invention of the microscope that researchers such as Leeuwenhoek and Havers were able to investigate the intricate microstructures of bone and cartilage (Hall, 2005). Polarized light microscopy and scanning electron microscopy studies made it possible to establish the architecture (de Visser et al., 2008). Adult articular cartilage has a zonal architecture that is determined by the alignment of its collagen fibers. The three zones of alignment are the superficial zone (closest to the articular surface) where the fibers are aligned parallel to the articular surface, the radial zone (closest to the bone) where they are normal to the surface, and the transitional zone (between the superficial and radial zones) where a continuous variation in average fiber orientation is observed. In addition, the chondrocytes change morphology from flattened and more aligned in the calcified zone at the bone interface to rounder in the superficial zone.

Muscles are organs of motion: folklore has it that although it takes 17 muscles to smile, it takes 43 to frown. They are mainly composed of muscle cells, which in turn contain myofibrils. The hierarchical structure means that these contain sarcomeres that are composed of the proteins actin and myosin. In most muscles, all the fibers are oriented in the same direction, running in a line from the origin to the insertion. They usually connect two or more anatomical structures, one of which is capable of moving: bone and skin, two bones, two parts of skin, two organs, etc. The function of muscles is intrinsically linked to ligaments, nerves, tendons, aponeurosis, and fascia, for example, these control the energy for the muscle contraction and its direction. All the muscles of facial expression are attached to skin, at least at the point of insertion. In addition, they are all superficial. Fried (1980) points out that they have the particular characteristic of coming in a wide range of sizes, shapes, and strengths. Apart from conveying emotions and displaying facial expressions, they fulfill many important functions such as closing the eyes and moving lips and mouth during mastication. Interestingly, the facial expression muscles work synergistically rather than independently. Hence, any repair is complicated by the fact that they are interdependent in many of their functions. Any material used to replace or repair bone, cartilage, or skin will clearly have some relationship with the muscles connected to them, not to mention interfaces (whether lesser or greater) with the ligaments.

AT is a specialized, loose, nonfibrous connective tissue composed primarily of fat cells called adipocytes. Its most important role is in energy homeostasis. In mammals, fats, usually in the form of triglycerides, are stored either in white adipose tissue (WAT) or (to a lesser extent in adults) brown adipose tissue (BAT). WAT contains a single lipid droplet, while BAT contains numerous smaller droplets and contains many more mitochondria. The latter are responsible for the brown color. BAT contains more capillaries due to its greater need for oxygen and is the main fat present in newborns. It is also responsible for generating body heat. Later, this is mostly replaced with WAT, although some BAT remains in the neck and intrascapular regions. WAT acts as an insulator and a protective layer around vital organs (think of the kidneys). In the face, it fulfills another vital function where it forms a cushioning layer with interfaces between bone and skin helping to shape and contour the facial features.

The skin that forms the external covering of the body is its largest organ, constituting 15%−20% of its total mass and a surface area of between 1.5 and 2 m^2 for the average human. The term "integumentary system" is used when derivatives such as mammary, sweat, and sebaceous glands are included.

Skin is broadly categorized as thick or thin, a reflection on its location, as it can vary from 1 mm to over 5 mm such as on the palms of the hands or soles of the feet. However, although these terms, which refer only to the epidermis, are really only of interest from a histological point of view, they are mentioned to illustrate the fact that skin can differ anatomically. Apart from the two main layers—the epidermis and the dermis—there is a third underlying hypodermis fatty (subcutaneous fat) layer (see Fig. 3.3). This layer is highly significant in facial repair and reconstruction (RR) and is relevant to this chapter.

The epidermis is composed of a keratinized, stratified squamous epithelium derived from the ectoderm. It grows continuously but, through the process of desquamation, maintains its regular thickness. One of its primary functions is as a protective barrier from the environment. The dermis is a dense connective tissue whose functions include imparting mechanical support and strength as well as thickness to the skin. Two distinct layers are clearly identified using light microscopy: the papillary and the reticular. Just beneath the reticular layer are found the layers of the hypodermis: lobules that contain varying amounts of AT separated by connective tissue septa, smooth muscle, and, in some sites, striated muscle. This is the layer that insulates inhabitants of cold climates. But it is also the layer that varies greatly from individual to individual, bestowing on each of us a large part of our "identity" based on how thin or not "quite so thin" we are. This is particularly true where the face is concerned.

3.2.2 Organs of special senses: eye, nose, and ear

Vision, hearing, equilibrium, and smell are all intrinsically linked to the anatomy of the head. While recognizing the importance of all of these, there does seem to be some

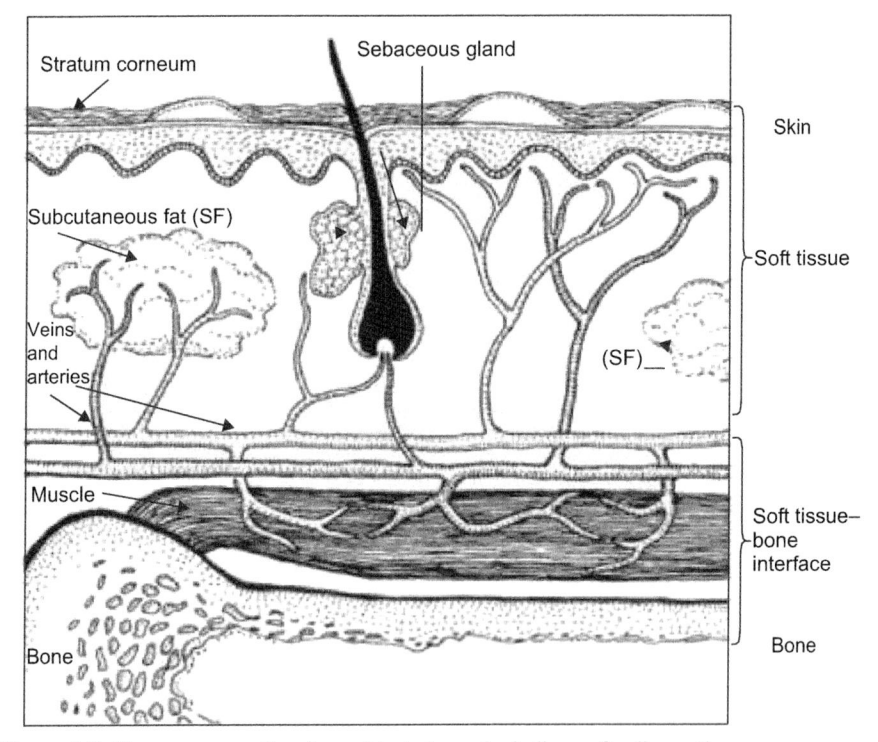

Figure 3.3 Tissue cross section from skin to bone, including soft adipose tissue.

consensus that vision is the most important of all the senses and its protection is vital. Although the "eye" itself can be considered as consisting of essential components such as the bulb and the accessory organs (muscles, eyelids and eyebrows, lachrymal system, etc.), where facial trauma is concerned, the bony orbit is critically important because this is where the eye is housed and protected. Broadly speaking, it consists of the orbital rim that is relatively thick and the orbital walls that are much more fragile. A recent book *"Biomaterials and Regenerative Medicine in Ophthalmology"* edited by Prof. Traian V. Chirila, Queensland Eye Institute, Australia (Chirila, 2010), covers many aspects of the repair of this particular system.

The external nose is mainly composed of bone, cartilage, skin, and mucosa. Nasal fractures are the most common facial trauma ($\sim 40\%$), and for whole body trauma, they rank third highest overall after wrist and clavicle (Dev, 2008). The history of medicine, and more particularly, the materials used to repair the body, makes fascinating reading, even if their historical accuracy is sometimes somewhat doubtful. One of the most interesting is the story of Tycho Brahe's artificial nose (1566) (Van Helden, 1995). After losing his nose (or part of it depending on which account one reads) in a duel, he had a gold/silver prosthesis fitted using an "adhesive balm" to keep it in place. This incident apparently kindled his interest in medicine, although he is best remembered for his astronomical discoveries. Another well-documented case of nasal repair is found in the Edwin Smith Papyrus, which is the only surviving copy of an ancient Egyptian book on trauma surgery. It consists exclusively of cases beginning with the head and working down. According to one report, of the 48 surviving case reports, only one resorted to "magic" (Wilkins, 1964). Another case described involved the "repositioning of the deviated nasal bones." Treatment included "the use of internal splints, firm external splints, and dressings made from linen and grease and honey." Further fascinating reading can be found in Lascaratos et al. (2003) "From the Roots of Rhinology: The Reconstruction of Nasal Injuries by Hippocrates." Returning to the present, Oeltjen and Hollier (2005) give an excellent overview of "Nasal and Naso—Orbital Ethmoid Fractures" in *"Head, Face and Neck Trauma"* edited by Stewart (2005).

In addition to hearing, the ear fulfills two other functions: equilibrium and cosmesis. The external ear consists of two major parts: the auricula or pinna and the external auditory meatus (Fried, 1980). Even focusing on the pinna and cosmesis issues, its repair is by no means trivial. It is composed of soft tissues, with cartilage being dominant. The chapter by Chang on "Auricular Trauma" gives a good overview of typical trauma, treatment options, and strategies (Stewart, 2005). Typically, microsurgical techniques have been used since the 1980s to reattach partially avulsed pinna. In cases where the total external ear is missing from birth, surgery (four operations over 2 years) is possible and most successful when the patient is 6—8 years old. When the loss is due to radical cancer surgery, amputation, burns, and/or congenital defects, then auricular prostheses are available (UT Southwestern, 2009).

3.3 Materials used in traditional interfacial repair

As in this chapter we are focusing principally on repair of the tissues underlying the skin and interfaced between it and the facial bones, as shown in Fig. 3.3, we now need to consider the materials and current approaches to facial soft tissue repair. There is a host of products used for soft tissue augmentation or replacement. Table 3.1 lists a selection of currently available materials. Our focus is on the three main classes of materials used either as fillers or for other soft tissue replacement applications. Depending on the genre of journal (or research), the descriptive words used in the classification of materials can vary. For example, the terms temporary, semipermanent, or permanent are often used, but in biomaterials science, the terms "bioresorbable" (degrades in vivo), "nonbiodegradable" (permanent), and "naturally derived" (sourced from living organisms) are preferred. Recently, it has been suggested by one surgeon that another approach could be to consider them "according to the reactions they induce within human tissue" (Nicolau, 2008). Seeking to control the rate at which bioresorbable materials degrade (or erode) in vivo is one that attracts an enormous amount of research as the processes involved greatly influence vascularization and/or new tissue growth, both of which in turn influence the extent of tissue regeneration. It should be pointed out here that, in general, naturally derived repair materials (e.g., AT) degrade very fast and often in an unpredictable manner. In contrast, the degradation rate of synthetic bioresorbable materials can often be tailored for a specific application.

Table 3.1 A selection of polymeric materials used as soft tissue fillers or in soft tissue repair.

Material	Description/use	Trade name	Manufacturer
Biostables			
Polytetrafluoroethylene (PTFE)	PTFE	Teflon	Dow Chemical/ Dupont
	PTFE-graphite	Proplast I	Vitek
	PTFE-aluminum oxide	Proplast II	Vitek
	Expanded PTFE	Gore-Tex	WL Gore & Associates
	Expanded PTFE tubing	SoftForm	Tissue Technologies
	Expanded PTFE tubing	UltraSoft	Tissue Technologies
	Expanded PTFE with silver and chlorhexidine (antibacterial)	MycroMesh Plus	WL Gore & Associates
	Expanded PTFE dual porosity	Advanta	Atrium Medical Products
	Expanded PTFE saline filled	Fulfil	Evera Medical
Polyacrylamide gels	5% PAA injectable gel	Aquamid	Contura International
	2.5% PAA injectable gel	Eutrophill	Lab Procytech
Polymethyl methacrylate	Beads suspended in bovine collagen gel	ArteFill	Artes Medical
Silicone	Silicone	Silastic	Dow Corning
	Silicone	Silikon 1000	Alcon Labs
	Silicone	Adatosil 5000	Bausch & Lomb
Bioresorbables			
Polylactide	Stimulatory filler	Sculptra/ NewFill	Sanofi Aventis
Naturally derived			
Collagen	Bovine collagen	Zyderm I	Allergan, Inc.
	Bovine collagen	Zyderm II	Allergan, Inc.
Cross-linked bovine collagen	Zyplast		Allergan, Inc.

Replacement materials for facial reconstruction at the soft tissue—bone interface 55

Table 3.1 Continued

Material	Description/use	Trade name	Manufacturer
Human collagen	Cosmo Derm I	Allergan, Inc.	
Human collagen	Cosmo Derm II	Allergan, Inc.	
Cross-linked human collagen	Cosmo Plast	Allergan, Inc.	
Human harvested autologous cells	Isolagen	Fibrocell Science, Inc.	
Human cadaver allogeneic collagen	Dermalogen	Collagenesis, Inc.	
Human harvested autologous cells	Autologen	Collagenesis, Inc.	
Porcine collagen	Evolence	ColBar LifeScience Ltd	
Hyaluronic acid	Rooster-derived hyaluronic acid	Hylaform	Allergan, Inc.
	Rooster-derived hyaluronic acid	Hylaform plus	Allergan, Inc.
	Bacterial or non-animal stabilized hyaluronic acid	Restylane	Medicis
	Bacterial or non-animal stabilized hyaluronic acid	Perlane	Medicis
	Bacterial or non-animal stabilized hyaluronic acid	Prevelle	Mentor
	Bacterial or non-animal stabilized hyaluronic acid	Juvederm	Allergan, Inc.
Gelatin	Gelfilm (absorbable gelatin film)	Gelfilm	Pharmacia & Upjohn Ophthalmic Division

Klein and Elson (2000) listed the requirements of a soft-tissue augmentation material as

- being potentially of high use,
- being cosmetically pleasing,

- having minimum undesirable reactions,
- having low "abuse" potential or abuse not leading to significant morbidity rates,
- being nonteratogenic,
- being noncarcinogenic,
- being nonmigratory,
- being capable of providing predictable, persistent, reproducible correction,
- being FDA-approved (presumably also other regulatory bodies) if nonautologous.

Because of the dominance of the "facial cosmetic surgery" market, even a superficial search reveals that there is a plethora of injectable materials (or fillers) described in a huge variety of journals that are used to "improve" facial appearances or rejuvenate aging features. Comprehensive lists can be found in various reviews such as "Soft Tissue Augmentation: A review" (Fernandez and Mackley, 2006) and "Long-lasting and Permanent Fillers: Biomaterial Influence over Host Tissue Response" (Nicolau, 2008). This "market" and the controversies surrounding its regulation in different countries are not of concern here. However, as many of the materials used also play an important role in the repair of facial defects, scars, and trauma injuries, these are being included. As Klein and Elson (as well as many others) pointed out currently, there are no repair materials that fulfill *all* the desired criteria (Klein and Elson, 2000; Nicolau, 2008). So, it is not surprising that it is also generally recognized that the "perfect" filler is still to be found. In the words of Hirsch in her recent review on "Soft Tissue Augmentation" (Hirsch and Cohen, 2006), "even a judicious selection of the perfect product for a given indication in a particular patient does not guarantee a perfect outcome." Time and time again in the literature, there are warnings about using particular products without having the necessary knowledge and expertise. However, one thing any surgeon or material scientist will agree on is that this is an ambitious "wish list" for any material to possess!

We shall now discuss the most frequently used materials in more detail. Table 3.1 lists a selection of soft tissue augmentation and filling materials. This unprejudiced list is by no means complete but is an example of representative materials for each of the classes to be discussed. We have included autologous materials such as fat and collagen but not botulinum toxin.

3.3.1 Naturally derived materials

Although two of the most frequently used naturally derived materials, "autologous fat tissue" and "collagen-based materials," lie outside the main focus of this chapter, we include a short overview of each. The oldest and most frequently used autograft is in fact "autologous fat tissue." Neuber first used it in whole graft form in 1893 (Neuber, 1893). Since the 1920s, it has been used in injectable form, but with the advent of liposuction, its use has greatly increased. There is much controversy surrounding its efficacy, in particular its longevity, and this is reflected even in the titles of some recent reviews: "Autologous Fat Transfer for Facial Recontouring: Is there Science behind the Art?" (Kaufman et al., 2007) and "Fat Grafting: Fact or Fiction?" (Calabria and Hills, 2005). It is used as a nonpermanent material to fill small defects and for scar repair. Because of the large degree of resorption (30%–60%), substantial overcorrection is necessary. The harvesting and tissue preparation techniques used are well documented.

Despite the many disadvantages, its use has persisted because its autologous nature means biocompatibility is not an issue.

Naturally derived materials come from a variety of sources including autologous (human) and allogenic sources such as bovine, porcine, avian, and bacterial. Collagen was one of the first naturally derived filler materials given FDA approval (Sarnoff et al., 2008). Bovine-derived collagen is one of the most popular injectable soft tissue fillers and is generally considered the gold standard against which all other materials are measured (Homicz and Watson, 2004; Klein and Elson, 2000). This is not really surprising, as collagen itself is a main structural element of both hard and soft tissues in mammals. Bovine collagen–based materials have been used for over 20 years, and the main disadvantages associated with their use are well documented: a propensity for hypersensitivity reactions and limited efficacy lifetimes. The need for skin testing requires multiple visits and hence delayed treatment times. Abscess formation, tissue necrosis, and granulations as a result of foreign body reactions have all been reported (Hanke et al., 1991). Some of the many autologous and allogenic collagen-based materials that have been developed over the years are expensive, customized products. One such commercially autologous product is Isolagen, which contains the patient's own living dermal fibroblast cells (Isolagen, 2007).

More recently, in their review, Homicz and Watson attest that hyaluronic acid–based materials such as Restylane and the Hylaform range of products appear to be a "safe alternative" to bovine collagen–based materials (Homicz and Watson, 2004). However, even though hyaluronic acid structures are believed to be identical across species (for example, it is isolated from rooster combs and streptococcus), traces of hyaluronic acid–associated proteins can cause allergic reactions (reactions currently reported as $<1\%$) (Lowe et al., 2005).

The rationale for the use of naturally derived polymers, e.g., collagen (Table 3.1), particularly when they are present in the patient's own system, appears to be logical. These polymers usually degrade in vivo enzymatically but many are also susceptible to hydrolysis. The degradation by-products are usually disposed of, or recycled, by the body through normal metabolic pathways. Furthermore, because of the chemical similarity between these polymers and ECM components already present in tissues, biocompatibility and integration would be expected to be enhanced. However, these polymers have very fast and uncontrollable degradation rates in vivo. In addition, to produce enough material, the crude polymer is usually sourced from a different species to the patient. As a result, there is concern over not only disease transmission but also the variable quality of these polymers, which often differs between batches.

3.3.2 Bioresorbable and nonbiodegradable materials

According to Eppley (2003) and others, the concept of an "alloplastic" material is synonymous with the term "synthetic" and means it originates from a nonbiological source. When used in the medical literature, the term alloplastic covers manufactured materials from "nonorganic, nonhuman, nonanimal sources" and encompass ceramics and metals as well as plastics (which to material chemists and material engineers means polymers). We are concentrating our discussion in this chapter on polymeric materials. Broadly speaking, polymeric biomaterials are divided into either nonbiodegradable (such as polytetrafluoroethylene [PTFE)]) and bioresorbable polymers (such as the poly(lactic acid)-based LactoSorb). Because different disciplines such as tissue engineering and materials chemistry use different definitions of the terms biodegradable, bioresorbable, bioabsorbable, and bioerodible, some confusion has arisen in the literature. So although the term biodegradable is the one most frequently used in its broadest sense, it is pertinent to define these terms because they can be important when discussing the chemical and physical properties of the kinds of polymers discussed in this review. Vert's definitions, given below for solid polymeric materials, are generally accepted by the materials and tissue engineering communities (Albertsson and Varma, 2003; Vert et al., 1992).

- Biodegradables break down due to macromolecular degradation. There is in vivo dispersion of the fragments/by-products but no proof of elimination from the body.
- Bioresorbables show bulk degradation and further resorb in vivo, i.e., the original foreign material and its breakdown products can be shown to be eliminated through the body's natural pathways.
- Bioerodibles show surface degradation and further resorb in vivo. Total elimination of low molecular weight by-products is inherent.
- Bioabsorbables can dissolve in body fluids in the absence of polymer chain cleavage or molecular mass loss, such as in the slow dissolution of water-soluble materials.

In addition to the many well-documented tissue RR applications for the materials shown in Table 3.1, another potential application (already mentioned in Section 3.1) that is very relevant to our topic is in HIV-LA surgery. However, some alloplastics are more used than others, and although many are approved for use in Europe, they are not always FDA-approved. Hence, we shall limit our discussion to two examples that are well studied and documented: polylactide (PLA) (bioresorbable) and expanded PTFE (ePTFE) (nonbiodegradable).

PLA belongs to a class of compounds known as the poly(α-esters). Because of their good biocompatibility and ability to be bioresorbed, aliphatic poly(α-esters), such as PLA, polyglycolide (PGA), and their copolymers, have been studied for biomedical purposes since the 1960s (Albertsson and Varma, 2003). The first commercially available product launched in 1962 was a PGA suture called Dexon (Tyco Healthcare Group, CT, USA). Since then, PLA has manifested itself in many forms for use in medical devices and in facial surgery applications. According to Moyle, PLA is the only one that has specific FDA approval for treatment of HIV-LA (Moyle et al., 2004, 2006). In a series of papers, He and others (Valantin et al., 2003; Burgess and Quiroga, 2005; Lafaurie et al., 2005; Barton et al., 2006; Cattelan et al., 2006; Lam et al., 2006) describe and evaluate injectable forms of PLA for use as a bioresorbable filler with a lifetime of 1−2 and sometimes up to 3 years. NewFill has been used since 1999 in Europe (Fernandez and Mackley, 2006). According to Hirsch, Sculptra, as it is known in the United States, was fast-tracked in 2004 for approval by the FDA for use in facial LA (Hirsch and Cohen, 2006).

One of the problems associated with the use of injectable PLA suspensions is the formation of subcutaneous papules and granulomas. As a result of various studies (Valantin et al., 2003; Woerle et al., 2004), suggestions as to how to avoid these side effects have been made. They are chiefly concerned with practical issues such as depth of injection, frequency of reapplications, and postinjection strategies (ice packs). However, from a fundamental biomaterials science perspective, the fact that these fillers are administered as suspensions of PLA microparticles means that, overall, they have a large surface area, and the body's immune response will be to reject or encapsulate each and every one of the "foreign body particles" injected. Similar side effects have been observed for nonresorbable fillers administered as microparticular suspensions, e.g., polymethyl methacrylate (PMMA) (Lemperle et al., 2003; Homicz and Watson, 2004). In his review, Nicolau (2008) addresses many relevant issues including the issue of particle size and surface area. This and other references cited therein merit reading as there appears to be some difference of opinion on what constitutes the "ideal" particle size. For example, in their 2002 paper, Morhenn et al. describe how PMMA particles less than 20 μm "appear" to have been phagocytosed by U-937 cells (human macrophage-type cells), while particles greater than 40 μm were not. This is in contrast to collagen-coated PLA microspheres 0.6—60 μm, which do "appear" to be all phagocytosed. Results for PMMA microspheres with Langerhans cells differ to some previously published results (Reis e Sousa et al., 1993), and it was concluded that "the different chemical nature of the microspheres and the type of Langerhans cells cultures may explain the differing results." On the other hand, it is well accepted that smooth, regularly shaped particles produce a much smaller inflammatory response than irregularly shaped ones (Nicolau, 2008 and refs therein). The many different commercial filler products all have subtle composition/structural differences, and hence, it becomes virtually impossible to predict the extent of an inflammatory response.

Another issue with "microsphere suspension fillers" is the fact that sometimes the microspheres can migrate, and this, of course, can lead to a new set of inflammatory responses. Again, it appears to be up to the individual surgeon to decide whether or not to use a particular filler for a specific application. For example, Nicolau (2008) states that he does not use the PLA product NewFill in the lips because they are known to displace and form unsightly nodules.

3.3.3 The polytetrafluoroethylenes

PTFE is a fully fluorinated unbranched polymer with a carbon backbone, analogous to the hydrocarbon polymer polyethylene (PE) (Sperati and Starkweather, 1961; Scheirs, 1997; Bunn and Howells, 1954). However, the conformation of PE is the common planar zigzag conformation, whereas PTFE is helical, Fig. 3.4 (Sperati and Starkweather, 1961; Scheirs, 1997). The comparatively large size of the fluorine atoms and the mutual repulsion between the adjacent fluorine atoms causes the polymer chains to twist to form compact helical rods with 13 fluorine atoms per 180° helical turn (Bunn and Howells, 1954; Drobny, 2000). Because it is easier to deform the C—C bond angle than stretch the C—C bond, each main-chain bond is rotated 20° from the next. This conformation creates a stiff rod shape, with the inner core of carbon completely encased within a sheath of electronegative fluorine atoms. This sheathing makes the rod conformation extremely stiff by inhibiting bending of the carbon backbone (Schiers, 1997). Consequently, the polymer chains are, of necessity, parallel

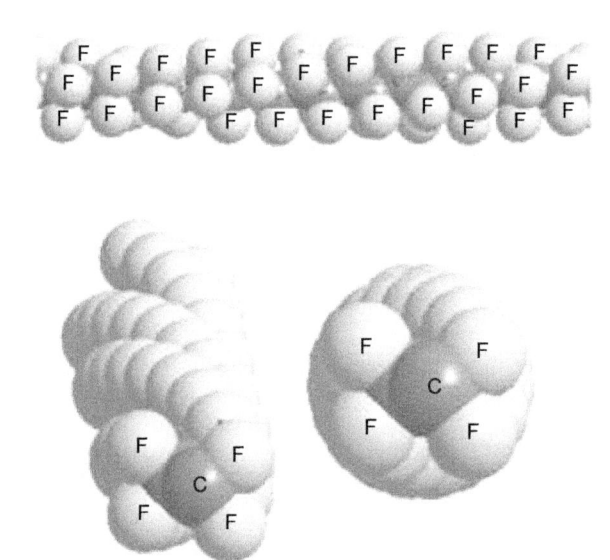

Figure 3.4 The helical structure of polytetrafluoroethylene.

packed, and this, combined with the high electronegativity of the outer sheath of fluorine atoms, allows the rods to slide past one another, culminating in the tendency of PTFE to cold flow (creep) (Schiers, 1997). It is likely that this phenomenon gives rise to the issues (discussed more fully later) of concern in implant surgery, namely shrinkage, migration, and micromovement.

The C−F bond is the strongest chemical bond found in polymers and confers on PTFE an exceptional chemical resistance (Sperati and Starkweather, 1961; Dargaville et al., 2003). It exhibits good electrical insulating abilities, a property that contributes to many commercially important applications. It is tough, nonadhesive, and has antifrictional properties combined with extreme hydrophobicity (Sperati, 1986; Gore, 1976a). The unique chemistry of PTFE gives it its unusual qualities of being both thermally and chemically inert. Although it is its inert nature that first made PTFE attractive as a biomedical material, it is also this property that leads to its lack of surface adhesion in vivo. This may necessitate aggressive suturing or anchoring in some applications, particularly in some facial "tight pockets."

PTFE's unique properties and versatile product range means that it has been used in medicine since the mid-20th century. The first use of the nonexpanded material was as an artificial heart valve, and later it became popular as a vascular graft replacement (Kannan et al., 2005). Voorhees and coworkers highlighted the importance of porosity in arterial grafts (Voor-hees et al., 1952). Until then, PTFE had been woven into a textile graft form. This was not ideal for the highly crystalline PTFE, which meant sacrificing some of its excellent features, and led to unraveling in the hostile in vivo environment. ePTFE is initially fibrillar and then expanded and has proven more favorable for medical applications than the nonexpanded versions. This is not only because it is nonbiodegradable but also because it has a low propensity to trigger thrombosis. These expanded materials have been described as having a "microstructure characterized by nodes interconnected by fibrils" (Gore, 1976a,b), as shown in Fig. 3.5. The size

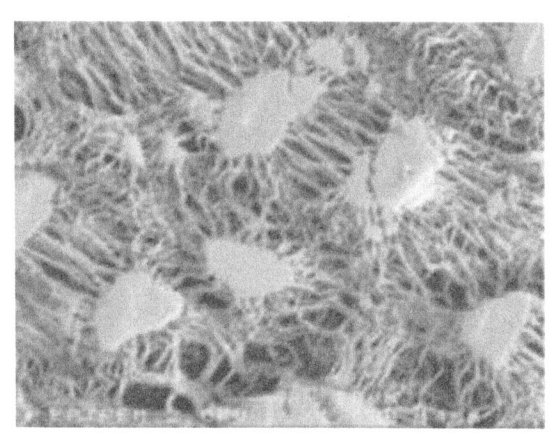

Figure 3.5 Scanning electron microscopy (SEM) image of expanded polytetrafluoroethylene (ePTFE) showing nodes and interconnecting fibrils.

of the nodes and the character of the fibrillar structures are dependent on the conditions used in the expansion process. Currently, ePTFE has a porosity of 70%–85%, with pore sizes ranging from 0.5 to 100 micron, depending on the source of the material. The porous microstructure allows ingrowth of tissue, which is essential for implant fixation, acceptance, and positive outcomes.

Expanded PTFE tubing produced as Gore-Tex was first used clinically in portal vein, thoracic vena cava, inferior vena cava, and external iliac veins in 1972 (Soyer et al., 1972). Since then, there has been a plethora of reports on the use of ePTFE, particularly Gore-Tex. Kannan et al. (2005) provide an excellent review of both models and clinical outcomes of prosthetic bypass and microvascular grafts reported in the literature from 1984 to the present. Reports indicate that both ePTFE and PTFE perform on a par with other synthetic grafts. However, they recommend their restriction to high-flow vessels. Smaller diameter grafts tend to fail due to occlusion caused by thrombosis and intimal hyperplasia (Bezuidenhout and Zilla, 2004). Guidoin et al. (1993) performed chemical analyses on 79 explanted arterial prostheses specimens removed after 6.5 years from human patients and concluded that ePTFE is stable for at least this duration in vivo.

PTFE's low friction coefficient has meant that it is an excellent coating material for devices that require ease of placement or ease of movement; for example, catheters and artificial joint coatings. PTFE-coated catheters were introduced during the late 1960s as their low friction properties minimize patient discomfort (Lawrence and Turner, 2005). The main longevity-related problem in total joint replacement is wear, particle-induced osteolysis around the acetabulum (cup) component. It has been demonstrated that using an ePTFE membrane as a physical seal prevents loosening and wear; subsequently, osteoblasts were found within the ePTFE seal (Bhumbra et al., 2000).

Periodontal applications were the first to exploit the principles of guided tissue regeneration (Zhao et al., 2000), later described as guided bone regeneration (GBR) for applications involving bone defects (Dahlin et al., 1991). Many periodontal concerns and treatments, such as periapical bone resorption (occurring at the base of the tooth apex), are applicable to GBR treatments elsewhere in the body, including maxillo- and craniofacial sites. In such cases, bone healing often fails to occur or occurs improperly. Large voids can occur either congenitally or through trauma, cysts, or natural resorption of the bone. Regeneration of bone at nonunion sites is dependent on patient age, bone structure, vascularization, the soft-tissue environment, and the size of the defect (Dahlin et al., 1994). Regeneration is often impaired by rapid ingrowth into the wound site by connective tissue from the surrounding areas.

Extensive research has taken place in the area of GBR using ePTFE as a nonresorbable membrane. In fact, in their review on the use of membranes for bone healing and neogenesis, Linde et al. (1993) state that virtually all investigations published on the promotion of bone regeneration by the membrane technique have utilized ePTFE membranes. There has been extensive experimental work using a variety of animal models from mice (Kidd et al., 2002) and rats (Dahlin et al., 1994; Kidd et al., 2002; Hagerty et al., 2000) to sheep (Paavolainen et al., 1993), pigs (Wiltfang et al., 1998), and monkeys (Hürzeler et al., 1998). These are only a small sample of both the model type and studies performed. One of the very early studies on rabbits is one in which Neel was the first (as far as we are aware) to suggest that these materials could be used in craniofacial applications (Neel, 1983). Furthermore, clinical periodontal radiographic studies were performed by Lorenzoni et al. (1999) to determine the regenerated bone response over 2 years to functional loading in vivo for 82 patients fitted with ePTFE GBR membranes in conjunction with prosthetic teeth. The regenerated bone appeared able to withstand functional loading. Periotest values were done at 6 and 24 months and revealed stable periimplant conditions and sustained osseointegration. Dahlin et al. (1998) tested the histological morphology of the ePTFE/tissue interface in oral implant patients. In their study, the membrane was penetrated by fibrous connective tissue that also separated the membrane from the bone. They concluded that fine micromovements of the implants might have caused this negative response. It is interesting to find that the results, both quantitative and qualitative, across all these studies showed no significant differences between the species.

Expanded PTFE is used as a soft tissue augmentation material for the face in the anterior nasal spine, glabella area, lips, malar area, mandible, mentum, mid-face region, nasal dorsum, nasolabial folds, and premaxilla (Mercandetti et al., 2008). It can be processed into a variety of forms, three-dimensional shapes, sheets, tubes, etc., which demonstrates its functional versatility. In particular, it was the ability to create these three-dimensional shapes that made ePTFE so valuable in facial soft-tissue suspension, augmentation, and correction (Panossian and Garner, 2004).

Early PTFE included Proplast I. This was a black PTFE—graphite composite and was initially designed as a coating material. It was replaced by Proplast II, a white PTFE—aluminum oxide composition that could be easily cut and shaped. The use of both Proplasts in these early applications was discontinued due to the many reports of gross malfunction. In some cases, these materials had been used in the temporomandibular joint and hence were put under considerable load. Their failure included significant wear, fragmentation, and perforation. The particulates formed in the immediate anatomical area triggered a foreign body response, thereby causing bone degeneration (including the glenoid fossa, mandibular condyle, and erosion into the cranial space). In some cases, the particulates were found to migrate to the lymph nodes where they could cause permanent hearing damage, chronic pain, loss of masticatory function, and motion (Food and Drug Admin, 1991). Until recently, the originators of PTFE marketed as Gore-Tex (W L Gore and Associates, Flagstaff, Arizona) were producing this material in a variety of products suited to facial soft tissue repair. These included solid tubes and patches in both preformed and customized shapes (Panos-sian and Garner, 2004; Chandler-Temple et al., 2008). Interestingly, although this product had been used since the early 1970s, it was not FDA approved for use for facial reconstruction and augmentation until 1993. In 1997, the FDA approved the use of the tubular implant SoftForm (Collagen Corporation, Palo Alto, California). SoftForm was used to treat deep facial furrows, such as nasolabial folds, as well as lip augmentation. The tubular SoftForm and the solid strand Gore-Tex used for the same purpose have similar porosities of 5—30 μm (Truswell, 2002). SoftForm was eventually replaced by tubular UltraSoft (Tissue Technologies Inc., San Francisco, California) in 2001. Cox (2005) reported on the advantages and disadvantages of permanent implants (ePTFE) versus nonpermanent (fillers). The permanent implants are easy to remove if they turn out to be malpositioned or if subsequently desired by the patient. However, they have limited filling capability and can feel undesirably firm to the touch. They can form elevated regions, extrude, contract, and migrate. The issue of shrinkage has been observed experimentally (Zhengbang et al., 2011). To address some of these issues (migration, shrinkage, and softness), Advanta (Atrium Medical Corporation, Hudson, NH, and Ocean Breeze Surgical, Amherst, NH) was developed and has been available since 2001 (Panossian and Garner, 2004; Redbord and Hanke, 2008). It consists of nonhollow strands of dual porosity: outer and inner core porosities of 40 and 100 μm, respectively. In late 2006, W L Gore withdrew their Gore-Tex SAMs series; however, W L Gore and Associates now supply materials for use as tissue space fillers in soft tissue reconstruction with products including Gore DUAL-MESH and GORE-TEX Soft tissue patch. Advanta was the only product of this type available commercially for a period (Redbord and Hanke, 2008). They continue to market and develop their MycroMesh Plus implant, which incorporates silver and chlorhexidine into the ePTFE matrix, producing a material with similar qualities to Gore-Tex. However, they report a greatly reduced infection rate (Malaisrie et al. 1998; Panossian and Garner, 2004).

Throughout the literature, there is an emphasis on the need for tissue ingrowth to assist in the biointegration and fixture of these implants. The nature of the tissue ingrowth needs to be carefully considered. Ideally, there should be enough specific cellular ingrowth to anchor the material, without the formation of a fibrous capsule, because this not only hinders the appropriate cellular integration but also leads to micromotion and ultimately more serious consequences.

This brief review of the products available for facial soft tissue augmentation demonstrates the ongoing development of new approaches to engineering materials to improve both material properties and implant outcomes. Tissue ingrowth has been improved as a result of pore size engineering. Tactility and antibacterial properties have also been improved by the customized engineering of this exceptional material.

However, these developments have not addressed a most critical aspect of the implant, namely, its surface. It is the material surface that first comes into contact with the patient's internal biological milieu, and the first response of the body's immune system is to the surface. Hence, the surface chemistry of implants and not just their gross morphologies should be considered.

3.4 Surface modification of facial membranes for optimal biointegration

As described in the previous section, ePTFE membranes have been used as soft tissue replacement materials at the soft tissue—bone interface. However, because of the inert nature of ePTFE, it does not integrate sufficiently with the underlying bone, and thus a number of research groups have been investigating methods to improve the bonding strength at this interface. Because PTFE is both thermally and chemically inert, its surface needs to be exposed to high energy processes to change its properties. Some of the approaches that have been used include ion implantation (Colwell et al., 2003), ArF excimer laser irradiation (Sato and Murahara, 2004), plasma activation using various carrier gases such as H_2O or N_2 (Oehr et al., 1999), O_2 (Kang et al., 2001), and Ar (Hsueh et al., 2003), electron beam preirradiation (He and Gu, 2003), and ^{60}Co gamma preirradiation (Mazzei et al., 2002). In addition, simultaneous grafting can be performed using plasma polymerization (Kühn et al., 2001), proton beam irradiation (Mazzei et al., 2003), as well as ^{60}Co gamma irradiation (Dargaville et al., 2003). Control of the induced changes can be achieved through judicious choice of reaction conditions, e.g., for gamma irradiation—induced grafting, these include monomer, concentration, dose, dose rate, and solvent. The surface properties of the modified polymers can differ substantially from those of the parent polymers, a fact that can be advantageously exploited to produce new materials with specific properties. Of the approaches listed above, the use of pretreatments (e.g., plasma or laser) before applying a calcium phosphate coat, as well as plasma treatment combined with ion implantation and gamma irradiation—induced grafting, has been explored for the improvement of ePTFE membranes (and PTFE) for applications in facial reconstruction. We will now describe these in some detail.

3.4.1 Calcium phosphate coatings

It is well established that a coat of a calcium phosphate phase, preferably HAP ($Ca_{10}(PO_4)_6(OH)_2$), on the surface of a biomaterial increases its ability to integrate with bone tissue in vivo. This has been extensively used for the fabrication of orthopedic metallic implants and has been shown to greatly improve the stability of the interface (Ratner et al., 2004). The method used to coat metallic implants, however, is not easily transferable to polymeric substrates. This is due to the fact that high reaction temperatures above the polymer melting point are required. A second issue when adding a coat to a substrate is the need for a strong interface between the polymer implant and the coating material itself. As PTFE is highly hydrophobic and has a very low friction coefficient, and HAP is highly hydrophilic, the resulting interfacial bond between them is very weak. In order therefore to produce a material with a well-bonded mineral coat on ePTFE membranes, it is necessary to first pretreat the membrane to decrease its hydrophobicity.

Early work in this area was conducted by Hontsu et al. (1998), who used an ArF excimer laser deposition technique to produce thin HAP films on PTFE substrates that had been chemically pretreated with a sodium—naphthalene complex. The chemical pretreatment introduced a mixture of oxygen-containing functional groups onto the polymer surface (e.g., −OH, −COOH), thus decreasing the hydrophobicity of the polymer. The HAP coat deposited at 310°C was highly amorphous but increased in crystallinity on annealing. The tensile strength of the interface was found to be 6 MPa. This was an improvement by an order of magnitude compared with a coating applied to a non-pretreated PTFE substrate.

Jansen and his research group used radio frequency (RF) magnetron sputtering to deposit HAP onto PTFE substrates (Feddes et al., 2004a,b,c). This is a technique for depositing a thin HAP coating that can also be applied to polymers as relatively low temperatures can be used. The coating composition was found to be dependent on a number of parameters such as discharge power (low power needed for polymers), sputter gas composition, gas pressure, and position of the sample in the chamber. In addition, the nature of the substrate and the thickness of the coat were also found to affect the coating composition. In their study on virgin PTFE (Feddes et al., 2004a), they found that the amount of calcium in the coating could be controlled by the Ar gas pressure: low levels were introduced at high pressures. In addition, UV irradiation caused detachment of F atoms from the polymer substrate. This was found to lead to a high Ca/P ratio (up to 5) as a result of calcium combining with the escaping F atoms. The result was the forma-tion of CaF_2 in the coating. At the same time, it was proposed that the removal of P occurred due to the formation of volatile PF_3 molecules. Despite this, the interfacial bond between the coating and the PTFE substrate was very strong (5.8 MPa) and displayed cohesive failure (i.e., within the PTFE substrate) during adhesion testing (Feddes et al., 2004b). In a subsequent study, various pretreatments including O_2 plasma pretreatment, Ar^+ ion gun treatment, and deposition of a Ti interlayer before coating deposition were evaluated (Feddes et al., 2004c). It was found that the presence of a Ti interlayer dramatically improved the adhesion of the coating (evaluated using the diamond scratch test). The other pretreatment methods, however, did not improve the adhesion of the coating layer onto the PTFE substrate. The reason proposed for the observed optimal adhesion for the Ti pretreatment was the ability of Ti to protect PTFE from UV irradiation—induced degradation. In addition, it was proposed that chemical bonding between the Ti interlayer and the PTFE substrate was an integral part of the system produced.

A mechanistic study was conducted into the use of single deposition runs by RF magnetron sputtering for producing HAP coatings (Surmenev et al., 2017). The study revealed the growth of calcium fluoride and calcium carbonate on the surface of the PTFE samples, while no HAP coating was deposited and these results agree with those obtained by Feddes et al. discussed above. The observed coating chemistry was attributed to the intensive PTFE substrate bombardment process by particles in the plasma, e.g., atoms and ions, causing fluorine ion sputtering thereby preventing HAP deposition. Further support of this mechanism came from the lack of HAP deposition on a Ti substrate situated in the same reaction chamber during the process; again, calcium fluoride and calcium carbonate were formed on the surface. Yamaguchi et al. (2008) applied a different process of electrophoretic deposition to produce an inorganic coating on PTFE. They applied an insulating photoresist mask to deposit patterns of wollastonite ($CaSiO_3$) onto a porous ePTFE substrate. Subsequent biomimetic HAP growth was achieved by immersion in simulated body fluid (SBF), leading to mineral nucleation at the wollastonite sites. The resulting HAP patterns were 200 μm in size and displayed distinct boundaries. To date, there has been no study reported on the adhesive strength of these coatings.

Although all the techniques described here are promising as methods for improving the interfacial bond of ePTFE with the underlying bone, much work still needs to be done to assess the biological response to the introduced coatings.

3.4.2 *Plasma treatment and ion implantation*

Plasma immersion ion implantation (PIII) is a combination of plasma treatment and ion implantation where the polymer sample is subjected to both treatments simultaneously. In has be utilized by the group of Chu and coworkers to enhance the cell attachment of osteoblast-like cells on PTFE biomaterials (Tong et al., 2010; Wang et al., 2010, 2012). In their work, they explored the use of oxygen in long-pulse, high-frequency versus short-pulse, low-frequency PIII treatments. Both treatment regimes resulted in surface confined oxidation, roughening, and increased hydrophobicity with the exact outcomes being depending on the treatment regime. These combined surface features were found to overall benefit cell attachment and proliferation both in MC3T3-E1 murine−derived osteoblastic cells (Tong et al., 2010), rat calvaria osteoblasts (Wang et al., 2010), and mesenchymal stem cells (MSCs) (Wang et al., 2012). The most recent study, expanded to investigate the use of different gases (oxygen, nitrogen, ammonia, and hydrogen) as well as two consecutive PIII processes with different gases (Wang et al., 2012). These studies found that the surface properties (chemistry, roughness, and wettability) were all affected by the different procedures. Interestingly, when the surface treatment involved a second ammonia PIII treatment, resulting in introduction of diverse chemistry including amine functionalities, the MSC displayed the best proliferation and osteogenic differentiation. Such enhanced performance of mixed functional group surfaces has also been shown for other biomaterial types (Yang et al., 2011).

The use of a different PIII treatment regime of PTFE biomaterials has, in separate studies by Bilek and coworkers, been shown to lead to different properties of a highly cross-linked subsurface as well as the introduction of free radicals (Bax et al., 2010, 2011). These free radicals can subsequently be used as linker-free immobilization of proteins. Particularly noteworthy is the enhanced bioactivity of the protein coatings compared with coatings of the same proteins produced by other immobilization methods. While this, so far, has not been explored for application in facial reconstruction, with judicious choice of proteins, the method has potential to enhance the bone-bonding ability of ePFTE membranes.

3.4.3 Gamma irradiation—induced grafting

Grafting techniques have been receiving increased attention in the area of biomaterials science due to their versatility. Simultaneous radiation-induced grafting involves the immersion of a polymer substrate in a monomer or monomer solution with simultaneous irradiation of the grafting system. This method produces radicals in the solution in the first instance and subsequently in the polymeric substrate (Wentrup-Byrne et al., 2005). The advantages of this method include its simplicity of execution and the versatility in tailoring the grafting outcome. Conversely, the formation of excess homopolymer may hinder efficient graft copolymerization and can thus be a disadvantage. Low doses are usually chosen due to the tendency of PTFE to undergo radiation-induced chain scission (rather than cross-linking).

Chapiro's seminal works studying radiation grafting of styrene and methyl methacrylate onto PTFE showed the remarkable propensity of the monomers to swell the substrate and hence lead to homogeneous grafting throughout (Chapiro, 1962). The mechanism for this phenomenon became known as the "grafting front mechanism" and has since been widely studied. It occurs when initial grafting on the substrate causes changes at the surface through swelling that subsequently allows the grafting to proceed gradually into the bulk polymer (Hegazy et al., 1992). Since these early advances, many studies have been performed grafting various monomers onto PTFE (Dargaville et al., 2003). There are, however, only a few reports of radiation graft polymerization onto ePTFE using monomers such as polyethylene oxide (Park et al., 2000), acrylic acid (Kepa et al., 2017; Hidzir et al., 2012, 2013, 2015, 2017; Grøndahl et al., 2003), and phosphates (Suzuki et al., 2005, 2015; Wentrup-Byrne et al., 2005, 2010; Chandler-Temple et al., 2010; Grøndahl et al., 2002). While the first of these studies aims at improving the nonfouling properties for use of ePTFE in blood vessels, the remaining works investigate the surface improvement of ePTFE membranes for facial applications.

The choice of acrylic acid, and in particular phosphate-containing monomers, for grafting onto ePTFE facial membranes originated from earlier studies that showed that these functional groups improve both the HAP nucleation process and growth, and therefore bone-bonding ability was inferred (often referred to as bioactivity). Indeed, the growth rate of apatite formation has been shown to decrease with the functional group in the order $PO_4H_2 > COOH >> CONH_2 \sim OH >> NH_2 >> CH_3 \sim 0$ (Tanahashi and Matsuda, 1997). In addition, grafting phosphate-containing monomers onto PE biomaterials resulted in doubling the amount of HAP growth on the modified material in SBF, whereas a smaller effect was observed for the acrylic acid grafted material (Tretinnikov and Ikada, 1997). Subsequent in vivo studies on the phosphate-modified PE biomaterial showed a significantly enhanced interface of the implant surface with the newly formed bone. This was attributed to the phosphate groups providing nucleation sites for HAP growth (Kamei et al., 1997).

The initial studies on modifying the surface properties of ePTFE facial membranes involved the acrylate monomer monoacryloxyethyl phosphate (Fig. 3.6(a)), using simultaneous grafting in water. This resulted in introduction of the phosphate-containing graft copolymer as evident from X-ray photoelectron spectroscopy (XPS) and in infrared spectroscopy investigations, albeit low graft yields

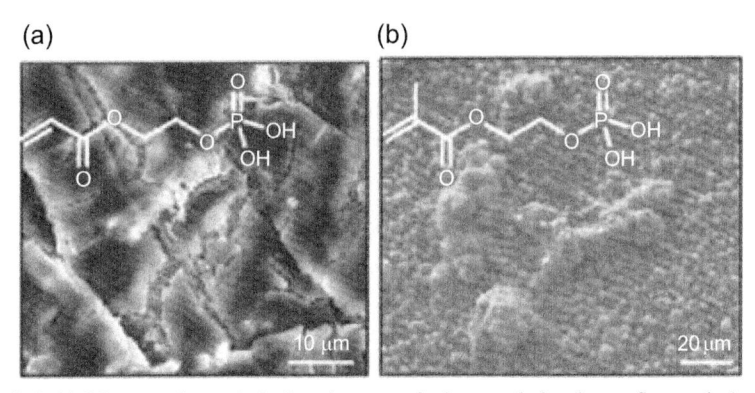

Figure 3.6 (a) Monoacryloxyethyl phosphate—grafted expanded polytetrafluoroethylene (ePTFE) and (b) methacryloxyethyl phosphate—grafted ePTFE showing brushite or monetite (a) and hydroxyapatite mineral phases (b), respectively.

(Grøndahl et al., 2002). Subsequent in vitro mineralization studies using SBF of the surface-modified ePTFE substrates revealed that not HAP but rather the more acidic calcium phosphate phase brushite or monetite formed predominantly, Fig. 3.6(a) (Grøndahl et al., 2003). A recent study observed that the addition of calcium ions to the grafting solution significantly increased the graft yield and surface coverage of the graft copolymer (Suzuki et al., 2015). This was attributed to the calcium ions enhancing the extent of grafting by increasing the local concentration of monomer near the surface during the grafting process. Furthermore, in vitro mineralization in SBF revealed that the mineral formed on these samples were brushite or monotite despite a globular mineral morphology. A key conclusion of the study was that the surface coverage of the graft copolymer, rather than the presence of calcium ions in the grafted layer, was linked to the amount of mineralization on these materials.

Subsequent studies involved graft polymerization of the methacrylate monomer methacryloxyethyl phosphate (MOEP, Fig. 3.6(b)) in various solvents (e.g., methanol and methyl ethyl ketone [MEK]) (Wentrup-Byrne et al., 2005, 2010; Suzuki et al., 2005; Chandler-Temple et al., 2010). It was found that the graft morphology varied with the solvent used (smooth morphology when grafted in methanol and globular morphology when grafted in MEK), and this was attributed to the solubility of the graft and homopolymers in the respective solvents (Wentrup-Byrne et al., 2005). Although, based on XPS examination, the chemistry of the two graft copolymers appeared to be very similar (i.e., they were both copolymers of MOEP and HEMA), the in vitro mineralization outcomes were very different (Suzuki et al., 2005). From infrared spectroscopic analysis of the mineralized layer, it was found that only the surface grafted in methanol actually induced nucleation and growth of HAP (Fig. 3.6(b)). Other modifications resulted in the growth of a mixture of calcium phosphate phases. It was later shown that this difference was in fact due to the degree of cross-linking in the graft copolymer (Suzuki et al., 2006). Depending on the grafting conditions, the nature of the graft copolymer varied (Fig. 3.7a or b), and it was shown that no cross-linking (Fig. 3.7(a)) was required for HAP to nucleate (Fig. 3.6(b)). In addition, it was found that the cross-linking reactions of the graft copolymer were due to the presence of large

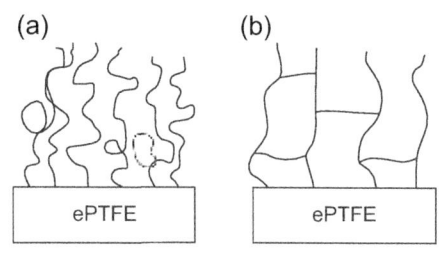

Replacement materials for facial reconstruction at the soft tissue–bone interface 69

Figure 3.7 Grafted polytetrafluoroethylene (PTFE) copolymer showing (a) no cross-linking and (b) some cross-linking.

amounts of a diene impurity in the monomers used in these studies. Subsequently, a method was devised to produce the nonbranched graft copolymer (and homopolymer) without incorporation of the diene impurity (Grøndahl et al., 2008). As a very nice conclusion to this mineralization study, it was found that the ePTFE membrane containing the nonbranched MOEP copolymer induced rapid and pure HAP growth in SBF (Wentrup-Byrne et al., 2010).

Because of the impurity present in the phosphate-containing monomers, a series of subsequent studies evaluated functionalization with carboxylate groups by grafting acrylic acid and/or itaconic acid onto ePTFE. Grafting was done in the presence of a homopolymer inhibitor (Mohr's salt) (Hidzir et al., 2013, 2015, Kepa et al., 2017) and a grafting enhancer (sulfuric acid) (Kepa et al., 2017) and demonstrated high graft yields and surface coverage. Enhancement of the carboxylate content was achieved when grafting with mixtures of acrylic and itaconic acid (Hidzir et al., 2015). In vitro mineralization studies in 1.5× SBF of carboxylate functionalized ePTFE membranes demonstrated that the surface coverage of the graft copolymer affect not only the mineral distribution but also the mineral phase that forms (Hidzir et al., 2017). Low surface coverage yielded coprecipitation of calcium phosphate and calcium carbonate, while high surface coverage resulted in carbonate substituted HAP formation. However, an enhanced carboxylate content did not enhance the mineralization outcome. Furthermore, as was observed for the phosphate-containing polymers described above, the topology of the graft copolymer influenced the type of calcium phosphate mineral that formed (Hidzir et al., 2017). Grafting of mixtures of amine and carboxylate functional monomers was explored as a means of "diluting" the carboxylate content on the surface of the graft copolymers (Kepa et al., 2017). A comparison was made between using a two-step consecutive grafting approach (yielding samples with high graft yield of grafted block copolymers) and grafting of monomer mixtures (yielding samples with moderate graft yield of grafted random copolymers). In vitro mineralization in 1.5× SBF yielded HAP mineralization of all samples with the mixed functional surfaces displaying enhanced mineralization compared with acrylic acid only grafted samples (Kepa et al., 2017). This result shows that surfaces with mixed functional groups have enhanced performance as was also pointed out for materials discussed in Section 3.4.2 above. Furthermore, the samples containing random copolymer grafts exhibited significantly larger amounts of mineral deposits than those containing block copolymer grafts despite of lower graft yield.

This results highlights that the architecture of grafted chains has a pronounced effect on the mineralization outcome (Kepa et al., 2017).

In addition to these promising in vitro mineralization studies, we have also shown that the grafted ePTFE membranes produced increased protein adsorption and osteoblast attachment, which are both important for increasing the interfacial strength (Suzuki et al., 2005). Furthermore, we have demonstrated that in vitro macrophage response is affected by the types of proteins that adsorb from serum and can be minimized by judicious choice of monomer and solvent combinations during the grafting process (Chandler-Temple et al., 2013). In conclusion, through selective modification of the material substrate surface, membranes can be improved (made more bioactive) and this results in stronger interfacing at the facial soft tissue–bone interface.

3.5 Future trends

It would appear that the most desirable materials are those that are both biostable and capable of integrating at the soft–hard tissue interface. As structurally graded PTFE materials are already commercially available, this concept is one that lends itself to further exploitation and development. Currently, there are plenty of ongoing fundamental materials research studies, including the chemical modification of surfaces of several well-defined polymers already used in facial repair and reconstruction. New developments should ideally combine both these aspects with the latest biomedical engineering technologies.

Acknowledgments

The authors (EW-B, LG, and AC-T) would like to thank everyone including their students who, over the years (in particular Dr. AC-T, Dr. Shuko Suzuki, and Dr. Norsyahidah Mohd Hidzir), worked on research projects, the results of which are included in this chapter; Dr. Richard Lewandowski for his generosity over many years in sharing his expertise in plastic surgery; and Dr. Bernard Môle and Dr. Pierre Nicolau (Paris), both of whom took the time to make a helpful contribution to our discussion on facial atrophy. We thank Sybil Curtis for her drawings in Figs 3.1 and 3.3, Dr. Oliver Locos for Fig. 3.4, and finally Patrick J. Lynch, medical illustrator, and C. Carl Jaffe, MD, cardiologist, for permission to use Fig. 3.2 (http://creativecommons.org/licenses/by/2.5/).

Thanks are also due to Professor Graeme George whose helpful discussions in the beginning helped to initiate this chapter and who proofread it at the end.

References

Afp, 2008. "China Face Transplant Patient Dies," the Age, 21 December. http://news.theage.com.au/world/china-face-transplant-patient-dies-20081221-72uc.html.

Albertsson, A.-C., Varma, I.K., 2003. Recent developments in ring opening polymerization of lactones for biomedical applications. Biomacromolecules 4 (6), 1466–1486. https://doi.org/10.1021/bm034247a.

Barton, S.E., Engelhard, P., Conant, M., 2006. Poly-L-lactic acid for treating HIV- associated facial lipoatrophy: a review of the clinical studies. Int. J. STD AIDS 17 (7), 429−435. https://doi.org/10.1258/095646206777689116.

Bax, D.V., Mckenzie, D.R., Weiss, A.S., Bilek, M.M., 2010. The linker-free covalent attachment of collagen to plasma immersion ion implantation treated polytetrafluoroethylene and subsequent cell-binding activity. Biomaterials 31, 2526−2534.

Bax, D.V., Wang, Y., Li, Z., Maitz, P.K.M., Mckenzie, D.R., Bilek, M.M., Weiss, A.S., 2011. Binding of the cell adhesive protein tropoelastin to PTFE through plasma immersion ion implantation treatment. Biomaterials 32, 5100−5111.

BBC NEWS, 2009. Woman Has First Face Transplant, 30 November 2005. http://news.bbc.co.uk/1/hi/health/4484728.stm.

Bezuidenhout, D., Zilla, P., 2004. Vascular grafts. In: Wnek, G.E., Bowlin, G.L. (Eds.), Encyclopedia of Biomaterials and Biomedical Engineering. Marcel Dekker Inc, New York, pp. 1715−1725.

Bhumbra, R.P.S., Walker, P.S., Berman, A.B., Emmanual, J., Barrett, D.S., Blunn, G.W., 2000. Prevention of loosening in total hip replacements using guided bone regeneration. Clin. Orthop. Relat. Res. 372, 192−204.

Bunn, C.W., Howells, E.R., 1954. Structures of molecules and crystals of fluoro- carbons. Nature 174, 549−551.

Burgess, C.M., Quiroga, R.M., 2005. Assessment of the safety and efficacy of poly- L-lactic acid for the treatment of HIV-associated facial lipoatrophy. J. Am. Acad. Dermatol. 52 (2), 233−239. https://doi.org/10.1016/j.jaad.2004.08.056.

Calabria, R., Hills, B., 2005. Fat grafting: fact or fiction? Aesthetic Surg. 25, 55. https://doi.org/10.1016/j.asj.2005.01.008.

Carr, A., Samaras, K., Burton, S., Law, M., Freund, J., Chisholm, D.J., Cooper, D.A., 1998. A syndrome of peripheral lipodystrophy, hyperlipidaemia, and insulin resistance in patients receiving HIV protease inhibitors. AIDS 12 (7), 51−58.

Cattelan, A.M., Bauer, U., Trevenzoli, M., Sasset, L., Campostrini, S., Facchin, C., Pagiaro, E., Gerzeli, S., Cadrobbi, P., Chiarelli, A., 2006. Use of polylactic acid implants to correct facial lipoatrophy in human immunodeficiency virus 1-positive individuals receiving combination antiretroviral therapy. Arch. Dermatol. 142 (3), 329−334.

Chandler-Temple, A., Wentrup-Byrne, E., Grøndahl, L., 2008. Expanded polytetrafluoroethylene: from conception to biomedical device. Chem. Aust. 75, 3−6.

Chandler-Temple, A., Wentrup-Byrne, E., Whittaker, A.K., Grøndahl, L., 2010. Graft copolymerisation of methoxyacrylethyl phosphate onto expanded poly(tetrafluoroethylene) facial membranes. J. Appl. Polym. Sci. 117, 3331−3339.

Chandler-Temple, A., Kingshott, P., Wentrup-Byrne, E., Whittaker, A.K., Cassidy, A.I., Grøndahl, L., 2013. Surface chemistry of grafted expanded poly(tetrafluoroethylene) membranes modifies the in vitro pro-inflammatory response in macrophages. J. Biomed. Mater. Res. A 101A (4), 1047−1058.

Chapiro, A., 1962. Preparation of graft copolymers with the aid of ionizing radiation. In: Chapiro, A. (Ed.), Radiation Chemistry of Polymeric Systems. John Wiley & Sons, New York, pp. 676−691.

Chirila, T.V. (Ed.), 2010. Biomaterials and Regenerative Medicine in Ophthalmology. Woodhead Publishing, Cambridge, UK.

Collins, E., Wagner, C., Walmsley, S., 2000. Psychosocial impact of the lipodystrophy syndrome in HIV infection. AIDS Read. 10, 546−550.

Colwell, J.M., Wentrup-Byrne, E., Bell, J.M., Wielunski, L.S., 2003. A study of the chemical and physical effects of ion implantation of micro-porous and nonporous PTFE. Surf. Coating. Technol. 168 (2–3), 216–222. https://doi.org/10.1016/S0257-8972(3)00204-4.

Cox, E., 2005. Who is still using expanded polytetrafluoroethylene? Dermatol. Surg. 31, 1613–1615.

Dahlin, C., Alberius, P., Linde, A., 1991. Osteopromotion for cranioplasty. J. Neurosurg. 74 (3), 487–491.

Dahlin, C., Sandberg, E., Alberius, P., Linde, A., 1994. Restoration of mandibular nonunion bone defects: an experimental study in rats using an osteopromotive membrane method. Int. J. Oral Maxillofac. Surg. 23 (4), 237–242.

Dahlin, C., Simion, M., Nanmark, U., Sennerby, L., 1998. Histological morphology of the ePTFE/tissue interface in humans subjected to guided bone regeneration in conjunction with oral implant treatment. Clin. Oral Implant. Res. 9 (2), 100–106.

Dargaville, T.R., George, G.A., Hill, D.J.T., Whittaker, A.K., 2003. High energy radiation grafting of fluoropolymers. Prog. Polym. Sci. 28 (9), 1355–1376. https://doi.org/10.1016/S0079-6700(03)00047-9.

Dev, V.R., Bryne, P., Tawfilis, A.R., Kim, D.W., 2008. Facial Trauma, Nasal Fractures. http://emedicine.medscape.com/article/1283709-overview.

De Visser, S.K., Bowden, J.C., Rintoul, L., Wentrup-Byrne, E., Bostrom, T., Pope, J.M., Momot, K.I., 2008. Anisotropy of collagen fibre alignment in bovine cartilage: comparison of polarised light microscopy and spatially-resolved diffusion-tensor measurements. Osteoarthr Cartil 16 (6), 689–697. https://doi.org/10.1016/j/joca.2007.09.015.

Drobny, J.G., 2000. Technology of Fluoropolymers, First ed. CRC Press, Florida.

Eppley, B.L., 2003. Alloplastic biomaterials for facial reconstruction. In: Booth, P.W., Eppley, B.L., Schmelzeisen, R. (Eds.), Maxillofacial Trauma and Esthetic Facial Reconstruction. Churchill Livingstone, Edinburgh, pp. 139–150.

Feddes, B., Vredenberg, A.M., Wolke, J.G.C., Jansen, J.A., 2004a. Bulk composition of rf magnetron sputter deposited calcium phosphate coatings on different substrates (polyethylene, polytetrafluoroethylene, silicon). Surf. Coating. Technol. 185 (2–3), 346–355. https://doi.org/10.1016/S0257-8972(03)01313-6.

Feddes, B., Wolke, J.G.C., Vredenberg, A.M., Jansen, J.A., 2004b. Adhesion of calcium phosphate ceramic on polyethylene (PE) and polytetrafluoroethylene (PTFE). Surf. Coating. Technol. 184 (2–3), 247–254. https://doi.org/10.1016/j.surfcoat.2003.10.013.

Feddes, B., Wolke, J.G.C., Weinhold, W.P., Vredenberg, A.M., Jansen, J.A., 2004c. Adhesion of calcium phosphate coatings on polyethylene (PE), polystyrene (PS), polytetrafluoroethylene (PTFE), poly(dimethylsiloxane) and poly-L-lactic acid (PLLA). J. Adhes. Sci. Technol. 18 (6), 655–672. https://doi.org/10.1163/156856104839347.

Fernandez, E.M., Mackley, C.L., 2006. Soft tissue augmentation: a review. J. Drugs Dermatol. 5 (7), 630–640.

Fialkov, J.A., Holy, C.E., Antonyshyn, O., 2000. Strategies for bone substitutes in craniofacial surgery. In: Davies, J.E. (Ed.), Bone Engineering, Toronto, EM Squared.

Food and Drug Administration, 1991. Clinical Information on Vitek TMJ Interpostional Implant and Vitek-Kent and Vitek-Kent I TMJ Implants. Public Health Advisory on Vitek Proplast Temporomandibular Joint Implants: A Letter to Practitioners. Issued September 1991. www.fda.gov/cdrh/safety/090191-tmj.pdf.

Fried, L.A., 1980. Anatomy of the Head, Neck, Face and Jaws, second ed. Lea & Febiger, Philadelphia.

Gonzalez, J., December 18, 2008. US Doctors Hail Near-Total Face Transplant. The Sydney Morning Herald. http://news.smh.com.au/world/us-doctors-hail-neartotal-face-transplant-20081218-716v.html.

Gore, R.W., 1976a. Process for Producing Porous Products. W L Gore & Associates, Inc, United States of America. United States Patent 3,953,566.
Gore, R.W., 1976b. Very Highly Stretched Polytetrafluoroethylene and Process Therefor. W. L. Gore & Associates, Inc., United States of America. United States Patent 3,962,153.
Grøndahl, L., Cardona, F., Chiem, K., Wentrup-Byrne, E., 2002. Preparation and characterization of the copolymers obtained by grafting of monoacryloxyethyl phosphate onto polytetrafluoroethylene membranes and poly(tetrafluoroethylene- co-hexafluoropropylene) films. J. Appl. Polym. Sci. 86 (10), 2550−2556.
Grøndahl, L., Cardona, F., Chiem, K., Wentrup-Byrne, E., Bostrom, T., 2003. Calcium phosphate nucleation on surface-modified PTFE membranes. J. Mater. Sci. Mater. Med. 14 (6), 503−510. https://doi.org/10.1023/A:1023403929496.
Grøndahl, L., Suzuki, S., Wentrup-Byrne, E., 2008. Influence of a diene impurity on the molecular structure of phosphate-containing polymers with medical applications. Chem. Comm. 3314−3316.
Guidoin, R., Maurel, S., Chakfe, N., How, T., Zhang, Z., Therrein, M., Formichi, M., Gosselin, C., 1993. Expanded polytetrafluoroethylene arterial prostheses in humans: chemical analysis of 79 explanted specimens. Biomaterials 14 (9), 694−704.
Hagerty, R.D., Salzmann, D.L., Kleinert, L.B., Williams, S.K., 2000. Cellular proliferation and macrophage populations associated with implanted expanded polytetrafluoroethylene and polyethylene terephthalate. J. Biomed. Mater. Res. 49 (4), 489−497.
Hall, B.K., 2005. In: Bones & Cartilage: Developmental and Evolutionary Skeletal Biology. Elsevier/Academic Press, London.
Hanke, C.W., Higley, H.R., Jolivette, D.M., Swanson, N.A., Stegman, S.J., 1991. Abscess formation and local necrosis after treatment with Zyderm or Zyplast collagen implant. J. Am. Acad. Dermatol. 25 (2), 319−326.
He, C., Gu, Z., 2003. Studies on the electron beam irradiated and acrylic acid grafted PET film. Radiat. Phys. Chem. 68 (5), 873−874.
Hegazy, E.-S., Dessouki, A.M., El-Assy, N.B., El-Sawy, N.M., El-Ghaffar, M.A., 1992. Radiation-induced graft polymerization of acrylic acid onto fluorinated polymers. 1. Kinetic study on the grafting onto poly(tetrafluoroethylene− ethylene) copolymer. J. Polym. Sci. Polym. Chem. Ed. 30, 1969−1976. https://doi.org/10.1002/pola.1992.080300920.
Hidzir, N.M., Hill, D.J.T., Martin, D., Grøndahl, L., 2012. Radiation-induced grafting of acrylic acid onto expanded poly(tetrafluoroethylene) membranes. Polymer 53, 6063−6071.
Hidzir, N.M., Hill, D.J.T., Taran, E., Martin, D., Grøndahl, L., 2013. Argon plasma treatment-induced grafting of acrylic acid onto expanded poly(tetrafluoroethylene) membranes. Polymer 54, 6536−6546.
Hidzir, N.M., Lee, Q., Hill, D.J.T., Rasoul, F., Grøndahl, L., 2015. Grafting of acrylic acid-co-itaconic acid onto ePTFE and characterisation of water uptake by the graft copolymers. J. Appl. Polym. Sci. 132, 41482.
Hidzir, N.M., Hill, D.J.T., Martin, D., Grøndahl, L., 2017. In vitro mineralisation of grafted ePTFE membranes carrying carboxylate groups. Bioact. Mater. 2, 27e34.
Hirsch, R.J., Cohen, J.L., 2006. Soft tissue augmentation. Cutis 78 (3), 165−172.
Holck, D.E., Ng, J.D., 2006. Evaluation and Treatment of Orbital Fractures: A Multidisciplinary Approach. Elsevier Saunders, Philadelphia.
Homicz, M.R., Watson, D., 2004. Review of injectable materials for soft tissue augmentation. Facial Plast. Surg. 20 (1), 21−29.
Hontsu, S., Nakamori, M., Kato, N., Tabata, H., Ishii, J., Matsumoto, T., Kawai, T., 1998. Formation of hydroxyapatite thin films on surface-modified polytetrafluoroethylene substrate. Jpn. J. Appl. Phys. 37, L1169−L1171.

Hsueh, C.-L., Peng, Y.-J., Wang, C.-C., Chen, C.-Y., 2003. Bipolar membrane prepared by grafting and plasma polymerization. J. Membr. Sci. 219, 1−13.

Hürzeler, M.B., Kohal, R.J., Naghshbandi, J., Mota, L.F., Conradt, J., Hutmacher, D., Caffesse, R.G., 1998. Evaluation of a new bioresorbable barrier to facilitate guided bone regeneration around exposed implant threads. An experimental study in the monkey. Int. J. Oral Maxillofac. Surg. 27 (4), 315−320.

Isolagen, 2007. The Isolagen Process. Isolagen Inc. http://www.isolagen.com/isolagen/default.htm.

Kamei, S., Tomita, N., Tamai, S., Kato, K., Ikada, Y., 1997. Histologic and mechanical evaluation for bone bonding of polymer surfaces grafted with a phosphate- containing polymer. J. Biomed. Mater. Res. 37 (3), 384−393. https://doi.org/10.1002/(SICI)1097-4636 (19971205)37:3<384::AID-JBM9>3.0.CO;2-H.

Kang, I.-K., Choi, S.-H., Shin, D.-S., Yoon, S.C., 2001. Surface modification of polyhydroxyalkanoate films and their interaction with human fibroblasts. Int J Biological Macromol 28 (3), 205−212. https://doi.org/10.106/S0141-8130(00)00165-3.

Kannan, R.Y., Salacinski, H.J., Butler, P.E., Hamilton, G., Seifalian, A.M., 2005. Current status of prosthetic bypass grafts: a review. J. Biomed. Mater. Res. B Appl. Biomater. 74B, 570−581. https://doi.org/10.1002/jbm.b.30247.

Kaufman, M.R., Miller, T.A., Huang, C., Roostaien, J., Wasson, K.L., Ashley, R.K., Bradley, J.P., 2007. Autologous fat transfer for facial recontouring: is there science behind the art? Plast. Reconstr. Surg. 119 (7), 2287−2296.

Kepa, K., Hill, D.J.T., Grøndahl, L., 2017. In vitro mineralization of dual grafted polytetrafluoroethylene membranes. Biointerphases 12, 02C413.

Kidd, K.R., Dal Ponte, D.B., Kellar, R.S., Williams, S.K., 2002. A comparative evaluation of the tissue responses associated with polymeric implants in the rat and mouse. J. Biomed. Mater. Res. 59 (4), 682−689.

Klein, A.W., Elson, M.L., 2000. The history of substances for soft tissue augmentation. Dermatol. Surg. 26 (12), 1096−1105. https://doi.org/10.1046/j.1524-4725.2000.t01-1-00512.x.

Kühn, G., Retzko, I., Lippitz, A., Unger, W., 2001. Homofunctionalized polymer surfaces formed by selective plasma processes. Surf. Coating. Technol. 142−144, 494−500.

Lafaurie, M., Dolivo, M., Porcher, R., Rudant, J., Madelaine, I., Molina, J.M., 2005. Treatment of facial lipoatrophy with intradermal injections of polylactic acid in HIV-infected patients. J. Acquir. Immune Defic. Syndr. 38 (4), 393−398.

Lam, S.M., Azizzadeh, B., Graivier, M., 2006. Injectable poly-L-lactic acid (Sculptra): technical considerations in soft-tissue contouring. Plast. Reconstr. Surg. 118 (3 Suppl.), 55S−63S.

Landau, T., 1989. Identity. In: About Faces. Doubleday, New York, pp. 36−43.

Lascaratos, J.G., Segas, J.V., Trompoukis, C.C., Assimakopoulos, D.A., 2003. From the roots of rhinology: the reconstruction of nasal injuries by Hippocrates. Ann. Otol. Rhinol. Laryngol. 112 (2), 159−162.

Lawrence, E.L., Turner, I.G., 2005. Materials for urinary catheters: a review of their history and development in the UK. Med. Eng. Phys. 27 (6), 443−453. https://doi.org/10.1016/j.medengphy.2004.12.013.

Lemperle, G., Romano, J.J., Busso, M., 2003. Soft tissue augmentation with Artecoll: ten-year history, indications, techniques, and complications. Dermatol. Surg. 29 (6), 573−587. https://doi.org/10.1046/j.1524-4725.2003.29140.x.

Linde, A., Alberius, P., Dahlin, C., Bjurstam, K., Sundin, Y., 1993. Osteopromotion: a soft-tissue exclusion principle using a membrane for bone healing and bone neogenesis. J. Periodontol. 64 (11), 1116−1128.

Lorenzoni, M., Pertl, C., Polansky, R., Wegscheider, W., 1999. 'Guided bone regeneration with barrier membranes — a clinical and radiographic follow-up study after 24 months'. Clin. Oral Implant. Res. 10 (1), 16—23.

Lowe, N.J., Maxwell, C.A., Patnaik, R., 2005. Adverse reactions to dermal fillers: review. Dermatol. Surg. 31 (s4), 1616—1625. https://doi.org/10.2310/6350.2005.31251.

Malaisrie, S.C., Malekzadeh, S., Biedlingmaier, J.F., 1998. In vivo analysis of bacterial biofilm formation on facial plastic bioimplants. Laryngoscope 108 (11/1), 1733—1738.

Mazzei, R.O., Smolko, E., Torres, A., Tadey, D., Rocco, C., Gizzi, L., Strangis, S., 2002. Radiation grafting studies of acrylic acid onto cellulose triacetate membranes. Radiat. Phys. Chem. 64, 149—160.

Mazzei, R., Tadey, D., Smolko, E., Rocco, C., 2003. Radiation grafting of different monomers onto PP foils irradiated with a 25 MeV proton beam. Nucl. Instrum. Methods Phys. Res. B 208, 411—415.

Mercandetti, M., Cohen, A.J., Chang, E.W., 2008. Implants, Soft Tissue. Gore-Tex. http://emedicine.medscape.com/article/878937-overview.

Morhenn, V.B., Lemperle, G., Gallo, R.L., 2002. Phagocytosis of different particulate dermal filler substances by human macrophages and skin cells. Dermatol. Surg. 28 (6), 484—490. https://doi.org/10.1046/j.1524-4725.2002.01273.x.

Moyle, G.J., Lysakova, L., Brown, S., Sibtain, N., Healy, J., Priest, C., Mandalia, S., Barton, S.E., 2004. A randomized open-label study of immediate versus delayed polylactic acid injections for the cosmetic management of facial lipoatrophy in persons with HIV infection. HIV Med. 5 (2), 82—87. https://doi.org/10.1111/j.1468-1293.2004.00190.x.

Moyle, G.J., Brown, S., Lysakova, L., Barton, S.E., 2006. Long-term safety and efficacy of poly-L-lactic acid in the treatment of HIV-related facial lipoatrophy. HIV Med. 7 (3), 181—185. https://doi.org/10.1111/j.1468-1293.2006.00342.x.

Neel, H.B., 1983. Implants of gore-tex. Arch. Otolaryngol. 109 (7), 427—433.

Neuber, G., 1893. Fettransplantation. Zentrabl Chir 22, 66.

Nicolau, P.J., 2008. The different products and their pharmacology. In: Ascher, B., Landau, M., Rossi, B. (Eds.), Injection Treatments in Cosmetic Surgery. Informa HealthCare, London & New York, pp. 413—417.

Oehr, C., Müller, M., Elkin, B., Hegermann, D., Vohrer, U., 1999. 'Plasma grafting — a method to obtain monofunctional surfaces'. Surf. Coating. Technol. 116—119, 25—35. https://doi.org/10.1016/S0257-5972(99)00201-7.

Oeltjen, J.C., Hollier, L., 2005. Nasal and naso-orbital ethmoid fractures. In: Stewart, M.G. (Ed.), Head, Face and Neck Trauma. Comprehensive Management, New York, Thieme, pp. 39—51.

Osmond, S.F., 2000. 'Rembrandt's Self-Portraits', the Arts, January 2000. http://www.worldandi.com/specialreport/rembrandt/rembrandt.html.

Paavolainen, P., Makisalo, S., Skutnabb, K., Holmstrom, T., 1993. Biologic anchorage of cruciate ligament prosthesis — bone ingrowth and fixation of the Gore- Tex® ligament in sheep. Acta Orthop. Scand. 64 (3), 323—328.

Panossian, A., Garner, W.L., 2004. Polytetrafluoroethylene facial implants: 15 years later. Plast. Reconstr. Surg. 113, 347—349. https://doi.org/10.1097/01. PRS.0000097285.32945.21.

Park, K., Shim, H.S., Dewanjee, M.K., Eigler, N.L., 2000. In vitro and in vivo studies of PEO-grafted blood-contacting cardiovascular prostheses. J. Biomater. Sci. Polym. Ed. 11 (11), 1121—1134. https://doi.org/10.1163/156856200744228.

Ratner, B.D., Hoffman, A.S., Schoen, F.J., Lemons, J.E. (Eds.), 2004. Biomaterials Science: An Introduction to Materials in Medicine, second ed. Elsevier Academic Press, Amsterdam; Boston.

Redbord, K.P., Hanke, C.W., 2008. Expanded polytetrafluoroethylene implants for soft-tissue augmentation: 5-year follow-up and literature review. Dermatol. Surg. 34 (6), 735−744. https://doi.org/10.1111/j.1524-4725.2008.34140.x.

Reis, E., Sousa, C., Stahl, P.D., Austyn, J.M., 1993. Phagocytosis of antigens by Langerhans cells in vitro. J. Exp. Med. 178, 509−519.

Rising, D., 2008. 'Real Mona Lisa's Identity Confirmed, University Says', National Geographic News, 15 January. http://news.nationalgeographic.com/news/pf/89566829.html.

Sarnoff, D.S., Saini, R., Gotkin, R.H., 2008. Comparison of filling agents for lip augmentation. Aesthet. Surg. J. 28 (5), 556−563. https://doi.org/10.1016/j.asj.2008.07.001.

Sato, Y., Murahara, M., 2004. Protein adsorption on PTFE surface modified by ArF excimer laser treatment. J. Adhes. Sci. Technol. 18 (13), 1545−1555. https://doi.org/10.1163/1568561042411259.

Scheirs, J., 1997. Structure/property considerations for fluoropolymers and fluoroelastomers to avoid in-service failure. In: Schiers, J. (Ed.), Modern Fluoropolymers: High Performance Polymers for Diverse Applications. John Wiley & Sons, Chichester.

Surmenev, R.A., Surmeneva, M.A., Grubova, I.Y., Chernozem, R.V., Krause, B., Baumbach, T., Lozac, K., Epple, M., 2017. RF magnetron sputtering of a hydroxyapatite target: a comparison study on polytetrafluorethylene and titanium substrates. Appl. Surf. Sci. 414, 335−344.

Siebel, M.J., Robins, S.P., Bilezikian, J.P., 2006. Dynamics of Bone and Cartilage Metabolism. Academic Press, San Diego.

Soyer, T., Lempinen, M., Cooper, P., Norton, L., Eiseman, B., 1972. A new venous prosthesis. Surgery 72 (6), 864−871.

Sperati, C.A., Starkweather, J.R., 1961. Fluorine-containing polymers. II. Polytetrafluoroethylene. Adv. Polym. Sci. 2, 465−495.

Sperati, C.A., 1986. Polytetrafluoroethylene: history of its development and some recent advances. In: Seymour, R.B., Kirshenbaum, G.S. (Eds.), High Performance Polymers: Their Origin and Development. Elsevier Science Publishing Co Inc., New York, pp. 267−278.

Stewart, M.G., 2005. Head, Face and Neck Trauma. Comprehensive Management, New York, Thieme.

Suzuki, S., Grøndahl, L., Leavesley, D., Wentrup-Byrne, E., 2005. In vitro bioactivity of MOEP grafted ePTFE membranes for craniofacial applications. Biomaterials 26, 5303−5312.

Suzuki, S., Whittaker, M.R., Grøndahl, L., Monteiro, M.J., Wentrup-Byrne, E., 2006. Synthesis of soluble phosphate polymers by RAFT and their in vitro mineralization. Biomacromolecules 7 (11), 3178−3187. https://doi.org/10.1021/bm060583q.

Suzuki, S., Wentrup-Byrne, E., Chandler-Temple, A., Shah, N., Grøndahl, L., 2015. Calcium ion-mediated grafting of a phosphate-containing monomer. J. Appl. Polym. Sci. 132, 42808.

Tanahashi, M., Matsuda, T., 1997. Surface functional group dependence on apatite formation on self-assembled monolayers in a simulated body fluid. J. Biomed. Mater. Res. 34, 305−315.

Tong, L., Kwok, D.T.K., Wang, H., Wu, L., Chu, P.K., 2010. Adv. Eng. Mater. B 12, 163−169.

Tretinnikov, O.N., Ikada, Y., 1997. Hydrogen bonding and wettability of surface grafted organophosphate polymer. Macromolecules 30, 1086−1090.

Truswell, W.H., 2002. Dural-porosity expanded polytetrafluoroethylene soft tissue implant. Arch. Facial Plast. Surg. 4, 92−97.

Ut Southwestern, 2009. Auricular Reconstruction (Ear Replacement Surgery). http://www8.utsouthwestern.edu/utsw/cda/dept28171/files/133537.html.

Valantin, M.A., Aubron-Olivier, C., Ghosn, J., Laglenne, E., Pauchard, M., Schoen, H., Bousquet, R., Katz, P., Costagliola, D., Katlama, C., 2003. Polylactic acid implants (New-Fill®) to correct facial lipoatrophy in HIV-infected patients: results of the open-label study VEGA. AIDS 17 (17), 2471−2477.

Van Helden, A., 1995. 'Tycho Brahe', the Galileo Project. http://galileo.rice.edu/sci/brahe.html.

Vert, M., Li, S.M., Spenlehauer, G., Guerin, P., 1992. Bioresorbability and biocompatibility of aliphatic polyesters. J. Mater. Sci. Mater. Med. 3 (6), 432−446. https://doi.org/10.1007/BF00701240.

Voorhees, A.B.J., Jaretzki Iii, A., Blakemore, A.H., 1952. The use of tubes constructed from Vinyon "N" cloth in bridging arterial defects. Ann. Surg. 135 (2), 332−336.

Walsh, W.R., 2006. Repair and Regeneration of Ligaments, Tendons and Joint Capsule. Humana Press, Totowa, NJ.

Wang, H., Kwok, D.T.K., Wang, W., Wu, Z., Tong, L., Zhang, Y., Chu, P.K., 2010. Biomaterials 31, 413−419.

Wang, H., Kwok, D.T.K., Xu, M., Shi, H., Wu, Z., Zhang, W., Chu, P.K., 2012. Adv. Mater. 24, 3315−3324.

Wentrup-Byrne, E., Grøndahl, L., Suzuki, S., 2005. Methacryloxyethyl phosphate- grafted expanded polytetrafluoroethylene membranes for biomedical applications. Polym. Int. 54, 1581−1588.

Wentrup-Byrne, E., Suzuki, S., Suwanasilp, J.J., Grøndahl, L., 2010. Novel phosphate-grafted ePTFE copolymers for optimum in vitro mineralization. Biomed. Mater. 5 (4), 045010.

Wilkins, R.H., 1964. 'Neurosurgical classic − XVII'. J. Neurosurg. 21, 240−244.

Wiltfang, J., Merten, H.-A., Peters, J.-H., 1998. Comparative study of guided bone regeneration using absorbable and permanent barrier membranes: a histologic report. Int. J. Oral Maxillofac. Implant. 13, 416−421.

Woerle, B., Hanke, C.W., Sattler, G., 2004. Poly-L-lactic acid: a temporary filler for soft tissue augmentation. J. Drugs Dermatol. 3 (4), 385−389.

Yamaguchi, S., Yabutsuke, T., Hibino, M., Yao, T., 2008. Generation of hydroxyapatite patterns by electrophoretic deposition. J. Mater. Sci. Mater. Med. 19, 1419−1424. https://doi.org/10.1007/s10856-006-0053-6.

Yang, Z., Tu, Q., Wang, J., Lei, X., He, T., Sun, H., Huang, N., 2011. Bioactive plasma-polymerized bipolar films for enhanced endothelial cell mobility. Macromol. Biosci. 11, 797−805.

Zhao, S.J., Pinholt, E.M., Madsen, J.E., Donath, K., 2000. Histological evaluation of different biodegradable and non-biodegradable membranes implanted subcutaneously in rats. J. Cranio-Maxillofacial Surg. 28 (2), 116−122.

Zhengbang, W., Haolin, T., Junsheng, L., Mu, P., 2011. Morphology change of biaxially oriented polytetrafluoroethylene membranes caused by solvent soakage'. J. Appl. Polym. Sci. 121, 1464−1468.

Tissue engineering of small-diameter vascular grafts

4

Kiran R. Adhikari, Bernabe S. Tucker, Vinoy Thomas
Polymers & Healthcare Materials/Devices, Department of Materials Science & Engineering, University of Alabama at Birmingham, Birmingham, AL, United States

Chapter outline

4.1 Background 79
4.2 Clinical significance 80
4.3 Tissue engineering approach 81
    4.3.1 Scaffold fabrication 81
    4.3.2 Materials and material properties 82
    4.3.3 Target architecture: native blood vessels 83
    4.3.4 Extracellular matrix 84
    4.3.5 Mimicry of the extracellular matrix: electrospinning 84
4.4 Surface modification approach 85
    4.4.1 Cold plasma–based techniques: chemical/physical modification 86
    4.4.2 Click chemistry 86
    4.4.3 Biological modification 87
4.5 Cellular interactions 88
    4.5.1 In vitro studies 88
    4.5.2 Stem cell studies 90
    4.5.3 Cell microintegration 91
    4.5.4 Core-shell method 93
    4.5.5 In vivo studies toward translational applications 93
4.6 Emerging perspectives 94
4.7 Conclusion 96
References 96

4.1 Background

Prosthetic vascular grafts trace their lineage back decades. Dacron or poly(ethylene terephthalate) was introduced as an option in 1957 (Ku and Allen, 1995). As a woven or knitted construct, a Dacron graft is resilient and total resorption occurs only after 30 years (Kannan et al., 2005). Teflon or poly(tetrafluoroethylene) (PTFE) began as a replacement heart valve material and has been employed as an inert prostheses in the form of expanded or ePTFE (Kannan et al., 2005). Together these materials are often considered the gold standard as vascular graft materials and remain commercially available as treatment options for operations requiring blood vessel replacement or bypass. However, they are only used for large caliber replacements (>6 mm) such as in aortic or peripheral bypass procedures (Kannan et al., 2005). Although they are among the most successful graft materials, thrombosis or clot formation still limits their efficacy for applications to replace smaller vessel members. Various other materials have also been used. The extensive list includes other polyesters, polyurethanes (PUs), and a plethora of additional polymers including bioresorbable types investigated as candidates for small caliber grafts (Kannan et al., 2005). As a tissue engineering subject, vascular grafts deserve further study for a full background and to frame their place in the broader field of biomaterials science. An extensive reading is highly recommended into the full scope of the use of biomaterials in medicine (Ratner et al., 2012) (Figs. 4.1–4.6).

4.2 Clinical significance

The World Health Organization has suggested that cardiovascular diseases (CVDs) are the leading cause of death globally (World Health Organization, 2017). In 2016 alone 17.9 million people died from CVDs, which accounts for 31% of all the deaths. The occurrence of CVDs is ubiquitous; not only the developed world but also the developing world is affected by this (National Center for Health Statistics, 2017). And these numbers are prone to grow each year and projected to be 23.3 million by year 2030, leading to the increase in the economic and health-care burdens (Antoniou et al., 2013). These rises in the numbers are due to the increasingly hostile cardiovascular environments, indicated by the change in diet, lack of exercises, socioeconomic, and other stresses (Gersh et al., 2010). Among the conditions common in CVD are blockages of coronary arteries, cerebrovascular, and peripheral arteries; all potentially leading to life-threatening conditions such as stroke or heart attack. Along with dietary modification, lifestyle changes, and medications, the treatment options for damaged blood vessels often may involve grafting a section of native vessel to replace the diseased portion known as autologous grafting, which is cited to be the gold standard (Pashneh-Tala et al. 2015). In United States alone, 1.4 million patients require arterial prosthetic each year (Browning et al., 2012; Hasan et al., 2014). These replacements could be autografts, allografts, or xenografts. Autografts, as they are transplanted from the patient's native tissues, suffer from limited supply and dimensional mismatch despite their superior compatibility. Allografts (from donors/cadavers) and xenografts (from animals) have also shown limited success due to immunogenic rejection of the body (Lee et al., 2007). In addition to that, there is a high chance of disease transfer from these grafts, and it takes longer time to incorporate in the patient's body. Also, these grafts have limited to no tunability in terms of diameter, physical properties, and manufacturing control.

4.3 Tissue engineering approach

Tissue engineering is one of the growing fields in the biomedical research and applications. The aim of the tissue engineering is to provide necessary environment in the targeted area for the regeneration of the tissues. The interest in the field is growing every day as the necessity of the regenerative medicine and the development of suitable material is in high demand. Many surgical procedures are done each day to regenerate, repair, or replace the damaged or degenerated organ, tissues, or body structures (O'brien, 2011). The National Institute of Health describes tissue engineering as "an interdisciplinary and multidisciplinary field that aims at the development of biological substitutes that restores, maintains, or improve tissue functions." It grew from the development of materials for biomaterials, and it indicates the areas of study that integrate cells, scaffolds, biological systems, and chemicals into the functional tissue structures. The main goal of tissue engineering is to provide the substitute for the native tissue structures while the repair is in place and to provide platform for the growth of natural system. Tissue engineering term itself was first introduced in the mid-1980s and growing ever since with the very ambitious goal of synthetically fabricating fully functional biological systems. Particularly, the realization of small-diameter tissue-engineered vascular grafts (TEVGs) represents a highly sought yet unreached goal for the field.

4.3.1 Scaffold fabrication

Scaffold fabrication and design is one of the major parts of biomaterial research; the research interest in the field of tissue engineering and thus in the development of suitable scaffold and the materials has grown immensely (Brahatheeswaran et al., 2011). While being the building structure for the repair of the damaged tissue, scaffolds can also act as drug, cell, or gene delivery vehicle. Depending on the function and purpose of the scaffolds, it could be a 3D porous matrix, a nanofiber mesh, a porous microsphere, or hydrogels or a combination of these. Scaffolds in general should have the following properties:

- Desirable mechanical properties, shape and strength, are sufficient to protect the growing tissue structure without hindering the passage of biochemical or biomechanical signals.
- Structured to facilitate the cell migration and cell contact and to support the cell adhesion and proliferation at desired rate.
- Biocompatible, nontoxic chemical composition.
- The dimension of the scaffold should match the native structure's dimensions.
- Stable physical dimension and physical and chemical structure over a desired period.
- A highly porous structure, i.e., high surface-to-volume ratio, is desired, with suitable high cell seeding density or drug loading.
- Reproducible bulk and microscopic structure.
- The scaffolds fabrication procedure should be relatively easy and should be capable of being produced in a sterile, stable product form. Polymers with wide processing flexibility are predominant in application and research.
- Cost effective and scalable.

Another important feature is that the developed scaffold should be a ready-to-use product. Clinicians typically prefer off-the-shelf product, which can be used in the procedures without further processing. There is a huge challenge in designing a scaffold that meets all the required properties. It starts from choosing the right materials to designing the scaffold into right shape and ultimately to assessing the effectiveness and impact in the human body.

4.3.2 Materials and material properties

Natural, synthetic, and combinations of both types of materials are currently being explored. The materials used should provide suitable and efficient environment for the tissue growth while providing the necessary support. The ultimate goal is to develop a scaffold that is suitable for in vivo uses so three important factor must be considered: biocompatibility, nontoxicity (material and its postdegradation products), and biomimicry (He and Lu, 2016; Galler et al., 2018).

Biocompatibility refers to the ability of the scaffolds/materials to perform desired task without causing a harmful response such as inflammation or rejection by immune system of the body. As the tissue growth takes place, the scaffold is absorbed in the body system. That brings to the second important factor toxicity. When scaffolds are implanted/absorbed in the body, it should neither trigger severe immune response from the body nor should the degraded product during the absorption process possess the toxicity. These are the most uncompromising factors for in vivo application and can be assessed during the in vitro studies. When the scaffolds are biologically active and nontoxic, they will be integrated in the surrounding tissues; if not they are encapsulated by the fibrous capsule, and surrounding tissue structure suffers damage and death due to its toxicity (Hench, 1998). The rate at which tissue regenerates in the body should match the rate of degradation of the scaffold materials. One of the major reasons for failure of scaffolds is that they fail to mimic the tissue structure and properties of the surrounding tissues. So, biomimicry is important for increasing the patency rate of the implanted scaffolds.

Using synthetic material gives a huge control over the fine-tuning of mechanical properties, shape, size, and degradation rate but that comes in expense of the favorable bioresponse. But on the other hand, using natural substances gives an excellent biochemical response from the body but lacks in the suitable mechanical properties, functionalization capabilities, and the consistency found from bulk materials (Wood et al., 2016). Natural materials such as proteins are inherently bioactive and biodegradable, making them suitable for numerous applications; however, the properties such as degradation are not quite controllable, thus making them difficult to use for applications where the tissue regeneration is desired to match the rate of degradation (Budhwani et al., 2016).

Therefore, different approaches of using a blend of natural and synthetic polymers to make scaffolds have been implemented (Thomas et al., 2006). By using the appropriate mixture of both natural and synthetic polymers, desired mechanical properties with favorable bioactivity can be achieved. A number of natural materials such as collagen (Han et al., 1999), chitosan (Suh and Matthew, 2000), alginate (Dar et al., 2002), silk fiber (Mandal and Kundu, 2010), gelatin (Kang et al., 1999), and cellulose (Müller et al., 2006) have shown promising results and have been widely used for many years. Synthetic polymers such as polyesters (Blanchemain et al., 2005), polyanhydrides (Lavik and Langer, 2004), polyphosphazenes (Carampin et al., 2007), PU (Uttayarat et al., 2010), polylactic acid (PLA) (Lavik and Langer, 2004), polyglycolic acid (Lavik and Langer, 2004), poly(caprolactone) (PCL) (Valence et al., 2012), Teflon (Zhu et al., 2017), among many others have shown tremendous potential and are actively studied.

4.3.3 Target architecture: native blood vessels

Blood vessels in the body carry the blood to and from the tissues and organs to facilitate the exchange of nutrients, metabolic wastes, gases, and signaling molecules. The shape and size of the vessels in the body varies in different sites within the circulatory system, but they are composed of three principal layers: the tunica intima, the tunica media, and the tunica adventitia (Song et al., 2011). The tunica intima, the innermost layer, consists of inner surface of smooth endothelium covered by the surface of elastic tissues. This is the layer that is in contact with blood flow, and this lining of endothelial cells (ECs) prevents the coagulation factors and also inhibits the attachments of the platelets to the inner surface of the vessels (Lusis, 2000). In addition to that, it acts as barrier for solvent and solutes in plasma and takes parts in regulating the vasomotor tone, the dilation, and constriction of the blood vessels. The tunica media is the middle layer of the vessels made up of mostly smooth muscle cells (SMCs), supported by connective tissues primarily made by elastic fibers and some collagen. It's usually thicker in the arteries compared to veins. This is very strong layer and the functions in controlling the caliber of the arteries and is important in maintaining the blood pressure in the arteries (Song et al., 2011). The outermost layer is tunica adventitia also known as tunica externa, which wraps the tunica media composed of connective tissues, mostly collagens. These collagens help anchoring vessels to nearby tissues. This layer by layer structure of a typical blood vessels is shown in figure 4.1 (Schmedlen et al., 2003).

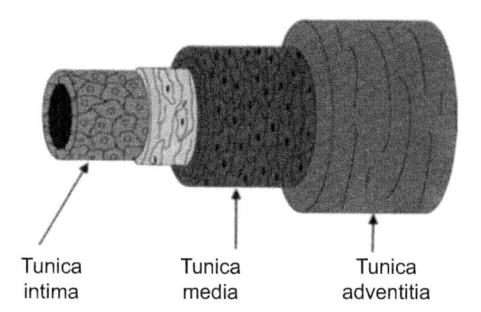

Figure 4.1 Hierarchy of structure in blood vessel members.
Adapted from Schmedlen, R.H., et al. 2003. Tissue engineered small-diameter vascular grafts. Clin. Plast. Surg. 30(4), 507–517.

4.3.4 Extracellular matrix

Extracellular matrix (ECM) is a three-dimensional network of collagen, different proteins macromolecule (elastin, fibronectin), proteoglycans or glycosaminoglycans, hyaluronic acid, and other soluble factors (Theocharis et al., 2016). ECM acts as substrate for cells to adhere. It not only is the foundation for the cell adhesion but also has the following major functions such as providing strength to 3D cell structure, providing guidance to the cell growth direction and alignment, facilitating the cell-to-cell contact establishment, mediating the signal transfer, and helping cell infiltrations. The cell surface has special receptor that binds with specific motif in the ECM proteins and that leads to the adhesion of cells to the ECM. Cells produce the components of the ECM and also enzymes to degrade it. The cells periodically modify the ECM topography and composition in response to the surrounding conditions. This process is called "matrix remodeling" and has a lot of significance in the development of tissue engineering scaffolds (Cox and Janine, 2011). The mechanical description of the ECM is quite complex. The matrix undergoes cyclic deformation; a single parameter such as Young's modulus cannot define the physical properties of an ECM matrix alone, but rather a set of descriptors are necessary (Akhmanova et al., 2015).

4.3.5 Mimicry of the extracellular matrix: electrospinning

Electrospinning process has gained much attention past couple of decades as it can produce a native ECM-like structure and microenvironment that is ideal for the process of fabricating TEVGs. It is capable of producing nanofibers in the range comparable with diameter of collagens. Fabrication of nanofiber gives high surface-to-volume ratio and high porosity, which is excellent for cell seeding and proliferation. All these parameters such as fiber density, porosity, surface area, fiber alignment, etc., in the scaffold can be adjusted either by using different materials or by varying the fabrication parameters for optimal microenvironmental properties. There is great interest in fabricating multilayered vascular grafts that mimic the anatomic multilayered structure of natural vessels. The layers are usually fabricated keeping degradation/regeneration rate, individual structure of each layer, and their mechanical properties in mind (Zhang et al., 2010).

In the electrospinning process, a polymer solution with suitable viscosity is subject to a high electrical potential difference. When the electric force is able to counteract the surface tension of the liquid droplet at the orifice (created by the pump system at specified rate) of a metallic needle, the droplet is stretched. And, at critical point, this stretched droplet erupts from the surface, initiating the formation of the fibers. This critical point at which rupturing of the droplet happens and a cone-like structure is formed called Taylor cone, and the shape and size of this cone is determined by the viscosity and the electrostatic repulsion. As the polymer jet is ejected, the rapid evaporation of the solvent occurs, which leads to the significant increase in the charge density at the surface. This increase in the surface density leading to the bending stability and thinning in the fibers produces the fibers in the nanometer range diameter. The fiber properties such as alignment, diameter, porosity, etc., dictate the overall effectiveness of the scaffold's application. These properties can be tuned by varying different parameters during the process of fabrication. Most importantly, diameter of the fiber produced guides the physical properties and degradation rate, so it is appropriate to discuss the effect of different experimental parameters in the diameter of the electrospun fibers (Deitzel et al., 2001). The following summarizes the important electrospinning parameters:

- Concentration—keeping molecular weight same; after some critical concentration (where beads are no longer observed), the increase in concentration leads to increase in the fiber diameter.
- Viscosity—molecular weight, concentration, and viscosity are in fact related to each other. Lower viscosity means surface tension of the liquid becomes dominant, and the formation of the beads is observed. But very high viscosity leads to hard ejection of the jet from the needle tip, and clogging of the system occurs due to early solidification of polymers before becoming a fiber.
- Conductivity—the solution used must have the appropriate conductivity. As fiber formation is dependent on the balance between the electric force and the surface tension of the polymer melts/solutions, less conductive solution is unable to create the enough surface charge density for the stretching of the fiber. Adding salt in the polymeric solution helps improving the conductivity of the solution, thus making smoother fibers.
- Voltage—increasing the voltage results in decrease in the fiber diameter.
- Distance between the collector and the needle tip—as the distance between the collector and the tip increases, the time for thinning of fiber also increases, resulting in fibers with smaller diameter. But one must keep in mind that electrostatic force decreases as the distance increases, so voltage might have to be adjusted in the electrospinning.
- Flow rate—increase in the flow rate of the polymer melt will also increase the fiber diameter.

Along with diameter, other morphological characteristics of fibrous mat structure can be changed by changing the collector geometry. For vascular graft application, using a rotating cylindrical collector gives a tubular structure, and the alignment of the fiber can be changed by changing the rotation speed of the collector. Randomly oriented fibrous mat can be collected either by rotating mandrel at low speed or using a stationary flat metallic collector. Detailed study in the fiber alignment can be found elsewhere (Yuan et al., 2017).

4.4 Surface modification approach

Polymeric scaffolds for TEVGs as fabricated are typically less than ideal for biological function. Broadly, there are several approaches taken for surface medication: physical, chemical, and biological. Physical modification to produce nanostructured and microstructured morphology can include coatings (additive) or surface etching (subtractive) to engineer specific physical characteristics. Chemical modification includes a wide range of techniques including surface treatment (acid, alkali, etc.), grafting (covalent modification), and click chemistry (thiol/alkene and alkyne/azide) among others. Biological modification entails the addition or immobilization of active biological molecules (peptides, growth factors, etc.,) to the surfaces (Jiao and Cui, 2007).

4.4.1 Cold plasma–based techniques: chemical/physical modification

Plasmas dominate as the overwhelming majority proportion of matter in the visible universe. Broadly, they are commonly known as a state of matter ranging from partially to fully ionized fluid that exhibits collective behavior. In the lab, cold plasmas have been used for decades as a material engineering technique for surface treatment to produce both physical and chemical modification. Thus, surface treatment by cold plasma has emerged as an interesting and reliable strategy for modification of biomaterials (Desmet et al., 2009). There are a variety of effects that result in both physical and chemical changes through the action of free electrons, radicals, charged molecular species, radiation, and bond rearrangement (Desmet et al., 2009, Chu et al., 2002).

The low-temperature plasma regime is characterized by a variety of generation methods and sources; overall, they can be described by their nonequilibrium thermal behavior, where the electron temperature greatly exceeds the neutral gas temperature (typically several orders of magnitude) (Conrads and Schmidt, 2000). The effect is a highly energetic and reactive medium that efficiently changes the surface properties while leaving the bulk portion of the substrates relatively unchanged; however, careful control of plasma parameters must be enacted to prevent destruction of the substrates (thermal decomposition, oxidation, melting, overetching). Control of cold plasma is accomplished by adjusting various parameters including pressure, electromagnetic field power, exposure time, feed gas composition, and feed gas flow rate (Desmet et al., 2009; Chu et al., 2002; Conrads et al., 2000). A variety of established techniques exist to generate cold plasma in the lab; these methods include DC glow, RF-coupled, microwave, dielectric barrier discharge (DBD), radiation-induced, and others discussed elsewhere (Conrads et al., 2000).

Electrospun vascular graft scaffolds have been successfully reported to be surface-modified with cold plasma operating at atmospheric pressure (Tucker et al., 2018). Atmospheric DBD plasma jets were directed through the inner surfaces of the electrospun polymer tubes and resulted in dramatic changes to the surface chemistry; in turn, the hydrophobic polymer surfaces exhibited hydrophilic behavior. The outstanding issue of aging remains, however, as plasma modification, notoriously, will be reversed as a function of time due presumably to rearranging and recovery (Brennan et al., 1991). Plasma modification offers a wide space of possibility through the introduction of new functional groups, activated surfaces, and even polymerized coatings and can serve as an excellent pretreatment for additional chemical/biological modification (Chu et al., 2002).

4.4.2 Click chemistry

Click chemistry was coined as a term to describe a new class of reactions in 2001 by Sharpless, Kolb, and colleagues (Kolb et al., 2001). The goals included high yields, wide scope, stereospecificity, innocuous by-products, nonchromatographic isolation, friendly solvent systems, and facile, rapid synthesis that would end up employing vast libraries of compounds (Kolb et al., 2001). Typical products of synthesis include the heteroatom (C-X-C) motif as the characteristic feature; this is formed from thermodynamically favorable "spring-loaded" reaction schemes. Since then, there has been tremendous exploitation of various reaction pathways in an effort to be labeled by the coveted click chemistry designation described by recent review, especially for biomaterials applications (Zou et al., 2018).

There are several important classes of relevant reaction schemes for constructing/modifying biomaterial scaffolds. Following are prominent examples:

- Copper catalyzed azide/alkyne cycloaddition (CuAAC) that results in the formation of stable triazole moieties (Song et al., 2013).
- Strain promoted azide/alkyne cycloaddition that offers an alternative to CuAAC to avoid toxic Cu(I) ions (Agard et al., 2004).
- Thiol/X reaction that can include thiol/alkene, thiol/alkyne, or thiol/Michael addition (Zou et al., 2018).
- Diels−Alder reaction that produces a stable cyclohexene product and can be thermally reversible for self-healing properties (Zhang et al., 2009).
- Oxime ligation that results in oxime or imine hydrazone bonds that are stable under physiologically relevant conditions (Kalia and Raines, 2008).

4.4.3 Biological modification

Bioactive compounds are attractive targets for surface functionalization in TEVGs. Several important classes include enzymes, peptide sequences, polysaccharides, and phospholipids (Goddard and Hotchkiss, 2007). Essentially, bioconjugation of these molecules offers an improved, biomimetic, biocompatible, and functional surface to increase the physiologically relevant biosurface interactions of tissue engineering scaffolds through either a barrier effect or through an active biological pathway.

A particularly promising example to increase hemocompatibility for blood-contact applications is by the immobilization of heparin or heparin-like analogues to reduce blood clotting, thus reducing thrombosis. Further reading into the scope and breadth of biosurface interaction is encouraged as the possible combinatorial approaches are extensive (Dee et al., 2002). An example of biofunctionalization route for engineering polymeric surfaces is depicted in figure 4.2 (Goddard et al., 2007).

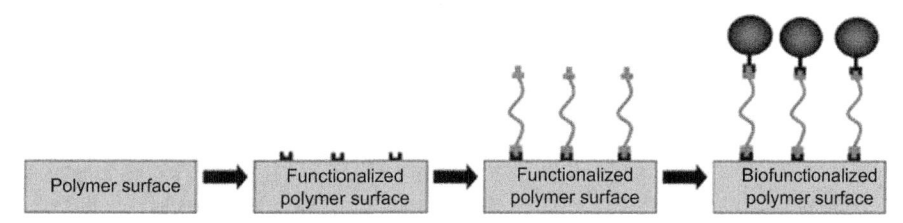

Figure 4.2 Graphical representation of the biofunctionalization route for engineering polymeric surfaces.
Adapted and reprinted with permission from Goddard, J.M., Hotchkiss, J.H., 2007. Polymer surface modification for the attachment of bioactive compounds. Prog. Polym. Sci. 32 (7), 698−725. doi: 10.1016/j.progpolymsci.2007.04.002.

4.5 Cellular interactions

Fabricating grafts with comparable mechanical properties and structure is not enough for developing a viable structure to be used for clinical purposes, it should be compatible for cell adhesion and proliferation. After implantation, it should integrate into the body by the formation of confluent endothelium in the lumen layer and a circumferentially aligned smooth muscle tissue in the outer layer. This means the scaffolds material should be a good host for the cells. A confluent endothelium layer is vital for the success of the graft implantation as it prevents the formation of thrombosis and the formation of SMCs layer outside of the endothelial layers as it gives the structural integrity and the mechanical strength to the TEVGs.

4.5.1 In vitro studies

In the electrospinning process, it is possible to fabricate the fibers with the desired orientation, pores sizes, and diameter, which can accelerate the cell matrix interaction necessary for the cell growth as shown in figure 4.4 (Zhang et al., 2010). The orientation of cell growth is favored by the fiber orientation reported by various researchers over the years (Zeltinger et al., 2001; Lowery et al., 2010). This ECM-mimicking fibrous mat fabricated by electrospinning polymer materials has shown to favor the cell—matrix interactions that are vital in the cell attachment and proliferation.

In vitro studies of the electrospun meshes are mainly focused on the growth and culture of ECs and SMCs in the inner and outer layers. Native endothelial layers in the

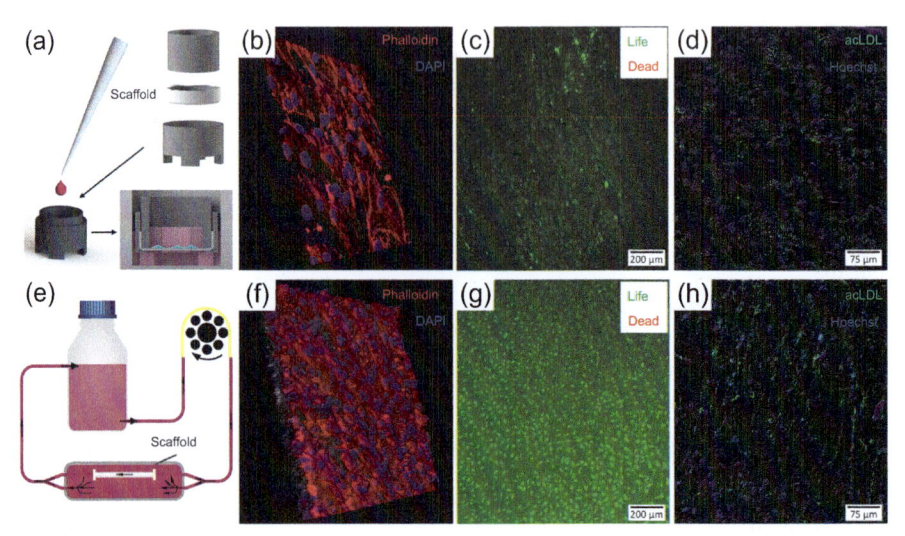

Figure 4.3 Cell seeding efficiency and viability under static and dynamic conditions. Panels (a, b, c, and d) correspond to static condition results and panels (e, f, g, and h) to dynamic. Adapted and reprinted with permission from Radakovic, D., et al. 2017. A multilayered electrospun graft as vascular access for hemodialysis. PLoS One 12 (10), e0185916.

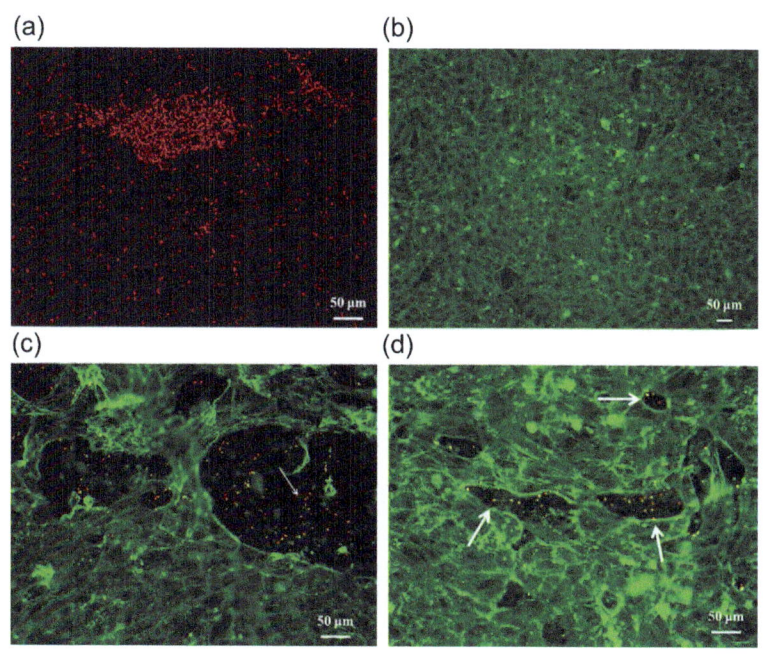

Figure 4.4 Fluorescent micrographs showing platelets (red) and actin cytoskeleton (green). (a) Platelets on bare scaffolds; (b) no platelet adhesion on fully endothelialized scaffolds; (c, d) platelet adhesion on exposed areas without coverage.

Adapted and reprinted with permission from Zhang, X., Thomas, V., Xu, Y., Bellis, S.L., Vohra, Y.K. 2010. An in vitro regenerated functional human endothelium on a nanofibrous electrospun scaffold. Biomaterials 15, 4376–4381.

vessels plays very important role in atherosclerosis and thrombosis. On vascular injury and repair process, a series of event takes place; endothelial layer plays important role in all of them. First, the damage is sealed off and followed by the prevention of excessive clotting. Endothelial layer that is in the interface between the blood and the surrounding tissue is capable of producing both antithrombotic and prothrombotic events depending on specific needs (Yau et al., 2015). Healthy ECs express agents that prevent the platelet aggregation and fibrin formation by release of nitric oxide; during endothelial dysfunction, ECs trigger fibrin formation and platelet adhesion, and finally, after repair, ECs release profibrinolytic agents that degrade the clot. So, in total, endothelial layer is vital in maintaining homeostasis in the vessels and preventing the thrombotic development.

Comparison of both static and dynamic conditions for cell proliferation using ECs has been accomplished (Radakovic et al., 2017). Interestingly, they reported successful endothelial coverage. After 7 days, the cells exhibited prolific metabolic activity.

Zhang et al., in 2010, reported fabrication of functional endothelium on the lumen of the electrospun graft. They reported using human aortic endothelial cells (HAECs), which adhered to and subsequently spread out on the scaffold after seeding as evidenced by florescent micrographs. Junctions were formed after 24 h, and the cells continued to become a monolayer in just 7 days. They also reported the release of antithrombotic factor PGI2 molecules by this endothelial layer in their study. The proliferation assay confirmed the HAEC proliferation continued for at least 11 days without cytotoxic effects and is depicted in figure 4.3 (Zhang et al. 2010).

4.5.2 Stem cell studies

Restoring the vital components in the native vessel structures, such as endothelial, smooth muscle layers in the fabricated graft, has shown to improve long-term patency of the implanted grafts (Glynn and Hinds, 2014). The formation of organized layer of ECs is vital not only in regulating the physiological processes such as angiogenesis, inflammation, and thrombosis but also in playing a critical role in the formation and maintenance of different tissues in the body (Levenberg et al., 2005). Stem cells and progenitor cells have shown promising results in tissue engineering and can be used as source for the required source cells. Stem cells can differentiate into other cell types in the body, making it attractive for use in researches such as gene therapy to tissue regeneration. Stem cells can be of various types and mainly can be divided into three categories based on the origin: embryonic stem cells (ESCs), which are harvested from the in vitro fertilization process from within 3—5 days of fertilization, non-ESCs (or adult stem cells), which are harvested from postnatal and adult human, consisting of two populations (hematopoietic and mesenchymal stem cells), and can be divided into certain types of cells depending on the location in the body, and the induced pluripotent stem cells (iPSC), which are adult stem cells reprogrammed to behave like the pluripotent stem cells. Detailed studies on development of iPSCs can be found in various research articles (Hamazaki et al., 2017; Yamanaka, 2012; Shi et al., 2017). As stem cell has a potential to divide into many kinds of cells or stem cells itself, it brings enormous opportunity in the field of regenerative medicine. The global interest and investment in the field of stem cells has increased tremendously, and in year 2016 alone, the investment grew almost seven times in North America (Mohtaram et al., 2017).

ESCs, being pluripotent cells that can develop into any cell structures and being able to self-renew, have some ethical concerns as they are harvested from the fertilized embryos. On the other hand, adult stem cells can be isolated from the patient itself and have less ethical concern but are limited by their ability to develop into the desired tissue structure. But the advantage of MSC is that it does not have the antigen to enable T-cell activation giving immunological advantages (Budhwani et al., 2016). This helps in the adoption of graft in the body in the long run.

Stem cells alone do not perform a specialized function in the metabolic or physiological activity as the body like other specialized tissues does. But instead they can divide into the specialized tissues, and this process of transformation of unspecialized stem cells into the specialized tissue structure is called differentiation (NIH, 2016). And this differentiation is facilitated by different signal, internal and external signals in the microenvironment. The external signals are chemicals secreted in the cells and physical contact in its vicinity, and the internal signals are controlled by the genes, and the information are coded in the DNA (NIH, 2016).

Researchers have reported the incorporation of stem cells during the process of electrospinning with the goal of in situ EC lining and SMC development. Budhwani et al. discuss the effect of fiber parameters in the cell biology of ESC, MSC, and iPSC. Cell growth is favored in the direction of fiber alignment, while the rough surface structure of the cell proliferation and presence of protein coating promote the differentiation of the cells. Also, increasing the conductivity of the fibers increases the differentiation of the seeded cells (Budhwani et al., 2016). Electrospun fibers not only help in orienting differentiation of the stem cells but also can be used as a tool to expand the stem cells (Mohtaram et al., 2017).

The process of loading of growth factors and the cells into the electrospun bioactive scaffold fibers is usually done by the following routes: physical adsorption, blend electrospinning, coaxial electrospinning, or covalent immobilization. Physical adsorption refers to the process of loading the biomolecules after the scaffold is prepared. The adsorption of biomolecules occurs simply though electrostatic forces (Zhang 2016). In this, the scaffold is dipped into the solution containing the aqueous solution or emulsion of biomolecules. The release of the adsorbed material, either proteins or genes, has shown to be faster than other approaches such as blend or coaxial electrospinning process in the system (Nie et al., 2008; Nie and Wang, 2007). In blend spinning, the loading is done during the process of creating the polymer solution that results in uniform distribution and sustained release of the biomolecules and gene in the system compared to physical adsorption. In coaxial spinning, polymer solution and biological solution/emulsion are spun through different capillary channels so that the fibers with distinct two layers consisting different material can be produced. This leads to controlled release of the biological material encapsulated into the polymer fibers. Covalent immobilization refers to binding of the chemicals in the surface of the scaffold fibers through reactions. This process can be used to modify the surface permanently. Detail reviews in methods of loading and release behavior can be found elsewhere (Ji et al., 2010).

Nieponice et al. in their 2010 study assessed the in vitro behavior of MSCs in compound electrospun vascular grafts. They reported the fabrication of scaffolds by combining thermally induced phase separation and electrospinning method using poly(ester urethane) urea (PEUU) as depicted in figure 4.5 (Nieponice et al., 2010). The cell seeding is done with muscle-derived stem cells, SMC-like layer was reported near the luminal surface, and the patency of the graft reported to improve in the rat models (Nieponice et al., 2010).

4.5.3 Cell microintegration

Oftentimes, because of the nature of the polymer used or the parameter used during the electrospinning, the porosity or the pore size does not allow cell migration at the desired rate. Researchers have been implementing different techniques to tackle this shortcoming to integrate cells into the fibers. This includes integration of the cells during the process of the electrospinning, which results in the cellularization of the vascular graft. In this section, we will discuss the integration of cells such as ECs and SMCs either by cospinning, cell-encapsulated spinning, or spraying over the

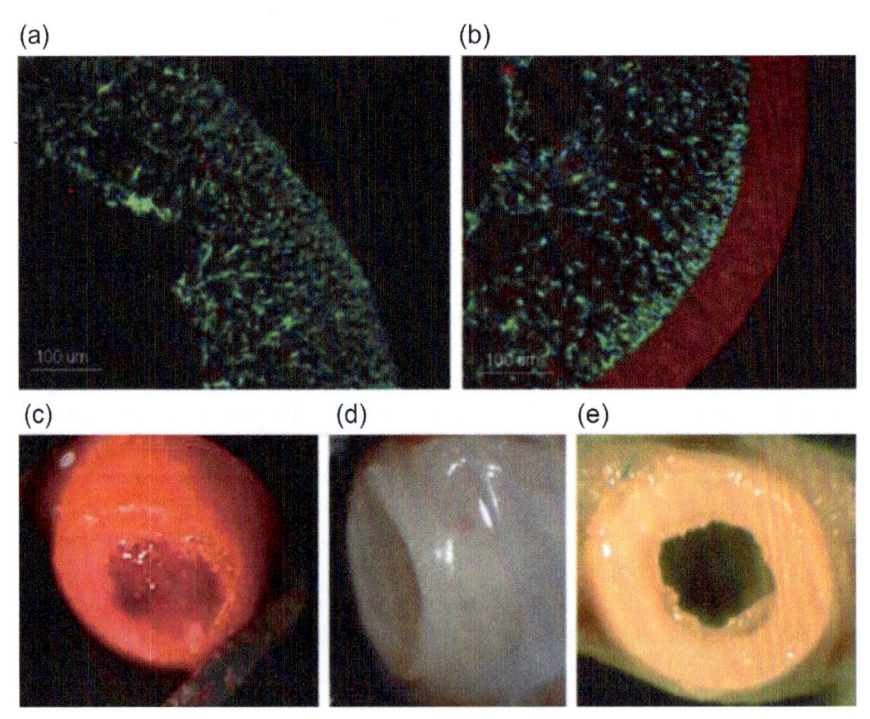

Figure 4.5 In vivo implantation of graft (a) thermally induced phase separation (TIPS) and (b) Electrospun enforced (ES-TIPS); preimplant section showing the muscle-derived stem cells populating the scaffolds through the thickness. Grafts after 8 week of implantation (c) intimal thrombosis in unseeded control (d) and (e) tissue-like aspects of explants for the grafts TIPS and ES-TIPS, respectively.
Adapted and modified from A. Nieponice, L. Soletti, J. Guan, Yi, H., B. Gharaibeh, T.M. Maul, J. Huard, W.R. Wagner, David A. Vorp. April 2010. Tissue Engineering Part A. (Ahead of print. http://doi.org/10.1089/ten.tea.2009.0427) and reprinted with permission from Mary Ann Liebert, Inc. Publishers.

mat structure, and the integration of the stem cells in the electrospinning will be discussed in a separate section later.

Stankus et al. described in their 2006 study that to increase the cellular proliferation and achieve better infiltration, simultaneously electrospraying the vascular SMCs along with electrospun PEUU from two different capillaries is necessary. The process of electrospraying is a method of electrohydrodynamic atomization in which the applied electric field deformed the charged droplet in the capillary opening, and eventually, this droplet breaks down to the smaller droplets which then deposited to the collector. The spraying of cells is done along with the media including gelatin so that the cells are protected from the physical or chemical stress during the pressurized spray process, while in the electrospraying process, the viability and proliferation of the SMC cells were not found to be affected. They reported higher cell densities in shorter time periods. Also, the orientation of the cells was found to be in the direction of the fiber orientation. For the cell viability study, the electrospun-electrosprayed constructs were maintained under static culture conditions, which allow the cell attachment, while preventing the washout of the cells found under dynamic culture conditions. In an another study, Lavielle et al. described this method applied toward composite membranes in which PLA was electrospun and microparticles of poly(ethylene glycol) (PEG) were electrosprayed into a fabricated hierarchical honeycomb-like structure. They reported selective leaching of the microparticles to achieve single-component membranes while retaining the structured membrane which is applicable in both biomedical and other applications (such as filtration) (Lavielle et al., 2013).

4.5.4 Core-shell method

Coaxial electrospinning has been well explored in the literature, examples of which are described by Zhang et al. (2004) and Wang et al. (2013). This method is very useful for producing fibers out of relatively less electrospinnable materials, embedding therapeutic materials for controlled release and other applications. These relatively less-spinnable materials can be embedded inside a fiber, which can later be released in a controlled manner to get a fiber out of those by simply dissolving the outer sheath. Yang et al. reported the stability and the release profile of the shell of electrospun poly(DL-lactide) ultrafine fibers. They reported the release of bovine serum albumin (BSA) along with methyl cellulose from the core of these fibers (Yang et al., 2007). They studied the release mechanism of the BSA from the nanofibers and reported that the diffusion is Fickian in nature. They also reported that the lower volume ration of aqueous to organic phase in the emulsion can reduce the initial burst release of protein. Another approach of core-shell electrospinning was implemented by a group Li et al. in their 2018 study, where the thin layer of drug material along with polyvinylpyrrolidone (PVP) K10 (low molecular weight PVP) is spun as outer ultrathin sheath layer and the core material is spun of either PVP K90 (high molecular weight PVP) or PCL (Li et al., 2018). This leads to the outer layer having thickness less than 10%, and the core is reported in their study. This ultrathin nature of the outer layer gives rise to fast release of embedded drug in the outer layer as reported in their in vitro dissolution test. It has been suggested that if one incorporates the same drug or other biomaterials in the bioabsorbable inner core, a system with variable release system can be designed. This process has shown promising prospects in various applications in biomedical fields and beyond.

4.5.5 In vivo studies toward translational applications

Many researchers have attempted to demonstrate significant improvements in TEVG during in vivo studies, thus a segue toward translational applications. Bai et al. (2019) reported in vivo results of biofunctionalized PS grafts (poly(ε-caprolactone)-b-poly(isobutyl-morpholine- 2,5-dione) electrospun with silk) in rats. The biofunctionalization is done by low-fouling PEG and two cell-adhesive peptide sequences (CREDVW and CAGW). The grafts were 2 cm long and 2 mm in diameter, which are implanted in the carotid artery position to assess the in vivo endothelialization and long-term patency. At 10 weeks, the biofunctionalized grafts shown patency while the as is graft developed restenosis. The SEM analysis of these grafts has shown a higher rate of endothelialization and is higher in the biofunctionalized grafts. Another approach combining the electrospinning with additive manufacturing is also getting attention in recent years. These methods include combining electrospinning with processes such as 3D printing, fused deposition, etc. Spadaccio et al. used grafts of 5 mm in diameter and 6 cm in length fabricated from the electrospinning process (heparin/PLLA) and outer armor made with PCL using fused deposition method and performed in vivo studies (Spadaccio et al., 2016). The heparin in the fiber from their previous studies showed increased patency due to its anticoagulant properties. They reported the patency of the grafts in the fourth week of implantation in the rabbit body. At explant, there was no evidence of the foreign body inflammation or sign of intraluminal thrombosis. Histological analysis had shown uniform cellularization, inner surface populated with elongated cells, very small cytoplasm/nuclei ratio, and nuclei protruding in the lumen, while outer surface was populated with spindle shaped cells. This difference in cell shapes different zones, suggesting the self-colonization of the scaffolds by the endogenous cells leading to different phenotypes within the scaffolds. Pektok et al. investigated the patency and structural integrity of electrospun PCL grafts and ePTFE graft in 15 rats each in their study. At the end, there is no reported case of stenosis in the groups of rats with PCL graft implanted, while there were 2 cases of stenosis lesions in the ePTFE grafts at 18 and 24 weeks (Pektok et al., 2008). Also, they reported the significantly better endothelial coverage and better cellular infiltration in the rats with PCL grafts implants. At the 12th week, the PCL grafts had confluent endothelial layer. In a different study with PCL scaffolds in rat mode in vivo, Tillman et al. have shown the confluent layer of sheep endothelial layer in the inner lumen of the electrospun PCL/collagen scaffolds and SMC formed a multi-layered structure on the exterior of the scaffolds, which is similar to the native blood vessel as shown in figure 4.6 (Tillman et al., 2009). This lining of ECs in the lumen of the scaffold materials prevents or resists the platelet adhesion. These grafts kept the structural integrity at retrieval, and the constant luminal diameter is also observed in the ultrasonography performed at 2 weeks interval.

4.6 Emerging perspectives

Small-diameter TEVGs are still just beyond reach but remain a clinically significant goal. The current literature is dominated by combinatorial approaches that integrate the strategies of tissue engineering, surface modification, and bioengineering. No single approach is likely to solve the problem; decades of research have been applied in attempts but none yet successful.

Silk is a sustainable material with economic importance since antiquity. Its inherent biocompatibility makes it a viable candidate for vascular graft engineering. Recently, electrospun silk vascular grafts of small caliber (1.5 mm inner diameter) have been demonstrated in rat models and are reported to achieve nearly full endothelialization within 6 weeks at a survival rate of 95%, outperforming Teflon as comparison

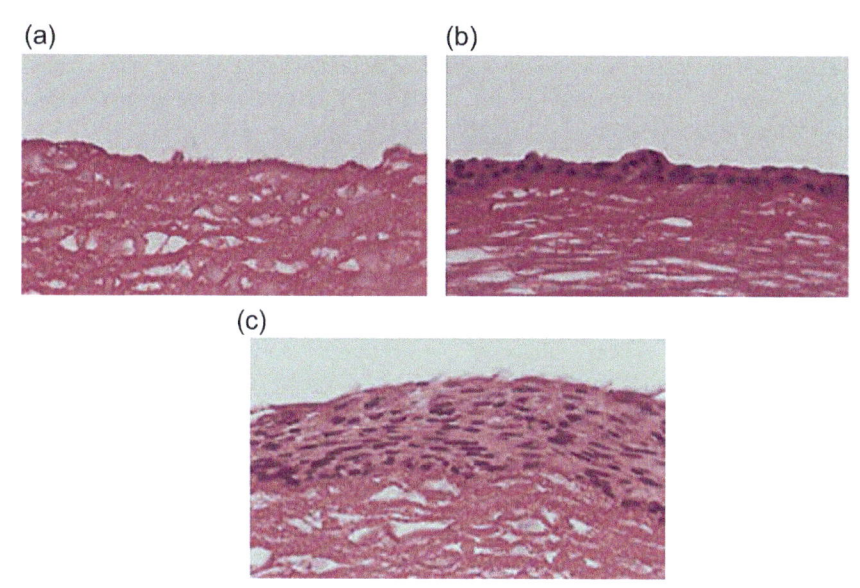

Figure 4.6 (a) Scaffolds with no cell seeded; (b) sheep endothelial adhered to inner lumen; (c) smooth muscle cell adhered to the exterior of the lumen (Tillman et al., 2009).
Adapted and reprinted with permission from Tillman B.W., Yazdani S.K., Lee S.J., Geary R.L., Atalan A. and Yoo J.J., The in vivo stability of electrospun polycaprolactone-collagen scaffolds in vascular reconstruction, Biomaterials 30 (4), 2009, 583−588.

(Felipe et al., 2018). Another promising approach reported with decellularized vascular scaffold coated with vascular endothelial growth factor (VEGF) resulted in functionality for up to 8 weeks in vivo attributed to the accelerated endothelialization due to the VEGF (Iijima et al., 2018). 3D printing combined with electrospinning has been reported as a method to produce a trilayered construct that mimics the native vessel architecture and has been suggested as an alternative to electrospinning alone (Huang et al., 2018) The dramatic increase in availability and decrease in price makes 3D printing an accessible route for rapid prototyping of tissue engineering substrates and will likely see further trending interest in future literature as the market for 3D printing and similar technology is expected to exceed 21 billion USD by the year 2020 (Ngo et al., 2018). Further approaches include hydrogels, self-assembly, and biohybrid scaffolds will also likely see extensive future exploration (Carrabba et al., 2018).

Nanotechnology and synthetic biology are still in their infancy but are suggested as the conjectural far future of medicine. Proteins and nucleic acids are the key building blocks, information storage, catalysts, and machinery of all known life. In principle, the ability to predict, manipulate, and program their behavior would constitute a significant leap forward in not only tissue engineering but also the whole of medical science and other fields (Ulijn and Jerala, 2018). Synthetic biology is currently relegated to the domain of prokaryotes due to significant challenges with eukaryotic cell modification; however, the potential to program the genome of a mammalian cell presents a fascinating proposition (Ceroni and Ellis, 2018). The need for vascular tissue engineering could be eliminated one day. The consequences of such unprecedented power over the basic unit of life are left to speculation and, for the moment, remain purely hypothetical.

4.7 Conclusion

The route to successful small-diameter TEVGs is likely within reach of the contemporary generation. The synergy of tissue engineering, surface modification, bioengineering, and emerging technology aims to provide a road map to guide researchers in their efforts, which bridges numerous disciplines. Overall, the key strategy is to provide the right combination of factors that promote rapid endothelialization; once this aspect is secured, the global impact will be monumental with the translational potential.

References

Agard, N.J., Prescher, J.A., Bertozzi, C.R., 2004. A strain-promoted [3 + 2] azide–alkyne cycloaddition for covalent modification of biomolecules in living systems. J. Am. Chem. Soc. 126, 15046–15047.

Akhmanova, M., Osidak, E., Domogatsky, S., Rodin, S., Domogatskaya, A., 2015. Stem Cells Int. 2015, 167025.

Antoniou, G.A., Chalmers, N., Georgiadis, G.S., Lazarides, M.K., Antoniou, S.A., Serracino-Inglott, F., Smyth, J.V., Murray, D., 2013. A meta-analysis of endovascular versus surgical reconstruction of femoropopliteal arterial disease. J. Vasc. Surg. 57, 242.

Bai, L., Zhao, J., Li, Q., Guo, J., Ren, X., Xia, S., Zhang, W., Feng, Y., 2019. Biofunctionalized electrospun PCL-PIBMD/SF vascular grafts with PEG and cell-adhevise peptides for endothelialization. Macromol. Biosci. 19 (2), 1800386.

Blanchemain, N., Haulon, S., Martel, B., Traisnel, M., Morcellet, M., Hildebrand, H.F., 2005. Vascular PET prostheses surface modification with cyclodextrin coating: development of a new drug delivery system. Eur. J. Vasc. Endovasc. Surg. 29, 628–632.

Brahatheeswaran, D., Yoshida, Y., Maekawa, T., Sakthi Kumar, D., 2011. Polymeric scaffolds in tissue engineering application: a review. Int. J. Polym. Sci. Article ID 290602, 19 pages https://doi.org/10.1155/2011/290602.

Brennan, W.J., Feast, W.J., Munro, H.S., Walker, S.A., 1991. Investigation of the ageing of plasma oxidized PEEK. Polymer 32 (8), 1527–1530. https://doi.org/10.1016/0032-3861(91)90436-M.

Browning, M., et al., 2012. Multilayer vascular grafts based on collagen-mimetic proteins. Acta Biomater. 8 (3), 1010–1021.

Budhwani, K.I., Wood, A.T., Gangrade, A., Sethu, P., Thomas, V., September 2016. Nanofiber and stem cell enabled biomimetic systems and regenerative medicine. J. Nanosci. Nanotechnol. 16 (9), 8923–8934, 12.

Carampin, P., Conconi, M.T., Lora, S., Menti, A.M., Baiguera, S., Bellini, S., Grandi, C., Parnigotto, P.P., March 1, 2007. Electrospun polyphosphazene nanofibers for in vitro rat endothelial cells proliferation. J. Biomed. Mater. Res. 80A (3), 661−668.

Carrabba, et al., April 17, 2018. Front. Bioeng. Biotechnol. 6 (41). https://doi.org/10.3389/fbioe.2018.00041.

Ceroni, F., Ellis, T., 2018. Nat. Rev. Mol. Cell Biol. 19, 481−482.

Chu, P.K., Chen, J.Y., Wang, L.P., Huang, N., 2002. Plasma-surface modification of biomaterials. Mater. Sci. Eng. R Rep. 36 (5−6), 143−206. https://doi.org/10.1016/S0927-796X(02)00004-9. Cited 958 times.

Conrads, H., Schmidt, M., 2000. Plasma generation and plasma sources. Plasma Sour. Sci. Technol. 9, 441−454.

Cox, T.R., Janine, T Erler, 2011. Remodeling and homeostasis of the extracellular matrix: implications for fibrotic diseases and cancer. Disease Models & Mechanisms 4 (2), 165−178.

Dar, A., Shachar, M., Leor, J., Cohen, S., 2002. Cardiac tissue engineering: optimization of cardiac cell seeding and distribution in 3D porous alginate scaffolds. Biotechnol. Bioeng. 80, 305−312.

Dee, K.C., Puleo, D.A., Bizius, R., 2002. An Introduction to Tissue-Biomaterial Interactions. Wiley-Liss, Hoboken, NJ.

Deitzel, J.M., Kleinmeyer, J., Harris, D., Beck Tan, N.C., 2001. The effect of processing variables on the morphology of electrospun nanofibers and textiles. Polymer 42, 261−272.

Desmet, T., Morent, R., De Geyter, N., Leys, C., Schacht, E., Peter, D., 2009. Nonthermal plasma technology as a versatile strategy for polymeric biomaterials surface modification: a review. Biomacromolecules 10 (9), 2351−2378. https://doi.org/10.1021/bm900186s.

Felipe, et al., 2018. JACC Basic Trans. Sci 3 (Issue 1), 38−53.

Galler, K.M., Brandl, F.P., Kirchhof, S., Widbiller, M., February 2018. Andreas Eidt, Wolfgang Buchalla, Achim Göpferich, and Gottfried Schmalz. Tissue Engineering Part A ahead of print. http://doi.org/10.1089/ten.tea.2016.0555.

Gersh, B.J., et al., 2010. The epidemic of cardiovascular disease in the developing world: global implications. Eur. Heart J. 31, 642−648. https://doi.org/10.1093/eurheartj/ehq030.

Glynn, J.J., Hinds, M.T., August 2014. Tissue Engineering Part B: Reviews ahead of print. http://doi.org/10.1089/ten.teb.2013.0285.

Goddard, J.M., Hotchkiss, J.H., 2007. Polymer surface modification for the attachment of bioactive compounds. Prog. Polym. Sci. 32 (7), 698−725. https://doi.org/10.1016/j.progpolymsci.2007.04.002.

Hamazaki, T., Rouby, N.El, Fredette, N.C., Santostefano, K.E., Terada, N., March 2017. Concise review: induced pluripotent stem cell research in the era of precision medicine. Stem Cells 35 (3), 545−550.

Han, B., Huang, L.L.H., Cheung, D., Cordoba, F., Nimni, M., 1999. Polypeptide growth factors with a collagen binding domain: their potential for tissue repair and organ regeneration. In: Zilla, P., Greisler, H.P. (Eds.), Tissue Engineering of Vascular Prosthetic Grafts. RG Landes, Austin, pp. 287−299.

Hasan, A., et al., 2014. Electrospun scaffolds for tissue engineering of vascular grafts. Acta Biomater. 10 (1), 11−25.

He, Y., Lu, F., 2016. Development of Synthetic and Natural Materials for Tissue Engineering Applications Using Adipose Stem Cells. Stem Cells International 2016, 5786257.

Hench, L.L., 1998. Bioceramics. J. Am. Ceram. Soc. 81, 1705−1727.

Huang, R., Gao, X., Wang, J., et al., 2018. Ann. Biomed. Eng. 46, 1254. https://doi.org/10.1007/s10439-018-2065-z.

Iijima, Aubin, H., Steinbrink, M., Schiffer, F., Assmann, A., Weisel, R.D., Matsui, Y., Li, R.K., Lichtenberg, A., Akhyari, P., January 2018. Bioactive coating of decellularized vascular grafts with a temperature sensitive VEGF conjugated hydrogel accelerates autologous endothelialization in vivo. Tissue Eng. Reg. Med. 12 (1), e513−e522.

Ji, W., Sun, Y., Yang, F., van den Beucken, J.J., Fan, M., Chen, Z., Jansen, J.A., 2010. Bioactive electrospun scaffolds delivering growth factors and genes for tissue engineering applications. Pharmaceut. Res. 28 (6), 1259−1272.

Jiao, Y.-P., Cui, F.-Z., November 26, 2007. Surface modification of polyester biomaterials for tissue engineering. Biomed. Mater. 2 (4).

Kalia, J., Raines, R.T., 2008. Hydrolytic stability of hydrazones and oximes. Angew. Chem. Int. Ed. 120, 7633−7636.

Kang, H.-W., Tabata, Y., Ikada, Y., July 1999. Fabrication of porous gelatin scaffolds for tissue engineering. Biomaterials 20 (14), 1339−1344.

Kannan, R.Y., Salacinski, H.J., Butler, P.E., Hamilton, G., Seifalian, A.M., July 2005. Current status of prosthetic bypass grafts: a review. J. Biomed. Mater. Res. B 74B (1), 570−581.

Kolb, H.C., Finn, M.G., Sharpless, K.B., 2001. Click chemistry: diverse chemical function from a few good reactions. Angew. Chem. Int. Ed. 40, 2004−2021, 10.1002/1521-3773(20010601)40:11<2004::AID-ANIE2004>3.0.CO;2-5.

Ku, D.N., Allen, R.C., 1995. Vascular grafts. In: Bronzino, J. (Ed.), The Biomedical Engineering Handbook. CRC Press, Boca Raton, FL, pp. 1871−1878.

Lavielle, N., Hebraud, A., Schlatter, G., Thony-Meyer, L., Rossi, R.M., Popa, A., 2013. Simultaneous electrospinning and electrospraying: a straightforward approach for fabricating hierarchically structured composite membranes. ACS Appl. Mater. Interfaces 5, 10090−10097.

Lavik, E., Langer, R., 2004. Appl. Microbiol. Biotechnol. 65, 1. https://doi.org/10.1007/s00253-004-1580-z.

Lee, S.J., et al., 2007. In vitro evaluation of electrospun nanofiber scaffolds for vascular graft application. J. Biomed. Mater. Res. A 83 (4), 999−1008.

Levenberg, S., Burdick, J.A., Kraehenbuehl, T., Langer, R., March 2005. Tissue Eng ahead of print. http://doi.org/10.1089/ten.2005.11.506.

Li, J.J., Yang, Y.Y., Yu, D.G., Du, Q., Yang, X.L., 2018. Fast dissolving drug delivery membrane based on the ultra-thin shell of electrospun core-shell nanofibers. Eur. J. Pharm. Sci. 122, 195−204.

Lowery, J.L., Datta, N., Rutledge, G.C., 2010. Effect of fiber diameter, pore size and seeding method on growth of human dermal fibroblasts in electrospun poly(ε-caprolactone) fibrous mats. Biomaterials 31 (3), 491−504. https://doi.org/10.1016/j.biomaterials.2009.09.072.

Lusis, A.J., 2000. Atherosclerosis. Nature 407, 233−241.

Mandal, B.B., Kundu, S.C., February 2010. Biospinning by silkworms: silk fiber matrices for tissue engineering applications. Acta Biomater. 6 (Issue 2), 360−371.

Mohtaram, N., Karamzadeh, V., Shafieyan, Y., et al., 2017. Commercializing electrospun scaffolds for pluripotent stem cell-based tissue engineering applications. Electrospinning 1 (1), 62−72. https://doi.org/10.1515/esp-2017-0003. Retrieved 26 Nov. 2018, from.

Müller, F.A., Müller, L., Hofmann, I., Greil, P., Wenzel, M.M., Staudenmaier, R., 2006. Cellulose-based scaffold materials for cartilage tissue engineering. Biomaterials 27, 3955−3963. https://doi.org/10.1016/j.biomaterials.2006.02.031.

National Center for Health Statistics Health, United States, 2016. With Chartbook on Long-Term Trends in Health. Hyattsville, MD. 2017. https://www.ncbi.nlm.nih.gov/pubmed/28910066.

Ngo, et al., June 15, 2018. Compos. B Eng. 143, 172−196.

Nie, H., Soh, B.W., Fu, Y.C., Wang, C.H., 2008. Three-dimensional fibrous PLGA/HAp composite scaffold for BMP-2 delivery. Biotechnol. Bioeng. 99 (1), 223−234. https://doi.org/10.1002/bit.21517.

Nie, H., Wang, C.H., 2007. Fabrication and characterization of PLGA/HAp composite scaffolds for delivery of BMP-2 plasmid DNA. J. Control. Release 120 (1−2), 111−121. https://doi.org/10.1016/j.jconrel.2007.03.018.

Nieponice, A., Soletti, L., Guan, J., Yi, H., Gharaibeh, B., Maul, T.M., Huard, J., Wagner, W.R., Vorp, David A., April 2010. Tissue Engineering Part A ahead of print. http://doi.org/10.1089/ten.tea.2009.0427.

NIH, 2016. Stem Cell Information Home Page. In Stem Cell Information [World Wide Web Site]. National Institutes of Health, U.S. Department of Health and Human Services, Bethesda, MD. Available at. http://stemcells.nih.gov/info/basics/2.htm.

O'brien, F., March 2011. Biomaterials & scaffolds for tissue engineering. Mater. Today 14 (3), 88−95. https://doi.org/10.1016/S1369-7021(11)70058-X.

Pashneh-Tala, S., MacNeil, S., Claeyssens, F., 2015. The tissue-engineered vascular graft-past, present, and future. Tissue Eng. B Rev 22 (1), 68−100. https://doi.org/10.1089/ten.teb.2015.0100. Advance online publication.

Pektok, E., Nottelet, B., Tille, J.C., Gurny, R., Kalangos, A., Moeller, M., Walpoth, B.H., 2008. Degradation and healing characteristics of small-diameter poly(ε-caprolactone) vascular grafts in the rat systemic arterial circulation. Circulation 118 (24), 2563−2570.

Radakovic, D., et al., 12 October 2017. A multilayered electrospun graft as vascular access for hemodialysis. PLoS One 12 (10), e0185916. https://doi.org/10.1371/journal.pone.0185916.

Ratner, B., Hoffman, A., Frederick Schoen, Lemons, J. (Eds.), 2012. Biomaterials Science, third ed. Academic Press.

Schmedlen, R.H., et al., 2003. Tissue engineered small-diameter vascular grafts. Clin. Plast. Surg. 30 (4), 507−517.

Shi, Y., Inoue, H., Wu, J.C., Yamanaka, S., 2017. Nat. Rev. Drug Discov. 16, 115−130.

Song, N., Ding, M., Pan, Z., Li, J., Zhou, L., Tan, H., Fu, Q., 2013. Construction of targeting-clickable and tumor-cleavable polyurethane nanomicelles for multifunctional intracellular drug delivery. Biomacromolecules 14, 4407−4419.

Song, Y., Feijen, J., Grijpma, D., Poot, A., 2011. Clin. Hemorheol. Microcirc. 49, 357−374.

Spadaccio, C., Nappi, F., De Marco, F., Sedati, P., Sutherland, F.W., Chello, M., Trombetta, M., Rainer, A., 2016. Preliminary in vivo evaluation of a hybrid armored vascular graft combining electrospinning and additive manufacturing techniques. Drug Target Insights 10 (1), 1−7.

Stankus, J.J., Guan, J., Fujimoto, K., Wagner, W.R., 2006. Microintegrating smooth muscle cells into a biodegradable, elastomeric fiber matrix. Biomaterials 27 (5), 735−744.

Suh, J.-K.F., Matthew, H.W.T., 2000. Application of chitosan-based polysaccharide biomaterials in cartilage tissue engineering: a review. Biomaterials 21, 2589−2598.

Theocharis, A.D., Skandalis, S.S., Gialeli, C., Karamanos, N.K., 2016. Adv. Drug Deliv. Rev. 97, 4−27.

Thomas, V., Dean, D.R., Vohra, Y.K., 2006. Nanostructured biomaterials for regenerative medicine. Curr. Nanosci. 2, 155. https://doi.org/10.2174/1573413710602030155.

Tillman, B.W., Yazdani, S.K., Lee, S.J., Geary, R.L., Atalan, A., Yoo, J.J., 2009. The in vivo stability of electrospun polycaprolactone-collagen scaffolds in vascular reconstruction. Biomaterials 30 (4), 583–588.

Tucker, B.S., Baker, P.A., Xu, K.G., Vohra, Y.K., Thomas, V., 2018. Atmospheric pressure plasma jet: a facile method to modify the intimal surface of polymeric tubular conduits. J. Vac. Sci. Technol. A 36, 04F404.

Ulijn, R.V., Jerala, R., 2018. Peptide and protein nanotechnology into the 2020s: beyond biology. Chem. Soc. Rev. 47, 3391–3394.

Uttayarat, P., Perets, A., Li, M., Pimton, P., Stachelek, S.J., Alferiev, I., et al., 2010. Micro-patterning of three-dimensional electrospun polyurethane vascular grafts. Acta Biomater 6, 4229–4237.

Valence, S. de, Tille, J.C., Mugnai, D., Mrowczynski, W., Gurny, R., Möller, M., Walpoth, B.H., 2012. Long term performance of polycaprolactone vascular grafts in a rat abdominal aorta replacement model. Biomaterials 33, 38–47.

Wang, T., Ji, X., Jin, L., Feng, Z., Wu, J., Zheng, J., Wang, H., Zu, Z., Guo, L., He, N., 2013. ACS Appl. Mater. Interfaces 5 (9), 3757–3763.

WHO, 2017. Cardiovascular Diseases (CVDs) ([online]). https://www.who.int/news-room/fact-sheets/detail/cardiovascular-diseases-(cvds).

Wood, A.T., Everett, D., Budhwani, K.I., Dickinson, B., Thomas, V., 2016. Wet-laid soy fiber reinforced hydrogel scaffold: fabrication, mechano-morphological and cell studies. Mater. Sci. Eng. C 63, 308–316. https://doi.org/10.1016/j.msec.2016.02.078.

Yamanaka, S., 2012. Induced pluripotent stem cells: past, present, and future. Cell Stem Cell 10 (6), 678–684. https://doi.org/10.1016/j.stem.2012.05.005.

Yang, Y., Li, X., Cui, W., Zhou, S., Tan, R., Wang, C., 2007. Structural stability and release profiles of proteins from core-shell poly(DL-lactide) ultrafine fibers prepared by emulsion electrospinning. J. Biomed. Mater. Res. A 86A (2), 374–385.

Yau, J.W., et al., 19 October, 2015. Endothelial cell control of thrombosis. BMC Cardiovasc. Disord. 15 (130) https://doi.org/10.1186/s12872-015-0124-z.

Yuan, H., Zhou, Q., Zhang, Y., 2017. Improving fiber alignment during electrospinning. Electrospun Nanofibers, pp. 125–147.

Zeltinger, J., Sherwood, J.K., Graham, D.A., Müeller, R., Griffith, L.G., October 2001. Tissue Engineering. http://doi.org/10.1089/107632701753213183.

Zhang, Y., Broekhuis, A.A., Picchioni, F., 2009. Thermally self-healing polymeric materials: the next step to recycling thermoset polymers? Macromolecules 42, 1906–1912.

Zhang, Y., Huang, Z.M., Xu, X., Lim, C.T., Ramakrishna, S., 2004. Preparation of core-shell structured PCL-r-gelatin Bi-component nanofibers by coaxial electrospinning. ACS Chem. Mater. 16 (18), 3406–3409.

Zhang, X., Thomas, V., Xu, Y., Bellis, S.L., Vohra, Y.K., 2010. An in vitro regenerated functional human endothelium on a nanofibrous electrospun scaffold. Biomaterials 15, 4376–4381. https://doi.org/10.1016/j.biomaterials.2010.02.017.

Zhu, A.P., Zhang, M., Shen, J., 2017. Blood compatibility of chitosan/heparin complex surface modified ePTFE vascular graft. Appl. Surf. Sci. 241 (3–4), 485–492, 2005.

Zou, Y., Zhang, L., Yang, L., Zhu, F., Ding, M., Lin, F., Wang, Z., Li, Y., 2018. "Click" chemistry in polymeric scaffolds: bioactive materials for tissue engineering. J. Control. Release 273, 160–179. https://doi.org/10.1016/j.jconrel.2018.01.023.

Clinical applications of mesenchymal stem cells

5

Rani James, Namitha Haridas, Kaushik D. Deb
DiponEd Biointelligence, Bangalore, India

Chapter outline

5.1 Introduction 102
    5.1.1 What are stem cells? 102
    5.1.2 Types of stem cells 102
    5.1.3 Stem cell differentiation 102
    5.1.4 Application of stem cells in tissue engineering 103
5.2 Mesenchymal stem cells 104
    5.2.1 Introduction of mesenchymal stem cells 104
5.3 Sources of mesenchymal stem cells 104
    5.3.1 Peripheral blood—derived mesenchymal stem cells 105
    5.3.2 Bone marrow—derived mesenchymal stem cells 105
    5.3.3 Umbilical cord—derived mesenchymal stem cells 105
    5.3.4 Adipose-derived mesenchymal stem cells 105
5.4 Properties of mesenchymal stem cells 106
    5.4.1 Differentiation 106
    5.4.2 Immune modulation 106
    5.4.3 Migratory capacity 107
5.5 Clinical applications of mesenchymal stem cells 107
    5.5.1 Bone and cartilage diseases 107
    5.5.2 Bone marrow transplant and graft versus host disease 107
    5.5.3 Cardiovascular diseases 108
    5.5.4 Autoimmune diseases 108
    5.5.5 Liver diseases 108
    5.5.6 Musculoskeletal diseases 108
    5.5.7 Rheumatoid arthritis 109
    5.5.8 Systemic lupus erythematosus 109
    5.5.9 Type 1 diabetes 110
    5.5.10 Multiple sclerosis 110
    5.5.11 Parkinson's disease 110
    5.5.12 Alzheimer's disease 111
5.6 Stem cell banking 111
    5.6.1 Introduction to stem cell banking 111
    5.6.2 Steps in mesenchymal stem cell banking 111
    5.6.3 Future orientation 112

References   112
Further reading   116

5.1 Introduction

5.1.1 What are stem cells?

Stem cells are special human cells that have the ability to develop into many different cell types, from muscle cells to brain cells. Stem cells are those cells stored as body's natural reservoir, replenishing stocks of specialized cells that have been used up or damaged. To maintain the body function, regeneration process is paramount. Some specialized cells, such as blood and muscle cells, are unable to make facsimiles of them through cell division. Stem cells are essential for the maintenance of tissues such as blood, skin, and gut, which undergo perpetual turnover (cell supersession), and muscle, which can be built up according to the body's needs and is often damaged during physical exertion.

Stem cells have three general properties: they are capable of dividing and renewing themselves for long periods; they are unspecialized; and they can give rise to specialized cell types.

5.1.2 Types of stem cells

Embryonic stem cells (ESCs)—derived from the blastocyst of a 5-day-old embryo; are pluripotent, i.e., they can differentiate into almost any cell type in the body (primary-like cells); can renew themselves indefinitely.

Adult stem cells (e.g., mesenchymal stem cells [MSCs], neural stem cells, adipose tissue—derived stem cells)—isolated from adult tissues, organs or blood, cord blood, etc.; are multipotent, i.e., can give rise to a number of related cell types; can renew themselves a number of times but not indefinitely induced. There are only small numbers of stem cells in these tissues, and they are more likely to generate only certain types of cells. For example, a stem cell derived from the liver will only generate more liver cells.

Pluripotent stem cells (induced pluripotent stem cells)—generated from reprogrammed somatic cells; similar or equivalent to ESCs, i.e., pluripotent and the ability to renew themselves indefinitely.

5.1.3 Stem cell differentiation

Stem cells are unspecialized. Stem cells can divide and generate more specialized cell types. This process is called differentiation. During the development of an organism, cells differentiate to become specialized. Specialized cells such as blood and muscle do not mundanely replicate themselves, which designates that when they are solemnly damaged by disease or injury, they cannot replace themselves. Stem cells from different tissues, and from different stages of development, they vary in the number and types of cells that they produce. According to the classical view, as an organism develops, the potential of a stem cell to generate any cell type in the body is gradually restricted. In regenerative medicine, undifferentiated cells are used to repair or replace damaged cells and tissues.

5.1.4 Application of stem cells in tissue engineering

Tissue engineering is an interdisciplinary field that combines principles of biology and medicine with engineering; design and construction of functional components that can be used for maintenance; and replacement or regeneration of damaged biological tissues.

Tissue engineering is first reported in 1933: Implantation of tumor cells wrapped in a polymer membrane into a pig to protect them from immune attack. Prosperously engendered artificial skin with fibroblasts seeded onto collagen scaffolds for the treatment of extensive burn injury which was developed in 1980 is still being used clinically.

Years later in 1993, when the concept of tissue engineering was introduced, mouse tumor cells encased in a biocompatible polymer membrane were implanted into the abdominal cavity of chick embryos. Decades later, researchers demonstrated that pancreatic beta cells from neonatal rats, cultured on synthetic capillaries and perfused with medium, relinquished insulin in replication to transmute in glucose concentration (Denham et al., 2005; Vats et al., 2005; Lajtha, 1967; Knight and Evans, 2004).

Majorly tissue engineering can be stratified into three categories:

1. Substitutive—essentially whole organ replacement.
2. Histoconductive—includes the replacement of missing or damaged components of an organ tissue with ex vivo constructs.
3. Histoinductive approaches—facilitate self-repair and may involve gene therapy utilizing DNA distribution via plasmid vectors or magnification factors; a number of criteria must be slaked to achieve efficacious, perennial rehabilitation of damaged tissues.
    (a) An adequate number of cells must be engendered to fill the defect; source can be autologous (from the patient), allogenic (from a human donor but not immunologically identical), or xenogenic (from a different species donor).
    (b) Cells must be able to differentiate into desired phenotypes.
    (c) Cells must adopt felicitous three-dimensional structural support/scaffold and produce extracellular matrix.
    (d) Engendered cells must be structurally and mechanically compliant with the native cell.
    (e) Cells must prosperously be able to integrate with native cells and overcome the jeopardy of immunological repudiation.
    (f) There should be minimal associated biological peril. Cell sources are further delineated into mature (nonstem) cells, adult stem cells or somatic stem cells, ESCs, and totipotent stem cells or zygotes. Mature cells are restricted in clinical application because of their low proliferative and differentiating potential.

(g) At least 20 major categories of somatic stem cells have been identified in the bone marrow (BM), blood, cornea and retina of the ocular perceiver, dental pulp, liver, skin, GI tract, and pancreas of mammals. Precisely, the MSCs, its self-renewal capacity, possess plasticity, and the possibility to utilize them as autologous cells renders MSCs opportunity for cell therapy and tissue engineering. Furthermore, it is kenned that MSCs engender and secrete a great variety of cytokines and chemokines that play benign paracrine actions when MSCs are utilized for tissue repair (Fuchs et al., 2001; Shieh and Vacanti, 2005; Stock and Vacanti, 2001; He et al., 2007).

5.2 Mesenchymal stem cells

5.2.1 Introduction of mesenchymal stem cells

MSCs are pluripotent progenitor cells that divide many times and whose progeny eventually gives rise to skeletal tissues. The main source of MSCs is the BM. MSCs are equivalent to stromal cells identified back in the 1960s by Dexter and colleagues. MSCs are isolated from BM mononuclear cells and are grown as adherence culture in tissue culture flasks. The International Society for Cellular Therapy defined the minimal criteria for defining multipotent mesenchymal stromal cells. The cells should be plastic-adherent and express markers such as CD105, CD73, and CD90 but not CD45, CD34, CD14, CD11b, CD79alpha, CD19, or HLA-DR. Besides, the cells must be differentiated to osteoblasts, adipocytes, and chondrocytes in vitro (Pittenger et al., 1999; Caplan and Bruden, 2001; Chang et al., 2006).

5.3 Sources of mesenchymal stem cells

There are different sources of MSCs in our body. The source of MSCs are bone marrow, adipose tissue, cord blood products, placenta, Wharton's jelly (WJ), amniotic fluid, and other tissues. MSCs can be easily isolated from adult tissues, and the main advantage is that isolation methods do not require ethical clearance.

Analysis of markers for chondrogenic and collagen gene expression are confirmed by quantitative polymerase chain reaction. Results showed that all three sources of MSCs such as bone, umbilical cord (UC), and adipose tissues presented a similar capacity for chondrogenic and osteogenic differentiation, but they differed in their adipogenic potential. So, it is crucial to predetermine the most appropriate MSC source for different applications. Besides, the MSCs also show regenerative potentials for a variety of conditions, such as graft versus host disease (GvHD), Crohn's disease, osteogenesis imperfecta (OI), cartilage damage, and myocardial infarction.

The MSCs can be isolated with 100% efficiency from the bone marrow and adipose tissues in comparison with MSCs isolated from adipose tissues. But the differentiation capacity was similar among MSCs from all sources as analyzed by flow results.

5.3.1 Peripheral blood–derived mesenchymal stem cells

MSCs are present in peripheral blood. PBMNC (peripheral blood mononuclear cell)-derived multipotent mesenchymal stromal cells circulate in low number and share most of the surface markers that are expressed in the bone marrow and adipose-derived MSCs. Studies reported that multipotent mesenchymal stromal cells possess diverse and complicated gene expression characteristics and are capable of differentiating even beyond mesenchymal cells lineages (Zvaifler et al., 2000; Ukai et al., 2007).

5.3.2 Bone marrow–derived mesenchymal stem cells

MSCs are present in peripheral blood. PBMNC-derived multipotent mesenchymal stromal cells circulate in low number and share most of the surface markers that are expressed in the bone marrow and adipose-derived MSCs. Studies reported that multipotent mesenchymal stromal cells possess diverse and complicated gene expression characteristics and are capable of differentiating even beyond mesenchymal cells lineages (Zvaifler et al., 2000; Ukai et al., 2007).

5.3.3 Umbilical cord–derived mesenchymal stem cells

MSCs can be isolated from umbilical arteries and umbilical vein, both embedded within a specific mucous connective tissue, known as WJ. UC is considered a medical waste, and the collection of UC-MSCs is noninvasive; furthermore, the access to UC-MSCs does not need encumbered ethical problems. UC-MSCs also express similar markers that are similar to BM− and adipose-derived MSC's. UC-derived MSCs have distinct self-renewal capacity and multipotency. Moreover, UC-MSCs show faster renewal rate that BM−derived MSC's, compared with other sources (Sunkomat et al., 2007; Seshareddy et al., 2008).

5.3.4 Adipose-derived mesenchymal stem cells

ADMSCs (adipose derived mesenchymal stem cells) are mainly isolated from adipose tissue and have the ability to differentiate into mesodermal tissue lineages, that is, bone, cartilage, muscle, and adipose. They have been incorporated into many different scaffold-based systems and play an established role in cartilage tissue engineering. The two major sources are abdominal fat and infrapatellar fat pad (IFP). Techniques and protocols for ADMSC harvest and isolation varies depending on different laboratory groups. Abdominal fat can be harvested from subcutaneous tissue via abdominoplasty or arthroscopy, which can be done easily under the expert guidance of a skilled doctor. IFP is a promising source of ADMSCs to produce cartilage lineage. Adipose tissue−derived MSCs express Oct4, Nanog, SOX2, alkaline phosphatase, and SSEA-4 markers (Brayfield et al., 2010; Bunnell et al., 2008; Freitag et al., 2017).

5.4 Properties of mesenchymal stem cells

5.4.1 Differentiation

MSCs have also been isolated from human first- and second-trimester fetal blood, liver, spleen, and BM. Although phenotypically similar, these culture-expanded MSCs showed heterogeneity in differentiation potential, which related to the tissue source. Taken together, these examples illustrate that mesenchymal precursor cells are phenotypically heterogeneous, and the relationship between traditional BM−derived MSCs and these other MSC-like populations remains to be fully clarified. MSCs can be isolated from different sources, for example, human adipose tissue is a source of multipotent stem cells called processed lipoaspirate (PLA) cells, which, like BM MSCs, can differentiate down several mesenchymal lineages in vitro. But there are some differences in the expressions of particular markers: CD49d is expressed on PLA cells but not MSCs, and CD106 is expressed on MSCs but not PLA cells. CD106 on MSCs in BM has been functionally associated with hematopoiesis, so the lack of CD106 expression on PLA cells is consistent with localization of these cells to a nonhematopoietic tissue. Unlike other stem cells such as hematopoietic stem cells (HSCs), which are identified by the expression of the CD34 surface marker, MSCs lack a unique marker. The CD105 surface antigen (endoglin) has been recently used to isolate hMSCs (human MSCs) from the BM, thus enabled the characterization of freshly isolated hMSCs before culture using flow cytometry. Peripheral blood−derived mesenchymal precursor cells have also been described in the blood of normal individuals, and these express many of the same markers similar to BM MSCs, as well as differentiating down the osteoblastic and adipogenic lineages. However, these appear to be a separate population from fibrocytes, which are mesenchymal precursor cells that circulate in the blood and can migrate into tissues. Adult human MSCs are reported to express intermediate levels of major histocompatibility complex (MHC) class I but do not express human leukocyte antigen (HLA) class II antigens on the cell surface, which makes the cells as an excellent source of immunosuppressive drug. Compared with other sources, the expression of HLA class I on fetal hMSCs is lower (Peister et al., 2004; Rogers et al., 2007).

5.4.2 Immune modulation

The clinical significance of MSCs mainly contribute to its potent immunosuppressive and antiinflammatory effects through the interactions between the lymphocytes associated with both the innate and adaptive immune systems. MSCs suppress activities such as T-cell proliferation, B-cell functions, natural killer cell proliferation, and cytokine production and prevent the differentiation, maturation, and activation of dendritic cells. Importantly, MSCs can suppress cells independently of the MHC identity between donor and recipient due to their low expression of MHC-II and other costimulatory molecules. While MSCs can exert immunosuppressive effects by direct cell to cell contact, their primary mechanism is production of soluble factors, including transforming growth factor-β, hepatocyte growth factor (HGF), nitric oxide, and indoleamine 2,3-dioxygenase (IDO) (Kode et al., 2009; Fibbe et al., 2007; Selmani et al., 2008; Zhao et al., 2008; Nasef et al., 2008)

5.4.3 Migratory capacity

The clinical significance of MSCs mainly contribute to its potent immunosuppressive and antiinflammatory effects through the interactions between the lymphocytes associated with both the innate and adaptive immune systems. MSCs suppress activities such as T-cell proliferation, B-cell functions, natural killer cell proliferation, and cytokine production and prevent the differentiation, maturation, and activation of dendritic cells. Importantly, MSCs can suppress cells independently of the MHC identity between donor and recipient due to their low expression of MHC-II and other costimulatory molecules. While MSCs can exert immunosuppressive effects by direct cell to cell contact, their primary mechanism is production of soluble factors, including transforming growth factor-β, HGF, nitric oxide, and IDO (Kode et al., 2009; Fibbe et al., 2007; Selmani et al., 2008; Zhao et al., 2008; Nasef et al., 2008)

5.5 Clinical applications of mesenchymal stem cells

MSCs have attracted attention because of their unique therapeutic properties.

5.5.1 Bone and cartilage diseases

MSCs have been used to treat various bone diseases as shown to be beneficial in treating bone disorders, such as OI and hypophosphatasia. The patients were transplanted with whole BM instead of MSCs alone. In a clinical study, patients who received HSCT were infused with the same donor MSCs. Infusion of MSCs showed further benefit, but this was limited in duration. MSC transplantation has been shown to produce significant clinical improvements with cartilage repair; but the mechanisms underlying cartilage regeneration are still unknown. The transplanted MSCs can be differentiated into chondrocytes, or it is also possible that MSCs produce soluble growth factors to induce other cells of the microenvironment to differentiate into cartilage.

5.5.2 Bone marrow transplant and graft versus host disease

MSCs show immunomodulatory activities that make the cells an attractive therapeutic approach during or after transplantation as their transplantation can minimize the toxicity of the conditioning regimens while inducing hematopoietic engraftment and decrease the incidence and severity of GvHD. In several clinical trials, MSCs were cotransplanted with HSCs to facilitate engraftment, but their mechanism of action remains unclear. Similarly, in another clinical trial, the infusion of third-party haploidentical MSCs during pediatric UC blood transplantation was shown to induce prompt hematopoietic recovery.

The European Group for Blood and Marrow Transplantation conducted a multicenter phase II study in which both pediatric and adult patients with steroid-resistant GvHD were treated with MSCs derived from various sources, including HLA-identical and haploidentical sibling donor BM or third-party mismatched donor BM that proved MSC infusion could prevent the development of GvHD (Zhao and Liu, 2016).

5.5.3 Cardiovascular diseases

Another area of MSC therapy is in the field of cardiovascular research. MSC therapy is an attractive candidate for cardiovascular repair because of its regenerative and immunomodulatory properties. Preclinical studies showed improved function of cardiac output after MSC administration. Clinical trials using MSCs improve cardiac function. There were no serious adverse events reported following MSC administration.

In addition to autologous MSCs, the efficacy of allogeneic MSCs has also been reported to improve cardiac functions. Allogeneic MSCs were well tolerated with a significant increase in left ventricular ejection fraction and lower incidences of arrhythmia and chest pain, as compared with the control group. Allogeneic MSCs derived from the placenta also resulted in significant clinical improvements that give the hope to preserve MSC stem cells for future use (Karantalis et al., 2012; Usunier et al., 2014; Hare et al., 2012).

5.5.4 Autoimmune diseases

Autoimmune diseases result from an inappropriate immune response of the body against normal cells and tissues. Based on their ability to modulate immune responses, MSCs are best stem cells as a treatment for autoimmune diseases. Patients suffering from severe autoimmune diseases do not respond to standard therapy and can be treated with autologous or allogenic MSC therapy. Besides patients with Crohn's disease, also known as inflammatory bowel disease, can be treated with MSC therapy.

5.5.5 Liver diseases

MSCs have been used to treat liver cirrhosis. Liver transplantation is often the only option in advanced stage patients; however, it is limited by lack of donors, surgical complications, and rejection. MSCs have the potential to be used for the treatment of liver diseases due to their regenerative potential and immunomodulatory properties. Furthermore, MSC therapy is minimally invasive compared with liver transplant (Tyndall, 2012; Kim & -Goo Cho, 2013).

5.5.6 Musculoskeletal diseases

Although the bones naturally restore without significant scarring, infections, trauma, and cancer could impair their functional restoration, causing several bone defects. Cell-based therapies need to isolate MSCs from the BM of the patient, to expand

and enrich the cells and to seed them into the most suitable three-dimensional scaffold and/or matrix. As an example, osteonecrosis is caused by femoral death due to poor blood supply: three patients were treated with MSCs infusion with TCP-treated matrix and good results were obtained. Similarly, Nöth et al. injected a preparation of MSCs into three patients and obtained encouraging results, as shown by radiographic and magnetic resonance imaging examination. MSCs were also successfully used for spinal fusion disease, so that phase I clinical trials arose. Patients affected by severe OI were injected systematically with purified allogenic MSCs: these cells were able to engraft into host bones, where they proliferated into osteoblasts and allowed an amelioration in the total bone mineral density (Nöth et al., 2007; Pneumaticos et al., 2011; Neen et al., 2006; Zhang et al., 2008; Horwitz et al., 2002).

5.5.7 Rheumatoid arthritis

Type II collagen (CII), one of the components of hyaline cartilage, acts as an autoantigen in rheumatoid arthritis (RA). When CII and the other antigenic peptides are recognized by T cells, they cause the uncontrolled activation of immune system cells, leading to destruction of the joints typical of RA patients. Zheng et al. demonstrated that MSCs isolated from RA patients exerted immunosuppressive functions, by inhibiting T-cell proliferation, blocking the secretion of several proinflammatory cytokines, and allowing the expression of antiinflammatory IL-10. They also obtained MSCs from chondrocytes and described that, following transplantation into RA joints, these cells not only suppressed the inflammation regulating the secretion of TGF-β1 but also prevented joints destruction. Similar to the previously described work of González et al. experimental data from Augello and Mao confirmed the positive effects of MSCs transplantation into animal model of collagen-induced arthritis while others did not describe any amelioration (Zheng et al., 2008; Augello et al., 2007; Mao et al., 2010).

5.5.8 Systemic lupus erythematosus

Systemic lupus erythematosus (SLE) is an autoimmune disease that can affect any part of the body. Recent findings demonstrated several defects in the hematopoietic system of SLE patients, probably due to unbalanced expression of cytokines and other growth factors. Interestingly, it was found that BM-derived CD34+ stem cells overexpressed different surface markers such as CD123 and CD166 that are closely related to T-cell development inflammation. Sun and colleagues determined a possible role of BM-derived MSCs in the hematological disorder typical of SLE patients and suggested that MSCs transplantation could be used to ameliorate the autoimmune progression of the disease (Borchers et al., 2004; Sun et al., 2007a,b; Kushida et al., 2001; Zhou et al., 2008).

5.5.9 Type 1 diabetes

Type 1 diabetes is an autoimmune disease mediated by the production of autoantibody directed against the β cells of the pancreas. As a consequence of the destruction of these cells, the quantity of insulin produced is not sufficient to control sugar blood level. Recently, it has been suggested that MSCs can overcome these problems as they can be differentiated into glucose-responsive, insulin-producing cells, and they possess immunomodulatory properties. It was hypothesized that resident pancreatic MSCs could be forced to adopt a pancreatic fate in vitro. Intriguing option comes from studies on umbilical cord blood (UCB)—derived MSCs that demonstrated the expression of pancreatic development genes in these cells. Recently, a population of UCB-derived cells was shown to behave like hES cells, recapitulating the same differentiation steps from early stages to β cells(Inverardi et al., 2003; Prabakar et al., 2011).

5.5.10 Multiple sclerosis

Multiple sclerosis is an important cause of neurological disability in young adults. Autoreactive T cells cause myelin destruction and secondary oligodendrocyte and axonal damage. Despite the efficacy of immunomodulatory or immunosuppressive drugs in controlling the number of relapses, no current therapy is effective to arrest the progressive phase of the disease. BM-derived MSCs have pronounced immune-modulating and immunosuppressive properties so as they were being tested in clinical trial for relapsing-remitting multiple sclerosis. MSCs promote self-repair by reducing scar formation, by stimulating the formation of new blood vessel, and by secreting growth and neuroprotective factors, such as superoxide dismutase-3. After intravenous injection, many cells entered the CNS and became widely distributed both in experimental models and in patients (Prockop and Youn Oh, 2012; Uccelli et al., 2007; Kemp et al., 2010; Gordon et al., 2010; Osaka et al., 2010; Zhang et al., 2006)

5.5.11 Parkinson's disease

Stem cell therapy, with the aim of replacing lost neurons, is the most promising strategy for this disease. It has been demonstrated that MSCs cells can enhance the levels of tyrosine hydroxylase and dopamine after transplantation in PD (parkinson's disease) animal models. Furthermore, it has been suggested that these cells contribute to neuroprotection by secreting trophic factors, such as EGF, VEGF, NT3, FGF-2, HGF, and BDNF, or through antiapoptotic signaling without differentiation in neuronal phenotype. For these reasons, new strategies, involving the genetic modification of hMSCs, arose, as a tool to induce the secretion of specific factors or to increase the percentage of DA (dopaminergic) cell differentiation (Wang et al., 2010; Wilkins et al., 2009; Moloney et al., 2010).

5.5.12 Alzheimer's disease

No treatment is currently able to stop the progression of Alzheimer's disease (AD). Recently, different studies tried to ameliorate neuropathological deficits in animal model of AD through stem cell therapy. In particular, Shin et al. focused on clearance of amyloid plaque, and they demonstrated that MSCs could enhance the cell autophagy pathway increasing neuron survival both in vitro and in vivo. Following promising results from MSCs treatment for autoimmune diseases, it was thought to modulate the inflammatory environment of AD. In particular, abnormalities of Tregs in cell number and/or function were observed, and it was shown that they could modulate microglial activation. Yang et al. demonstrated that UC-derived MSCs activated Tregs in vitro, and once transplanted in AD animal model, Tregs modulated microglia activation, increasing neuron survival (Shin et al., 2014; Ma et al., 2013; Saresella et al., 2010).

In conclusion, before MSCs' clinical application in various indications, further studies are needed to improve MSC-based protocols and above all to expand our knowledge on MSC immunogenicity in an HLA-mismatched context.

5.6 Stem cell banking

5.6.1 Introduction to stem cell banking

Stem cells are those master cells of the body that can be transformed into any cell of the body. Salient features of stem cells as follow: (a) they have immense regenerative capacities; (b) they can be used to become tissue- or organ-specific cells; (c) they are responsible for repair of body; d) they can be stored for long term and can remain undifferentiated for a long time.

Stem cell banking is the process of preserving your stem cells at temperatures several degrees below freezing point ($-196°C$) (cryopreservation). These stem cells can possibly provide a cure for many conditions and offer opportunities for improved quality of life.

Stem cell banking through long-term storage of different stem cells represents a basic source to store original features of stem cells for patient-specific clinical applications. Stem cells can heal the body, promote recovery, and offer an enormous amount of therapeutic potential (Hanna and Hubel, 2009).

5.6.2 Steps in mesenchymal stem cell banking

(1) Bone marrow, dental pulp, adipose, or cord tissue aspiration is performed at a medical center; (2) shipping of the tissue sample at $4°C$ in temperature-controlled packaging to the processing facility; (3) MSC isolation and culturing in laminar air flow chambers/biosafety cabinets (manual production) or bioreactor; (4) characterization for fresh MSCs; (5) aliquoting of MSC samples with cryopreservation medium for future repeated doses; (6) freezing and storage at $-196°C$ in liquid nitrogen freezer as

per regulatory guidelines; (7)slow thawing at 37°C in water bath; (8) characterization for thawed MSCs; (9) expansion of thawed MSCs in incubator for processing into repeated doses; (10) activation of MSC's into final cell therapy product; (11) packaging and shipping of final product in optimized, minimized loss of cell viability and potency to medical center at 4°C in temperature-controlled packaging with temperature recorder; (12) injection of the MSC into patient (Cooper and Chandra, 2011).

5.6.3 Future orientation

MSCs have a wide range of clinical implications because of their ability to differentiate into multiple lineages, secrete growth factors related to immune regulation, and migration to inflammatory sites. The results of different areas of research application and clinical trials results using MSCs have been promising and also highlight many critical challenges that need to be addressed in the future. The area of MSC research requires much extensive research in the field of therapeutic efficiency (Ullah et al., 2015).

References

Augello, A., Tasso, R., Negrini, S.M., Cancedda, R., Pennesi, G., 2007. Cell therapy using allogeneic bone marrow mesenchymal stem cells prevents tissue damage in collagen-induced arthritis. Arthritis Rheum. 56 (4), 1175–1186 (View at Publisher · View at Google Scholar · View at Scopus).

Borchers, A.T., Keen, C.L., Shoenfeld, Y., Gershwin, M.E., 2004. Surviving the butterfly and the wolf: mortality trends in systemic lupus erythematosus. Autoimmun. Rev. 3 (6), 423–453 (View at Publisher · View at Google Scholar · View at Scopus).

Brayfield, C., Marra, K., Rubin, J.P., 2010. Adipose stem cells for soft tissue regeneration. Plast. Chir. 42, 124–128.

Bunnell, B.A., Estes, B.T., Guilak, F., Gimble, J.M., 2008. Differentiation of adipose stem cells. Mol. Biol. 456, 155–171.

Caplan, A.I., Bruden, S.P., 2001. Mesenchymal stem cells: building blocks for molecular medicine in the 21th century. Trends Mol. Med. 7, 259–264.

Chang, C.J., Yen, M.L., Chen, Y.C., Chien, C.C., Huang, H.I., Bai, C.H., Yen, B.L., 2006. Placenta-derived multipotent cells exhibit immunosuppressive properties that are enhanced in the presence of interferon-gamma. Stem Cells 24 (11), 2466–2477.

Cooper, K., Chandra, V., 2011. Establishment of a mesenchymal stem cell bank. Stem Cell. Int 2011, 905621.

Denham, M., Conley, B., Olsson, F., Cole, T.J., Mollard, R., 2005. Stem cells: an overview. Curr Protoc. Cell Biol. 23, 23.1.

Fibbe, W.E., Nauta, A.J., Roelofs, H., 2007. Modulation of immune responses by mesenchymal stem cells. Ann. N. Y. Acad. Sci. 1106, 272–278 (View at Publisher · View at Google Scholar · View at Scopus).

Freitag, J., Shah, K., James, W., Boyd, R., Abi Tenen, et al., 2017. The effect of autologous adipose derived mesenchymal stem cell therapy in the treatment of a large osteochondral defect of the knee following unsuccessful surgical intervention of osteochondritis dissecans — a case study. Freitag et al. BMC Muscoskelet. Disord. 18, 298. https://doi.org/10.1186/s12891-017-1658-2.

Fuchs, J.R., Nasseri, B.A., Vacanti, J.P., 2001. Tissue engineering: a 21st century solution to surgical reconstruction. Ann. Thorac. Surg. 72, 577−591.

Gordon, D., Pavlovska, G., Uney, J.B., Wraith, D.C., Scolding, N.J., 2010. Human mesenchymal stem cells infiltrate the spinal cord, reduce demyelination, and localize to white matter lesions in experimental autoimmune encephalomyelitis. JNEN (J. Neuropathol. Exp. Neurol.) 69 (11), 1087−1095 (View at Publisher · View at Google Scholar · View at Scopus).

Hanna, J., Hubel, A., 2009. Preservation of stem cells. Organogenesis 5 (3), 134−137.

Hare, J.M., Fishman, J.E., Gerstenblith, G., et al., 2012. Comparison of allogeneic vs autologous bone marrow-derived mesenchymal stem cells delivered by transendocardial injection in patients with ischemic cardiomyopathy: the poseidon randomized trial. J. Am. Med. Assoc. 308, 2369−2379 ([PMC free article][PubMed]).

He, Q., Wan, C., Li, G., 2007. Concise review: multipotent mesenchymal stromal cells in blood. Stem Cells 25, 69−77.

Horwitz, E.M., Gordon, P.L., Koo, W.K.K., et al., 2002. Isolated allogeneic bone marrow-derived mesenchymal cells engraft and stimulate growth in children with osteogenesis imperfecta: implications for cell therapy of bone. Proc. Natl. Acad. Sci. U. S. A 99 (13), 8932−8937.

Inverardi, L., Kenyon, N.S., Ricordi, C., 2003. Islet transplantation: immunological perspectives. Curr. Opin. Immunol. 15 (5), 507−511 (View at Publisher · View at Google Scholar ·View at Scopus).

Karantalis, V., Balkan, W., Schulman, I.H., Hatzistergos, K.E., Hare, J.M., 2012. Cell-based therapy for prevention and reversal of myocardial remodeling. Am. J. Physiol. Heart Circ. Physiol. 303, H256−H270 ([PMC free article] [PubMed]).

Kemp, K., Hares, K., Mallam, E., Heesom, K.J., Scolding, N., Wilkins, A., 2010. Mesenchymal stem cell-secreted superoxide dismutase promotes cerebellar neuronal survival. J. Neurochem. 114 (6), 1569−1580 (View at Publisher · View at Google Scholar · View at Scopus).

Kim, N., Goo Cho, S., July 2013. Clinical applications of mesenchymal stem cells. Korean J. Intern. Med. 28 (4), 387−402.

Knight, M.A., Evans, G.R., 2004. Tissue engineering: progress and challenges. Plast. Reconstr. Surg. 114, 26E−37E. 2.

Kode, J.A., Mukherjee, S., Joglekar, M.V., Hardikar, A.A., 2009. Mesenchymal stem cells: immunobiology and role in immunomodulation and tissue regeneration. Cytotherapy 11 (4), 377−391 (View at Publisher · View at Google Scholar · View at Scopus).

Kushida, T., Inaba, M., Hisha, H., et al., 2001. Intra-bone marrow injection of allogeneic bone marrow cells: a powerful new strategy for treatment of intractable autoimmune diseases in MRL/lpr mice. Blood 97 (10), 3292−3299 (View at Publisher · View at Google Scholar · View at Scopus).

Lajtha, L.G., 1967. Stem cells and their properties. Proc. Can. Cancer Conf. 7, 31−39.
   External Resources
   • Pubmed/Medline (NLM)
   • Chemical Abstracts Service (CAS)

Ma, T., Gong, K., Ao, Q., et al., 2013. Intracerebral transplantation of adipose-derived mesenchymal stem cells alternatively activates microglia and ameliorates neuropathological deficits in Alzheimer's disease mice. Cell Transplant. 22 (Suppl. 1), S113−S126 (View at Publisher · View at Google Scholar).

Mao, F., Xu, W.-R., Qian, H., et al., 2010. Immunosuppressive effects of mesenchymal stem cells in collagen-induced mouse arthritis. Inflamm. Res. 59 (3), 219−225 (View at Publisher · View at Google Scholar · View at Scopus).

Moloney, T.C., Rooney, G.E., Barry, F.P., Howard, L., Dowd, E., 2010. Potential of rat bone marrow-derived mesenchymal stem cells as vehicles for delivery of neurotrophins to the Parkinsonian rat brain. Brain Res. 1359, 33−43.

Nasef, A., Mazurier, C., Bouchet, S., et al., 2008. Leukemia inhibitory factor: role in human mesenchymal stem cells mediated immunosuppression. Cell. Immunol. 253 (1−2), 16−22 (View at Publisher · View at Google Scholar · View at Scopus).

Neen, D., Noyes, D., Shaw, M., Gwilym, S., Fairlie, N., Birch, N., 2006. Healos and bone marrow aspirate used for lumbar spine fusion: a case controlled study comparing healos with autograft. Spine 31 (18), E636−E640 (View at Publisher · View at Google Scholar · View at Scopus).

Nöth, U., Reichert, J., Reppenhagen, S., et al., 2007. Cell based therapy for the treatment of femoral head necrosis. Der Orthopäde 36 (5), 466−471 (View at Publisher · View at Google Scholar · View at Scopus).

Osaka, M., Honmou, O., Murakami, T., et al., 2010. Intravenous administration of mesenchymal stem cells derived from bone marrow after contusive spinal cord injury improves functional outcome. Brain Res. 1343, 226−235 (View at Publisher · View at Google Scholar · View at Scopus).

Peister, A., Mellad, J.A., Larson, B.L., et al., 2004. Adult stem cells from bone marrow (MSCs) isolated from different strains of inbred mice vary in surface epitopes, rates of proliferation, and differentiation potential. Blood 103, 1662−1668.

Pittenger, M.F., Mackay, A.M., Beck, S.C., Jaiswal, R.K., Douglas, R., Mosca, J.D., Moorman, M.A., Limoneti, D.W., Carig, S., Marshak, D.R., 1999. Multilineage potential of adult human mesenchymal stem cells. Science 284, 143−147.

Pneumaticos, S.G., Triantafyllopoulos, G.K., Chatziioannou, S., Basdra, E.K., Papavassiliou, A.G., 2011. Biomolecular strategies of bone augmentation in spinal surgery. Trends Mol. Med. 17 (4), 215−222 (View at Publisher · View at Google Scholar · View at Scopus).

Prabakar, K.R., Domínguez-Bendala, J., Molano, R.D., et al., 2011. Generation of glucose-responsive, insulin-producing cells from human umbilical cord blood-derived mesenchymal stem cells. Cell Transplant. 21 (6), 1321−1339.

Prockop, D.J., Youn Oh, J., 2012. Mesenchymal stem/stromal cells (MSCs): role as guardians of inflammation. Mol. Ther. 20 (1), 14−20 (View at Publisher · View at Google Scholar · View at Scopus).

Rogers, I., et al., 2007. Identification and analysis of in vitro cultured CD45-positive cells capable of multi-lineage differentiation. Exp. Cell Res. 313, 1839−1852.

Saresella, M., Calabrese, E., Marventano, I., et al., 2010. PD1 negative and PD1 positive CD4$^+$ T regulatory cells in mild cognitive impairment and Alzheimer's disease. J. Alzheimer's Dis. 21 (3), 927−938 (View at Publisher · View at Google Scholar · View at Scopus).

Selmani, Z., Naji, A., Zidi, I., et al., 2008. Human leukocyte antigen-G5 secretion by human mesenchymal stem cells is required to suppress T lymphocyte and natural killer function and to induce CD4$^+$CD25^{high}FOXP3$^+$ regulatory T cells. Stem Cells 26 (1), 212−222 (View at Publisher View at Google Scholar View at Scopus).

Seshareddy, K., et al., 2008. "Methods to isolate mesenchymal-like cells from Wharton's jelly of umbilical cord. Methods *Cell Biol.* 86, 101−119.

Shieh, S.J., Vacanti, J.P., 2005. State-of-the-art tissue engineering: from tissue engineering to organ building. Surgery 137, 1−7, 4.

Shin, J.Y., Park, H.J., Kim, H.N., et al., 2014. Mesenchymal stem cells enhance autophagy and increase beta-amyloid clearance in Alzheimer disease models. Autophagy 10 (1), 32−44 (View at Publisher View at Google Scholar).

Stock, U.A., Vacanti, J.P., 2001. Tissue engineering: current state and prospects. Annu. Rev. Med. 52, 443−451.

Sun, L.Y., Zhang, H.Y., Feng, X.B., Hou, Y.Y., Lu, L.W., Fan, L.M., 2007a. Abnormality of bone marrow-derived mesenchymal stem cells in patients with systemic lupus erythematosus. Lupus 16 (2), 121−128 (View at Publisher View at Google Scholar View at Scopus).

Sun, L.-Y., Zhou, K.-X., Feng, X.-B., et al., 2007b. Abnormal surface markers expression on bone marrow $CD34^+$ cells and correlation with disease activity in patients with systemic lupus erythematosus. Clin. Rheumatol. 26 (12), 2073−2079 (View at Publisher View at Google Scholar View at Scopus).

Sunkomat, et al., 2007. Cord blood-derived MNCs delivered intracoronary contribute differently to vascularization compared to CD34+ cells in the rat model of acute ischemia. J. Mol. Cell. Cardiol. 42, S97.

Tyndall, A., 2012. Application of autologous stem cell transplantation in various adult and pediatric rheumatic diseases. Pediatr. Res. 71 (4 Pt 2), 433−438.

Uccelli, A., Pistoia, V., Moretta, L., 2007. Mesenchymal stem cells: a new strategy for immunosuppression? Trends Immunol. 28 (5), 219−226 (View at Publisher View at Google Scholar ·View at Scopus).

Ukai, R., Honmou, O., Harada, K., Houkin, K., Hamada, H., Jeffery, D., March 2007. Kocsis. Mesenchymal stem cells derived from peripheral blood protects against ischemia. J. Neurotrauma 24 (3), 508−520.

Ullah, I., Baregundi Subbarao, R., Rho, G.J., 2015. Human mesenchymal stem cells - current trends and future prospective. Biosci. Rep. 35 (2), e00191.

Usunier, B., Benderitter, M., Tamarat, R., Chapel, A., 2014. Management of fibrosis: the mesenchymal stromal cells breakthrough. Stem Cell. Int. 2014, 340257 ([PMC free article] [PubMed]).

Vats, A., Bielby, R.C., Tolley, N.S., Nerem, R., Polak, J.M., 2005. Stem cells. Lancet 366, 592−602.

External Resources
- Pubmed/Medline (NLM)
- Crossref (DOI)
- Chemical Abstracts Service (CAS)
- ISI Web of Science

Wang, F., Yasuhara, T., Shingo, T., et al., 2010. Intravenous administration of mesenchymal stem cells exerts therapeutic effects on parkinsonian model of rats: focusing on neuroprotective effects of stromal cell-derived factor-1α. BMC Neurosci. 11 article 52, View at Publisher View at Google Scholar View at Scopus.

Wilkins, A., Kemp, K., Ginty, M., Hares, K., Mallam, E., Scolding, N., 2009. Human bone marrow-derived mesenchymal stem cells secrete brain-derived neurotrophic factor which promotes neuronal survival in vitro. Stem Cell Res. 3 (1), 63−70 (View at Publisher View at Google Scholar View at Scopus).

Zhang, P., Gan, Y.-K., Tang, J., et al., 2008. Clinical study of lumbar fusion by hybrid construct of stem cells technique and biodegradable material. Zhonghua Wai Ke Za Zhi 46 (7), 493−496 (View at Google Scholar View at Scopus).

Zhang, J., Li, Y., Lu, M., et al., 2006. Bone marrow stromal cells reduce axonal loss in experimental autoimmune encephalomyelitis mice. J. Neurosci. Res. 84 (3), 587−595 (View at Publisher).

Zhao, Z.-G., Li, W.-M., Chen, Z.-C., You, Y., Zou, P., 2008. Immunosuppressive properties of mesenchymal stem cells derived from bone marrow of patients with chronic myeloid leukemia. Immunol. Investig. 37 (7), 726–739 (View at Publisher View at Google Scholar View at Scopus).

Zhao, K., Liu, Q., 2016. The clinical application of mesenchymal stromal cells in hematopoietic stem cell transplantation. J. Hematol. Oncol. 9, 46.

Zheng, Z.H., Li, X.Y., Ding, J., Jia, J.F., Zhu, P., 2008. Allogeneic mesenchymal stem cell and mesenchymal stem cell-differentiated chondrocyte suppress the responses of type II collagen-reactive T cells in rheumatoid arthritis. Rheumatology 47 (1), 22–30 (View at Publisher View at Google Scholar View at Scopus).

Zhou, K., Zhang, H., Jin, O., et al., 2008. Transplantation of human bone marrow mesenchymal stem cell ameliorates the autoimmune pathogenesis in MRL/lpr mice. Cell. Mol. Immunol. 5 (6), 417–424.

Zvaifler, N.J., Marinova-Mutafchieva, L., Adams, G., et al., 2000. Mesenchymal precursor cells in the blood of normal individuals. Arthritis Res. 2, 477–488.

Further reading

Campioni, D., Rizzo, R., Stignani, M., et al., 2009. A decreased positivity for CD90 on human mesenchymal stromal cells (MSCs) is associated with a loss of immunosuppressive activity by MSCs. Cytometry B: Clin. Cytomet. 76 (3), 225–230 (View at Publisher · View at Google Scholar · View at Scopus).

Chan, J., Waddington, S.N., O'Donoghue, K., et al., 2007. Widespread distribution and muscle differentiation of human fetal mesenchymal stem cells after intrauterine transplantation in dystrophic mdx mouse. Stem Cells 25 (4), 875–884 (View at Publisher · View at Google Scholar · View at Scopus).

Jones, B.J., Brooke, G., Atkinson, K., McTaggart, S.J., 2007. Immunosuppression by placental indoleamine 2,3-dioxygenase: a role for mesenchymal stem cells. Placenta 28 (11–12), 1174–1181 (View at Publisher · View at Google Scholar · View at Scopus).

MacKenzie, T.C., Kobinger, G.P., Kootstra, N.A., et al., 2002. Efficient transduction of liver and muscle after in utero injection of lentiviral vectors with different pseudotypes. Mol. Ther. 6 (3), 349–358 (View at Publisher).

Pan, X.-N., Zheng, L.-Q., Lai, X.-H., 2014. Bone Marrow-Derived Mesenchymal Stem Cell Therapy for Decompensated Liver Cirrhosis: A Meta-Analysis. World J. Gastroenterol. 20 (38), 14051–14057. https://doi.org/10.3748/wjg.v20.i38.14051.

Soler Rich, R., Munar, A., Soler Romagosa, F., Peirau, X., Huguet, M., et al., 2015. Treatment of knee osteoarthritis with autologous expanded bone marrow mesenchymal stem cells: 50 cases clinical and MRI results at one year follow-up. J. Stem Cell Res. Ther. 5, 285. https://doi.org/10.4172/2157-7633.1000285.

Part Two

Tissue regeneration

Cardiac regeneration

6

Raghav Murthy[1], Aditya Sengupta[2]

[1]Assistant Professor, Assistant Director of Pediatric Cardiac Transplantation, Department of Cardiovascular Surgery, Mount Sinai Hospital, Icahn School of Medicine at Mount Sinai, New York, NY, United States; [2]Cardiothoracic Surgery Resident, Department of Cardiovascular Surgery, Mount Sinai Hospital, Icahn School of Medicine at Mount Sinai, New York, NY, United States

Chapter outline

- **6.1 Introduction** 120
- **6.2 The controversy** 120
- **6.3 Mechanisms** 121
  - 6.3.1 Survival and protection 122
  - 6.3.2 Inflammation reduction 122
  - 6.3.3 Cell–cell communication 122
  - 6.3.4 Angiogenesis/vascularization 124
  - 6.3.5 Cardiomyogenesis 124
  - 6.3.6 Molecular regulation of proliferation, the cell cycle, and commitment 124
  - 6.3.7 Cardiac aging 124
- **6.4 Stem cell therapies** 125
- **6.5 Barriers in stem cell therapy** 127
- **6.6 Tissue engineering** 130
- **6.7 Cellular reprogramming** 132
- **6.8 Stem cell–derived exosomes and small vesicles** 132
- **6.9 Hydrogels** 132
- **6.10 Cardiac regeneration in children** 133
- **6.11 Valves** 133
- **6.12 Biointegration** 136
- **6.13 Conclusion** 138
- **References** 138

6.1 Introduction

The human heart has been an object of awe and mystery from ancient times. This hollow, muscular, pumping chamber has become the obsession of many people. The average heart beats 3,363,840,000 times in a lifetime. Eight out of every 100 children born have a heart defect. Perhaps most importantly, cardiovascular disease is the leading global cause of death, accounting for more than 17 million deaths per year. According to the American Heart Association, this number is expected to grow to more than 23 million by 2030 (Benjamin et al., 2017). In 2011, nearly 787,000 people died from heart disease, stroke, and other cardiovascular disorders in the United States (Terashvili and Bosnjak, 2018). They account for about 3.9 million deaths per year in Europe (Muller et al., 2018). Evidently, the wear and tear that this organ suffers is immense. With regard to the pediatric population, the cardiac structures most afflicted by birth defects are the valves and great vessels. The advantages of having tissue-engineered structures that can grow, repair themselves, and remodel as the child grows are substantial. On the other hand, adult structures are fully grown, but the ability to self-regenerate and self-repair with the aid of biointegrated materials would be a game changer. The current treatments for cardiovascular disease do not address the underlying issue of permanent cardiomyocyte loss. Innovative therapies in cardiac regeneration aim to alter this. This chapter looks into both the regenerative capacity of the heart, and the breadth and depth of cardiac tissue regeneration that can be performed in the pediatric and adult worlds.

6.2 The controversy

A fundamental issue concerning the ability of the heart to sustain injury is whether myocardial regeneration occurs in the adult organ or whether this growth and adaptation are restricted to prenatal life, thus limiting the response of the heart to pathologic loads (Anversa et al., 2006). The concept of the heart as a terminally differentiated organ (postmitotic organ) unable to replace working myocytes has been at the center of cardiovascular research and therapeutic developments for the past 50 years (MacLellan and Schneider, 2000). The dogmatic view has been that the heart reacts to an increase in workload only by increasing myocyte size or hypertrophy. Heart failure ensues when this phenomenon can no longer be sustained. The possibility that the heart can regenerate has been dismissed, and the concept of myocardial repair is viewed with skepticism (Anversa et al., 2006). However, several studies have indicated that myocyte regeneration occurs after myocardial infarction and pressure overload (Anversa et al., 2006; Beltrami et al., 2001; Urbanek et al., 2003, 2005; Linke et al., 2005; Olivetti et al., 1994).

The concept that myocytes cannot divide originated from a difficulty in identifying mitotic figures within these cells (Anversa et al., 2006). It was believed that the number of cardiomyocytes a person would have during his or her lifetime was fixed, and any that died would decrease the total number of cardiac muscle cells present in any individual organ. But, numerous studies have provided evidence that myocytes die and that new ones are constantly being formed (Anversa et al., 2006; Kang and Izumo, 2000). Over the years, evidence has been gathered from studies of sex-mismatched cardiac transplants of the existence of a true cardiac stem cell (Anversa et al., 2006; Quaini et al., 2002). Confocal microscopy has further demonstrated evidence of myocyte turnover in the human heart (Anversa et al., 2006). Thus, the heart must be considered a dynamic organ.

The ability of adult stem cells to generate cells beyond their own tissue boundary constitutes a process called developmental plasticity or transdifferentiation (Anversa et al., 2006). Bone marrow cells are the most versatile of all cells and therefore have been targeted to develop strategies for tissue regeneration. They have been injected into infarcted myocardium and have been used to treat nonischemic cardiomyopathy (Anversa et al., 2006). These cells became the center of several preliminary studies aimed at rescuing the infarcted myocardium. The fact that infarcted myocardium does not heal to become functional contractile tissue, but rather forms scar, is not in contention. Rather, the issue at hand is devising a means of promoting translocation of cardiac stem cells into infarcted areas of the heart such that they can "set up shop," repair the damaged tissue, and replace it with fully functional and contractile cardiomyocytes. This process also requires the development of new and viable blood supply to the area (Anversa et al., 2006). Several cell line–based strategies have been implemented experimentally to repair the infarcted heart. They include use of fetal cardiomyocytes, skeletal myoblasts, embryonic-derived endothelial cells, bone marrow–derived immature myocytes, fibroblasts, smooth muscle cells, endothelial progenitor cells, and bone marrow–derived cells (BMCs) (Anversa et al., 2006; Anversa, 2005). These approaches had a rather uniform outcome in animal studies but have shown variable degrees of improvement in cardiac performance in human studies. This was most likely due to the formation of a passive graft that reduced negative remodeling by decreasing the stiffness of the scarred portion of the ventricular wall. An active graft, which dynamically contributes to myocardial contractility, has been observed in only a few cases (Anversa et al., 2006; Anversa, 2005; Orlic et al., 2001a,b; Kajstura et al., 2005; Yoon et al., 2005; Kawada et al., 2004). However, the implanted cells may exert a paracrine effect, activating a growth response of resident progenitor cells (Anversa et al., 2006; Yoon et al., 2005; Behfar et al., 2002). Myocardial regeneration necessitates the administration of a more primitive cell that is multipotent and can differentiate into the main cardiac cell lineages, namely myocytes, vascular smooth muscle cells, and endothelial cells (Anversa et al., 2006).

6.3 Mechanisms

Desperate need for new treatment strategies for cardiovascular diseases has prompted several clinical approaches to cardiac regeneration. Unfortunately, the outcomes have been less than desirable. To promote comprehensive discussion of cardiovascular regenerative medicinal products, an international collaboration, known as TACTICS

(Transnational Alliance for Regenerative Therapies in Cardiovascular Syndromes), was formed. TACTICS recently produced a consensus statement prelude to topic-specific articles asserting the need for a thorough understanding of mechanisms involved in myocardial repair and regeneration (Broughton et al., 2018; Fernandez-Aviles et al., 2017). Fig. 6.1 below provides an overview of the mechanisms involved in cardiac regeneration and repair.

6.3.1 Survival and protection

Myocardial preservation postinjury, including cardiomyocyte survival and limitation of scar formation, is vital to mitigate the long-term effects of cardiac dysfunction (Broughton et al., 2018). Multiple signaling pathways are involved in this process. One of the most extensively studied is the Akt pathway (Matsui et al., 2001; Mohsin et al., 2012). Survival is also enhanced by activation of BCL2 interacting protein 3 (Broughton et al., 2018; Dong et al., 2010). Antiapoptotic signaling is the key and involves tumur necrosis factor (TNF) α, Bcl-2-associated-X protein (BAX), BH-3 proteins, and p53 (Broughton et al., 2018).

6.3.2 Inflammation reduction

Inflammatory responses in injured myocardium include mononuclear cell infiltration, interleukins (IL) that modulate inflammation, mast cell accumulation in the healing scar, fibroblast and extracellular matrix (ECM) remodeling, and temporal regulation of angiogenesis in the evolving and subsequently healing infarct (Broughton et al., 2018). Various cytokines and growth factors are associated with this process, including compliment 5a (C5a), tumor growth factor beta 1, Monocyte chemoattractant protein 1 (MCP-1), TNFα, IL-6, intercelluar adhlesion molecule (ICAM), macrophage colony-stimulating factor, IL-10, tissue inhibitor of metalloproteinases (TIMP), stem cell factor, vascular endothelial growth factor (VEGF), platelet-derived growth factor (PDGF), and IL-8 to name a few (Broughton et al., 2018; Kukielka et al., 1995). There is usually a combined cell mediated and humoral inflammatory reaction at the site of injury.

6.3.3 Cell–cell communication

Cellular cross talk is essential, and cardiomyocytes have been shown to communicate through multiple routes including gap junctions, adhesion complexes, and secretome factors (Broughton et al., 2018). Fibroblasts and other cells that reside in the interstitium of cardiomyocytes are highly contributory to cellular communication. The regulation and proliferation of these cells determines the degree of ensuing fibrosis.

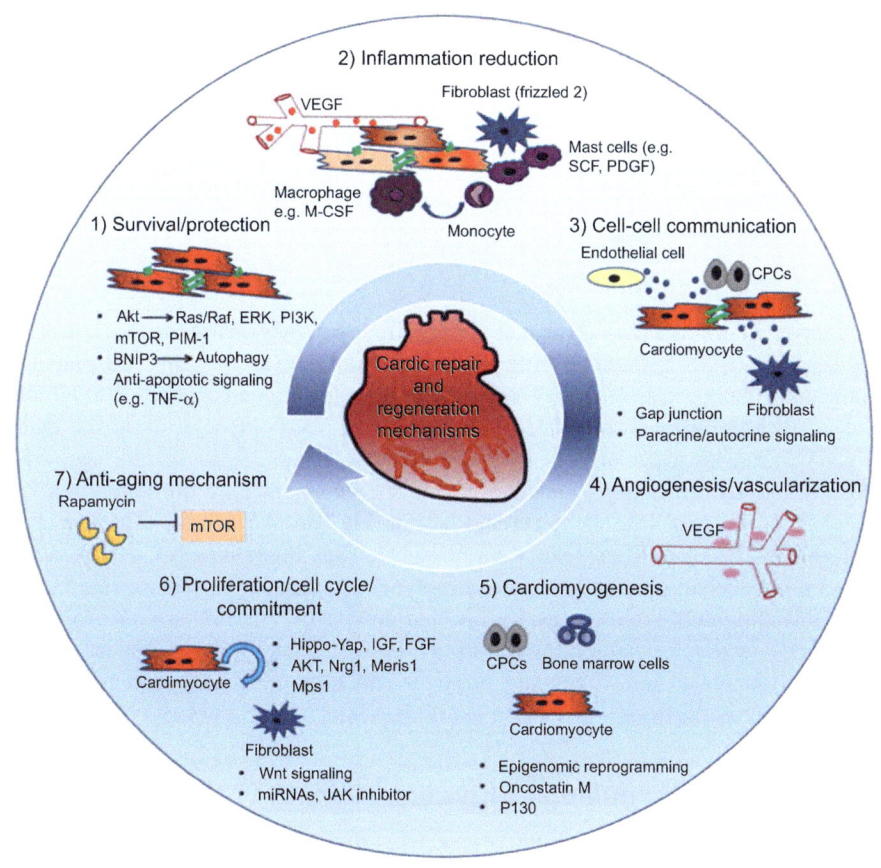

Figure 6.1 Cardiovascular repair and regeneration involve multiple mechanisms. Representative categories and selected examples of processes to enhance heart repair and regeneration are covered in this review. Mechanisms work independently on a molecular level to collectively mediate the concurrent cellular actions of survival, repair, and regenerative responses. *AKT*, protein kinase B; *BNIP3*, BCL2 interacting protein 3; *CPC*, cardiac progenitor cell; *ERK1/2*, extracellular signal-related kinase-1/2; *FGF*, fibroblast growth factor; *IGF*, insulin-like growth factor; *JAK*, janus kinase; *M-CSF*, macrophage colony-stimulating factor; *Meis-1*, Meis homobox 1; *Mps-1*, monopolar spindle 1; *mTOR*, mammalian target of rapamycin; *Nrg-1*, neuregulin 1; *PDGF*, platelet-derived growth factor; *PIM-1*, protooncogene proviral integration site for Moloney murine leukemia virus (PIM) kinase 1; *PI3K*, phosphoinositide 3-kinase; *SCF*, stem cell factor; *TNF*, tumor necrosis factor; and *VEGF*, vascular endothelial growth factor (Broughton et al., 2018).
Reproduced with permission from Wolters Kluwer Health, Inc.

6.3.4 Angiogenesis/vascularization

Functional development of blood vessels is requisite for the proper delivery of oxygen, nutrients, and metabolites to the site of injury and also for the removal of waste products generated as part of the regenerative process. Endothelial progenitor cells seem to play a key role in this process of angiogenesis, which probably involves a combination of cellular and paracrine processes (Broughton et al., 2018).

6.3.5 Cardiomyogenesis

Cardiomyocyte mitosis during the first year of life in humans contributes to 0.04% of total cardiomyocytes present at birth; this number drops to 0.009% in a 20-year-old young adult heart as assessed by phospho-histone 3 immunolabeling (Broughton et al., 2018; Mollova et al., 2013). Annual human cardiomyocyte turnover rates over a lifespan are calculated to be 1.9% in adolescents, 1% in middle age, and 0.45% in old age; by age 50, the cardiomyocytes remaining from birth are approximately 55%, whereas 45% are generated later in life (Broughton et al., 2018; Bergmann et al., 2009). There are clearly substantial limitations of endogenous regenerative mechanisms intrinsic to cardiomyocytes in the adult mammalian heart. Albeit this extremely limited capacity for cardiomyocyte renewal, an understanding of the mechanism(s) of cardiomyocyte proliferation and cell cycle arrest is necessary to develop strategies for stimulating turnover and promoting cardiac regeneration (Broughton et al., 2018).

6.3.6 Molecular regulation of proliferation, the cell cycle, and commitment

Genetic triggers for cell cycle reactivation to drive cardiomyocyte mitosis in the adult heart have been advanced as potential therapeutic targets for heart regeneration. The Hippo-YAP signaling pathway is critical for intrinsic cellular regulation of cardiomyocyte proliferation. Other candidates include Meis homobox 1 and neuregulin 1 (Broughton et al., 2018). Despite decades of targeted and focused efforts, the adult mammalian cardiomyocyte remains remarkably refractory to molecular interventions intended to promote reentry into the cell cycle and proliferation. It is also clear that the neonatal and lower vertebrate myocardium is composed of a very different molecular and cellular milieu that permits the manipulation of cardiomyocyte proliferation relative to the adult heart (Broughton et al., 2018; Crippa et al., 2016).

6.3.7 Cardiac aging

Aging is a heterogeneous process characterized by increased levels of reactive oxygen species, genomic DNA damage, epigenetic modifications, and telomere shortening. Consequences of aging pursuant to these deleterious changes include defective protein homeostasis, progressive loss of quality control processes, and accumulation of dysfunctional organelles that directly impact cardiomyocyte, fibroblast, and stem cell populations (Broughton et al., 2018). Cardiac progenitor cells are especially affected by the aging process, thus impairing reparative and regenerative potential.

6.4 Stem cell therapies

Although the existing therapies for ischemic heart disease have been successful in decreasing early mortality rates, preventing additional damage to the heart muscle, and reducing the risk of additional heart attacks, most patients are still likely to have a worse quality of life, including more frequent hospitalizations. Therefore, there is an ultimate need for a treatment strategy that improves clinical conditions by either replacing the damaged heart cells and/or improving cardiac performance. Currently available surgical treatment modalities, such as coronary bypass surgery, left ventricular assist devices, and valve replacement, prevent further deterioration of function and allow for remodeling of the diseased ventricle to some extent. The gold standard treatment for end-stage heart failure continues to be orthotropic heart transplantation. However, the severe shortage of donor organs limits the ability to extend this treatment widely. Thus, cardiac tissue regeneration with the use of stem cells, or their exosomes, may be an effective therapeutic option (Terashvili and Bosnjak, 2018; Fernandez-Aviles et al., 2017).

There are many different kinds of stem cells. The following, however, describes the current key players (Terashvili and Bosnjak, 2018):

(1) Embryonic stem cells, or ESCs, are obtained from the inner cell mass of the blastocyst that forms 3—5 days after an egg cell is fertilized by a sperm. They can give rise to every cell type in the fully formed body, but not the placenta and umbilical cord.
(2) Tissue-specific stem cells (somatic or adult stem cells) are more specialized than ESCs. Their primary roles are to maintain and repair tissue.
(3) Mesenchymal stem cells (MSCs) are multipotent stromal cells that can be isolated from the bone marrow. They are nonhematopoietic, multipotent stem cells with the capacity to differentiate into mesodermal lineage cells such as bone, cartilage, muscle, and fat cells.
(4) Induced pluripotent stem cells are cells taken from any tissue (usually skin or blood) and are genetically modified to behave like an ESC. They are pluripotent.
(5) Umbilical cord blood stem cells are collected from the umbilical cord at birth, and they can produce all of the blood cells in the body.

Apart from directly affecting regenerative efforts, stem cells can also secrete vital cytokines and growth factors. The cardioprotective panel of stem cell—derived factors is numerous and includes, but are not limited to, basic fibroblast growth factor (bFGF)/f, IL-1β, IL-10, vascular endothelial growth factor, hepatocyte growth factor, insulin-like growth factor-1, stromal cell—derived factor 1, thymosin-β4, Wnt5a, angiopoietin-1 and -2, macrophage inflammatory protein 1, erythropoietin, and PDGF. Fig. 6.2 demonstrates the various mechanisms involved in stem cell therapy.

Human induced pluripotent stem cells (hiPSCs) have considerable therapeutic potential because they are patient-specific stem cells that do not face the immunologic barrier, in contrast to ESCs. Furthermore, hiPSCs can be reprogrammed from easily

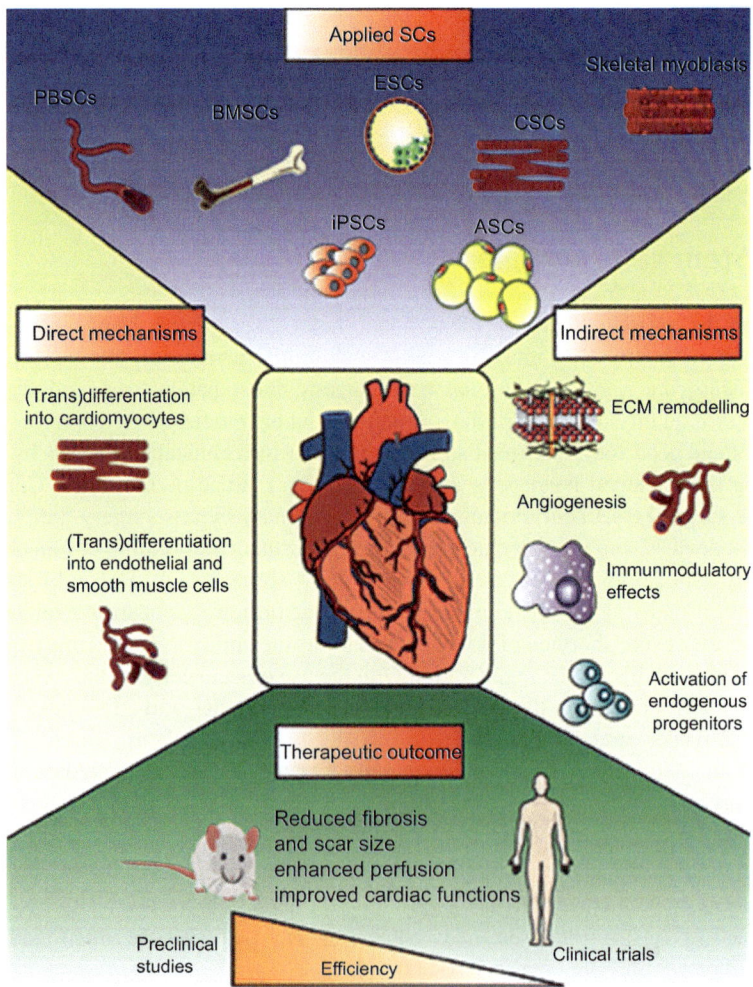

Figure 6.2 Stem cell (SC) therapy for cardiac regeneration. Multipotent adult SCs from various tissues (bone marrow [BM], peripheral blood [PB], cardiac and adipose tissue), skeletal myoblasts, and pluripotent SCs (induced pluripotent SCs [iPSCs], embryonic SC [ESCs]) have been applied to stimulate cardiac regeneration of the adult heart in preclinical and clinical studies. SCs possess the capability to differentiate into cardiomyocytes, endothelial, and smooth muscle cells, supporting regeneration. However, numerous in vitro and in vivo studies have identified indirect paracrine pathways as the main mediators of the beneficial effects of SC therapy. The release of soluble factors positively influences the remodeling of the extracellular matrix (ECM) in the injured tissue. Similarly, the formation of new blood vessels and various immunomodulatory effects are stimulated by paracrine signaling of SCs. A large number of preclinical studies demonstrated significant therapeutic outcomes with SC treatment, including reduced fibrosis and infarction size, enhanced perfusion, and improved cardiac performance. In contrast, functional data obtained in clinical trials are inconsistent with and did not confirm the pronounced benefits of SC therapy observed in various animal models. *ASC*, adipose-derived stromal/stem cell; *CSC*, cardiac stem cell.
Adapted, with permission-open access article. The final, published version of this article is available at http://www.karger.com/?doi=http://doi.org/10.1159/000492704. Muller, P., Lemcke, H., David, R. 2018. Stem cell therapy in heart diseases — cell types, mechanisms and improvement strategies. Cell. Physiol. Biochem 48 (6), 2607−2655.

accessible tissue, such as the donor's skin, fat, or blood. Their use may avoid common legal and ethical problems that arise from the use of ESCs. In addition, they can differentiate into functional cardiomyocytes and are now one of the most promising cell sources for cardiac regenerative therapy (Terashvili and Bosnjak, 2018).

There are two main techniques for cell administration: intramyocardial delivery and intracoronary injection. Direct muscle injections into the endocardium can be performed surgically or percutaneously. Coronary injections can be done antegrade or retrograde through the coronary sinus. When stem cells are injected in an antegrade coronary fashion, they have to permeate the capillary wall to get into tissue, and thus there exists concern over the possibility of microplugging and microinfarctions. Direct myocardial injections avoid these issues.

The landmark clinical trials using bone marrow–derived MSCs to treat acute myocardial infarction and heart failure are listed in Table 6.1 (Terashvili and Bosnjak, 2018). The net result of these studies has been disappointing. The Cochrane Reviews for heart attack patients have also found no benefit in primary outcomes such as morbidity, mortality, quality of life, and ejection fraction (Fisher et al., 2015, 2016).

Compared with bone marrow stem cell treatments, cord blood stem cell treatments offer the following advantages: increased tolerance of the human leukocyte antigen mismatching, decreased risk of graft-versus-host disease, and enhanced proliferative ability (Thomas, 1999). However, there is usually only a small quantity of cord cells that can be harvested. A number of groups have used such cells to treat heart failure with varying degrees of success (Bartolucci et al., 2017). But as mentioned previously, the overall inconsistent results obtained from clinical studies have been disappointing. A recent metaanalysis of cell therapy studies in heart failure and acute myocardial infarction elaborates this issue (Gyongyosi et al., 2018). There exists a gap between in vitro and in vivo studies, and the key to the "next generation" of stem cell therapeutics resides in fixing this "gap." The proposed strategies for this are shown in Fig. 6.3.

6.5 Barriers in stem cell therapy

There are several concerns with injecting stem cells into living beings. The following are a few (Fig. 6.4):

(1) Tumorigenicity: This is a salient concern. The division and differentiation of injected stem cells cannot be controlled, and there are reports of tumor formation from injecting adult stem cells (Tang et al., 2018; Heslop et al., 2015; Andrews et al., 2005). As long as live cells are used, this risk can never be eliminated. Various alternatives have been proposed to decrease the risk of tumor formation, including cell-free agents, extracellular vesicles, microvesicles, and exosomes (Tang et al., 2018).
(2) Immunogenicity: Autologous stem cells should not generate an immune reaction and should have immune tolerance. However, these are expensive to generate. Stem cells manufactured from other sources could generate immunogenicity (Tang et al., 2018; Heslop et al., 2015).
(3) Retention/engraftment: The engraftment and survival rates of implanted stem cells are currently poor. This hampers the potential long-term effects of these cells. Ischemia-reperfusion injury and secondary injury from inflammation are the primary culprits

Table 6.1 Selected list of landmark clinical trials that used mostly bone marrow—derived mesenchymal stem cells to treat acute myocardial infarction and heart failure.

Study	N	Patient Type	Cell Type/ Delivery	Follow-Up (Months)	Outcome (LVEF Change, %)
Janssens et al. (Janssens et al., 2006)	67	AMI	BMD/ intracoronary	4	NS
BOOST (Meyer et al., 2009)	60	AMI	BMD/ intracoronary	61	NS
ASTAMI (Lunde et al., 2006)	100	AMI	BMD/ intracoronary	36	NS
REGENT (Tendera et al., 2009)	200	AMI	BMD/ intracoronary	6	NS
REPAIR-AMI (Assmus et al., 2010)	204	AMI	BMD/ intracoronary	24	5
MAGIC (Menasche et al., 2008)	97	HF	SMB	6	NS
FOCUS-CCTRN (Perin et al., 2012)	92	HF	BMD/ transendocardial	6	NS
POSEIDON (Hare et al., 2012)	30	ICM	BMD/ transendocardial	12	NS
SCIPIO (Chugh et al., 2012)	33	IHF	CPCs/ intracoronary	12	8
CADUCEUS (Malliaras et al., 2014)	25	AMI	CDCs/ intracoronary	12	NS
TOPCARE-AMI (Leistner et al., 2011)	55	AMI	BMD/ intracoronary	60	11
TAC-HFT (Heldman et al., 2014)	65	ICM	BMD/ transendocardial	12	NS
MSC-HF (Mathiasen et al., 2015)	55	IHF	BMD/ transendocardial	6	5

Table 6.1 Continued

Study	N	Patient Type	Cell Type/ Delivery	Follow-Up (Months)	Outcome (LVEF Change, %)
MESAMI (Guijarro et al., 2016)	10	ICM	BMD/ transendocardial	12	6
SWISS-AMI (Surder et al., 2013)	200	AMI	BMD/ intracoronary	4	NS
REGENERATE-AMI (Choudry et al., 2016)	100	AMI	BMD/ intracoronary	12	NS
PreSERVE-AMI (Quyyumi et al., 2017)	161	AMI	BMD/ intracoronary	6	NS
MiHeart-AMI (Nicolau et al., 2018)	121	AMI	BMD/ intracoronary	6	NS
RIMECARD (Bartolucci et al., 2017)	30	HF	US-MSC/ intravenous	12	5
TIME (Traverse et al., 2018)	120	AMI	BMD/ intracoronary	24	NS

AMI, acute myocardial infarction; *BMD*, bone marrow–derived; *CDC*, cardiosphere-derived cell; *CPC*, cardiac progenitor cell; *HF*, heart failure; *ICM*, ischemic cardiomyopathy; *IHF*, ischemic heart failure; *LVEF*, left ventricular ejection fraction; *NS*, not significant; *SMB*, skeletal myoblast; *UC-MSC*, umbilical cord–derived mesenchymal stem cell.
Reproduced with permission from Elsevier.

(Tang et al., 2018; Zeng et al., 2007; Hong et al., 2013). One potential solution to this problem is to administer repeated doses, but this implies higher costs along with the increased risks associated with invasive procedures. Other proposed therapies to avoid washout include covering cells with biomaterials or using injectable hydrogels. The direct application of cells, either by a patch of stem cells on the infarcted heart or spraying cells, has also been proposed (Tang et al., 2018).

(4) Tissue targeting: A sure way of delivering cells into the myocardium is by directly injecting them into the heart. This often requires open surgery. The alternative, injecting cells into the coronary arteries, involves loss of cells in the systemic circulation. Cells injected into the coronaries also need to cross the capillary membrane and engraft into host tissue, a process that can be rendered ineffective by the poor blood supply to the infarcted myocardium.

(5) Storage/shipping stability: Off-the-shelf availability of live stem cells is limited. Thus, the effects of freezing, thawing, storage, and shipping could affect their efficacy (Tang et al., 2018).

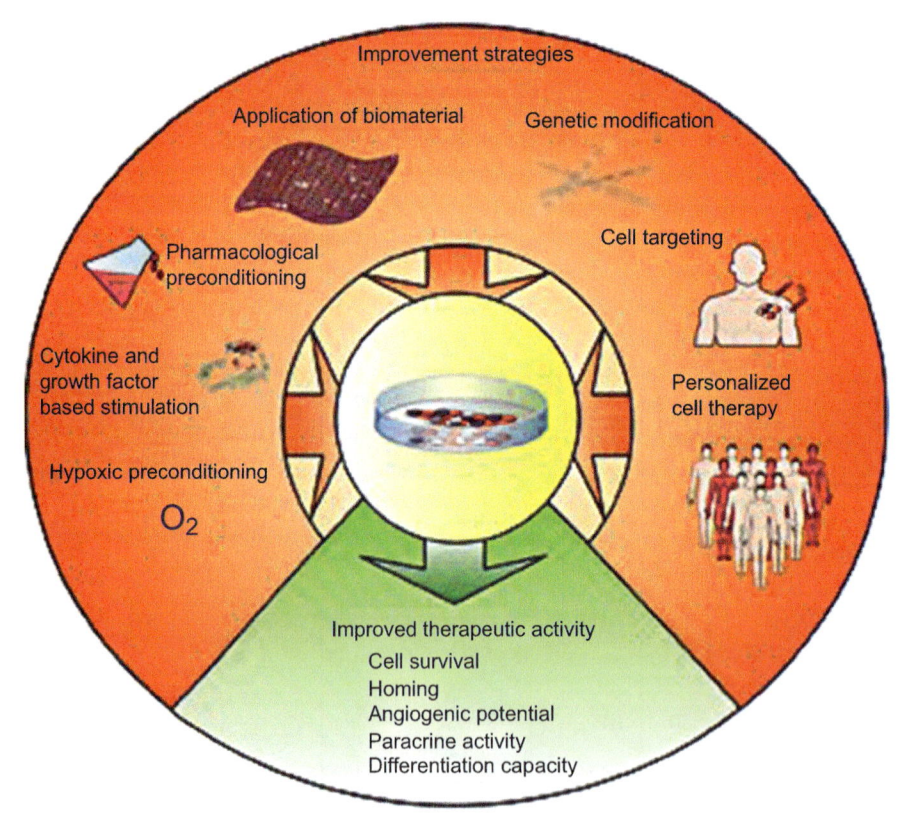

Figure 6.3 Strategies for improving stem cell (SC) efficiency in the treatment of cardiovascular diseases. Genetic modifications are based on alterations of the cellular genome (genome/DNA editing) or on posttranscriptional gene regulation (miRNA, siRNA). Nongenetic strategies include preconditioning with environmental factors such as temperature and oxygen content, pharmacological agents, and cytokines/growth factors. Biomaterials, such as cell patches or biodegradable scaffolds, can also greatly enhance therapeutic effects of transplanted SCs. Cell targeting represents another strategy to augment SC efficiency, for instance, by tagging cells with magnetic particles. In addition to modifying transplanted SCs, proper patient selection is another novel approach used to optimize SC therapy. This concept of personalized cell therapy relies on targeted therapeutic treatments that are tailored to a patient's particular pathology. These applied strategies improve cell survival and enhanced homing on transplantation into the injured heart.

Adapted, with permission-open access article. The final, published version of this article is available at http://www.karger.com/?doi=http://doi.org/10.1159/000492704.

6.6 Tissue engineering

In addition to injection therapy, stem cell tissue patches, both cardiogenic and noncardiogenic, have been used in the treatment of myocardial infarction. Animal studies have shown promise, and data are emerging for human studies as well. Implantation of engineered myoblast sheets over an infarction site have yielded improved neovascularization, attenuated left ventricular dilation, decreased fibrosis, improved

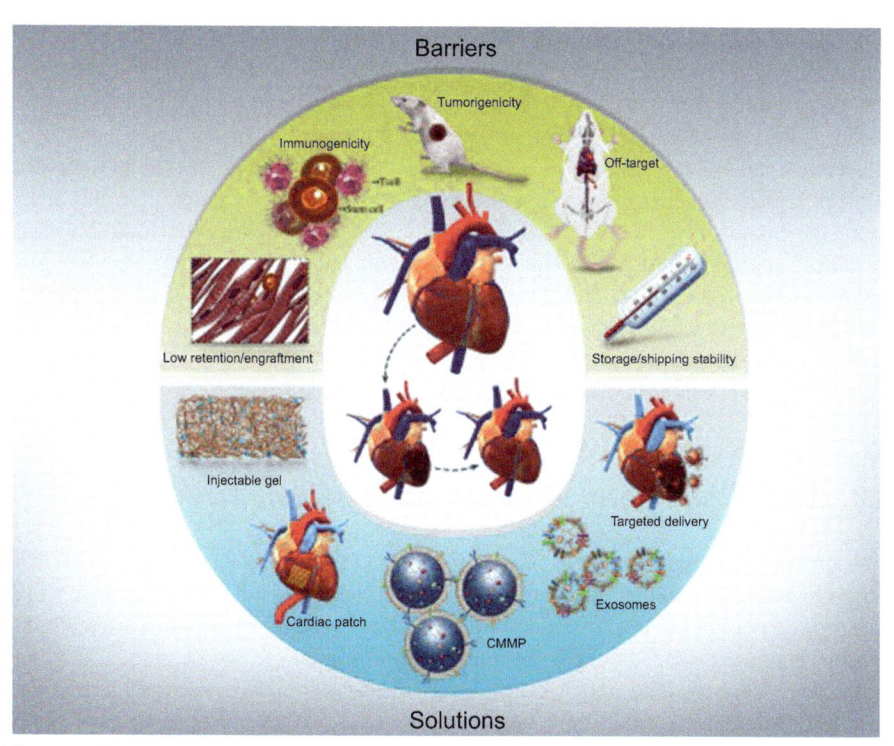

Figure 6.4 Challenges to the field of cardiac cell therapy and new emerging solutions. Open access article, reproduced under Creative Commons License. Tang JN, Cores J, Huang K, Cui XL, Luo L, Zhang JY, et al. 2008. Concise review: is cardiac cell therapy dead? Embarrassing trial outcomes and new directions for the future. Stem Cells Transl. Med. 7 (4), 354–359.

fractional shortening, and prolonged animal survival compared with the delivery of the same number of myoblasts by cell injection (Terashvili and Bosnjak, 2018; Bursac, 2009).

In a recently published study, Nummi et al. reported that during on-pump coronary artery bypass graft (CABG) surgery, part of the right atrial appendage can be excised on insertion of the right atrial cannula of the heart-lung machine and that the removed tissue can be easily cut into micrografts for transplantation. Appendage tissue is harvested during cannulation of the right atrium, and therefore no additional procedure is needed. Cell isolation and matrix preparation for transplantation are done simultaneously with the bypass surgery in the operating room, and thus the perfusion and aortic clamp times are not increased. After the anastomoses are completed, the atrial appendage sheet is placed on the myocardium with 3–4 sutures, thereby allowing the myocardium to contract without interference. The authors concluded that atrial appendage—derived cell therapy administered during CABG surgery will have an effect on patient treatment in the future (Nummi et al., 2017).

6.7 Cellular reprogramming

The adult human heart lacks sufficient ability to replenish damaged cardiac muscle because the rate of cardiomyocyte renewal activity is less than 1% per year. The mechanical and electrical engraftment of injected cardiomyocytes is largely not feasible at the scale that would be necessary for cardiac improvement. On the other hand, the human heart contains a large population of fibroblasts that may be used for direct reprogramming. As such, direct fibroblast reprogramming in vivo has emerged as a possible approach to cardiac regeneration. However, even though we have a much better understanding of the various molecular mechanisms underlying fibroblast reprogramming, this technology still lacks sufficient efficacy for use in human cells (Terashvili and Bosnjak, 2018).

6.8 Stem cell–derived exosomes and small vesicles

The survival of transplanted stem cells is dismal, and the beneficial effects of stem cell therapies are not due to the differentiation of new cardiomyocytes, but rather arise from a temporary pool of secreted exosomal growth factors. Therefore, despite early cellular demise, treatment with stem cells has a number of limited cardiac benefits, including decreased cardiomyocyte apoptosis, reduced fibrosis, enhanced neovascularization, and improved left ventricular ejection fraction. It is for these reasons that exosome therapy recapitulates the benefits of stem cell therapy (Kishore and Khan, 2017), and many studies have shown that the activation of cardioprotective pathways involved in stem cell therapy can be reproduced by the injection of exosomes produced by stem cells (Davidson et al., 2017). The lack of tumor-forming potential is an additional benefit of using exosomes for cardioprotection and regeneration. However, the mechanisms underlying stem cell or hiPSC-derived exosome therapy are still unclear. Despite this, numerous scientific investigations have identified recent applications of exosomes in the development of molecular diagnostics, drug delivery systems, and therapeutic agents (Terashvili and Bosnjak, 2018; Kishore and Khan, 2017; Davidson et al., 2017).

6.9 Hydrogels

Recently, the concept of injecting hydrogels into scar tissue after a heart attack, or applying patches onto the surface of ventricular myocardium, has been tested. The basic premise of injecting these hydrogels is to limit the extension and expansion of the ischemic area of the heart. They are also useful as vehicles for the delivery of cells and drugs. The optimal material should be easy to manufacture, inexpensive, biocompatible, biodegradable, nontoxic, nonimmunogenic, and should possess mechanical properties similar to that of cardiac tissue. The materials that have been tested in clinical trials include collagen, gelatin, hyaluronic acid, fibrin, alginate, agarose, chitosan, keratin, Matrigel, decellularized ECM, polyacrylic acid derivatives, polyethylene glycol, polyethylene oxide, polyvinyl alcohol, and polyphosphazene (Di Franco et al., 2018). While animal data has been encouraging, human studies have been limited to rare clinical trials. A detailed report of all the available hydrogels and patches has been reviewed by Pena et al. (Pena et al., 2018).

6.10 Cardiac regeneration in children

The incidence of heart failure in the pediatric population is low, but the mortality is extremely high. About 2.5 million children are born every year with congenital heart disease, and about 15%−25% of them develop heart failure (Pavo and Michel-Behnke, 2017). Neonates and children seem to have a higher regenerative capacity compared with adults for reasons discussed before. Some of the relevant clinical reports in pediatric cardiac regeneration are shown in Table 6.2 (Pavo and Michel-Behnke, 2017).

As noted from Table 6.2, transient improvements in heart function in children have been mostly limited to anecdotal experience. However, there are several ongoing clinical trials in patients with cardiomyopathy and hypoplastic left heart syndrome (Pavo and Michel-Behnke, 2017).

6.11 Valves

In the United States, valvular heart disease (VHD) affects about 5 million adults and 44,000 children annually (Cheung et al., 2015; Go et al., 2013). Cardiac regeneration has a lot to offer in the treatment of VHD, especially with regard to valve substitutes. The ideal valve substitute should be able to grow with the patient, last a lifetime, be infection-resistant, require no anticoagulation, be resistant to degeneration, be compatible with tissue and blood components (thereby not eliciting any inflammatory reactions), have minimal transvalvular gradients with no valvular regurgitation, and be affordable. However, none of the currently available synthetic or biologic substitutes in the market have all of these properties. For instance, mechanical heart valves are very durable, but they require anticoagulation to reduce the risk of thrombosis and thromboembolism. These limitations almost invariably subject children and adults to several operations over a lifetime.

Creating a tissue-engineered cardiac valve involves developing tissue composed of living cells that will, at some point, form and remodel their own ECM and thus provide durability and growth potential. The biological and engineering challenge lies in providing structural organization, support, and mechanical integrity for these cells until they are capable of forming their own mature ECM. The variables involved in the creation of a living "tissue-engineered" valve include choice or manipulation of cell phenotype, induction of appropriate ECM formation by mechanical and/or biochemical signals, and provision of structural integrity and cellular organization

Table 6.2 Published clinical studies with pediatric cell-based cardiac regeneration.

Ref.	Study Type	Diagnosis	N	Mean Age (Months)	Sex	Stem Cell Type	Cell Application	Follow-Up	Results
Lacis and Erglis (2011)	Case report	Dilated CMP	1	3.5	F	BM-MNC	IM	4 months	Increase in LVEF from 20% to 41%
Rupp et al. (2012)	Case report	Dilated CMP	9	4 months – 16 years	NA	BM-MNC	IC	1–52 months	Three Patients HTX, one Death, Others Improved
Ishigami et al. (2015)	Controlled Study	HLHS	7 Treated, 7 Controls	<6 years	NA	CDC	IC	18 months	Increase in RVEF from 46.9% to 52.1% in treated patients
Rupp et al. (2010)	Case report	HLHS	1	11	M	BMC	IC	14 months	RVEF from 22% to 44%
Rupp et al. (2009)	Case report	Dilated CMP	1	24	M	BMC	IC	6 months	EF from 24% to 45%, BNP and NYHA Decreased
de Lezo et al. (2009)	Case report	Post-AMI	1	7	NA	BM-MNC	IC	14 months	LVEF from 20% to 43%

Olgunturk et al. (2010)	Case report	Dilated CMP	2	6 and 9 years	M, F	PBSC After GCSF	IC	8 Weeks, 6 months	Patient LVEF from 16% to 39%; Patient LVEF from 34% to 54%
Limsuwan et al. (2010)	Case report	Post-AMI	1	9 years	F	BMV After GCSF	IC	3 months	LVEF form 30% to 47%
Zeinaloo et al. (2011)	Case report	Dilated CMP	1	11 years	M	BM-MSC	IC	12 months	LVEF from 20% to 42%
Rivas et al. (2011)	Case report	Dilated CMP	2	3 and 4	M	PBSC After GCSF	IC	4 months	EF from <30% to >40%
Bergmane et al. (2013)	Case report	Dilated CMP	7	4 months – 17 years	NA	BMC	IM	12 months	Six Patients Controlled, LVEF from 33.5% to 54%
Burkhart et al. (2015)	Case report	HLHS	1	3	NA	Umbilical Cord Blood-Derived Cells	IM	3 months	EF Increased to 45%

BMC, Bone marrow cells; *BM-MNC*, Bone marrow mononuclear cell; *BNP*, Brain natriuretic peptide; *CDC*, Cardiosphere-derived cells; *CMP*, Cardiomyopathy; *EF*, Ejection fraction; *F*, Female; *FUP*, Follow-up; *GCSF*, Granulocyte colony-stimulating factor; *HLHS*, Hypoplastic left heart syndrome; *HTX*, Heart transplantation; *IC*, Intracoronary; *IM*, Intramyocardial; *LV*, Left ventricle; *M*, Male; *NA*, Data not available; *NYHA*, New York Heart Association Classification; *PBSC*, Peripheral blood stem cell; *RV*, Right ventricle. Pavo and Michel-Behnke, 2017.

by the scaffold until new "native" ECM formation occurs. This is discussed further in the next section.

Tissue-engineered heart valves will one day revolutionize the field of cardiac surgery with advancements in growth and biological integration. Although this technology is not currently commercially available, there have been promising results, reviewed in detail by Cheung et al. (Cheung et al., 2015).

6.12 Biointegration

Biointegration refers to the process of coalescing the mechanical and functional properties of a given biomaterial, such as an implant, with living tissue. This integration, or biologization, provides a scaffold on which native tissue is allowed to regenerate, thus essentially converting biomaterials into a dynamic component of living beings that react and adapt to the environment (Amling et al., 2006). This has been a significant paradigm shift in tissue engineering and has seen tremendous progress in the past few decades. With regard to cardiovascular disease, decellularized matrices have come to the forefront. These are biocompatible, three-dimensional scaffolds without a cellular component, but with an intact ECM, that have the capacity to promote endothelial cell migration and proliferation, along with recruitment of vascular smooth muscle cells and pericytes (Gilbert et al., 2006). Although still in its infancy, decellularized matrices have seen wide applications in heart, valve, and vessel engineering with implications for the future treatment of heart failure and valvular and vascular diseases (Moroni and Mirabella, 2014).Before biointegration, approaches to heart regeneration had undergone a number of iterations. For instance, Zimmermann et al. described a method of creating a patch consisting of scaffold substrates seeded with cardiac muscle cells in vitro (Zimmermann et al., 2006). To combat the issues of inadequate cell migration into the scaffolds and to minimize the inflammatory response to scaffold degradation, Shimuzu and colleagues utilized layered cardiomyocyte cell sheets that precluded the need for an artificial scaffold (Shimizu et al., 2003). However, the problem of creating a geometrical structure capable of supporting the metabolic demands of cardiac muscle cells at depths greater than approximately 100 μm from the surface persisted. A method of constructing whole-heart scaffolds was needed.

In 2008, Ott et al. generated bioartificial hearts by perfusing rat hearts with sodium dodecyl sulfate (SDS), thus effectively decellularizing the cadaveric hearts but simultaneously maintaining the underlying ECM. These were then recellularized with cardiac and endothelial cells and maintained by coronary perfusion for up to 4 weeks; by day 8, pump function equal to ~2% of adult heart function could be generated (Ott et al., 2008). Since then, various groups have engineered techniques of decellularizing—recellularizing cardiac tissue from various substrates, including human pericardium from cadaveric donors, porcine hearts, and human ventricular myocardium (Moroni and Mirabella, 2014). Common decellularization protocols involve the use of SDS, Triton X-100 detergents, and trypsin/

ethylenediaminetetraacetic acid. Results have been variable, but most groups have reported preservation of glycosaminoglycan content (Mirsadraee et al., 2006; Wainwright et al., 2010; Oberwallner et al., 2014). Thus, it may now be possible to engineer whole hearts and/or cardiac tissue parts using the techniques and principles of biointegration, but there are drawbacks. For one, donor pathology often limits the quality of the final product. Furthermore, whole-heart studies are currently limited by the constraints of heterotopic transplantation models (Lop et al., 2017).

Similar concepts have also been applied to the creation of biointegrated heart valves. Surgical valve replacement/repair remains the mainstay of therapy for most severe forms of VHD. However, current prosthetic valves have a number of drawbacks, including hemorrhagic and embolic events for mechanical valves and structural valve degeneration for bioprosthetic ones (Chikwe et al., 2010). Here, the future of valvular tissue engineering is headed toward creating a decellularized scaffold seeded with suitable cell types that is durable, nonimmunogenic, and mimics the mechanical properties of native heart valves. The protocols used are similar to those utilized in cardiac regeneration; common substrates include porcine aortic and pulmonary valves and aortic homograft leaflets. Decellularization can be achieved using trypsin, Triton X-100, or deoxycholic acid, and recellularization is often accomplished by seeding with cardiac mesenchymal stromal cells (Dainese et al., 2012). However, further validation is needed as most human studies have shown conflicting results regarding the outcomes of such biointegrated valves (Dohmen, 2012).

Finally, vascular biology has also benefited from biointegration. A significant number of arterial bypass operations performed in the United States involve placement of synthetic grafts that often fail due to thrombosis (Klinkert et al., 2004). Decellularized matrices can be very advantageous here as they are nonimmunogenic and provide a functional framework for endothelial cell migration and growth (Moroni and Mirabella, 2014). As with bioartificial hearts and biointegrated valves, the most common decellularization protocols involve use of Triton X-100/sodium deoxycholate and SDS; porcine aorta and carotid arteries and human umbilical arteries/veins have been used as substrates (Pellegata et al., 2012). Recently, two single-arm phase 2 trials were conducted to evaluate the efficacy of a biointegrated acellular vessel as an alternative to polytetrafluoroethylene arteriovenous grafts in patients with end-stage renal disease. Sixty patients were implanted, 63% (95% CI, 47–72) and 28% (95% CI 17–40) of who had primary patency at 6 and 12 months, respectively. Although promising, larger, randomized-controlled studies are needed to determine if such implants can be used as conduits for dialysis access (Lawson et al., 2016). Moving forward, the concepts and practices of biointegration have to be refined before such technology can be used to treat cardiovascular disease. For one, decellularization and recellularization strategies have to be optimized so as to generate efficient cell lysis and simultaneously promote sustainable cell migration, growth, and differentiation. Part of this process mandates a more complete understanding of recellularization and engraftment at the molecular and cellular levels. Furthermore, the immune response to ECM-based scaffolds has yet to be completely elucidated (Brown et al., 2012). Despite these current shortcomings, biointegration will invariably be an integral part of the future of cardiac tissue engineering and regeneration.

3D and 4D bioprinting of the myocardium are the newer methods of cardiac tissue engineering. 3D bioprinting refers to the creation of a physical object in 3 dimensions by the deposition of material in successive layers. This may or may not use a baseline scaffold as a backbone. The incorporation of the dimension of time into this concept is 4D bioprinting. Various cell lines can be printed over scaffolds such as alginate, collagen, hylarounic acid, gelatin, or decellularized ECM to create complex structures, such as sheets and tubes of contractile tissue. This concept can also be reproduced without using an underlying scaffold, thus eliminating the disadvantages of a scaffold (cost, immunogenicity, byproduct creation, and elimination produced by disintegration of the scaffold). The hurdles to bioprinting currently remain in the establishment of thickness and maintaining tissue viability. This barrier can be overcome by vascularization of the tissue. The concept of introducing micro- and microvasculature into a bioprinted object has several limitations and challenges. 4D bioprinting includes instances in which objects can change their shape based on the presence of external stimuli. This would be a game changer (Ong et al., 2018).

6.13 Conclusion

The heart is a magnificent organ! A complete understanding of its structural and functional properties holds the key to the future development of cardiac regenerative therapies. Despite the current limitations in transferring promising preclinical results to the bedside, tremendous knowledge has been gained over the past few decades. We have come a long way from the days of mice studies where injection of stem cells induced microtubule formation and are now in the era of clinical trials where patients receiving stem cell therapy have shown improved myocardial function. Although these effects may be more from a paracrine effect rather than regenerating cardiomyocytes, numerous novel avenues for therapeutic intervention have opened up as the techniques and practices of biointegration mature. The future of this field is indeed very exciting.

References

Amling, M., Schilling, A.F., Pogoda, P., Priemel, M., Rueger, J.M., 2006. Biomaterials and bone remodeling: the physiologic process required for biologization of Bone Substitutes. Eur. J. Trauma Emerg. Surg. 32 (2), 102–106.

Andrews, P.W., Matin, M.M., Bahrami, A.R., Damjanov, I., Gokhale, P., Draper, J.S., 2005. Embryonic stem (ES) cells and embryonal carcinoma (EC) cells: opposite sides of the same coin. Biochem. Soc. Trans 33 (Pt 6), 1526–1530.

Anversa, P., Leri, A., Kajstura, J., 2006. Cardiac regeneration. J. Am. Coll. Cardiol 47 (9), 1769–1776.

Anversa, P.L.A., 2005. Myocardial regeneration. In: Zipes, D.P., Libby, P., Bonow, R.O., Braunwald, E. (Eds.), Braunwald's Heart Disease, pp. 1911–1924. Philadelphia, PA.

Assmus, B., Rolf, A., Erbs, S., Elsasser, A., Haberbosch, W., Hambrecht, R., et al., 2010. Clinical outcome 2 years after intracoronary administration of bone marrow-derived progenitor cells in acute myocardial infarction. Circ. Heart Fail 3 (1), 89–96.

Bartolucci, J., Verdugo, F.J., Gonzalez, P.L., Larrea, R.E., Abarzua, E., Goset, C., et al., 2017. Safety and efficacy of the intravenous infusion of umbilical cord mesenchymal stem cells in patients with heart failure: a phase 1/2 randomized controlled trial (RIMECARD trial [randomized clinical trial of intravenous infusion umbilical cord mesenchymal stem cells on cardiopathy]). Circ. Res 121 (10), 1192–1204.

Behfar, A., Zingman, L.V., Hodgson, D.M., Rauzier, J.M., Kane, G.C., Terzic, A., et al., 2002. Stem cell differentiation requires a paracrine pathway in the heart. FASEB J 16 (12), 1558–1566.

Beltrami, A.P., Urbanek, K., Kajstura, J., Yan, S.M., Finato, N., Bussani, R., et al., 2001. Evidence that human cardiac myocytes divide after myocardial infarction. N. Engl. J. Med 344 (23), 1750–1757.

Benjamin, E.J., Blaha, M.J., Chiuve, S.E., Cushman, M., Das, S.R., Deo, R., et al., 2017. Heart disease and stroke statistics-2017 update: a report from the American heart association. Circulation 135 (10), e146–e603.

Bergmane, I., Lacis, A., Lubaua, I., Jakobsons, E., Erglis, A., 2013. Follow-up of the patients after stem cell transplantation for pediatric dilated cardiomyopathy. Pediatr. Transplant 17 (3), 266–270.

Bergmann, O., Bhardwaj, R.D., Bernard, S., Zdunek, S., Barnabe-Heider, F., Walsh, S., et al., 2009. Evidence for cardiomyocyte renewal in humans. Science 324 (5923), 98–102.

Broughton, K.M., Wang, B.J., Firouzi, F., Khalafalla, F., Dimmeler, S., Fernandez-Aviles, F., et al., 2018. Mechanisms of cardiac repair and regeneration. Circ. Res. 122 (8), 1151–1163.

Brown, B.N., Ratner, B.D., Goodman, S.B., Amar, S., Badylak, S.F., 2012. Macrophage polarization: an opportunity for improved outcomes in biomaterials and regenerative medicine. Biomaterials 33 (15), 3792–3802.

Burkhart, H.M., Qureshi, M.Y., Peral, S.C., O'Leary, P.W., Olson, T.M., Cetta, F., et al., 2015. Regenerative therapy for hypoplastic left heart syndrome: first report of intraoperative intramyocardial injection of autologous umbilical-cord blood-derived cells. J. Thorac. Cardiovasc. Surg. 149 (3), e35–e37.

Bursac, N., 2009. Cardiac tissue engineering using stem cells. IEEE Eng. Med. Biol. Mag. 28 (2), 80, 2, 4-6, 8-9.

Cheung, D.Y., Duan, B., Butcher, J.T., 2015. Current progress in tissue engineering of heart valves: multiscale problems, multiscale solutions. Expert Opin. Biol. Ther 15 (8), 1155–1172.

Chikwe, J., Filsoufi, F., Carpentier, A.F., 2010. Prosthetic valve selection for middle-aged patients with aortic stenosis. Nat. Rev. Cardiol 7 (12), 711–719.

Choudry, F., Hamshere, S., Saunders, N., Veerapen, J., Bavnbek, K., Knight, C., et al., 2016. A randomized double-blind control study of early intra-coronary autologous bone marrow cell infusion in acute myocardial infarction: the REGENERATE-AMI clinical trialdagger. Eur. Heart J 37 (3), 256–263.

Chugh, A.R., Beache, G.M., Loughran, J.H., Mewton, N., Elmore, J.B., Kajstura, J., et al., 2012. Administration of cardiac stem cells in patients with ischemic cardiomyopathy: the SCIPIO trial: surgical aspects and interim analysis of myocardial function and viability by magnetic resonance. Circulation 126 (11 Suppl. 1), S54–64.

Crippa, S., Nemir, M., Ounzain, S., Ibberson, M., Berthonneche, C., Sarre, A., et al., 2016. Comparative transcriptome profiling of the injured zebrafish and mouse hearts identifies miRNA-dependent repair pathways. Cardiovasc. Res 110 (1), 73−84.

Dainese, L., Guarino, A., Burba, I., Esposito, G., Pompilio, G., Polvani, G., Rossini, A., 2012. Heart valve engineering: decellularized aortic homograft seeded with human cardiac stromal cells. J. Heart Valve Dis 21, 125−134.

Davidson, S.M., Takov, K., Yellon, D.M., 2017. Exosomes and cardiovascular protection. Cardiovasc. Drugs Ther 31 (1), 77−86.

de Lezo, J.S., Pan, M., Herrera, C., 2009. Combined percutaneous revascularization and cell therapy after failed repair of anomalous origin of left coronary artery from pulmonary artery. Cathet. Cardiovasc. Interv 73 (6), 833−837.

Di Franco, S., Amarelli, C., Montalto, A., Loforte, A., Musumeci, F., 2018. Biomaterials and heart recovery: cardiac repair, regeneration and healing in the MCS era: a state of the "heart". J. Thorac. Dis. 10 (Suppl. 20), S2346−S2362.

Dohmen, P.M., 2012. Clinical results of implanted tissue engineered heart valves. HSR Proc. Intensive Care Cardiovasc. Anesth. 4, 225−231.

Dong, Y., Undyala, V.V., Gottlieb, R.A., Mentzer Jr., R.M., Przyklenk, K., 2010. Autophagy: definition, molecular machinery, and potential role in myocardial ischemia-reperfusion injury. J. Cardiovasc. Pharmacol. Ther 15 (3), 220−230.

Fernandez-Aviles, F., Sanz-Ruiz, R., Climent, A.M., Badimon, L., Bolli, R., Charron, D., et al., 2017. Global position paper on cardiovascular regenerative medicine. Eur. Heart J 38 (33), 2532−2546.

Fisher, S.A., Zhang, H., Doree, C., Mathur, A., Martin-Rendon, E., 2015. Stem cell treatment for acute myocardial infarction. Cochrane Database Syst. Rev 9, CD006536.

Fisher, S.A., Doree, C., Mathur, A., Taggart, D.P., Martin-Rendon, E., 2016. Stem cell therapy for chronic ischaemic heart disease and congestive heart failure. Cochrane Database Syst. Rev 12, CD007888.

Gilbert, T.W., Sellaro, T.L., Badylak, S.F., 2006. Decellularization of tissues and organs. Biomaterials 27, 3675−3683.

Go, A.S., Mozaffarian, D., Roger, V.L., Benjamin, E.J., Berry, J.D., Borden, W.B., et al., 2013. Heart disease and stroke statistics—2013 update: a report from the American Heart Association. Circulation 127 (1), e6−e245.

Guijarro, D., Lebrin, M., Lairez, O., Bourin, P., Piriou, N., Pozzo, J., et al., 2016. Intramyocardial transplantation of mesenchymal stromal cells for chronic myocardial ischemia and impaired left ventricular function: results of the MESAMI 1 pilot trial. Int. J. Cardiol 209, 258−265.

Gyongyosi, M., Haller, P.M., Blake, D.J., Martin Rendon, E., 2018. Meta-analysis of cell therapy studies in heart failure and acute myocardial infarction. Circ. Res 123 (2), 301−308.

Hare, J.M., Fishman, J.E., Gerstenblith, G., DiFede Velazquez, D.L., Zambrano, J.P., Suncion, V.Y., et al., 2012. Comparison of allogeneic vs autologous bone marrow-derived mesenchymal stem cells delivered by transendocardial injection in patients with ischemic cardiomyopathy: the POSEIDON randomized trial. JAMA 308 (22), 2369−2379.

Heldman, A.W., DiFede, D.L., Fishman, J.E., Zambrano, J.P., Trachtenberg, B.H., Karantalis, V., et al., 2014. Transendocardial mesenchymal stem cells and mononuclear bone marrow cells for ischemic cardiomyopathy: the TAC-HFT randomized trial. J. Am. Med. Assoc 311 (1), 62−73.

Heslop, J.A., Hammond, T.G., Santeramo, I., Tort Piella, A., Hopp, I., Zhou, J., et al., 2015. Concise review: workshop review: understanding and assessing the risks of stem cell-based therapies. Stem Cells Transl. Med. 4 (4), 389−400.

Hong, K.U., Li, Q.H., Guo, Y., Patton, N.S., Moktar, A., Bhatnagar, A., et al., 2013. A highly sensitive and accurate method to quantify absolute numbers of c-kit+ cardiac stem cells following transplantation in mice. Basic Res. Cardiol. 108 (3), 346.

Ishigami, S., Ohtsuki, S., Tarui, S., Ousaka, D., Eitoku, T., Kondo, M., et al., 2015. Intracoronary autologous cardiac progenitor cell transfer in patients with hypoplastic left heart syndrome: the TICAP prospective phase 1 controlled trial. Circ. Res. 116 (4), 653−664.

Janssens, S., Dubois, C., Bogaert, J., Theunissen, K., Deroose, C., Desmet, W., et al., 2006. Autologous bone marrow-derived stem-cell transfer in patients with ST-segment elevation myocardial infarction: double-blind, randomised controlled trial. Lancet 367 (9505), 113−121.

Kajstura, J., Rota, M., Whang, B., Cascapera, S., Hosoda, T., Bearzi, C., et al., 2005. Bone marrow cells differentiate in cardiac cell lineages after infarction independently of cell fusion. Circ. Res. 96 (1), 127−137.

Kang, P.M., Izumo, S., 2000. Apoptosis and heart failure: a critical review of the literature. Circ. Res. 86 (11), 1107−1113.

Kawada, H., Fujita, J., Kinjo, K., Matsuzaki, Y., Tsuma, M., Miyatake, H., et al., 2004. Non-hematopoietic mesenchymal stem cells can be mobilized and differentiate into cardiomyocytes after myocardial infarction. Blood 104 (12), 3581−3587.

Kishore, R., Khan, M., 2017. Cardiac cell-derived exosomes: changing face of regenerative biology. Eur. Heart J. 38 (3), 212−215.

Klinkert, P., Post, P.N., Breslau, P.J., van Bockel, J.H., April 2004. Eur. J. Vasc. Endovasc. Surg. 27 (4), 357−362.

Kukielka, G.L., Smith, C.W., Manning, A.M., Youker, K.A., Michael, L.H., Entman, M.L., 1995. Induction of interleukin-6 synthesis in the myocardium. Potential role in post-reperfusion inflammatory injury. Circulation 92 (7), 1866−1875.

Lacis, A., Erglis, A., 2011. Intramyocardial administration of autologous bone marrow mononuclear cells in a critically ill child with dilated cardiomyopathy. Cardiol. Young 21 (1), 110−112.

Lawson, J.H., Glickman, M.H., Ilzecki, M., et al., 2016. Bioengineered human acellular vessels for dialysis access in patients with end-stage renal disease: two phase 2 single-arm trials. Lancet 387, 2026−2034.

Leistner, D.M., Fischer-Rasokat, U., Honold, J., Seeger, F.H., Schachinger, V., Lehmann, R., et al., 2011. Transplantation of progenitor cells and regeneration enhancement in acute myocardial infarction (TOPCARE-AMI): final 5-year results suggest long-term safety and efficacy. Clin. Res. Cardiol. 100 (10), 925−934.

Limsuwan, A., Pienvichit, P., Limpijankit, T., Khowsathit, P., Hongeng, S., Pornkul, R., et al., 2010. Transcoronary bone marrow-derived progenitor cells in a child with myocardial infarction: first pediatric experience. Clin. Cardiol. 33 (8), E7−12.

Linke, A., Muller, P., Nurzynska, D., Casarsa, C., Torella, D., Nascimbene, A., et al., 2005. Stem cells in the dog heart are self-renewing, clonogenic, and multipotent and regenerate infarcted myocardium, improving cardiac function. Proc. Natl. Acad. Sci. U. S. A 102 (25), 8966−8971.

Lop, L., Sasso, E.D., Menabo, R., Lisa, F.D., Gerosa, G., 2017. The rapidly evolving concept of whole heart engineering. Stem Cells Int. 8920940.

Lunde, K., Solheim, S., Aakhus, S., Arnesen, H., Abdelnoor, M., Egeland, T., et al., 2006. Intracoronary injection of mononuclear bone marrow cells in acute myocardial infarction. N. Engl. J. Med. 355 (12), 1199−1209.

MacLellan, W.R., Schneider, M.D., 2000. Genetic dissection of cardiac growth control pathways. Annu. Rev. Physiol. 62, 289−319.

Malliaras, K., Makkar, R.R., Smith, R.R., Cheng, K., Wu, E., Bonow, R.O., et al., 2014. Intracoronary cardiosphere-derived cells after myocardial infarction: evidence of therapeutic regeneration in the final 1-year results of the CADUCEUS trial (CArdiosphere-Derived aUtologous stem CElls to reverse ventricUlar dySfunction). J. Am. Coll. Cardiol. 63 (2), 110−122.

Mathiasen, A.B., Qayyum, A.A., Jorgensen, E., Helqvist, S., Fischer-Nielsen, A., Kofoed, K.F., et al., 2015. Bone marrow-derived mesenchymal stromal cell treatment in patients with severe ischaemic heart failure: a randomized placebo-controlled trial (MSC-HF trial). Eur. Heart J. 36 (27), 1744−1753.

Matsui, T., Tao, J., del Monte, F., Lee, K.H., Li, L., Picard, M., et al., 2001. Akt activation preserves cardiac function and prevents injury after transient cardiac ischemia in vivo. Circulation 104 (3), 330−335.

Menasche, P., Alfieri, O., Janssens, S., McKenna, W., Reichenspurner, H., Trinquart, L., et al., 2008. The Myoblast Autologous Grafting in Ischemic Cardiomyopathy (MAGIC) trial: first randomized placebo-controlled study of myoblast transplantation. Circulation 117 (9), 1189−1200.

Meyer, G.P., Wollert, K.C., Lotz, J., Pirr, J., Rager, U., Lippolt, P., et al., 2009. Intracoronary bone marrow cell transfer after myocardial infarction: 5-year follow-up from the randomized-controlled BOOST trial. Eur. Heart J. 30 (24), 2978−2984.

Mirsadraee, S., Wilcox, H.E., Korossis, S.A., Kearney, J.N., Watterson, K.G., Fisher, J., Ingham, E., 2006. Development and characterization of an acellular human pericardial matrix for tissue engineering. Tissue Eng 12, 763−773.

Mohsin, S., Khan, M., Toko, H., Bailey, B., Cottage, C.T., Wallach, K., et al., 2012. Human cardiac progenitor cells engineered with Pim-I kinase enhance myocardial repair. J. Am. Coll. Cardiol. 60 (14), 1278−1287.

Mollova, M., Bersell, K., Walsh, S., Savla, J., Das, L.T., Park, S.Y., et al., 2013. Cardiomyocyte proliferation contributes to heart growth in young humans. Proc. Natl. Acad. Sci. USA 110 (4), 1446−1451.

Moroni, F., Mirabella, T., 2014. Decellularized matrices for cardiovascular tissue engineering. Am. J. Stem Cells 3 (1), 1−20.

Muller, P., Lemcke, H., David, R., 2018. Stem cell therapy in heart diseases − cell types, mechanisms and improvement strategies. Cell. Physiol. Biochem. 48 (6), 2607−2655.

Nicolau, J.C., Furtado, R.H.M., Silva, S.A., Rochitte, C.E., Rassi Jr., A., Moraes Jr., J., et al., 2018. Stem-cell therapy in ST-segment elevation myocardial infarction with reduced ejection fraction: a multicenter, double-blind randomized trial. Clin. Cardiol 41 (3), 392−399.

Nummi, A., Nieminen, T., Patila, T., Lampinen, M., Lehtinen, M.L., Kivisto, S., et al., 2017. Epicardial delivery of autologous atrial appendage micrografts during coronary artery bypass surgery-safety and feasibility study. Pilot Feasibility Stud. 3, 74.

Oberwallner, B., Brodarac, A., Choi, Y.H., Saric, T., Anic, P., Morawietz, L., Stamm, C., 2014. Preparation of cardiac extracellular matrix scaffolds by decellularization of human myocardium. J. Biomed. Mater. Res A 102 (9), 3263−3272.

Olgunturk, R., Kula, S., Sucak, G.T., Ozdogan, M.E., Erer, D., Saygili, A., 2010. Peripheric stem cell transplantation in children with dilated cardiomyopathy: preliminary report of first two cases. Pediatr. Transplant 14 (2), 257−260.

Olivetti, G., Melissari, M., Balbi, T., Quaini, F., Sonnenblick, E.H., Anversa, P., 1994. Myocyte nuclear and possible cellular hyperplasia contribute to ventricular remodeling in the hypertrophic senescent heart in humans. J. Am. Coll. Cardiol. 24 (1), 140–149.

Ong, C.S., Nam, L., Ong, K., Krishnan, A., Huang, C.Y., Fukunishi, T., Hibino, N., April 22, 2018. 3D and 4D Bioprinting of the myocardium: current approaches, challenges, and future prospects. BioMed Res. Int. 2018, 6497242. https://doi.org/10.1155/2018/6497242. eCollection 2018. Review.

Orlic, D., Kajstura, J., Chimenti, S., Jakoniuk, I., Anderson, S.M., Li, B., et al., 2001a. Bone marrow cells regenerate infarcted myocardium. Nature 410 (6829), 701–705.

Orlic, D., Kajstura, J., Chimenti, S., Limana, F., Jakoniuk, I., Quaini, F., et al., 2001b. Mobilized bone marrow cells repair the infarcted heart, improving function and survival. Proc. Natl. Acad. Sci. USA 98 (18), 10344–10349.

Ott, H.C., Matthiesen, T.S., Goh, S.K., Black, L.D., Kren, S.M., Netoff, T.I., Taylor, D.A., 2008. Perfusion-decellularized matrix: using nature's platform to engineer a bioartificial heart. Nat. Med 14, 213–221.

Pavo, I.J., Michel-Behnke, I., 2017. Clinical cardiac regenerative studies in children. World J. Cardiol 9 (2), 147–153.

Pellegata, A.F., Asnaghi, M.A., Zonta, S., Zerbini, G., Mantero, S., 2012. A novel device for the automatic decellularization of biological tissues. Int. J. Artif. Organs 35, 191–198.

Pena, B., Laughter, M., Jett, S., Rowland, T.J., Taylor, M.R.G., Mestroni, L., et al., 2018. Injectable hydrogels for cardiac tissue engineering. Macromol. Biosci 18 (6), e1800079.

Perin, E.C., Willerson, J.T., Pepine, C.J., Henry, T.D., Ellis, S.G., Zhao, D.X., et al., 2012. Effect of transendocardial delivery of autologous bone marrow mononuclear cells on functional capacity, left ventricular function, and perfusion in chronic heart failure: the FOCUS-CCTRN trial. JAMA 307 (16), 1717–1726.

Quaini, F., Urbanek, K., Beltrami, A.P., Finato, N., Beltrami, C.A., Nadal-Ginard, B., et al., 2002. Chimerism of the transplanted heart. N. Engl. J. Med 346 (1), 5–15.

Quyyumi, A.A., Vasquez, A., Kereiakes, D.J., Klapholz, M., Schaer, G.L., Abdel-Latif, A., et al., 2017. PreSERVE-ami: a randomized, double-blind, placebo-controlled clinical trial of intracoronary administration of autologous $CD34^+$ cells in patients with left ventricular dysfunction post STEMI. Circ. Res 120 (2), 324–331.

Rivas, J., Menendez, J.J., Arrieta, R., Alves, J., Romero, M.P., Garcia-Guereta, L., et al., 2011. Usefulness of intracoronary therapy with progenitor cells in patients with dilated cardiomyopathy: bridge or alternative to heart transplantation? An. Pediatr. 74 (4), 218–225.

Rupp, S., Bauer, J., Tonn, T., Schachinger, V., Dimmeler, S., Zeiher, A.M., et al., 2009. Intracoronary administration of autologous bone marrow-derived progenitor cells in a critically ill two-yr-old child with dilated cardiomyopathy. Pediatr. Transplant 13 (5), 620–623.

Rupp, S., Zeiher, A.M., Dimmeler, S., Tonn, T., Bauer, J., Jux, C., et al., 2010. A regenerative strategy for heart failure in hypoplastic left heart syndrome: intracoronary administration of autologous bone marrow-derived progenitor cells. J. Heart Lung Transplant 29 (5), 574–577.

Rupp, S., Jux, C., Bonig, H., Bauer, J., Tonn, T., Seifried, E., et al., 2012. Intracoronary bone marrow cell application for terminal heart failure in children. Cardiol Young 22 (5), 558–563.

Shimizu, T., Yamato, M., Kikuchi, A., Okano, T., 2003. Cell sheet engineering for myocardial tissue reconstruction. Biomaterials 24, 2309–2316.

Surder, D., Manka, R., Lo Cicero, V., Moccetti, T., Rufibach, K., Soncin, S., et al., 2013. Intracoronary injection of bone marrow-derived mononuclear cells early or late after acute myocardial infarction: effects on global left ventricular function. Circulation 127 (19), 1968−1979.

Tang, J.N., Cores, J., Huang, K., Cui, X.L., Luo, L., Zhang, J.Y., et al., 2018. Concise review: is cardiac cell therapy dead? Embarrassing trial outcomes and new directions for the future. Stem Cells Transl. Med 7 (4), 354−359.

Tendera, M., Wojakowski, W., Ruzyllo, W., Chojnowska, L., Kepka, C., Tracz, W., et al., 2009. Intracoronary infusion of bone marrow-derived selected CD34+CXCR4+ cells and non-selected mononuclear cells in patients with acute STEMI and reduced left ventricular ejection fraction: results of randomized, multicentre Myocardial Regeneration by Intra-coronary Infusion of Selected Population of Stem Cells in Acute Myocardial Infarction (REGENT) Trial. Eur. Heart J. 30 (11), 1313−1321.

Terashvili, M., Bosnjak, Z.J., 2019 Jan. Stem cell therapies in cardiovascular disease. J. Cardiothorac. Vasc. Anesth 33 (1), 209−222.

Thomas, E.D., 1999. A history of haemopoietic cell transplantation. Br. J. Haematol 105 (2), 330−339.

Traverse, J.H., Henry, T.D., Pepine, C.J., Willerson, J.T., Chugh, A., Yang, P.C., et al., 2018. TIME trial: effect of timing of stem cell delivery following ST-elevation myocardial infarction on the recovery of global and regional left ventricular function: final 2-year analysis. Circ. Res 122 (3), 479−488.

Urbanek, K., Quaini, F., Tasca, G., Torella, D., Castaldo, C., Nadal-Ginard, B., et al., 2003. Intense myocyte formation from cardiac stem cells in human cardiac hypertrophy. Proc. Natl. Acad. Sci. USA 100 (18), 10440−10445.

Urbanek, K., Torella, D., Sheikh, F., De Angelis, A., Nurzynska, D., Silvestri, F., et al., 2005. Myocardial regeneration by activation of multipotent cardiac stem cells in ischemic heart failure. Proc. Natl. Acad. Sci. USA 102 (24), 8692−8697.

Wainwright, J.M., Czajka, C.A., Patel, U.B., Freytes, D.O., Tobita, K., Gilbert, T.W., Badylak, S.F., 2010. Preparation of cardiac extracellular matrix from an intact porcine heart. Tissue Eng. C Methods 16, 525−532.

Yoon, Y.S., Wecker, A., Heyd, L., Park, J.S., Tkebuchava, T., Kusano, K., et al., 2005. Clonally expanded novel multipotent stem cells from human bone marrow regenerate myocardium after myocardial infarction. J. Clin. Investig. 115 (2), 326−338.

Zeinaloo, A., Zanjani, K.S., Bagheri, M.M., Mohyeddin-Bonab, M., Monajemzadeh, M., Arjmandnia, M.H., 2011. Intracoronary administration of autologous mesenchymal stem cells in a critically ill patient with dilated cardiomyopathy. Pediatr. Transplant 15 (8), E183−E186.

Zeng, L., Hu, Q., Wang, X., Mansoor, A., Lee, J., Feygin, J., et al., 2007. Bioenergetic and functional consequences of bone marrow-derived multipotent progenitor cell trans-plantation in hearts with postinfarction left ventricular remodeling. Circulation 115 (14), 1866−1875.

Zimmermann, W.H., Melnychenko, I., Wasmeier, G., Didie, M., Naito, H., Nixdorff, U., Hess, A., Budinsky, L., Brune, K., Michaelis, B., Dhein, S., Schwoerer, A., Ehmke, H., Eschenhagen, T., 2006. Engineered heart tissue grafts improve systolic and diastolic function in infarcted rat hearts. Nat. Med. 12, 452−458.

Tissue-based products

Umashankar P.R., Priyanka Kumari
Bioemedical Technology Wing, Sree Chitra Tirunal Institute for Medical Sciences and Technology, Trivandrum, Kerala, India

Chapter outline

7.1 Introduction 145
7.2 Acellular tissue products 147
7.3 Chemically cross-linked tissue products 158
7.4 Tissue-derived products 159
7.5 Host response to tissue products 171
7.6 Sterilization of tissue-based/tissue-derived products 174
    7.6.1 Ethylene oxide treatment 174
    7.6.2 Gamma and electron beam irradiation 175
    7.6.3 Ultraviolet irradiation 175
    7.6.4 Ethanol treatment 175
    7.6.5 Cryopreservation and freeze-drying 175
    7.6.6 Antibiotic regimen 176
    7.6.7 Aqueous glutaraldehyde sterilization 176
7.7 Risk management of tissue-based products 176
7.8 Conclusion 177
References 177
Further reading 185

7.1 Introduction

In medical science, six approaches are typically used to treat a disease or defect at organ scale. They are transplantation, autografting, implantation of a permanent prosthesis, use of stem cells, in vitro synthesis of organs, and induced regeneration (Yannas et al., 2001). Tissue-based products form a major component for the surgical treatment of disease or defect at organ scale either by replacement or by induced regeneration.

They are made from nonviable or rendered nonviable animal/human tissue or its derivatives, either in chemically cross-linked form or in an acellular form. Acellular form is achieved through decellularization. The term nonviable means having no potential for metabolism or multiplication, and "derivative" indicates noncellular substance extracted from tissue or cells through a manufacturing process.

The purpose of chemical cross-linking or decellularization of tissues is to reduce the host immune response to an insignificant level so as to have reasonable durability without manifesting early structural failure, calcification, or unacceptable inflammatory response. Apart from the immune response, the biggest risk for tissue-based products will be contamination with infectious agents such as prions, virus, bacteria, yeast, and molds apart from pyrogens, residual chemicals of processing, which may cause toxicological reactions.

Tissue for manufacturing medical devices can be sourced from any vertebrates or invertebrates including amphibians, arthropods such as crustacea, birds, or coral. Examples of animal tissue products/derivatives can be pericardium from bovine, porcine kangaroo, equine, buffalo, porcine aortic valve, bone, chitosan, small intestinal submucosa (SIS), forestomach, gelatin, collagen, tallow and its derivatives, chondroitin sulfate, hyaluronic acid, etc. Examples of device technologies manufactured from such animal tissue can be pericardial patch, bioprosthetic valve, absorbable/nonabsorbable meshes/sutures, gelatin sealed grafts, transcatheter valves, dura grafts, resorbable nerve products, wound dressing, powders, matrices, hemostats, bone substitutes, cartilage repair devices, bone void fillers, resorbable dental membranes, etc.

There are potential hazards associated with use of tissue-based products. ISO 22442 parts 1, 2, and 3 deal with the risk management, sourcing, collection and handling, storage, and transport of animal tissue. The Medical Devices Regulation (2017/745/EU) or CFR21 part 1271 covers devices manufactured utilizing derivatives of tissues or cells of human origin, which are nonviable or are rendered nonviable. Examples of nonviable human tissue are decellularized human dermis, allograft tendons, demineralized bone, cartilage, pericardium, acellular cornea, fascia lata, etc.

At present, the tissue or its derivatives are regulated only when it is nonviable or rendered nonviable. Transplantation tissues or cells do not come under this. All the devices manufactured utilizing tissues or cells of human or animal origin or their derivatives, which are nonviable and rendered nonviable are in class III (FDA) or Class D, Medical Devices rules 2017, Government of India, unless such devices are intended to come in contact with intact skin only.

In this chapter, tissue products are categorized into acellular or decellularized tissue products, chemically cross-linked tissue products, and tissue-derived products. This is followed by sections on tissue response to tissue-derived products, sterilization methods, and their risk management.

7.2 Acellular tissue products

Acellular tissue is an excellent material for surgical grafting because of its optimum mechanical properties on account of its natural architecture. Although clinically used, acellular or decellularized tissue represents a biomaterial whose biological performance is not well understood. Acellular tissue matrices are produced by selective removal of cellular components that are believed to promote immunogenicity and calcification. These acellular matrices promote induced regeneration through remodeling of the prosthesis by neovascularization, recellularization, and laying of new extracellular matrix (ECM) by the host (Yannas et al., 1989). Native matrix architecture of decellularized tissue may provide structural and chemical cues for cell interactions (Badylak et al., 2009). Decellularized tissue can be remodeled by cellular enzymes present in human host, which makes them an ideal regenerative matrix (Gilbert et al., 2006). Hence, it is considered regenerative.

Interestingly, mildly cross-linked decellularized tissue such as bovine pericardium was shown to have better biological response in in vitro and in vivo studies (Umashankar et al., 2011). It showed better vascular healing compared with glutaraldehyde tissue in chronic pig aortic patch model (Umashankar et al., 2017). It was also observed that decellularized pericardium had better host collagen deposition and improved mechanical properties compared with glutaraldehyde-treated pericardium following subcutaneous implantation in rats (Vishnu, 2015, Sreelakshmi, 2017). It was also shown that even mild cross-linking adversely affects regenerative potential of decellulairzed bovine pericardium (Umashankar et al., 2013).

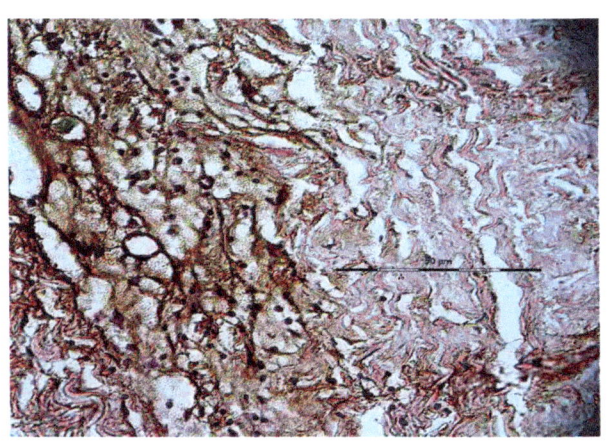

Histochemistry of decellularized pericardium at 60 days showing two-third areas of neocollagenization (pink in color) and nearly one-third (dark drown) areas of scaffold collagen (streptavidin-peroxidase stain X 400) (Vishnu, 2015).

Comparison of the host collagenization following implantation of processed bovine pericardium at different duration in rat subcutaneous implantation model (Vishnu, 2015).

Treatment	Mean ± SE		
	15D	30D	60D
Decellularized	11.17 ± 1.13	36.05 ± 1.59	72.40 ± 2.81
Glutaraldehyde-treated	1.88 ± 0.23	22.95 ± 2.42	39.38 ± 2.59

Mean ± SE bearing superscripts in a column differ significantly at 5% level (p<0.05).

Comparison of tensile strength (MPa) of processed bovine pericardium following explanation in rat subcutaneous implantation model at different duration (Sreelakshmi et al 2017)

Material	0th day	7 days	15 days	30 days
Decellularized	13.14 ± 4.32	0.99 ± 0.11	5.07 ± 1.39	5.92 ± 0.58
Glutaraldehyde te	13.17 ± 2.92	2.49 ± 0.56	2.94 ± 0.47	2.79 ± 0.25

Mean ± SE bearing different superscripts in a column differ significantly (p<0.05).

Immunogenicity of decellularized tissue is a concern as residual xenogeneic and allogeneic cellular antigens are recognized as foreign by the host organism and therefore elicit an inflammatory or rejection response. However, components of ECM are generally conserved among species and hence well tolerated even by xenogenic recipients (Bernad et al 1983). Methods are developed to produce completely acellular matrices from allogenic or xenogenic tissues by specifically removing the cellular components leaving a material composed of ECM components (Crapo et al., 2011). The ECM is a complex mixture of structural and functional proteins, glycoproteins, and proteoglycans arranged in a unique, tissue-specific three-dimensional ultrastructure serving as structural support and a reservoir of growth factors and cytokines (Badylak, 2002; Crapo et al., 2011). The ECM increases local concentration of these factors and presents them efficiently to resident cell surface receptors as well as protects them from degradation and modulates their synthesis (Bonewald, 1999). Collagen is identified as the most abundant protein within mammalian ECM and is mainly constituted by collagen type I, which is also widely used for therapeutic applications (Vanderrest and Garrone, 1991). Fibronectin is second only to collagen in ECM and exists in soluble and tissue forms. It has ligands for adhesion of many cell types (Miyamoto et al., 1998). Laminin is another adhesion molecule primarily seen in basement membrane ECMs (Schwarzbauer, 1999). Laminin is very important for vascularization of scaffolds as it has a prominent role in formation and maintenance of vascular structure due to its favorable interaction with endothelial cells (Ponce et al., 1999). Glycosaminoglycans are another important component of ECM and play important role in binding of growth factors and cytokines, water retention, and the gel properties of ECM. The heparin-binding properties of numerous cell surface receptors and of many growth factors such as VEGF make the heparin-rich GAGs in ECM extremely important (Hodde et al., 1996). The list of growth factors present within ECM although in small quantities includes vascular endothelial growth factor (VEGF), beta fibroblast growth factor (bFGF), epidermal growth factor (EGF), Transforming growth factor beta (TGFβ), keratinocyte growth factor (KGF), hepatocyte growth factor (HGF), and platelet derived growth factor (PDGF) (Badylak, 2002).

The most commonly utilized method of decellularization of tissues involves a combination of physical and chemical methods. Physical methods used are snap freezing (Jackson et al., 1987) and mechanical force or agitation (Dahl et al., 2003). Chemical methods consists of either alkaline or acid treatment (Probst et al., 1997), nonionic detergents such as Triton X-100 (Gulati, 1988), ionic detergents such as sodium dodecyl sulfate, sodium deoxycholate, and Triton X-200 (Rieder et al., 2004), zwitterionic detergents such as CHAPS, sulfobetaine 10 and 16, and tri (n-butyl)phosphate, hypotonic/hypertonic solutions (Dahl et al., 2003), EDTA/EGTA (Bader et al., 1998), and enzymes such as trypsin and endonucleases (Courtman et al., 1994). The effectiveness of decellularization and the alterations to the ECM vary depending on the source of tissue, the composition of the tissue, the tissue density, and other factors (Gilbert et al., 2006). Decellularized tissues such as the pericardium (Courtman et al., 1994), heart valves (Bader et al., 1998), blood vessels (Conklin et al., 2002), skin (Chen et al., 2004), nerves (Hudson et al., 2004), skeletal muscle (Borschel et al., 2004), tendons (Cartmell and Dunn, 2000), ligaments (Woods and Gratzer, 2005), SIS (Badylak et al., 1989), urinary bladder(Chen and Field, 1995), and liver (Lin et al., 2004) have been studied for tissue engineering application. Biological scaffolds derived from decellularized tissue are successfully used in human clinical applications (Badylak, 2004).

Despite many studies and its successful clinical use, it is still not possible to truly predict biocompatibility of one decellularized material over another (Badylak et al., 2009). It was observed that more detailed investigation of host immune response, the ECM constituents that affect the response, and the effect of these factors on scaffold remodeling and outcomes is warranted (Badylak and Glibert, 2008). Hence, this category of tissue products is considered as emerging.

Acellular tissue products

Product	Company	Description	Available forms	Indications	Sterilization	Tests conducted	References
MIRODERM Biologic Wound Matrix	Miromatrix Medical Inc.	Non–cross-linked acellular wound matrix derived from the porcine liver that is processed.	Wound matrix	Indicated for the management of wounds including partial- and full-thickness wounds, indicated for various types of ulcers undermined wounds, trauma wounds, drainage wounds, and surgical wounds.	Sterilized with electron beam irradiation.	—	https://www.miromatrix.com/miroderm
SURGISIS Ocular Graft	Innovative Ophthalmic Products Inc.	Derived from porcine small intestinal submucosa (SIS). It is supplied sterile in a dry lyophilized state sealed in a double peel pouch system.	Sheets of various sizes	Intended for implantation to reinforce and support the reconstruction of the soft tissue of the eyelid and eyelid spacer graft.	Ethylene oxide	Undergone extensive biocompatibility testing, viral inactivation testing, and mechanical testing.	https://www.accessdata.fda.gov/cdrh_docs/pdf5/K053622.pdf
KeraSys Bioengineered Lamellar Patch Graft	Innovative Ophthalmic Products Inc.	It is constructed from four layers of laminated porcine SIS.	Lamellar patch graft	Intended for implantation to reinforce sclera and aid the physical reconstruction of the ocular surface.	Ethylene oxide	Undergone extensive biocompatibility testing, viral inactivation testing, and mechanical testing.	https://www.accessdata.fda.gov/cdrh_docs/pdf9/K090078.pdf

STRATTICE Reconstructive Tissue Matrix Perforated	Lifecell Corporation	Surgical mesh derived from porcine skin, which is processed and preserved in a patented phosphate-buffered aqueous solution containing matrix stabilizers.	Tissue patch	Intended for use as a soft tissue patch to reinforce soft tissue where weakness exists and for the surgical repair of damaged or ruptured soft tissue membranes. Includes repair of hernias and/or body wall defects.	Sterilized by electron beam irradiation.	—	https://www.hcp.stratticetissuematrix.com/hcp
Alpha Chondro Shield	Swiss Biomed Orthopaedics	Cell-free cartilage implant for cartilage regeneration, soft, albeit highly resilient, tear-resistant, absorbable matrix, that serves as tissue replacement.	Matrix	Indicated for underlying tissue protection. It can protect regenerating cartilage against mechanical shear and compression effects in articular cartilage surgery.	Sterilized using gamma radiation.	—	https://www.swissbiomedortho.com/downloads/package_insert.pdf
TissueMend Soft Tissue Repair Matrix	TEI Biosciences Inc., Boston, MA	An acellular, collagen (fetal bovine dermis) membrane used to reinforce soft tissues where weakness exists. Composed of pure, nondenatured collagen and is for augmentation of tendon repair surgery.	Flat, dry sheet of uniform thickness and color	Indicated for tendon repair surgery, including reinforcement of the rotator cuff, patellar, Achilles, biceps, quadriceps, or other tendons. Can be used in both open and arthroscopic applications. TissueMend scaffold helps in healing of the tendon and fortifies the tissue until the healing process has been completed.	Validated terminal sterilization process: sterility assurance level (SAL) 10^{-6}.	Undergone biocompatibility testing, virus inactivation testing, and has been tested free of any transmissible spongiform encephalopathies.	https://www.accessdata.fda.gov/cdrh_docs/pdf6/k060989.pdf

Continued

Continued

Product	Company	Description	Available forms	Indications	Sterilization	Tests conducted	References
EpiFix	MiMedx Group, Inc.	Bioactive tissue matrix allograft composed of dehydrated human amnion/chorion membrane that preserves and contains multiple extracellular matrix proteins, growth factors, cytokines, and other specialty proteins.	Membrane	Intended for homologous use in the treatment of acute and chronic partial- and full-thickness wounds, modulates inflammation, acts as barrier membrane, enhances healing, and reduces scar tissue formation.	Terminally sterilized for enhanced patient safety.	Processed through the PURION process.	https://mimedx.com/epifix
AmnioCord	MiMedx Group, Inc.	Dehydrated, nonviable cellular umbilical cord allograft for homologous use.	Connective tissue matrix	Provides protective environment for healing, provides a connective tissue matrix to replace/supplement damaged/inadequate integumental tissue.	Terminally sterilized for enhanced patient safety.	Processed using the PURION PLUS process.	https://mimedx.com/amniocord/
EpiBurn	MiMedx Group, Inc.	Bioactive tissue matrix allograft comprising dehydrated human amnion/chorion membrane that preserves and contains multiple extracellular matrix proteins, growth factors, cytokines, and other specialty proteins.	Bioactive tissue matrix	Intended for homologous use to reduce scar tissue formation, modulates inflammation, enhances healing, and acts as a barrier membrane.	Terminally sterilized.	Processed through the PURION process.	https://mimedx.com/epiburn/

AmnioFill	MiMedx group, Inc.	Nonviable cellular tissue matrix allograft composed of multiple extracellular matrix proteins, growth factors, cytokines, and other specialty proteins present in placental tissue.	Human collagen matrix	Homologous use as a placental connective tissue matrix to replace or supplement damaged or inadequate integumental tissue, for acute and chronic wounds and modulates inflammation; enhances healing and reduces scar tissue formation.	Terminally sterilized for enhanced patient safety.	PURION processed for an effective allograft with excellent characteristics.	https://mimedx.com/amniofill/
EpiCord	MiMedx Group, Inc.	Dehydrated, nonviable cellular umbilical cord allograft for homologous use.	Connective tissue matrix	Provides protective environment for healing process; provides a connective tissue matrix to replace/supplement damaged or inadequate integumental tissue.	Terminally sterilized for enhanced patient safety.	Processed using the PURION PLUS process.	https://mimedx.com/epicord/
AlloPatch HD	MTF Biologics	Human allograft skin minimally processed to remove epidermal and dermal cells. The process utilized preserves the extracellular matrix of the dermis.	Acellular dermis	Used for replacement of damaged or inadequate integumental tissue/for the repair, reinforcement/ supplemental support of soft tissue defects.	Aseptically processed and is not terminally sterilized.	—	https://www.mtfbiologics.org/
Cardiograft	LifeNet Health	Cryopreserved (−120°C or below) cardiac allograft bioimplant (aortic and pulmonary heart valves), processed from human donated tissue.	Bioprosthetic heart valve	Intended for replacement or reconstruction of diseased, damaged, malformed, or malfunctioning native or prosthetic heart valves.	Subjected to an antibiotic regimen.	—	https://www.lifenethealth.org/sites/default/files/files/68-40-083.pdf

Continued

Continued

Product	Company	Description	Available forms	Indications	Sterilization	Tests conducted	References
Decellularized Cardiograft	LifeNet Health	Decellularized pulmonary artery patch allograft, processed from human donated tissue.	Cryopreserved pulmonary artery tissue	Repair of the right ventricular outflow tract.	Subjected to an antibiotic regimen.	—	https://www.cryolife.com/Products/Cardiac Allografts.
CryoValve's SG Pulmonary Human Heart Valve (and Conduit)	CryoLife, Inc.	Human heart valve aseptically recovered from qualified donors.	Human heart valve	Indicated for the replacement of diseased, damaged, malformed, or malfunctioning native or prosthetic pulmonary valves. Pulmonary heart valve allografts are used to repair both congenital and acquired valvular lesions.	Valve is treated with an antimicrobial solution and is cryopreserved in a tissue culture medium, containing a cryoprotectant.	—	https://www.accessdata.fda.gov/cdrh_docs/pdf10/K101866.pdf
CryoPatch SG Pulmonary Human Cardiac Patch	CryoLife, Inc.	Derived from human pulmonary valve and artery tissue aseptically recovered from qualified donors.	Pulmonary human cardiac patch	Indicated for repair or reconstruction of the right ventricular outflow tract.	Patch treated with an antimicrobial solution and is cryopreserved in a tissue culture medium containing cryoprotectant.	—	https://www.lifenethealth.org/sites/default/.../aortoiliac_spec_sheet_68-60-147v7.pdf
Angiograft Human Aortoiliac Artery	LifeNet Health	A safe and efficacious alternative to synthetic grafts, in abdominal aortoiliac procedures.	Arterial graft	Intended for infected synthetic graft replacement, trauma, mycotic aneurysm, aortoenteric fistula.	Processing involves dissection, antibiotic disinfection, and cryopreservation.	—	https://www.biotissue.com/downloads/amniograft.../amniograft-insert_PI-BT-001E_V2.pdfbiotissue

AmnioGraft	Bio-Tissue, Inc.	Human amniotic membrane product classified as a "human cell- and tissue-based product" derived from donated human tissue.	Biotissue	Designated by the FDA as a homologous graft for ocular wound repair and healing. On the ocular surface, AmnioGraft acts as an antiscarring, antiinflammatory, and antiangiogenic agent and supports epithelial adhesion and differentiation.	—	Aseptically processed as per good tissue practice (GTP), microbiological testing, donor screening, and infectious disease testing done.	www.biotissue.com/products/prokera.aspx
PROKERA	Bio-Tissue, Inc.	Self-retaining biologic corneal bandage, comprised of a cryopreserved amniotic membrane graft fastened to an ophthalmic conformer.	Biologic corneal bandage	Intended for use in eyes in which ocular surface cells are damaged or underlying stroma is inflamed or scarred. Acts as a self-retaining biologic corneal bandage; treats superficial corneal surface diseases, inserted between the eyeball and the eyelid to maintain space in the orbital cavity and to prevent closure or adhesions.	—	Aseptically processed as per current good tissue practices (CGTP) and current good manufacturing practices (CGMP), microbial testing performed.	https://www.katena.com 〉 Tutoplast Fascia Lata
Tutoplast Processed Fascia Lata	RTI Surgical, Inc.	It is dehydrated, processed fascia lata from donated human tissue.	Processed fascia lata	Homologous use for the repair, replacement, reconstruction, or augmentation of soft tissues. This includes connective tissue graft replacement during ophthalmic surgery.	Low-dose gamma irradiation applied terminally to dry implant so as to achieve minimum SAL 10^{-6}.	Microbial testing done, potential pathogens removed.	www.bioteck.com/images/PDF/Brochures/NEURO_HeartDM_EN.pdf

Continued

Continued

Product	Company	Description	Available forms	Indications	Sterilization	Tests conducted	References
Heart DM	Bioteck	Biological matrix for dural replacement and repair, obtained from equine pericardium (native equine collagen matrix). Pericardium decellularized using Zymo-Teck process.	Membrane	Inert, biological scaffold supports tissue repair process without undesired reactions in surrounding tissues, acts as matrix for fibroblast infiltration, substrate for new collagen deposit, and used for strengthening and covering dural grafts.	Beta rays sterilization	—	https://ecatalog.baxter.com/ecatalog/loadResource.blob?bid=55435
DuraGuard	Synovis Life Technologies, Inc.	Prepared from bovine pericardium that is cross-linked with glutaraldehyde.	Repair patch	Indicated for use as a dura substitute for the closure of dura mater during neurosurgery.	Chemically sterilized using ethanol and propylene oxide, has been treated with 1 M sodium hydroxide for 60–75 min at 20–25°C.	—	https://www.lifenethealth.org/sites/default/files/product/68-60-084-02.pdflifenethealth

Oracell	LifeNet Health	Bioimplants decellularized using MatrACELL (proprietary and patented technology), which removes $\geq 97\%$ of the DNA without compromising desired biomechanical/biochemical properties, which allows the matrix to retain its growth factors, native collagen scaffold, and elastin required for healing.	Dermal matrix	Acellular dermal matrix intended for soft tissue repair and reconstruction.	Terminal sterilization, with a SAL of 10^{-6}.	—	alliqua.com/wp-content/uploads/2014/10/Biovance-Package-Insert.pdf
BIOVANCE	Alliqua BioMedical	Decellularized, dehydrated human amniotic membrane with a preserved natural epithelial basement membrane and an intact extracellular matrix structure with its biochemical components. The epithelial basement membrane and extracellular matrix of this allograft provide a natural scaffold that allows cellular attachment or infiltration and growth factor storage. BIOVANCE provides a protective cover and supports the body's wound healing processes.	Membrane	An allograft intended for use as a biological membrane covering that provides the extracellular matrix while supporting the repair of damaged tissue. As a barrier membrane, BIOVANCE is intended to protect the underlying tissue and preserve tissue plane boundaries with minimized adhesion or fibrotic scarring.	Aseptically processed product and is terminally sterilized with e-beam irradiation.	—	https://www.miromatrix.com/miroderm

7.3 Chemically cross-linked tissue products

Glutaraldehyde is the most successfully used chemical cross-linking agent for tissue-based products. Glutaraldehyde-stabilized tissue on accounts of its durability is clinically used since 1970s as surgical implants or bioprostheses. While glutaraldehyde cross-linking rendered xenografts immunologically inert, it also made them nonresorbable and nonamenable to in vivo remodeling on account of its resistance to matrix metalloproteinases. This has resulted in a graft that fails to remodel in vivo, which is required for structural and physiological integration into host tissue. Although glutaraldehyde-treated tissue has this disadvantage, with the advent of more sophisticated postprocessing antimineralization treatments, devices such as glutaraldehyde-treated bovine pericardial valves were reported to function more than 17 years in patients (Bourguingnon et al., 2015), proving its excellent durability.

Glutaraldehyde can introduce stable cross-links into collagens fibers in comparison with other aldehydes (Nimni, 1968; Carpentier and Dubost, 1972). The specific chemistry of glutaraldehyde fixation of collagen is not fully understood. It is assumed that intermolecular—intramolecular covalent bonds are formed in two ways, formation of Schiff bases by reaction of an aldehydes group with an amino group of lysine or hydroxylysine or an aldol condensation between two adjacent aldehydes (Jayakrishnan and Jameela, 1996). Besides amino groups, glutaraldehyde can also interact with carboxy, imido, and other groups of protein (Bowers and Cater, 1966; Anderson, 1967; Blauer et al., 1975). Fixation time and concentration of glutaraldehyde used are important. Slower time-dependent cross-linking of glutaraldehyde occurs at lower concentration, and at higher concentration, cross-linking occurs only at surface of the collagen fibers due to rapid polymerization and impairment of glutaraldehyde molecule access to the interstitium of the larger collagen fibers by steric hindrance (Nimni et al., 1987). The degree of cross-linking determines the degradation rate of chemically cross-linked tissue, and as a consequence, it can significantly affect its regeneration pattern (Liang et al., 2004). Highly cross-linked matrices are more stable to degradation and therefore less likely to be remodeled as part of tissue morphogenesis.

Glutaraldehyde-treated xenografts such as bovine/porcine/equine pericardium or porcine aortic/pulmonary valve have been extensively used as medical implant in the form of cardiovascular patch or are used in the fabrication of bioprosthetic heart valve. However, it has been shown that glutaraldehyde-treated tissues have propensity for calcification and are susceptible for early mechanical failure (Schoen and Levy, 2005). Calcification hence accounts for the main reason for failure of glutaraldehyde cross-linked tissue when used in the fabrication of bioprosthetic valves, especially when implanted in younger patients. Besides this, glutaraldehyde-treated pericardium and its extract produced cytotoxic effect and caused release of inflammatory cytokines from macrophages in vitro. It also produced periimplant necrosis, infiltration with chronic and acute inflammatory cells, and calcification in rat subcutaneous implantation model (Umashankar et al., 2012).

The mechanism of calcification on glutaraldehyde-treated pericardium is not well understood on account of its complexity. It is reported that pericardial tissue residual antigens through immunological mechanisms, free aldehyde groups of glutaraldehyde, and tissue phospholipids through host calcium binding are the reasons causing tissue calcification (Schoen and Levy, 2005). Several strategies have been investigated to tackle the above issues. Suppression of residual antigenicity was achieved through decellularization and glutaraldehyde fixation. Free aldehyde groups on account of glutaraldehyde fixation were removed through postfixation treatments with amino acids. As an example, the use of 2-amino oleic acid reduced residual and unbound aldehyde groups on glutaraldehyde-fixed bovine pericardium and significantly decreased the risk of calcification (Chen et al., 1994). Treating with ethanol has been shown as a method to remove tissue phospholipids, thus preventing calcification (Vyavahare et al., 1997). Several other techniques such as treatment with aluminum−ethanol combination (Ogle et al., 2003) lyophilization, treatments with heparin (Chanda, 1997) or sulphonated poly(ethylene oxide [EtO]), which block side effects of glutaraldehyde residues (Lee et al., 2001), grafting of hyaluronic acid to glutaraldehyde-treated bovine pericardium have been tried for preventing calcification of glutaraldehyde-treated pericardium (Ohri et al., 2004).

Besides this, several proprietary techniques such as L-Hydro process (Nina et al., 2005), ADAPT technique (Neethling et al., 2004), Edward Carpentier Xenologix process, and ThermaFix process (Carpentier et al., 1998) have been utilized for prevention of bioprosthetic calcification and are exploited commercially.

Many of the patented methods for reducing graft calcification have utilized trivalent cations such as Fe3, Al3, and Sn3 by simple incubation of glutaraldehyde-treated tissue in the above salt solutions or in combination with anticalcification agents such as diphosphonates or ethanol.

Different tissue products cross-linked with aldehydes meant for applications ranging from wound matrix to heart valve substitute are presented below.

7.4 Tissue-derived products

Tissue-derived products consists of noncellular substance extracted from human or animal tissue or cells through a manufacturing process. However, the final substance used in the device should not contain any cells or tissues. It is a material obtained from animal or human tissue, through one or more treatment, transformations, or steps of processing. Tissue-derived products are sourced from human or animal tissue and are available alone or in combination with synthetic material. Human tissue−derived products such as lyophilized amniotic fluid, recombinant human platelet−derived growth factor, and demineralized human bone matrix are some of the examples. Examples of animal tissue−derived products include porous bovine bone mineral matrix, purified bovine collagen type I, purified porcine collagen, bovine thrombin, chondroitin sulfate from shark, chytosan from crustaceans, and absorbable suture material from animal gut submucosa. Injectable device−drug combination, which has bovine collagen, beta calcium phosphate, as well as recombinant human protein PDGF meant for bone fusion application, is an example for combination product.

Examples of tissue-derived products have been discussed.

Chemically cross-linked tissue products

Product	Company	Description	Available forms	Indications	Sterilization	Tests conducted	References
EZ Derm	Mölnlycke Health Care AB	Porcine Xenograft. Porcine dermis chemically cross-linked with an aldehyde to provide durability and storage.	Wound matrix	Indicated for use on partial-thickness skin loss, donor sites, skin ulcerations and abrasions; also used as temporary covering for full-thickness skin loss, Toxic epidermal necrolysis and meshed autograft protection.	—	Clinically tested, Latex-friendly, noncytotoxic.	https://www.molnlycke.ae/products-solutions/ez-derm/
Medtronic CoreValve system	Medtronic CoreValve LLC	Consists of a transcatheter aortic valve (referred to as the CoreValve), a delivery catheter, and a compression loading system. CoreValve is made of natural tissue obtained from the porcine heart and attached to a flexible, self-expanding, nickel-titanium (Nitinol) frame for support.	Aortic Valve	Indicated for relief of aortic stenosis in patients with symptomatic heart disease due to severe native calcific aortic stenosis, indicated for use in patients with symptomatic heart disease due to failure of a surgical bioprosthetic aortic valve.	Sterilized with glutaraldehyde solution.	—	https://www.accessdata.fda.gov/cdrh_docs/pdf13/P130021S033C.pdf

Edwards SAPIEN 3 Transcatheter Heart Valve with the Edwards Commander Delivery System	Edwards Lifesciences	Comprises a balloon-expandable, radiopaque, cobalt-chromium frame, trileaflet bovine pericardial tissue valve, and polyethylene terephthalate (PET) fabric skirt. The Edwards Commander delivery system (useable length 105 cm) used for delivery of the Edwards SAPIEN 3 transcatheter heart valve.	Bioprosthetic Heart Valve	Indicated for relief of aortic stenosis in patients with symptomatic heart disease due to severe native calcific aortic stenosis.	Sterilized with glutaraldehyde solution.	—	https://www.edwards.com/gb/devices/heart-valves/transcatheteredwards; https://www.accessdata.fda.gov/cdrh_docs/pdf14/P140031c.pdf
Edwards SAPIEN XT Transcatheter Heart Valve with the Ascendra + Delivery System	Edwards Lifesciences	Comprises a balloon-expandable, radiopaque, cobalt-chromium frame, trileaflet bovine pericardial tissue valve, and a PET fabric skirt.	Bioprosthetic Heart Valve	Indicated for relief of aortic stenosis in patients with symptomatic heart disease due to severe native calcific aortic stenosis and with native anatomy appropriate for the 23, 26, or 29 mm valve system.	Sterilized with glutaraldehyde solution. The delivery system is supplied sterilized with ethylene oxide (EtO) gas.	Nonclinical testing has demonstrated that the Edwards SAPIEN XT THV is MR conditional.	https://www.edwards.com/gb/devices/heart-valves/transcatheteredwards; https://www.accessdata.fda.gov/cdrh_docs/pdf13/P130009d.pdf

Continued

Continued

Product	Company	Description	Available forms	Indications	Sterilization	Tests conducted	References
Melody™ Transcatheter Pulmonary Valve (TPV) and Ensemble Transcatheter Valve Delivery System	Medtronic, Inc.	TPV consists of a heterologous (bovine) jugular vein valve sewn within a laser-welded, platinum-iridium stent with gold brazing of the welds. The Ensemble transcatheter valve delivery system consists of a balloon-in-balloon catheter with a retractable polytetrafluoroethylene sheath large enough to cover the valve after crimping.	Transcatheter Pulmonary Valve	Patients born with heart defects often receive a pulmonary valve conduit (an artificial graft with a valve inside that connects the heart to the lungs) to correct the defects. The Melody TPV is used to treat a failing conduit.	Melody TPV sterilant contains 1% glutaraldehyde and 20% isopropyl alcohol, the Ensemble transcatheter valve delivery system is sterilized with EtO gas.	Nonclinical testing and modeling has demonstrated that Melody TPV is MR conditional.	https://www.edwards.com/gb/devices/heart-valves/transcatheteredwards; https://www.accessdata.fda.gov/cdrh_docs/pdf14/P140017d.pdf
Perceval Sutureless Aortic Heart Valve	Sorin Group USA Inc.	A bioprosthetic valve designed to replace a diseased native/malfunctioning prosthetic aortic valve through an open-heart surgery. It consists of a tissue component made from bovine pericardium and a self-expandable Nitinol stent that supports the valve and fixes it in place.	Bioprosthetic valve	Indicated for the replacement of diseased, damaged, or malfunctioning native or prosthetic aortic valves.	Preserved in a buffered aldehyde-free sterile solution.	Investigated in three clinical studies: • the Perceval Pilot Trial (V10601); • the Perceval Pivotal Trial (V10801); and • the Perceval CAVALIER Trial (TPS001).	https://www.accessdata.fda.gov/cdrh_docs/pdf15/P150011d.pdf

Edwards Pericardial Aortic Bioprosthesis	Edwards Lifesciences	A stented trileaflet valve comprising RESILIA bovine pericardial tissue that is mounted on a flexible cobalt-chromium metal frame.	Bioprosthetic heart valve	Intended for use as a heart valve replacement.	Preservation with glycerol.	The clinical safety and effectiveness was established based on the outcome data of the COMMENCE trial.	www.accessdata.fda.gov/cdrh_docs/pdf15/P150048C.pdf
VASCU-GUARD Peripheral Vascular Patch	Synovis Life Technologies, Inc. and Baxter	Prepared from bovine pericardium that is cross-linked with glutaraldehyde; VASCU-GUARD is treated with 1 molar sodium hydroxide for 60–75 min at 20–25°C.	Vascular Patch	Intended for use in peripheral vascular reconstruction including the carotid, renal, iliac, femoral, profunda, and tibial blood vessels and arteriovenous access revisions.	VASCU-GUARD is chemically sterilized using ethanol and propylene oxide.	Safety and performance was evaluated through biocompatibility, bench testing, and animal studies.	https://www.accessdata.fda.gov/cdrh_docs/pdf14/K142461.pdf; http://ecatalog.baxter.com
Edwards Bovine Pericardial Patch	Edwards Lifesciences LLC	Comprises a rectangular sheet of glutaraldehyde-treated bovine pericardium, the pericardial patch is in the form of a 10 × 15 cm size and may be tailored during surgery to meet the individual needs.	Rectangular sheet	Intended for use as a surgical patch open-heart surgery; intracardiac defects; septal defects and annulus repairs; cardiac and vascular reconstruction and repairs; peripheral vascular reconstruction and repairs; great vessel reconstruction and repairs; and suture-line buttressing.	Preserved with glutaraldehyde solution.	Functional and safety testing conducted.	https://www.edwards.com/devices/bovine-pericardial-patches/cardiacedwards

Continued

Continued

Product	Company	Description	Available forms	Indications	Sterilization	Tests conducted	References
Peri-Guard Repair Patch	Synovis Life Technologies, Inc.	Biologic tissue prepared from bovine pericardium cross-linked with glutaraldehyde.	Pericardial patch	Intended for repair of pericardial structures and for use as a prosthesis for the surgical repair of soft tissue deficiencies. Also intended for use as a patch material for intracardiac defects, great vessel, septal defect, and annulus repair, and suture-line buttressing.	Chemically sterilized using ethanol and propylene oxide. It is treated with 1 molar sodium hydroxide for 60–75 min at 20–25°C.	Clinically tested.	https://www.accessdata.fda.gov/cdrh_docs/pdf/K983162.pdf; http://ecatalog.baxter.com
SJM Biocor Bovine Pericardial Patch	St. Jude Medical, Inc.	Bovine pericardial patch glutaraldehyde cross-linked and treated with a proprietary anticalcification treatment, has favorable characteristics of durability and thrombogenicity.	Pericardial patch	Indicated for use as a biological patch graft in cardiovascular surgery (used in closure of intercavitary faults, widening of VD outlet, widening of the aortic ring in valve replacements, treatment of aneurysms of the aorta and left ventricle, atrial and ventricular reconstitution, mustard surgery, arterioplasties and others). In thoracic surgery, used in closing of the bronchial stump and diaphragmatic hernias, treatment of thoracic wall faults and others.	Preservative used is formaldehyde.	–	https://www.sjm.com

Tissue-derived products

Product	Company	Description	Available forms	Indications	Sterilization	Tests conducted	References
AUGMENT Injectable	BioMimetic Therapeutics, LLC	Device/drug combination product for use in bone fusion of the foot/ankle. First part granules of beta-tricalcium Phosphate (β-TCP). Second part from tissue from cow hides (bovine collagen). Third part recombinant human platelet–derived growth factor (rhPDGF-BB).	Injectable	For alternative to autograft in arthrodesis Intended for patients who need supplemental graft material.	Provided as two sterile trays: Matrix tray containing 10 mL polypropylene syringe containing either 0.5/1.0 gm of a milled β-TCP/ bovine Type I collagen matrix is sterilized by gamma irradiation. Vial tray sterilized by ethylene oxide (EtO).	—	www.wright.com/.../AUGMENT-Injectable-Surgical-Technique-and-Package-Insert_J.
OrthoFlo	MiMedx Group, Inc.	Lyophilized amniotic fluid allograft, intended for homologous use.	Human Amniotic Fluid Allograft	Protects and provides cushion to the joint, provides lubrication for enhanced mobility, modulates inflammation.	Terminally sterilized	—	https://mimedx.com/orthoflo/

Continued

Continued

Product	Company	Description	Available forms	Indications	Sterilization	Tests conducted	References
OsteoTape	Impladent, Limited	An OsteoGen' Collagen Resorbable Bone Graft Matrix (OsteoTape) is a resorbable bone grafting substitute comprising highly purified Type I bovine collagen, used as a carrier derived from bovine Achilles tendon, combined with crystals of the product OsteoGen', a synthetic bioactive resorbable graft of the nonceramic hydroxylapatite category.	Sterile cubes and strip forms	Indicated for periodontal and maxillofacial use in surgical procedures.	Complies with radiation sterilization	Safety evaluation by standardized tests, undergone viral inactivation, and biocompatibility testing both in vivo and in vitro.	https://www.accessdata.fda.gov/cdrh_docs/pdf9/K090794.pdf
PuraPly Antimicrobial Wound Matrix	Organogenesis, Inc.	Purified collagen matrix (derived from a porcine source) combined with polyhexamethylenebiguanide hydrochloride.	Native collagen matrix	Indicated for the management of wounds including: partial- and full-thickness wounds, indicated for various types of ulcers undermined wounds, trauma wounds, drainage wounds, surgical wounds.	—	Clinically tested, latex free.	https://organogenesis.com/products/puraply-antimicrobial-surgical.html

Omnigraft (also known as Integra Dermal Regeneration Template)	Integra LifeSciences Corporation	Two-layered skin repair product: The layer next to the wound is collagen from cattle and chondriotin-6-sulfate from sharks. The top layer is silicone.	Dermal Regeneration template	Indicated for use in the treatment of partial- and full-thickness diabetic foot ulcers that are greater than 6 weeks in duration.	Gamma irradiation	Clinically tested on humans.	Integra LifeSciences Corp. Integra Matrix Wound Dressing [website]. Plainsboro, NJ: Integra LifeSciences, 2008. https://www.accessdata.fda.gov/cdrh_docs/pdf/P900033S042c.pdf
Plain Catgut Absorbable Surgical Suture	Ethicon US, LLC	Surgical sterile, absorbable suture material from purified sheep gut submucous fibrous tissue (mostly collagen).	Suture thread	Indicated for use in general soft tissue approximation and/or ligation, used in ophthalmic procedures, but not used in cardiovascular and neurological tissues.	EtOEO and gamma radiation	–	https://www.ethicon.com/na/products/wound.../surgical-gut-suture-plain-and-chromic C

Continued

Continued

Product	Company	Description	Available forms	Indications	Sterilization	Tests conducted	References
HEMOBLAST Bellows	Biom'up USA Inc.	A handheld device used to achieve hemostasis (control bleeding) during surgical procedures. Includes a dry, sterile powder made of highly purified porcine collagen (with glucose), chondroitin sulfate, and thrombin.	Dry, sterile powder	Indicated in surgical procedures as an adjunct to hemostasis when control of minimal, mild, and moderate bleeding by conventional procedures is ineffective in neurosurgical, ophthalmic, and urological procedures.	Sterilized using gamma sterilization	Single-arm, pilot, clinical studies conducted.	https://www.accessdata.fda.gov/cdrh_docs/pdf17/P170012C.pdf
Lyoplant	Aesculap, Inc.	Implant of pure collagen derived from bovine pericardium a Type I collagen that is known for its low propensity to cause immunological reactions.	Bioabsorbable dura substitute	Used for replacement and or augmentation of dura mater during cranial and spinal procedures.	EtO	Manufactured from bovine pericardium which is bovine spongiform encephalitis free, it is treated with a purifying NaOH solution during processing, to further reduce any risk.	http://itmedica.com/wp-content/uploads/2017/01/LYOPLANT.pdf

DBX Inject Putty	MTF Biologics	Demineralized bone matrix (DBM) with osteoinductive potential and is osteoconductive. Contains demineralized bone from human donors, with sodium hyaluronate as carrier. It is nonhemolytic.	Available as putty, mix, strips, and injectable putty	Osteoinductive bone void filler, extender in the spine with autograft or allograft, treatment of osseous defects, used with bone marrow, used in pelvis and extremities, can be used as extender in pelvis and extremities with autograft or allograft.	Aseptically processed and passes USP <71> sterility tests. DBX inject tissue is not terminally sterilized. The plastic syringe provided is terminally sterilized by gamma radiation.	Tested and validated in vitro and in vivo (athymic mouse model), undergone viral clearance inactivation.	https://www.mtf.org/documents/PI_-70_Rev_5.pdf
OrthoBlast II Demineralized Bone Matrix	IsoTis Ortho-Biologics, Inc.	Bone graft substitute composed of DBM and a poloxamer reverse phase medium (RPM) carrier with the additional benefit of cancellous bone (supports tissue and vascular growth).	Available in putty and paste forms	As an autograft extender (extremities, spine, and pelvis) and as a bone void filler (extremities and pelvis). Packed gently into bony defects (surgically created or resulting from a trauma) of the skeletal system.	Electron beam (e-beam) sterilization	Tested in vivo in skeletally mature sheep model.	https://www.seaspine.com/products/orthoblast-2/seaspine; https://www.accessdata.fda.gov/cdrh_docs/pdf7/K070751.pdf
DynaGraft II Demineralized Bone Matrix	IsoTis Ortho-Biologics, Inc.	Bone graft substitute composed of DBM and a poloxamer RPM, a biocompatible carrier.	Available in putty and paste forms	For orthopedic applications as filler for gaps or voids. Indicated to be packed gently into bony gaps in the skeletal system as a bone graft extender (extremities, spine, and pelvis) and as bony void filler of the extremities and pelvis.	E-beam sterilization	Tested in vitro	https://www.accessdata.fda.gov/cdrh_docs/pdf4/K043573.pdf

Continued

Continued

Product	Company	Description	Available forms	Indications	Sterilization	Tests conducted	References
OCS-B	NIBEC Co., Ltd.	Sterile, porous bone mineral matrix produced by the removal of organic compounds from bovine bone. It is supplied as cancellous (spongiosa) or cortical granules in a single use container, packaged in a secondary thermoform blister.	Porous, biocompatible bone grafts	Used as an adjective therapy in restoring bony defects.	Sterile by Gamma irradiation	Undergone biocompatibility testing, viral inactivation testing.	https://www.accessdata.fda.gov/cdrh_docs/pdf11/K113246.pdf
SurFuse Gel, SurFuse Putty, ExFuse Gel, ExFuse Putty	Hans Biomed Corporation	Resorbable bone void filler, primary component of SurFuse and ExFuse, is demineralized particle bone that is derived from human donor cortical bone. The additional bone powder in the ExFuse is derived from human donor cancellous bone. CMC is added to enhance the cohesiveness of the composition.	Bone void filler available in gel and putty form	Indicated for bony voids or gaps that are not intrinsic to the stability of the bone structure. They are intended to be gently packed into bony voids or gaps of the skeletal system as a bone void filler in the extremities and pelvis.	Sterilized	Undergone biocompatibility testing, serological testing, viral inactivation testing, also tested in vivo in the athymic (nude) rat muscle pouch model.	https://www.accessdata.fda.gov/cdrh_docs/pdf17/K171568.pdf

7.5 Host response to tissue products

The source of biological scaffold, decellularization method, and methods of terminal sterilization vary widely, and each of these variables affects host response to the tissue products (Gilbert et al., 2006; Reing et al., 2010). Residual processing chemicals such as detergents in the decellularized scaffold can potentially cause adverse tissue reaction (Cebotari et al., 2010). Host response as reported (Valentin et al., 2006) for five different commercially available tissue products such as GraftJacket, Restore, CuffPatch, TissueMend, and Permacol is presented in the table below.

Scaffold	Tissue and process	Host response
GraftJacket	Human dermis, proprietary cryogenic processing	Elicited most intense acute cell response, which was not predictive of an adverse remodeling outcome. Cellular response was predominantly mononuclear response. Multinucleate giant cell was observed. The device was replaced with fibrous connective tissue and a persistent low-grade chronic inflammatory response.
Restore	Porcine small intestinal submucosa (SIS), minimally processed	Elicited most intense acute cell response, which was not predictive of an adverse remodeling outcome. Cellular response was predominantly mononuclear response. Biological scaffold was replaced with a mixture of muscle cells and organized connective tissue.
CuffPatch	Porcine SIS, carbodiimide cross-linked	Multinucleate giant cell was observed. Cellular response was predominantly neutrophilic response throughout the entire study. Device showed accumulation of dense collagenous tissue, a persistent foreign body response, and relatively slower remodeling.
TissueMend	Fetal bovine skin, proprietary process	Cellular response was predominantly mononuclear response, low-grade chronic inflammation, minimal scaffold degradation, and fibrous encapsulation.
Permacol	Porcine dermis, cross-linked with isocyanate	Cellular response was predominantly mononuclear response. Multinucleate giant cell was observed. Low-grade chronic inflammation, minimal scaffold degradation, and fibrous encapsulation.

Studies on decellularized porcine Matrix P valve revealed a foreign body—type reaction accompanied by severe fibrosis and massive neointima formation around decellularized porcine valve wall. There was low recellularization of decellularized matrix, and neovascularization was observed only in the neointima and scar tissue. Inflammatory infiltrates, composed mainly of T cells, B cells, and plasma cells, as well as the presence of dendritic cells, macrophages, and mast cells, were detected in the tissue surrounding the porcine matrix. In the fibrous tissue, overexpression of connective tissue growth factor was observed. The above observation was attributed to incomplete decellularization of porcine matrix, which may have contributed to increased immunogenicity of these conduits (Cicha et al., 2010). The immunogenicity of un—cross-linked decellularized xenograft was observed as tissue over growth, inflammatory cell infiltration, and incidence of aneurismal dilation by Hilbert et al. (2004) through comparing various reported decellularization protocols in long-term sheep implantation model. Even positive results found in short-term animal trials of an un—cross-linked decellularized xenograft valve had culminated in catastrophic failure during clinical trial (Simon et al., 2003). However, nondetergent-based decellularization with mild cross-linking was shown to produce decellularized bovine pericardium with better healing response in rat subcutaneous model (Parvathy et al., 2013). Pulmonary valve conduit fabricated from the above material survived 6 months in sheep pulmonary artery position without aneurysmic dialataion, minimum clacification, and host tissue incoroporation, which was limited to graft wall (Umashankar et al. unpublished work).

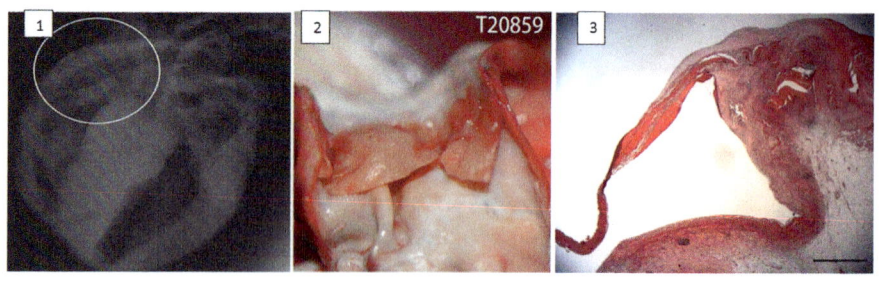

Explanted mildly cross-linked decellurized bovine pericardial pulmonary valved conduit after 6 months implantation in sheep pulmonary artery position. 1. X-ray of heart with valved conduit. 2. Opened valved conduit showing thin valve leaflets. 3. Micrograph of explanted pulmonary graft wall with leaflet showing neointimal growth at laeaflet base.

Immunogenicity of tissue products is another aspect to be considered. It is reported that the immune response against incompletely decellularized xenograft was seen much more prominent compared with isografts or even allografts (Rossini et al., 1999; O'Brien et al.,1984). Even fully decellularized xenograft is reported to elicit specific acquired immunity in vivo as demonstrated by immunoblot analysis (O'Brien et al., 1984). The residual immunogenicity of decellularized xenograft was reported as tissue over growth, inflammatory cell infiltration, and incidence of aneurismal dilation by Hilbert et al. (2004), through comparing various reported decellularization protocols in long-term sheep implantation model. It was also observed that although un—cross-linked decellularized bovine pericardium showed chronic inflammatory response evidenced by macrophage and lymphocyte infiltration and elicited cell mediated and humoral immune response, it also induced skeletal muscle formation within graft at 90 days in rat abdominal defect model (Umashankar et al., 2013).

The presence of Gal epitope on the surface of vascular endothelium is the primary cause of rejection in xenogeneic organ transplants (Collins et al., 1994; Cooper et al., 1993; Galili et al., 1985; Oriol et al., 1993). Humans produce large amounts of anti-Gal antibodies (1% of circulating IgG) including IgG, IgM, and IgA as a result of constant exposure to intestinal bacteria that carry Gal epitope (Gabrielli et al., 1991; Galili et al., 1984; Goldberg et al., 1995; Koren et al., 1992). The presence of Gal epitope in biologic scaffolds is reported on glutaraldehyde cross-linked porcine bioprosthetic heart valves (Konakei et al., 2005), porcine anterior cruciate ligament and cartilage (Stone et al., 2007; Stone et al., 1998), and porcine SIS ECM (McPherson et al., 2000). An immunologic response to the xenoantigens other than α-Gal (non-Gal antigens) has also been reported (Stone et al., 2007; Baumann et al., 2007; Zhu and Hurst, 2002). Residual DNA in bioscaffolds is also directly correlated to adverse host reactions (Nagata et al., 2010; Zheung et al., 2005). Despite the universal presence of DNA remnants in commercially available ECM devices, the clinical efficacy of these devices for their intended application has been largely positive, and hence, it appears unlikely that the remaining DNA fragments contribute to any adverse host response or a cause for concern (Badylak and Gilbert, 2008). Decellularized porcine valves were shown to have considerable amount of residual proteins with different molecular weights and have significantly higher potential for inciting monocyte migratory response (Reider et al., 2005). Complete elimination of cells and remnants could not be achieved with different processing techniques, and decellularized porcine heart valve matrix has the potential to attract inflammatory cells and to induce platelet activation (Kasimir et al., 2006).

In long-term pig aortic implantation study, glutaraldehyde-treated bovine pericardium showed calcification and became isolated with respect to surrounding tissue and it was seen progressively pushed adluminally, resulting in a weak neointima bearing the pulsatile mechanical load at the grafted region (Umashankar et al., 2017). It also showed significantly thicker neointima at longer periods compared with decellularized bovine pericardium indicating prolonged state of inflammation compared with decellularized bovine pericardium (Rocco et al., 2014). Rat subcutaneous implantation studies on glutaraldehyde-treated bovine pericardium also showed similar findings such as calcification and thick capsule around the implant, with periimplant necrosis. Angiogenesis was seen only in the periphery of the implant. Moderate to severe inflammation evidenced as mononuclear cell infiltration could be seen at the periphery of implant. The implant interior was remarkably acellular (Umashankar et al., 2011).

7.6 Sterilization of tissue-based/tissue-derived products

Tissues are widely used as substitutes for synthetic grafts in the form of tissue-based or tissue-derived products for the repair or reconstruction of damaged or injured parts of the body. Allograft from human or animal donors is considered as good alternative to synthetic grafts, but the problem associated with them is the bioburden level that can increase the risk of transmission of infectious diseases.

Sterilization of these biomaterials is of utmost importance so that the associated bioburden levels are under control, thereby rendering successful clinical application of the product. Also some biomaterials might show unwanted response due to their nature toward the sterilization techniques used, thus it is important to have knowledge of the various sterilization techniques and their effects on the biomaterials.

Sterilization is any physical or chemical process that renders a product free of contamination from all living forms of microorganisms, which includes bacteria, spores, yeasts, and viruses by either killing them or inactivating them (Kim et al., 2007; Heseltine, 2001), and thereby preventing allograft associated life-threatening infections (Dziedzic-Goclawska et al., 2005). Various sterilization techniques viz. chemical processing (Prolo et al., 1980), antibiotic regimens (Tom and Rodeo, 2002), gamma irradiation (Nguyen et al., 2007), EtO treatment (Arizono et al., 1994), electron beam sterilization, use of cryoprotectants, etc., are widely used to minimize/get rid of the bioburden levels, hence rendering the product sterile.

Sterility assurance level (SAL) is defined as the possibility of a product being non-sterile after undergoing a valid sterilization method (https://my.aami.org, 2016), which is derived from the kinetic studies on inactivation of microorganisms. Implantable biomaterials are sterilized at such a high dose so as to achieve a SAL of $10-6$, which implies that the probability of a microorganism surviving the treatment is less than 1 in 1,000,000.

The following are the different sterilization methods used on tissue products, which can attain a SAL value of $10-6$.

7.6.1 Ethylene oxide treatment

It is a chemical sterilization method widely employed for over 40 years and is commercially used for sterilization of tissue allografts, heat- and moisture-sensitive medical devices (Parisi, 1991), and health-care medicinal and clinical products. EtO acts by causing irreversible alkylation of cell molecules containing amino, carboxyl, thiol, hydroxyl, and amide groups, thereafter suppressing cell metabolism and division permanently (Gogolewski and Mainil-Varlet, 1997). It effectively renders the product sterile, i.e., free of bacteria (Phillips and Kaye, 1949) and viruses (Jordy et al., 1975; Mcquillan et al., 1999), when used at appropriate concentration due to its low-temperature requirements and wide spectrum of antimicrobial (bactericidal and virucidal) activity (Hsiao et al., 2012). EtO sterilization is known to effectively inactivate vegetative Gram-negative and Gram-positive bacteria, fungal, spores, (DNA, RNA, enveloped and naked), and viruses (Moore et al., 2004).

7.6.2 Gamma and electron beam irradiation

Ionizing radiation viz. gamma (γ) irradiation is commercially used for sterilization of biomaterials. Electromagnetic gamma (γ) rays are generally obtained from 60Co or 137Cs sources, and a dose of 25 kGy is considered effective (Leonard et al., 2006) for sterilization of medical products.

While electron beam irradiation is a process that utilizes high energy accelerated electron stream for sterilization (Kim et al., 2011), both gamma (γ) and electron beam irradiation disrupt the DNA and RNA strands, thereby generating reactive oxygen species (ROS) and damaging cell components (Cox and Battista, 2005; Baird, 2004), and the ROS cause degradation of the DNA molecules by cleaving the phosphodiester backbones of DNA molecules (Baird, 2004). Gamma and electron beam irradiation are established procedures for inactivation of bacteria, i.e., both Gram-positive and Gram-negative, some bacterial and fungal spores, yeasts, and most of the viruses (Spoto et al., 2000; Smith et al., 2001; Grieb et al., 2005), except some endospores capable of withstanding high doses of ionizing irradiation (Prescott et al., 2004).

7.6.3 Ultraviolet irradiation

Ultraviolet (UV) irradiation usually possessing a wavelength in the range of 200–280 nm (Prescott et al., 2004) intertwined with a specific time duration (Fischbach et al., 2001; Yixiang et al., 2008) is gaining importance for sterilizing biodegradable scaffolds. In this method, production of photoproducts by excitation of electrons results in damage to DNA and its molecules, thereby affecting DNA replication leading to microbial inactivation (Setlow, 2006). UV irradiation effectively kills vegetative bacteria and enveloped viruses, while naked viruses show more resistance (Setlow, 2006; Watanabe et al., 1989).

7.6.4 Ethanol treatment

Ethanol in the concentration of 60%–80% is employed for inactivation of bacteria, i.e., Gram-positive, Gram-negative, and acid-fast bacteria and lipophilic viruses (Fendler et al., 2002). Ethanol acts by denaturing the proteins, leading to dehydration of cellular components and dissolution of cell membrane lipids, leading to microbial inactivation (Prescott et al., 2004; Marreco et al., 2004).

7.6.5 Cryopreservation and freeze-drying

Cryopreservation of tissue transplants with the use of cryoprotectants viz. glycerol and dimethyl sulfoxide is extensively used for preserving biodegradable scaffolds.

Freeze-drying (lyophilization) enacts by protein and enzyme denaturation by incorporating low temperature (Privalov, 1990). This method disrupts the membranous structure of the microorganisms and thus removes the bound water by means of dehydration finally leading to microbial inactivation.

7.6.6 Antibiotic regimen

Limited information on the use of antibiotics for sterilization of biodegradable scaffolds is available. Antibiotic sterilization act by interference in DNA replication, synthesis of cell walls, and protein synthesis, which result in bacterial inactivation (Hancock, 2005). This method has been noted to be ineffective against viruses, fungi, and molds.

7.6.7 Aqueous glutaraldehyde sterilization

Glutaraldehyde available as aqueous solution possesses strong biocidal activity when activated by acquiring the required pH. Glutaraldehyde is also known to possess excellent polymerization properties. Complete immersion of biological products in the glutaraldehyde-sterilizing solution is a must followed by a thorough rising of the product so as to remove the residual solution (Rutala and Weber, 2014).

7.7 Risk management of tissue-based products

Typical hazards of medical devices manufactured utilizing animal tissues or derivatives are contamination by bacteria, molds, or yeast, contamination by viruses or transmissible agents such as pathogenic entities or agents causing spongiform encephalopathies, prions, and similar entities (bovine spongiform encephalitis [BSE], scrapie), and undesired pyrogenic, immunologic, and toxicological reactions.

Of these, more importance has been given to contamination with transmissible agents such as agents causing spongiform encephalopathies considering their resistance to most of the existing sterilization methods.

For ensuring the safety of medical devices based on animal tissues, two complimentary approaches are adapted to control the potential contamination of tissues. They are

1. Selecting source material for minimal contamination with agents by way of implementing ISO 22442-1 and ISO 22442-2. Here emphasis is on collection of animal tissue from a "low-risk herd" or "well-monitored herd" in which for at least the previous 6 years is given. "The low-risk herd" will have
   * A documented veterinary monitoring;
   * There has been no case of BSE;
   * There has been no feeding of mammalian-derived protein;
   * There is a fully documented breeding history;
   * Each animal is traceable;
   * Genetic material has been introduced only from herds with the same BSE-free status.
2. Testing the ability of the production processes to remove or inactivate agents by implementing the standard ISO 22442-3.

FDA differs in its approach with respect to tissue sourced from bovine and other animal sources. For example, all materials in a device, which are derived from a bovine source or exposed to it, are to be identified, as well as the bovine material should be sourced from cattle, which have not originated from or resided in a country where

BSE has been diagnosed or which presents a significant risk of introducing BSE. Each lot of bovine material as well as the product should be traceable, with records indicating the country of origin and residence of the animals. In case the bovine-derived material is only available from a country where BSE is known to exist, then the manufacturer should provide evidence to indicate that the BSE agent is inactivated during the manufacturing process.

For tissues from animal sources other than bovine, such as porcine heart valves, corneal shields, porcine blood vessels used in vascular grafts, porcine collagen used in wound dressings, and hyaluronic acid from rooster combs used in viscoelastic fluids, have to be identified by tissue type, species of origin, and country of origin/residence. Products derived from human tissue are covered under 21 CFR 1270, Human Tissue Intended for Transplantation, for additional requirements.

7.8 Conclusion

Tissue-derived products are made from nonviable or rendered nonviable animal/human tissue or its derivatives. They are available in acellular, chemically cross-linked, or tissue-derived forms. Different sterilization methods are available for tissue-derived products, which are effective. There are various national and international guidelines that elaborate on safe use of tissue-derived products.

References

alliqua.com/wp-content/uploads/2014/10/Biovance-Package-Insert.pdf.
AlloPatch HD®. Available online: https://www.mtfbiologics.org/.
Anderson, P.J., 1967. Purification and quantification of glutaraldehyde and its effect on several enzyme activities in skeletal muscle. J. Histochem. Cytochem 15, 652−661.
Arizono, T., Iwamoto, Y., Okuyama, K., Sugioka, Y., 1994. Ethylene oxide sterilization of bone grafts: residual gas concentration and fibroblast toxicity. Acta Orthopaedica Scandinavica 65, 640−642. https://doi.org/10.3109/17453679408994621.
Bader, A., Schilling, T., Teebken, O.E., Brandes, G., Herden, T., Steinhoff, G., 1998. Tissue engineering of heart valves- human endothelial cell seeding of detergent acellularised porcine valves. Eur. J. Cardiothorac. Surg 14, 279−284.
Badylak, S.F., 2002. The extracellular matrix as a scaffold for tissue reconstruction. Semin Cell Dev. Biol 13, 377−383.
Badylak, S.F., 2004. Xenogenic extracellular matrix as a scaffold for tissue reconstruction. Transplant. Immunol 12, 367−377.
Badylak, S.F., Freytes, D.O., Gilbert, T.W., 2009. Extracellular matrix as a biological scaffold material: structure and function. Acta Biomater 5 (1), 1−13.
Badylak, S.F., Gilbert, T.W., 2008. Immune response to biologic scaffold materials. Semin. Immunol 20, 109−116.
Badylak, S.F., Lantz, G.C., Coffey, A., Geddes, L.A., 1989. Small intestinal submucosa as a large diameter vascular graft in the dog. J. Surg. Res 47, 74−80.

Baird, R.M., 2004. "Sterility assurance: concepts, methods and problems," *Russell, Hugo & Ayliffe's Principles and Practice of disinfection*. Preserv. Steriliz 526−539. https://doi.org/10.1002/9780470755884.ch16.

Baumann, B.C., Stussi, G., Huggel, K., Rieben, R., Seebach, J.D., 2007. Reactivity of human natural antibodies to endothelial cells from Gala(1,3)Gal-deficient pigs. Transplantation 83, 193−201.

Baxter, Peri-Guard. Available online: http://ecatalog.baxter.com.

Bernad, M.P., Chu, M.L., Myers, J.C., Ramirez, F., Eikenberry, E.F., Prockop, D.J., 1983a. Nucleotide sequences of complimentary deoxyribonucleic acids for the proalpha 1 chain of human type1 procollagen: statistical evaluation of structures that are conserved during evolution. Biochemistry 22, 5212−5223.

Blauer, G., Harmatz, E., Meir, D., Swenson, M.K., Zvilichowsky, B., 1975. The interaction of glutaraldehyde with poly(alpha-L-lysine), N-butyl amine and collagen: the primary photon release in aqueous media. Biophysics 14, 2585−2598.

Bonewald, L.F., 1999. Regulation and regulatory activities of transforming growth factor beta. Crit. Rev. Eukaryot. Gene Expr 9, 33−44.

Borschel, G.H., Dennis, R.G., Kuzon, J.R.W.M., 2004. Contractile skeletal muscle tissue engineered on an acellular scaffold. Plast. Reconstr. Surg 113, 595−602.

Bourguignon, T., El khoury, R., Candolfi, P., Loardi, C., Mirza, A., Boulanger-lothion, J., Bouquiaux-stablo-duncan, A.L., Espitalier, F., Marchand, M., Aupart, M., 2015. Very long term outcomes of the Carpentier-Edwards perimount aortic valve in patients aged 60 or younger. Ann. Thorac. Surg 100, 853−859.

Bowers, J.H., Cater, C.W., 1966. The reaction of glutaraldehyde with proteins and other biological materials. J. Microsc. Soc 85, 193−200.

CarpentieR, S.M., Chen, L., Shen, M., Fornes, P., Martinet, B., Quintero, L.J., Witzel, T.H., Carpentier, A.F., 1998. Heat treatment mitigates calcification of valvular bioprostheses. Ann. Thorac. Surg 66, S264−S266.

Carpentier, A., Dubost, C., 1972. From xenograft to bioprosthesis: evolution of concepts and techniques of valvular xenograft. In: Ionescu, M.I., Ross, D.N., Wooler, G.H. (Eds.), Biological Tissue in Heart Valve Replacement. Butterworth &Co. Ltd., London, p. 515.

Cartmell, J.S., Dunn, M.G., 2000. Effect of chemical treatments on tendon cellularity and mechanical properties. J. Biomed. Mater. Res 49, 134−140.

Cebotari, S., Tudorache, I., Jaekel, T., Hilfiker, A., Dorfman, S., Ternes, W., Haverich, A., Lichtenberg, A., 2010. Detergent decellularization of heart valves for tissue engineering: toxicological effects of residual detergents on human endothelial cells. Artificial Organs 34 (3), 206−210.

Chanda, J., 1997. Heparin in calcification prevention of porcine pericardial bioprostheses. Biomaterials 18 (16), 1109−1112.

Chen, N., Field, E.H., 1995. Enhanced type 2 and diminished type 1 cytokines in neonatal tolerance. Transplantation 59, 933−934.

Chen, R.N., Ho, H.O., Tsai, Y.T., Sheu, M.T., 2004. Process development of an acellular dermal matrix for biomedical applications. Biomaterials 25, 2679−2686.

Chen, W., Schoen, F.J., levy, R.J., 1994. Mechanism of efficacy of 2-amino oleic acid for inhibition of calcification of glutaraldehyde-pretreated porcine bioprosthetic heart valves. Circulation 90, 323−329.

Cicha, I., Rüffer, A., Cesnjevar, R., Glöckler, M., Abbas, A., Werner, G.D., Christoph, D.G., 2010. Early obstruction of decellularised xenogenic valves in pediatric patients: involvement of inflammatory and fibroproliferative processes. Cardiovasc. Pathol. https://doi.org/10.1016/j.carpath.2010.04.006.

Collins, B.H., Chari, R.S., Magee, J.C., Harland, R.C., Lindman, B.J., Logan, J.S., Bollinger, R.R., Meyers, W.C., Platt, J.L., 1994. Mechanisms of injury in porcine livers perfused with blood of patients with fulminant hepatic failure. Transplantation 58, 1162−1171.

Conklin, B.S., Richter, E.R., Kreutziger, K.L., Zhong, D.S., Chen, C., 2002. Development and evaluation of a novel decellularised vascular xenograft. Med. Eng. Phys 24, 173−183.

Cooper, D.K., Good, A.H., Koren, E., Oriol, R., Malcolm, A.J., Ippolito, R.M., Neethling, F.A., Ye, Y., Romano, E., Zuhdi, N., 1993. Identification of apha galactosyl and other carbohydrate epitopes that are bound by human anti-pig antibodies: relevance to discordant xenografting in man. Transpl. Immunol 1, 198−205.

Courtman, D.W., Pereira, C.A., Kashef, V., Mccomb, D., Lee, J.M., Wilson, G.J., 1994. Development of a pericardial acellular matrix biomaterial: biochemical and mechanical effects off cell extraction. J. Biomed. Mater. Res 28, 655−666.

Cox, M.M., Battista, J.R., 2005. Deinococcus radiodurans—the consummate survivor. Nat. Rev. Microbiol 3, 882.

Crapo, M.P., GilberT, T.W., Badylak, S.F., 2011. An overview of tissue and whole organ decellularization process. Biomaterials 32, 3233−3243.

Dahl, S.L., Koh, J., Prabhakar, V., Niklason, L.E., 2003. Decellularised native and engineered arterial scaffolds for transplantation. Cell Transplant 12, 659−666.

DuraGuard. Available online: https://ecatalog.baxter.com/ecatalog/loadResource.blob?bid=55435.

Dziedzic-Goclawska, A., Kaminski, A., Uhrynowska-tyszkiewicz, I., Stachowicz, W., 2005. Irradiation as a Safety Procedure in Tissue Banking. Cell Tissue Bank 6, 201−219. https://doi.org/10.1007/s10561-005-0338-x.

Fendler, E.J., Ali, Y., Hammond, B.S., Lyons, M.K., Kelley, M.B., Vowell, N.A., 2002. The impact of alcohol hand sanitizer use on infection rates in an extended care facility. Am. J. Infect. Control 30, 226−233. https://doi.org/10.1067/mic.2002.120129.

Fischbach, C., Tessmar, J., Lucke, A., Schnell, E., Schmeer, G., Blunk, T., Göpferich, A., 2001. Does UV irradiation affect polymer properties relevant to tissue engineering? Surface Sci 491, 333−345. https://doi.org/10.1016/S0039-6028(01)01297-3.

Gabrielli, A., Candela, M., Ricciatti, A.M., Caniglia, M.L., Wieslander, J., 1991. Antibodies to mouse laminin in patients with systemic sclerosis(Scleroderma) recognizes galactosyl (alpha1-3) galactose epitopes. Clin. Exp. Immunol 86, 367−373.

Galili, U., Macher, B.A., Buehler, J., Shohet, S.B., 1985. Human natural anti-alpha galactosyl IgGII. The specific recognition of alpha (1-3) linked galactose residues. J. Exp. Med 162, 573−582.

Galili, U., Rachmilewitz, E.A., Peleg, A., Fletchner, I., 1984. A unique natural human IgG antibody with anti-alpha-galactosyl specificity. J. Exp. Med 160, 1519−1531.

Gilbert, T.W., Sellaro, T.L., Badylak, S.F., 2006. Decellularisation of tissues and organs. Biomaterial 27 (19), 3675−3683.

Gogolewski, S., Mainil-Varlet, P., 1997. Effect of thermal treatment on sterility, molecular and mechanical properties of various polylactides: 2. Poly (l/d-lactide) and poly (l/dl-lactide). Biomaterials 18, 251−255. https://doi.org/10.1016/S0142-9612(96)00132-9.

Goldberg, L., Lee, J., Cairns, T., Cook, T., Lin, C.K., Palmer, A., Simpson, P., Taube, D., 1995. Inhibition of human antipig xenograft reaction with soluble oligosaccharides. Transplant. Proc 27, 249−250.

Grieb, T.A., Forng, R.Y., Stafford, R.E., Lin, J., AlmeidA, J., Bogdansky, S., Ronholdt, C., Drohan, W.N., Burgess, W.H., 2005. Effective use of optimized, high-dose (50 kGy) gamma irradiation for pathogen inactivation of human bone allografts. Biomaterials 26, 2033−2042. https://doi.org/10.1016/j.biomaterials.2004.06.028.

Gulati, A.K., 1988. Evaluation of acellular and cellular nerve grafts in repair of rat peripheral nerve. J. Neurosurg 68, 117−123.

Hancock, R.E., 2005. Mechanisms of action of newer antibiotics for Gram-positive pathogens. Lancet Infect. Dis 5, 209−218. https://doi.org/10.1016/S1473-3099(05)70051-7.

Heseltine, P., 2001, 504 pages. In: Block, S.S. (Ed.), Disinfection, Sterilization and Preservation, Infection Control & Hospital Epidemiology, vol. 23. Lippincott Williams & Wilkins, Philadelphia, p. 1. https://doi.org/10.1017/S0195941700084289, 109-109.

Hilbert, S., Yanagida, R., Krueger, P., Jones, L.A., Wolfinbarger, L., Hopkins, R., 2004. A comparison of explant pathology findings of anionic and non-anionic detergent decellularised heart valve conduits. In: Nerem, R.M. (Ed.), Cardiovascular Tissue Engineering. From Basic Biology to Cell-Based Therapies. Hilton Head. Georgia Institute of Technology, SC, p. 2043.

Hodde, J.P., Badylak, S.F., Brightman, A.O., Voytik-Harbin, S.L., 1996. Glycosaminoglycan content of small intestinal submucosa: a bioscaffold for tissue replacement. Tissue Eng 2, 209−217.

Hsiao, C.Y., Liu, S.J., Ueng, S.W., Chan, E.C., 2012. The influence of γ irradiation and ethylene oxide treatment on the release characteristics of biodegradable poly (lactide-co-glycolide) composites. Polym. Degrad. Stab 97, 715−720. https://doi.org/10.1016/j.polymdegradstab.2012.02.015.

http://itmedica.com/wp-content/uploads/2017/01/LYOPLANT.pdf.
https://my.aami.org/aamiresources/previewfiles/14160_2016preview.pdf.
https://mimedx.com/amniocord/.
https://mimedx.com/amniofill/.
https://mimedx.com/epiburn/.
https://mimedx.com/epicord/.
https://mimedx.com/epifix/.
https://mimedx.com/orthoflo/.
www.miromatrix.com/miroderm.
https://organogenesis.com/products/puraply-antimicrobial-surgical.html
https://www.accessdata.fda.gov/cdrh_docs/pdf13/P130009d.pdf.
https://www.accessdata.fda.gov/cdrh_docs/pdf15/P150011d.pdf.
https://www.accessdata.fda.gov/cdrh_docs/pdf17/P170012C.pdf.
https://www.accessdata.fda.gov/cdrh_docs/pdf14/P140017d.pdf.
https://www.accessdata.fda.gov/cdrh_docs/pdf5/K053622.pdf.
https://www.accessdata.fda.gov/cdrh_docs/pdf14/P140031c.pdf.
https://www.accessdata.fda.gov/cdrh_docs/pdf13/P130021S033C.pdf.
https://www.accessdata.fda.gov/cdrh_docs/pdf/P900033S042c.pdf.
https://www.accessdata.fda.gov/cdrh_docs/pdf11/K113246.pdf.
https://www.accessdata.fda.gov/cdrh_docs/pdf7/K070751.pdf.
https://www.accessdata.fda.gov/cdrh_docs/pdf14/K142461.pdf.
https://www.accessdata.fda.gov/cdrh_docs/pdf/K983162.pdf.
https://www.accessdata.fda.gov/cdrh_docs/pdf10/K101866.pdf.
https://www.accessdata.fda.gov/cdrh_docs/pdf17/K171568.pdf.
https://www.accessdata.fda.gov/cdrh_docs/pdf4/K043573.pdf.
https://www.accessdata.fda.gov/cdrh_docs/pdf9/K090078.pdf.
https://www.accessdata.fda.gov/cdrh_docs/pdf6/k060989.pdf.
https://www.accessdata.fda.gov/cdrh_docs/pdf9/K090794.pdf.

https://www.biotissue.com/downloads/amniograft.../amniograft-insert_PI-BT-001E_V2.pdf.
https://www.edwards.com [homepage on the Internet]. transcatheter heart valve. Available from: https://www.edwards.com/gb/devices/heart-valves/transcatheter.
https://www.edwards.com/devices/bovine-pericardial-patches/cardiac.
https://www.hcp.stratticetissuematrix.com/.
https://www.lifenethealth.org/sites/default/files/files/68-40-083.pdf.
https://www.lifenethealth.org/sites/default/.../aortoiliac_spec_sheet_68-60-147v7.pdf.
https://www.lifenethealth.org/sites/default/files/product/68-60-084-02.pdf.
https://www.molnlycke.ae/products-solutions/ez-derm/
https://www.mtf.org/documents/PI_-70__Rev_5.pdf.
https://www.seaspine.com/products/orthoblast-2/.
https://www.swissbiomedortho.com/downloads/package_insert.pdf

Hudson, T.W., Zawko, S., Deister, C., Lundy, S., Hu, C.Y., Lee, K., 2004. Optimized acellular nerve graft is immunologically tolerated and supports regeneration. Tissue Eng 10, 1642−1651.

Integra LifeSciences Corp, 2008. Integra Matrix Wound Dressing [website]. Integra LifeSciences, Plainsboro, NJ.

Jackson, D.W., Grood, E.S., Arnoczky, S.P., Butler, D.L., Simon, T.M., 1987. Cruciate reconstruction using freeze dried anterior cruciate ligament allograft and a ligament augmentation device: an experimental study in a goat model. Am. J. Sports Med 15, 528−538.

Jayakrishnan, A., Jameela, S.R., 1996. Glutaraldehyde as fixative in bioprosthesis and drug delivery matrices. Biomaterials 17, 471−484.

Jordy, A., Hoff-jorgensen, R., Flagstad, A., Lund, E., 1975. Virus inactivation by ethylene oxide containing gases. Acta Veterinaria Scandinavica 16, 379−387.

ST. Jude medical. SJMTM Pericardial Patch with EnCapTM AC Technology. Available online: https://www.sjm.com.

Kasimir, M.T., Rieder, E., Seebacher, G., Nigisch, A., Dekan, B., Wolner, E., Weigel, G., Simon, P., 2006. Decellularisation does not eliminate thrombogenicity and inflammatory stimulation in tissue-engineered porcine heart valves. J. Heart Valve Dis 15 (2), 278−286.

Kim, S.M., Eo, M.Y., Kang, J.Y., Myoung, H., Lee, J.H., Cho, H.J., Yea, K.H., Lee, B.C., 2011. Bony Regeneration Effect of Electron-Beam Irradiated Hydroxyapatite and Tricalcium Phosphate Mixtures With 7 to 3 Ratio in the Caravel Defect Model of Rat.

Kim, M.S., Khang, G., Lee, H.B., 2007. Method and Techniques for Scaffold Sterilisation. A Manual For Biomaterials/Scaffold Fabrication Technology, pp. 239−249. https://doi.org/10.1142/9789812772114_0023.

Konakei, K.Z., Bohle, B., Blumer, R., Hoetzenecker, W., Rothe, G., Moser, B., Boltznitulescu, G., Gorhitzer, M., KlepetkO, W., Wolner, E., Ankersmit, H.G., 2005. Alpha Gal on bioprosthesis:Xenograft immune response in cardiac surgery. Eur. J. Clin. Invest 35, 17−23.

Koren, E., Neethling, F.A., Ye, Y., Niekrasz, M., Baker, J., Martin, M., Zuhdi, N., Cooper, D.K., 1992. Transplant. Proc 24, 598−601.

Lee, W.K., Park, K.D., Kim, Y.H., Suh, H., Park, J.C., Lee, J.E., Sun, K., Baek, M.J., Kim, H.M., Kim, S.H., 2001. Improved calcification resistance and biocompatibility of tissue patch grafted with sulphonated PEO or heparin after glutaraldehyde fixation. J. Biomed. Mater. Res 58 (1), 27−35.

Leonard, D., Buchanan, F., Farrar, D., 2006. Investigation into depth dependence of effect of E-beam radiation on mechanical and degradation properties of polylactide. Plast Rubber Compos 35, 303−309. https://doi.org/10.1179/174328906X143840.

Liang, H.C., Chang, Y., Hsu, C.K., Lee, M.H., Sung, H.W., 2004. Effects of cross linking degree of an acellular biological tissue on it tissue regeneration pattern. Biomaterials 25, 3541−3552.

Lin, P., Chan, W.C., Badylak, S.F., Bhatia, S.N., 2004. Assessing porcine liver derived biomatrix for hepatic tissue engineering. Tissue Eng 10, 1046−1053.

Marreco, P.R., Moreira, P.D., Genari, S.C., Moraes Â, M., 2004. Effects of different sterilization methods on the morphology, mechanical properties, and cytotoxicity of chitosan membranes used as wound dressings. J. Biomed. Mater. Res. B Appl. Biomater. Off. J. Soc Biomater 71, 268−277. https://doi.org/10.1002/jbm.b.30081. The Japanese Society for Biomaterials, and The Australian Society for Biomaterials and the Korean Society for Biomaterials.

Mcpherson, T.B., Liang, H., Record, R.D., Badylak, S.F., 2000. Gal alpha (1-3) Gal epitope in porcine small intestinal submucosa. Tissue Eng 6, 233−239.

Mcquillan, G.M., Coleman, P.J., Kruszon-moran, D., Moyer, L.A., Lambert, S.B., Margolis, H.S., 1999. Prevalence of hepatitis B virus infection in the United States: the national health and nutrition examination surveys, 1976 through 1994. Am. J. Public Health 89, 14−18. https://doi.org/10.2105/AJPH.89.1.14.

MEDICAL DEVICES RULES, 2017. Ministry of Health and Family Welfare. Department of Health and Family Welfare) Government of India.

Miyamoto, S., Katz, B.Z., Lafrenie, R.M., Yamada, K.M., 1998. Fibronectin and integrins in cell adhesion signaling and morphogenesis. Ann. NY Acad. Sci 857, 119−129.

Moore, T.M., Gendler, E., Gendler, E., 2004. Viruses adsorbed on musculoskeletal allografts are inactivated by terminal ethylene oxide disinfection. J. Orthop. Res 22, 1358−1361. https://doi.org/10.1016/j.orthres.2004.05.002.

Nagata, S., Hananyama, R., Kawane, K., 2010. Autoimmunity and clearance of dead cells. Cell 140 (5), 619−630.

Neethling, W.M.L., Ross, G., Hodge, A.J., 2004. ADAPT-treated porcine valve tissue (Cusp and wall) versus medtronic freestyle and prima plus: crosslink stability and calcification behavior in the subcutaneous rat model. J Heart Valve Dis 13, 689−696.

Nguyen, H., Morgan, D.A., Forwood, M.R., 2007. Sterilization of allograft bone: effects of gamma irradiation on allograft biology and biomechanics. Cell Tissue Bank 8, 93−105. https://doi.org/10.1007/s10561-006-9020-1.

Nimni, M.E., 1968. A defect in the intramolecular and intermolecular crosslinking of collagen caused by penicillamine.I. Metabolic and functional abnormalities in soft tissues. J. Biol. Chem 243, 1457−1466.

Nimni, M.E., Cheung, D., Strates, B., Kodama, M., Sheik, K., 1987. Chemically modified collagen: natural biomaterial for tissue replacement. J. Biomed. Mater. Res 21, 741−771.

Nina, V.J., Pomerantzeff, P.M., Casagrande, I.S., Cheung, D.T., Brandão, C.M., Oliveira, S.A., 2005. Comparative study of the L-hydro process and glutaraldehyde preservation, 13 (3), 203−207.

Ogle, M.F.1, Kelly, S.J., Bianco, R.W., Levy, R.J., 2003. Calcification resistance with aluminium-ethanol treated porcine aortic valve bioprostheses in juvenile sheep. Ann. Thorac. Surg 75 (4), 1267−1273.

Ohri, R., Hahn, S.K., Hoffman, A.S., Stayton, P.S., Giachelli, C.M., 2004. Hyaluronic acid grafting mitigates calcification of glutaraldehyde fixed bovine pericardium. J. Biomed. Mater. Res A 70 (2), 328−333.

O'brien, T.K., Gabbay, S., parkes, A.C., Knight, R.A., Zalesky, P.J., 1984. Immunological reactivity to a new glutaraldehyde tanned bovine pericardial heart valve. Trans. Am. Soc. Arti. Intern. Organs 30, 440−444.

Oriol, R., Ye, Y., Koren, E., Cooper, D.K., 1993. Carbohydrate antigens of pig tissue reacting with human natural antibodies as potential targets for hyperacute vascular rejection in pig to man organ xenotransplantation. Transplantation 56, 1433–1442.

Parisi, A.N., 1991. "Sterilization with Ethylene Oxide and Other Gases," Disinfection, Sterilization and Preservation.

Parvathy, T., Divakaran, N., Lalithakunjamma, R., Vijayan, N., Syam, V., Umashankar, R., 2013. Pathological effects of processed bovine pericardial scaffolds—a comparative in vivo evaluation. Artificial Organs 37 (7), 600–605.

Phillips, C.R., Kaye, S., 1949. The sterilizing action of gaseous ethylene oxide. Am. J. Hyg 50, 270–279.

Ponce, M., Nomizu, M., Delgado, M.C., Kuratomi, Y., Hoffman, M.P., Powell, S., Yamada, Y., Kleinmann, H.K., Malinda, K.M., 1999. Identification of endothelial cell binding sites on laminin gamma-1 chain. Circ. Res 84, 688–694.

Prescott, L., Harley, J., Klein, D., 2004. Microbiology, sixth ed. McGrawHillScience/Engineering/Math, New York, NY.

Privalov, P.L., 1990. Cold denaturation of protein. Crit. Rev. Biochem. Mol. Biol 25, 281–306. https://doi.org/10.3109/10409239009090612.

Probst, M., Dahiya, R., Carrier, S., Tanagho, E.A., 1997. Reproduction of functional smooth muscle tissue and partial bladder replacement. Br. J. Urol 79, 505–515.

Prolo, D.J., Pedrotti, P.W., White, D.H., 1980. Ethylene oxide sterilization of bone, dura mater, and fascia lata for human transplantation. Neurosurgery 6, 529–539. https://doi.org/10.1227/00006123-198005000-00006.

Reider, E., Seebacher, G., Kasimir, M.T., Eichmair, E., Winter, B., Dekan, B., Wolner, E., Simno, P., Weigel, G., 2005. Decellularised porcine and human valve scaffolds differ importantly in residual potential to attract monocytic cells. Circulation 111, 2792–2797.

Reing, J.E., Brown, B.N., Daly, K.A., Freund, J.M., Glibert, T.W., Hsiong, S.X., Huber, A., Kullas, K.E., Tottey, S., Wolf, M.T., Badylak, S.F., 2010. The effects of processing methods upon mechanical and biologic properties of porcine dermal extracellular matrix scaffolds. Biomaterials 31, 8626–8633.

Rieder, E., Kasimir, M.T., Siberhumer, G., Seebacher, G., Wolner, E., Simon, P., 2004. Decellularisation protocols of porcine heart valves differ importantly in efficiency of cell removal and susceptibility of the matrix to recellularisation with human vascular cells. J. Thorac. Cardiovasc. Surg 127, 399–405.

Rocco, K.A., Maxfield, M.W., Best, C.A., Dean, E.W., Bruer, C.K., 2014. In vivo application of tissue engineered vasculat graft: a review. Tissue Eng. B 20 (6), 628–640.

Rossini, A.A., Greiner, D.L., Mordes, J.P., 1999. Induction of immunological tolerance for transplantation. Physiol. Rev 79 (1), 101–106.

Rutala, W.A., Weber, D.J., 2014. Selection of the ideal disinfectant. Infect. Control Hospital Epidemiol 35, 855–865. https://doi.org/10.1086/676877.

Schoen, F.J., Levy, R.J., 2005. Calcification of tissue heart valve substitutes: progress toward understanding and prevention. Ann. Thorac. Surg 79, 1072–1080.

Schwarzbauer, J.E., 1999. Basement membranes: putting up the barriers. Curr. Biol. 9, R242–R244.

Setlow, P., 2006. Spores of Bacillus subtilis: their resistance to and killing by radiation, heat and chemicals. J. Appl Microbiol 101, 514–525. https://doi.org/10.1111/j.1365-2672.2005.02736.x.

Simon, P., Kasimir, M.T., Seebacher, G., Weigel, G., Ullrich, R., Salzer-muhar, U., Reider, E., Wolner, E., 2003. Early failure of the tissue engineered porcine heart valve SYNERGRAFT™ in pediatric patients. Eur. J. Cardiothorac. Surg 23, 1002–1006.

Smith, R.A., Ingels, J., Lochemes, J.J., Dutkowsky, J.P., Pifer, L.L., 2001. 'Gamma irradiation of HIV-1'. J. Orthopaedic. Res 19, 815−819. https://doi.org/10.1016/S0736-0266(01)00018-3.

Spoto, M.H., Gallo, C.R., Alcarde, A.R., Gurgel, M.S., Blumer, L., Walder, J.M., Domarco, R.E., 2000. Gamma irradiation in the control of pathogenic bacteria in refrigerated ground chicken meat. Scientia Agricola 57, 389−394. https://doi.org/10.1590/S0103-90162000000300003.

Sreelakshmi, M., 2017. Biomechanical and Histopathological Assessment of in Vivo Remodeling Response of Decellularized Bovine pericardium. MVSc Thesis. Kerala Veterinary and Animal Sciences University, Pookode.

Stone, K.R., Ayala, G., Goldstein, K., Hurst, R., Walgenbach, A., Galili, U., 1998. Porcine cartilage transplants in cynomolgus monkey.III Transplantation of porcine alpha galactosidase treated porcine cartilage. Transplantation 65, 1577−1583.

Stone, K.R., Walgenbach, A.W., Turek, T.J., Somers, D.L., Wicomb, W., Galili, U., 2007. Anterior cruciate ligament reconstruction with a porcine xenograft: a serologic, histologic, and biomechanical study in primates. Arthroscopy 23, 411−419.

Tom, J.A., Rodeo, S.A., 2002. Soft tissue allografts for knee reconstruction in sports medicine. Clin. Orthopaedic. Related Res 402, 135−156. https://doi.org/10.1097/01.blo.0000026965.51742.57.

Umashankar, P.R., Arun, T., Kumari, T.V., 2011. Short duration glutaraldehyde crosslinking of decellularised bovine pericardium improves biological response. J. Biomed. Mater. Res. A 97 (3), 311−320.

Umashankar, P.R., Arun, T., Kumari, T.V., 2013. Effect of chronic Inflammation and immune response on regeneration induced by decellularised bovine pericardium. J. Biomed. Mater. Res. A 101A, 2202−2209.

Umashankar, P.R., Mohanan, P.V., Kumari, T.V., 2012. Glutaraldehyde treatment elicits toxic response compared to decellularisation in bovine pericardium. Toxicol. Int 19 (1), 51−58.

Umashankar, P.R., Sabareeswaran, A., Sachin, J.S., 2017. Long term healing of mildly crosslinked decellularised bovine pericardial aortic patch. J. Biomed. Mater. Res. B Appl. Biomater 105 (7), 2145−2152.

US FDA, CFR- Code of Federal Regulations Title 21 Part 1271. Human Cells, Tissues and Cellular and Tissue Based Products.

Valentin, J.E., Badylak, J.S., Mccabe, J.P., Badylak, S.F., 2006. Extracellular matrix bioscaffolds for orthopedic applications: a comparative histologic study. J. Bone Joint. Surg. Am 88 (12), 2673−2686.

Vanderrest, M., Garrone, R., 1991. Collagen family of proteins. FASEB J. 5, 2814−2823.

Vascu-Guard. Available online: http://ecatalog.baxter.com.

Vishnu, S., 2015. Effect of differently processed bovine pericardium on tissue remodeling in rat subcutaneous model. MVSc thesis. Kerala Veterinary and Animal Sciences University, Pookode.

Vyavahare, N., Hirsch, D., Lerner, E., Baskin, J.Z., Schoen, F.J., Bianco, R., Kruth, H.S., Zand, R., Levy, R.J., 1997. Prevention of bioprosthetic heart valve calcification by ethanol pre-incubation: efficacy and mechanisms. Circulation 95, 479−488.

Watanabe, Y., Miyata, H., Sato, H., 1989. Inactivation of laboratory animal RNA-viruses by physicochemical treatment. Exp. Anim 38, 305−311. https://doi.org/10.1538/expanim1978.38.4_305.

Woods, T., Gratzer, P.F., 2005. Effectiveness of three extraction techniques in the development of a decellularised bone anterior cruciate ligament-bone graft. Biomaterials 26, 7339−7349.

www.accessdata.fda.gov/cdrh_docs/pdf15/P150048C.pdf.
www.bioteck.com/images/PDF/Brochures/NEURO_HeartDM_EN.pdf.
www.biotissue.com/products/prokera.aspx.
www.ethicon.com/na/products/wound.../surgical-gut-suture-plain-and-chromicC.
www.katena.com ⟩ Tutoplast® Fascia Lata
www.wright.com/.../AUGMENT-Injectable-Surgical-Technique-and-Package-Insert_J.
Yannas, I.V., Lee, E., Orgill, D.P., Skrbut, E.M., Murphy, G.F., 2001. Synthesis and Characterization of a model extracellular matrix that induces partial regeneration of adult mammalian skin. Proc. Natl. Acad. Sci. USA 86, 933−937.
Yixiang, D., Yong, T., Liao, S., Chan, C.K., Ramakrishna, S., 2008. Degradation of electrospun nanofiber scaffold by short wave length ultraviolet radiation treatment and its potential applications in tissue engineering. Tissue Eng. A 14, 1321−1329. https://doi.org/10.1089/ten.tea.2007.0395.
Zheung, M.H., Chen, J., Kirilak, Y., Willers, C., Xu, J., Wood, D., 2005. Porcine small intestine submucosa is not an acellular collagenous matrix and contains porcine DNA: possible implications in human implantation. J. Biomed. Mater. Res. B Appl. Biomater 73 (1), 61−67.
Zhu, A., Hurst, R., 2002. Anti-N-glycolylneuraminic acid antibodies identified in healthy human serum. Xenotransplantation 9, 376−381.

Further reading

CDRH BSE WORKING GROUP, 1998. Guidance For FDA Reviewers and Industry Medical Devices Containing Materials Derived From Animal Sources (Except For In Vitro Diagnostic Devices). U.S. Department of Health and Human Services, Food and Drug Administration, Center for Devices and Radiological Health.
https://www.accessdata.fda.gov/cdrh_docs/pdf9/K092021.pdf.
https://www.fda.gov/medicaldevices/deviceregulationandguidance/overview/classifyyourdevice/.
ISO 22442-1:2015: Medical Devices Utilizing Animal Tissues and Their Derivatives — Part 1: Application of Risk Management.
ISO 22442-2:2015: Medical Devices Utilizing Animal Tissues and Their Derivatives — Part 2: Controls on Sourcing, Collection and Handling.
ISO 22442-3:2007: Medical Devices Utilizing Animal Tissues and Their Derivatives — Part 3: Validation of the Elimination And/or Inactivation of Viruses and Transmissible Spongiform Encephalopathy (TSE) Agents.
REGULATION (EU) 2017/745 of the EUROPEAN PARLIAMENT and of the COUNCIL of 5 April 2017.
www.cryolife.com ⟩ Products ⟩ Cardiac Allografts

Tendon Regeneration

Jeffery D. St. Jeor[1], Donald E. Pfeifer[1], Krishna S. Vyas[2]
[1]Mayo Clinic School of Medicine, Rochester, MN, United States; [2]Department of Plastic Surgery, Mayo Clinic, Rochester, MN, United States

Chapter outline

- 8.1 Tendon cells and composition 188
- 8.2 Internal architecture 189
- 8.3 Importance of the complex three-dimensional structure 190
- 8.4 Tendon to bone insertion 191
- 8.5 Pure dense fibrous connective tissue 192
- 8.6 Uncalcified fibrocartilage 192
- 8.7 Tidemark 192
- 8.8 Calcified fibrocartilage 193
- 8.9 Bone 193
- 8.10 Supporting structures 193
- 8.11 Blood supply 194
- 8.12 Biomechanical properties 195
- 8.13 Impacting factors 196
- 8.14 Effects of aging 197
    - 8.14.1 Biochemical effects 197
    - 8.14.2 Biomechanical effects 198
- 8.15 Effects of exercise 198
    - 8.15.1 Biochemical response 198
    - 8.15.2 Biomechanical effects 199
    - 8.15.3 Overuse 199
- 8.16 Effects of immobilization 199
- 8.17 Tendon injury 199
- 8.18 Types of injury 200
- 8.19 Tendon healing 200
- 8.20 Mechanisms of healing 201
- 8.21 Surgical intervention 202
- 8.22 Tendon regeneration 202
- 8.23 Utilization of growth factors in tendon healing 203
    - 8.23.1 Transforming growth factor beta 203
    - 8.23.2 CTGF/CCN2 204
    - 8.23.3 Bone morphogenic protein family 204
    - 8.23.4 bFGF/FGF-2 205
        - 8.23.4.1 Insulin-like growth factor 1 205
        - 8.23.4.2 Platelet-derived growth factor 205

        8.23.4.3 VEGF 206
   8.23.5 Other growth factors of interest 206
   8.23.6 Autologous growth factor sources, platelet-rich plasma 207
8.24 Stem cell–based approaches to tendon healing 207
   8.24.1 Bone marrow–derived mesenchymal stem cells 208
   8.24.2 Adipose-derived mesenchymal stem cells 209
   8.24.3 Embryonic stem cells 209
   8.24.4 Induced pluripotent stem cells 209
   8.24.5 Tendon-derived stem cells 210
8.25 The role of biologic and synthetic scaffolds in tendon healing 210
   8.25.1 Collagen-based constructs 210
   8.25.2 Tissue-based constructs 211
   8.25.3 Synthetically engineered constructs 212
8.26 The role of gene transfer in tendon healing 213
8.27 Future of tendon regeneration 214
References 214

8.1 Tendon cells and composition

Tendon composition consists primarily of water (55%–70%), associated with a variety of highly organized and densely packed extracellular matrix (ECM) proteins and connective tissue cells. Elastin, proteoglycans (i.e., aggrecan and decorin), and glycoproteins (i.e., tenascin C and fibronectin) constitute 4%, 4%, and 2% of the matrix, respectively (Bordoni and Morabito, 2018). Proteoglycans in the ECM are highly associated with the water component of tendons. The proteoglycans hold water and resist compression, which gives the viscoelastic nature of tendons. Glycoproteins contribute to the mechanical stability and tendon healing (Bordoni and Morabito, 2018; Wang et al., 2012; Kjaer, 2004; Kannus, 2000; Benjamin et al., 2008). These ECM elements are produced by two types of fibroblasts in the tendon tissue: tenocytes and tenoblasts (Kannus, 2000). Tenoblasts represent 90%–95% of these cells. Fibroblasts themselves comprise 90%–95% of the tendon's cells with the other 5%–10% consisting of chondrocytes, which can be found at pressure and insertion points; synovial cells, which make up the tendon's sheath; and vascular cells. Fibroblasts are the cells responsible for the secretion of the ECM and therefore collagen assembly and turnover. These cells are typically arranged in longitudinal rows, in close proximity to the collagen fibrils (Bordoni and Morabito, 2018; Wang et al., 2012; Benjamin et al., 2008). Tendon collagen is mostly type I (60%). Along with type I collagen there are several other types of collagen in tendons, but they are in much smaller amounts, a few of these include II, III, V, X, and XII. Although they are found in much smaller amounts, these collagen types still play an important function for the tendons. Type II collagen is found at the osteotendinous junctions. Type III collagen plays a role in the healing process of tendons by forming rapid cross-links at repair sites to help stabilize the tendon. Type V collagen, in conjunction with type I, regulates the fibril diameter, a structural unit of collagen.

Type X collagen facilitates endochondral ossification by regulating the mineralization of the matrix. Type XII collagen provides lubrication between collagen fibers (Bordoni and Morabito, 2018; Wang et al., 2012; Kjaer, 2004; Kannus, 2000). There is also a type of tendon cell, termed tendon stem cell, that has the ability to self-renew and differentiate into tenocytes. This ability means that tendon stem cells play a crucial role in tendon maintenance and repair. Moreover, studies suggest that tendon stem cells may also be responsible for the development of tendinopathy by undergoing nontenocyte differentiation due to excessive mechanical loading conditions (Wang et al., 2012; Zhang and Wang, 2010; Zhou et al., 2010).

8.2 Internal architecture

Tendon collagen molecules are made up of polypeptide chains. Three polypeptide chains combine together to form tropocollagen, which is a densely packed helical and soluble molecule. Five tropocollagen molecules cross-link to make an insoluble microfibril. An aggregate of microfibrils forms fibrils. Fibrils are grouped into fibers, fibers into fiber bundles, and fiber bundles into fascicles (Kannus, 2000; Benjamin et al., 2008). The collagen fibers can be seen running longitudinally, transversely, and horizontally along the course of the tendon. Additionally, the longitudinal fibrils not only run only parallel but also cross each other in a variety of ways, forming spirals (plaits). Along the whole length of tendons, the ratio of longitudinally to transversely (or horizontally) running fibers ranges between 10:1 and 26:1 (Kannus, 2000; Jozsa et al., 1991). Along with the cellular divisions mentioned, cells in the tendon are linked to each other via gap junctions, which were shown by immunolabeling for connexin32 and connexin43. Connexin32 represents contact between the cell bodies, and connexin43 represents the meeting of cell processes as well as where cell bodies meet. This architecture of the fibroblasts in the tendon and their interconnection provides a three-dimensional network that surrounds the collagen fibrils and provides a basis for cell-to-cell interactions (Kjaer, 2004; Ralphs et al., 1998; McNeilly et al., 1996). This structural organization and the properties of the individual tendon components are what allow the tendon to withstand high tensile forces, while still maintaining a degree of compliance. This complex hierarchical division of structural components also ensures that minor damage inflicted does not spread to the entire tendon, rather is contained within a given structural area (Kjaer, 2004).

Within the tendon, the bundles of collagen fibers are not straight but rather show a wavy pattern with periodic changes of direction. This characteristic pattern is known as crimping. However, along the course of individual fibers or between the fibers of a fascicle, crimping is a rather varying phenomenon. In the same tendon, the crimps may differ in both size and geometry, appearing as isosceles or scalene triangles. This may be due to varying contribution of the proteoglycan cross-linking to the crimp (Bordoni and Morabito, 2018; Kannus, 2000). The ability of the tendons to transmit the force of muscle contraction is also closely related to tendon crimping. The greater load a tendon is subjected to, the greater the angle at the base of the crimps. When the tendon is stretched, the crimps gradually tend to become flat. The crimping helps provide a buffer for which longitudinal elongation can occur without fibrous damage. They also act as a shock absorber along the length of the tendon during the early stages of pulling and allow the tendon to recover its form when the applied force has ended (Bordoni and Morabito, 2018; O'Brien, 2005).

Surrounding the tendon unit is a membrane called the epitenon, which functions to reduce friction with adjacent tissues, acting like a synovium (Kirkendall and Garrett, 1997). Within the epitenon, the collagen fibrils are found with different orientations, transversely as well as longitudinally and obliquely. Occasionally, the epitenon fibrils appear to be fused with the superficial tendon fibrils. On its inner surface, the epitenon is continuous with the endotenon, a thin membrane of loose connective tissue that covers the individual tendon fibers and groups them in larger units represented by bundles of fibers of various order. The function of endotenon is to circumscribe the various orders of bundles and to allow the penetration and the capillary distribution of neurovascular bundles and lymphatic structures inside the tendon (Bordoni and Morabito, 2018; Kirkendall and Garrett, 1997). Such a hierarchical structure aligns all structural levels parallel to the long axis of the tendon, making it ideal for carrying and transmitting large tensile mechanical loads (Wang et al., 2012).

8.3 Importance of the complex three-dimensional structure

Tendons typically join muscle to bone; however, there are some tendons that connect one muscle belly to another. Each muscle essentially has two tendons: a proximal and a distal tendon. Tendons are capable of resisting high tensile forces from muscle contraction. The force applied to the tendon then leads to three possible outcomes:

(1) Tendon compliance and changes in tendon length lead to difficulty in holding a joint steady (i.e., movement). However, fine motor movements are made easier because the changes in muscle length lead to reduced force needed for change (Kannus, 2000; Benjamin et al., 2008; Cutts et al., 1991).

(2) The now-stretched tendons are able to store energy capable of reducing the work of the muscle by releasing the stored energy on recoil (Kannus, 2000; Benjamin et al., 2008; Cutts et al., 1991).

(3) To stretch the tendon, the muscle must shorten, possibly more than it normally would. This can be both advantageous (i.e., in situations when the tendon is acting like a spring) or disadvantageous (i.e., when the energy is transferred to an external system). The complex macro- and microstructure of tendons, and tendon fibers make this possible (Kannus, 2000; Benjamin et al., 2008; Cutts et al., 1991).

The muscles used during fine and delicate movements have long, thin tendons (hand flexor tendons), whereas those muscles that are used during power or endurance movements have short, robust tendons (Achilles tendon). These short tendons have a greater tensile strength than long tendons; this is shown in the larger load that is required to rupture shorter tendons of the same diameter. Along with tensile force, tendons are also subjected to shear and compression as tendons move and interact with surrounding bone (Benjamin et al., 2008). Longer tendons can undergo greater deformation before rupturing, when compared with shorter tendons, showing that strength and resistance are two separate properties of a tendon and they depend on the tendon's diameter and length (Bordoni and Morabito, 2018). However, tendons have a low resistance to shear forces due to their primary function of transmitting forces while losing little energy to deformation (Kirkendall and Garrett, 1997). The mechanical function and properties of a tendon is related to the shape of the tendon and the magnitude of forces applied to the tendon (Bordoni and Morabito, 2018). The above-described three-dimensional internal structure of the fibers forms a buffer medium against forces from various directions, thus preventing damage and disconnection of the fibers (Kannus, 2000).

8.4 Tendon to bone insertion

The enthesis is defined as the region where tendon, ligament, or joint capsule inserts into bone (i.e., an "attachment site," "insertion site," or "osteotendinous junction"). The entheses are sites of stress concentration and act to transmit tensile load from soft tissues to bone. Entheses are critical, as they allow for the proper transmission of contractile forces from the muscle belly to the respective skeletal attachment, while simultaneously dissipating force away from the enthesis itself, from tendon into bone (Apostolakos et al., 2014; Benjamin et al., 2006; Benjamin et al., 2002). While transferring forces, it is important to consider that the enthesis connects two mechanically different materials: tendon (a compliant, tough, protein-rich material) and bone (a hard, mineralized tissue). When two materials with differing mechanical properties and a sharp interface are exposed to externally applied loads, they are prone to stress concentrations and will exhibit nonuniform deformation. This mismatch in deformation between the two phases will cause a stress singularity to arise locally, increasing the risk of failure. Consequently, they are commonly subject to overuse injuries (Benjamin et al., 2006; Deymier et al., 2017).

This is explained by a general engineering principle that stress concentrates at interfaces between structures with differing mechanical properties. To avoid these stress concentrations and allow for effective stress transfer, the enthesis must be able to balance the differing elastic moduli of both tendon and skeletal tissue. This is accomplished by a complex hierarchical structure. At the tissue level, tendons attach to the bone with a splayed morphology and a large attachment area to diffuse stress at the interface. At the cellular level, the mineral and collagen content and organization help create a load-sharing mechanism (Benjamin et al., 2002; Deymier et al., 2017).

Entheses can be further described according to the type of tissue present at the skeletal attachment site, either dense fibrous connective tissue or fibrocartilage. Fibrous and fibrocartilaginous entheses have also been referred to as "periosteal-diaphyseal" and "chondroapophyseal" or "indirect" and "direct," respectively (Apostolakos et al., 2014; Benjamin et al., 2002; Benjamin and Ralphs, 2001).

At fibrous entheses, the tendon attaches either directly to the bone or periosteum through fibrous tissue, which resembles the tendon midsubstance. The bone or periosteum insertions are used to classify fibrous entheses as either "bony" or "periosteal," respectively (Apostolakos et al., 2014; Benjamin and Ralphs, 2001). Fibrous entheses are common in tendons that attach muscles to the metaphysis and diaphysis of long bones (i.e., deltoid). Fibrous entheses insertions typically occur over large surface areas and are characterized by perforating mineralized collagen fibers (Apostolakos et al., 2014; Lu and Thomopoulos, 2013). These entheses are less common and suffer overuse injuries less often than fibrocartilaginous entheses (Apostolakos et al., 2014; Benjamin et al., 2006; Benjamin et al., 2002; Lu and Thomopoulos, 2013).

Fibrocartilaginous entheses, as their name suggests, attach to bone through a layer of fibrocartilage, which acts as a transition from the fibrous tendon tissue to bone (Apostolakos et al., 2014; Benjamin et al., 2006). In contrast to fibrous enthesis insertions, fibrocartilaginous insertions are found on the epiphysis and apophysis (i.e., rotator cuff). Fibrocartilaginous entheses are also sites where chondrogenesis occurs, thus creating four distinct tissue zones that act as a structural gradient from uncalcified tendon to calcified bone. The four zones are pure dense fibrous connective tissue, uncalcified fibrocartilage, calcified fibrocartilage, and bone (Apostolakos et al., 2014; Benjamin et al., 2006).

8.5 Pure dense fibrous connective tissue

Pure dense fibrous connective tissue is composed primarily of pure tendon. Similar to tendons, pure dense fibrous connective tissue composition consists of fibroblasts, type I and III collagen, elastin, and proteoglycans. Because of the similarity to tendon midsubstance, the mechanical properties of the pure dense fibrous connective tissue zone are similar to the properties of tendons (Apostolakos et al., 2014; Benjamin and Ralphs, 2001).

8.6 Uncalcified fibrocartilage

Uncalcified fibrocartilage, as the name suggests, is an uncalcified avascular zone. This zone's composition is made up of fibrochondrocytes, aggrecan, and types I, II, and III collagen. On a functional note, this zone acts as a force damper to dissipate stress generated by bending collagen fibers during movement. This unique characteristic leads to differing amounts of uncalcified fibrocartilage present, based on the degree of force and movement that an enthesis is subjected to (Apostolakos et al., 2014; Benjamin and Ralphs, 2001).

8.7 Tidemark

Although not mentioned as one of the four zones, the tidemark is a basophilic line that is seen separating the uncalcified and calcified fibrocartilage zones. It acts as the mechanical boundary between a soft and hard tissue (Benjamin and Ralphs, 2001; Angeline and Rodeo, 2012). The tidemark is a relatively straight demarcation, which suggests that the mineralization process taking place in calcified fibrocartilage produces a flat surface, which is important clinically because it reduces the risk of damage to soft tissues during joint movement (Apostolakos et al., 2014; Angeline and Rodeo, 2012).

8.8 Calcified fibrocartilage

Calcified fibrocartilage is an avascular and calcified zone that is less cellular compared with its uncalcified counterpart. Calcified fibrocartilage consists of fibrochondrocytes, mostly type II collagen with type I and X, and aggrecan. This zone represents the junction between tendon and bone, and unlike the relatively flat tidemark, this transition is highly irregular. However, this irregularity is as important as the uniformity in the tidemark. The irregularity is from the interlocking attachments of the calcified fibrocartilage and bone; these attachments are what provide the mechanical integrity of the enthesis (Apostolakos et al., 2014; Angeline and Rodeo, 2012; D'Agostino, 2010; McGonagle and Benjamin, 2015).

8.9 Bone

The bone zone consists of osteoclasts, osteocytes, and osteoblasts, with a type I collagen and carbonate apatite mineral matrix (Apostolakos et al., 2014). The bone at the site of an enthesis is closely integrated with the nearby cancellous bone. This close relationship, along with the fibrocartilage and boney attachments mentioned above, is what gives the tendon anchorage at the site of attachment (D'Agostino, 2010; McGonagle and Benjamin, 2015).

8.10 Supporting structures

The main task of the structures surrounding the tendon is to facilitate tendon sliding and prevent it from deviating course during movement (Bordoni and Morabito, 2018). These structures can be divided into five categories:

(1) Fibrous sheath (or retinacula): There are channels or grooves through which tendons, usually long ones (i.e., extensors and flexors of the hand and feet), glide during their course of movement. The sliding of tendons on neighboring tissues could be seriously impaired by the resulting friction if not for these boney grooves and notches. These boney channels and grooves generally have a fibrocartilage floor and are covered with a fibrous sheath or retinaculum (Bordoni and Morabito, 2018; Kannus, 2000; O'Brien, 2005).
(2) Synovial sheaths: Synovial sheaths are associated with the fibrous retinaculum, found directly below the fibrous layer, and facilitate sliding of the tendon inside the retinaculum. The synovial sheath consists of two thin serous sheets: the parietal sheet and visceral sheet. The parietal layer covers the wall of the fibrous sheath, and the visceral layer covers the tendon surface. These two sheets essentially form a closed duct and contain a peritendinous liquid that serves as lubrication. It is important to note, however, that synovial sheaths are not ubiquitous with tendons; they are only found in areas where tendons experience a sudden change in direction or are subjected to increased friction that require efficient lubrication (Bordoni and Morabito, 2018; Kannus, 2000; O'Brien, 2005).
(3) Reflection pulleys: Reflection pulleys serve as the anatomic reinforcements for fibrous sheaths of tendons that have curves along their course. They are a circumscribed thickening of dense fibrillary tissue whose task is to keep the tendon inside the sliding bed (Bordoni and Morabito, 2018; Kannus, 2000; O'Brien, 2005).

(4) Tendon bursae: Tendon bursae are small serous vesicles and serve to reduce friction between tendons and surrounding structures. They are particularly found at sites where bony prominences might otherwise compress and put wear and tear on the tendon. Typical examples include subacromial, infrapatellar, and retrocalcaneal bursae (Bordoni and Morabito, 2018; Kannus, 2000).
(5) Paratenon: In some tendons that do not have a true synovial sheath, there can be a peritendinous sheet to reduce friction. The paratenon is composed of type I and III collagen and thin elastic fibers. The presence of elastic fibers allows for stretch, and the paratenon as a whole helps reduce friction and acts as an elastic sleeve that provides free movement with respect to the surrounding structures. The characteristic example of a well-defined paratenon is the Achilles tendon (Bordoni and Morabito, 2018; Kannus, 2000; Kirkendall and Garrett, 1997).

8.11 Blood supply

Early in development, tendons are highly vascular and metabolically active, receiving their nutrition through vascular perfusion from a dense capillary network (Peacock, 1959; Fenwick et al., 2002). Mature tendons, however, are poorly vascularized and are much less metabolically active, relying on nutrition through diffusion from synovial fluid (Fenwick et al., 2002). Generally, a number of different vessels are responsible for the blood supply of a tendon. These blood vessels originate from three different areas: the musculotendinous junction, the osteotendinous junction, or vessels from surrounding connective tissue such as the paratenon (Peacock, 1959; Fenwick et al., 2002). The blood vessels of tendons are typically arranged in a longitudinal orientation within the tendon. Because tendons are moving tissue and subjected to mechanical load and large degrees of movement, the tendon vasculature must be able to comply to these forces as well. That is why in tendons that are subjected to a large degree of movement (i.e., tendons of the hand), their blood supply is composed of tortuous vessels that allow for straightening during tendon movement (Fenwick et al., 2002; Brockis, 1953; Tempfer and Traweger, 2015). This demand for compliance from tendon vasculature leads to some position-dependent filling of vascular beds known as watershed areas or critical zones (Tempfer and Traweger, 2015; Rathbun and Macnab, 1970). It is at these zones that you see that tendons are prone to inflammatory episodes and/or rupture (Tempfer and Traweger, 2015).

There are significant differences between the vascular networks of different tendons: sheathed or unsheathed. Sheathed tendons have a well-defined vasculature. The blood vessels only enter the tendon at specific points along the tendon, whereas in unsheathed tendons, vessels may pass through the surrounding paratenon into the tendon at any point (Fenwick et al., 2002; Tempfer and Traweger, 2015). These two vascular patterns between the sheathed and unsheathed tendons have been referred to as "avascular tendons" and "vascular tendons," respectively, with major implications on healing potential (Tempfer and Traweger, 2015; Chaplin, 1973).

8.12 Biomechanical properties

In addition to their high stiffness, tendons, like many tissues in the body, exhibit viscoelastic, or time-dependent, behavior. This means that when the tendon is held at a constant strain level, the stress in the tendon decreases, a phenomenon known as stress relaxation. Conversely, when tendons are held at a constant stress level, the strain in the tendon increases; this is known as creep. What this means practically is that tendons at low strain rates absorb more mechanical energy but are less effective in carrying mechanical loads. At high strain rates, tendons become stiffer and more effective in transmitting large muscular loads to bone (Wang et al., 2012; Duenwald et al., 2009). This viscoelasticity characteristic is thought to be a function of the tendon structure and composition (i.e., collagen, proteoglycans, glycoproteins, and water) (Duenwald et al., 2009).

The time-dependent properties of tendons affect their ability to convert muscle contraction into skeletal movement, as well as positional stability of the body. The viscoelastic character of tendons leads to a nonlinear stress–strain curve under quasi-static loading conditions (Fig. 8.1).

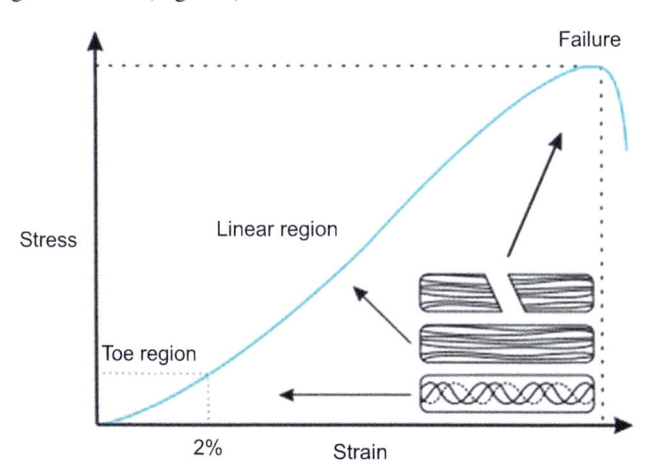

Figure 8.1 Strain to stress relationship and failure of tendons. From Kelc R., Naranda J., Kuhta M., Vogrin M., May 15, 2013. The physiology of sports injuries and repair processes. In: Michael H., Nick D., Yaso K. (Eds.), Current Issues in Sports and Exercise Medicine, IntechOpen, https://doi.org/10.5772/54234. Available from: https://www.intechopen.com/books/current-issues-in-sports-and-exercise-medicine/the-physiologyof-sports-injuries-and-repair-processes.

The non–linear stress–strain curve consists of three distinct regions:

(1) Toe region: This is where "stretching out" or "uncrimping" of collagen fibers occurs from mechanically loading the tendon up to 2% strain. This region is responsible for non–linear stress/strain curve because the slope of the toe region is not linear (Kannus, 2000; Screen, 2008; Kelc & et al., 2013).

(2) Linear region: This is the physiological upper limit of tendon strain, whereby the collagen fibrils orient themselves in the direction of tensile mechanical load and begin to stretch. The majority of tendon extension occurs through the intermolecular sliding of collagen triple helices. If strain is less than 4%, the tendon will return to its original length when unloaded, therefore this portion is elastic and reversible, and the slope of the curve represents the Young's modulus (Kannus, 2000; Screen, 2008; Kelc et al., 2013).

(3) Yield and failure region: This is where the tendon stretches beyond its physiological limit, and intramolecular cross-links between collagen fibers fail. If microfailure continues to accumulate, stiffness is reduced and the tendon begins to fail, resulting in irreversible plastic deformation. If the tendon stretches beyond 10%—15% strain, the tendon fails through fiber pullout (Kannus, 2000; Screen, 2008; Kelc et al., 2013). More on tendon injury will be discussed later in this chapter.

To accurately understand the behavior and mechanism of injury of tendons, it is important to understand the viscoelasticity of tendons. So not only does this mean understanding stress relaxation, creep, and the nonlinear stress—strain relations as mentioned above but also understanding the recovery behavior of a tendon when the load has been removed. Tendons, because of their viscoelastic properties, experience different loading and unloading curves in the range of 5%—10%, where energy is dissipated as heat; this phenomenon is known as hysteresis (Screen, 2008; Kelc et al., 2013). The postload recovery helps us understand not only the properties and characteristics of an unloaded tendon but also how the tendon will respond to subsequent loadings, especially if the tendon has not fully recovered from the previous load (Duenwald et al., 2009; Screen, 2008). Tendons with a slow postload recovery will likely not achieve full recovery before subsequent loading. This means it is more likely to deform from its original length under similar loads. This causes a slightly longer tendon, which reduces the efficiency of the muscle—tendon contraction (Duenwald et al., 2009).

These characteristics become important when trying to understand, quantify, and treat injuries in tendons. For example, when considering the use of a tendon graft, understanding the viscoelastic and postload recovery of the predamaged tendon is important to ensure proper joint kinematics and to maximize success in joint reconstruction. This will be discussed in greater detail later in the chapter (Duenwald et al., 2009; Screen, 2008).

8.13 Impacting factors

As previously discussed, the force generated by muscle contraction is transmitted via tendons to the bone and produces joint moments. Tendons experience much higher stress during locomotion than any other component in the musculoskeletal system. The greater the force generated by the muscle, the greater the stress that is transmitted through the tendon to the bone. Historically, tendons have been thought of as relatively inert structures; however, we now know that tendons are capable of changing their composition, structure, and mechanical properties in response to the external and internal factors they are subjected to (Wang et al., 2012; Kjaer, 2004; Svensson et al., 2016; Christensen et al., 2008; Tardioli et al., 2012). However, even with these tendon adaptations, it still has poor healing ability, and both acute and chronic injuries remain a clinical challenge.

As expected from differing functions and demands, different types of tendons vary in their mechanical and biomechanical properties. For example, the Young's modulus varies greatly from tendon to tendon. Young's modulus is classically defined as the modulus of elasticity of a material, calculated by the rate of change of stress with strain, and is an intrinsic property and is used along with stiffness to measure the elastic properties of a tendon (Ensey et al., 2009). The patellar tendon has a Young's modulus around 660 MPa, whereas the tibialis anterior tendon is about 1200 MPa (Wang et al., 2012; Maganaris and Paul, 1999; Johnson et al., 1994). Another difference seen among tendons is that, as a general rule, extensor tendons are more flattened, while flexor tendons are more rounded or oval. Some of the longest tendons are those in the hands and feet. Here, the tendons help to modulate the speed at which the distal elements can move. This is accomplished by having their attachment sites either closer to or father from the axis of movement (Benjamin et al., 2008).

8.14 Effects of aging

8.14.1 Biochemical effects

During early development, the tendon fibrils are small and uniform in diameter, but from adolescence onward, they become more variable in size. However, aging has been shown to cause a decrease in mean fibril diameter, possibly regulated by type V collagen. The largest mean fibril diameters of tendons are reported to occur between 20 and 29 years old, and the average diameter then decreases with increasing age (Benjamin et al., 2008). As humans age, the tendon cells' morphology change from the relatively round cells in the young/immature tendon, to flatter, very long spindle-shaped cells. They become less numerous, with low amounts of cytoplasm and reduced organelle activity, and their long, thin cytoplasmic projections shorten and diminish in number (Benjamin et al., 2008; Svensson et al., 2016). There is also a change in both cell density and cellular activity seen with an increase in age. The drop in cell density is believed to be from age, as well the large expansion of ECM. Age-related changes in cellular function are seen with in vitro studies on primary tendon cell cultures. Tendon stem cell activity, cell proliferation, and migration have been reported to decrease with age after maturation has been reached (Svensson et al., 2016; Kohler et al., 2013). There is also very limited tissue turnover in aging adult tendons, which is to be expected with the slowing in growth after tendon maturity has been reached (Svensson et al., 2016).

The changes to the structure and composition seen with aging are thought to impact the tendon mechanical function as the tissue ages. Both enzymatic and nonenzymatic cross-linking reactions that occur in between collagen molecules of tendons can be affected by age. With maturation, there is a drastic change in enzymatic cross-links—divalent cross-links are replaced by trivalent cross-links (Svensson et al., 2016; McCrum et al., 2018). The nonenzymatic cross-linking reaction is a glycation process and is mostly unregulated. This process involves sugar molecules continuously attaching themselves to collagen molecules, leading to increasing amounts of advanced glycation end products (AGEs) as we age. AGE accumulation is a result of collagen turnover rates; as tendons have a very low turnover rate, there is a great deal of AGE accumulation. This process is thought to have an important role with aging (Svensson et al., 2016; McCrum et al., 2018). The accumulated AGEs increase the distance between collagen molecules within fibrils, which affects both the molecular structure and properties of tendons. AGEs likely also contribute to the loss of water, as the glycation reaction causes dehydration of collagen. Using magnetic resonance imaging (MRI) of human tendon in vivo, it has been shown that the MRI signal intensity is altered with aging, reflecting a change in the internal milieu of the tendon (Svensson et al., 2016; Carroll et al., 2008).

8.14.2 Biomechanical effects

Mechanical properties of tendons, such as tension and Young's modulus, influence the overall performance of the muscle—tendon complex. During development, there is an increase in tendon strength until maturation. As we age, it is no surprise that in conjunction with the changes mentioned above, tendons also experience degradation in their mechanical properties, which can affect the rate of force development, elastic energy return, and electromechanical delay (Svensson et al., 2016; Ensey et al., 2009; McCrum et al., 2018). An example of this was shown when examining patellar tendon strength from donors. The Young's modulus of young donors (29—50 years old) was 660 MPa, whereas in tendons from old donors (64—93 years old), Young's modulus was 504 MPa (Wang et al., 2012; Maganaris and Paul, 1999). The lower figures of Young's modulus seen on older donors are indicative of intrinsically weaker tendon structures. A decrease in tendon stiffness in older tendons is an indication that the tendon has become more compliant and in older adults, that translates to slower transmission of force, slower torque development, and decreased performance (Ensey et al., 2009; McCrum et al., 2018). These parameters can create problems with balance and mobility, which is often seen in elderly people.

8.15 Effects of exercise

8.15.1 Biochemical response

There are a number of effects that long-term exercise and physical training have on tendons. Long-term training causes connective tissue remodeling and increased tendon collagen turnover (Tardioli et al., 2012; Brumitt and Cuddeford, 2015). There is a three- to sevenfold increase in blood flow during exercise. This increase in blood flow is in response to changes in peritendinous pressure and release of prostaglandins, bradykinin, and adenosine (Tardioli et al., 2012; Brumitt and Cuddeford, 2015). There is a change in the metabolic activity of tendons with an increase of glucose uptake and enzymatic function. Exercise also influences peritendinous lactate and inflammatory markers such as prostaglandin-E2, thromboxane-B2, and interleukin (IL)-6 turnover

(Tardioli et al., 2012). In addition, proteoglycan content appears to increase, along with enzymatic cross-linking. Conversely, AGEs accumulation can be reduced in the tendons due to increased tendon turnover (Svensson et al., 2016).

8.15.2 Biomechanical effects

The mechanical properties of tendons are determined by the microstructural parameters, including collagen fiber content, fiber orientations, and cross-link density. Tendon fibroblasts change these parameters in response to exercise by increasing their biosynthetic activity, causing an increase in the number of collagen fibrils, an increase in collagen fibril diameter, an increase in fibril packing density, an increase in tendon cross-sectional area, and aligning fibers along the direction of the tensile strain (Brumitt and Cuddeford, 2015; Wren et al., 2000). Tendons have also shown an increase of up to 20% in stiffness as a response to exercise (Wren et al., 2000; Buchanan and Marsh, 2002). Along with an increase in stiffness, there is an increase in both tendon strength and tendon Young's modulus (Wren et al., 2000).

8.15.3 Overuse

However, exercise can cause negative effects to the tendon. Repetitive loading of human tendons can lead to an overuse injury because of inadequate time for the tendon tissue to return to its normal nonloaded state. In overuse injuries, tendons experience a change within the tendon substance itself. These changes are either from primary alterations in the biochemical composition or gradually developing degenerative changes (Kjaer, 2004).

8.16 Effects of immobilization

Chronic inactivity (20–90 days bed rest) results in a reduction in tendon stiffness and collagen synthesis. Immobilization may also lead to a decrease in cross-sectional area of the tendon due to a lack of tendon loading. Immobilization of tendons also causes catabolic effects on tendons. Overall reduced habitual loading due to decreased physical activity and muscle strength can lead to loss of tendon functionality (Wang et al., 2012; Christensen et al., 2008; McCrum et al., 2018).

Overall, it is important to note that the various changes mentioned above are somewhat controversial, and a number of studies have been done and are ongoing to add clarity to the tendon response to intrinsic and extrinsic factors.

8.17 Tendon injury

Because of their key role in locomotion, tendons are at risk for overuse, traumatic, and degenerative injury. When tendons are subjected to strain levels above the tissue's tensile capabilities, microtrauma or macrotrauma will result. There have been some predisposing factors identified, which influence the frequency of tendon injuries. Some of these predisposing factors include genetic, chronic disease, and drug use (Kjaer, 2004; Wu et al., 2017). Other factors such as high body weight, leg length inequality, foot abnormalities, and low flexibility of joint, tendon, or muscle have been identified as important associations, not necessarily causations, for developing tendon injury (Kjaer, 2004; Wu et al., 2017).

8.18 Types of injury

Tendon injuries, like most injuries, can be broken up into two major classes: acute and chronic. Acute injuries include tendonitis injuries (i.e., peritenonitis, tenosynovitis, and tenovaginitis), lacerations, and ruptures. Tendonitis injuries are accompanied by pain and inflammation (Wang et al., 2012; Brumitt and Cuddeford, 2015; Wu et al., 2017). Chronic injuries are tendinosis, or sometimes called tendinopathy. Tendinosis is degenerative in nature and usually lacks any signs of inflammation but rather is associated with the formation of damaging lipids, proteoglycans, and calcified tissue in tendons (Wang et al., 2012; Brumitt and Cuddeford, 2015; Wu et al., 2017). Tendon ruptures most frequently occur in regions of low vascularization and tend to follow tendinopathy injuries (Fenwick et al., 2002; Wu et al., 2017).

Tendon degeneration occurs when there is a maladaptation remodeling in response to chronic mechanical load and stressors. There is an imbalance between the matrix breakdown and synthesis. This imbalance leads to some molecular and structural changes: higher type III collagen in relation to type I collagen in the ECM and increased levels of proteoglycans and glycoproteins (Fenwick et al., 2002; Wu et al., 2017).

8.19 Tendon healing

Tendons have the capability of repair. This process is controlled by the tendon cells and the components found in the ECM. The healing process is broken up into three distinct but overlapping stages: tissue inflammation, cell proliferation (reparative), and ECM remodeling (consolidation and maturation). The phase duration is largely dependent on the severity and location of the injury (Wu et al., 2017; Docheva et al., 2015).

(1) Inflammatory stage: This stage occurs shortly after injury and begins with the formation of a hematoma. Proinflammatory cytokines then recruit cellular components and inflammatory cells such as neutrophils, monocytes, and platelets (Brumitt and Cuddeford, 2015; Wu et al., 2017; Docheva et al., 2015). Secreted angiogenic factors then create a vascular network at the injury site. This newly formed network is responsible for stabilization and survival of the newly forming fibrous tissue. Angiogenesis is a key step in healing because lack of a blood supply has been shown to impair healing (Docheva et al., 2015).
(2) Reparative stage: During this stage, synthesis of ECM components takes place and occurs a few days after injury. In particular, type III collagen synthesis increases due to recruited fibroblasts, but there is also an increase of other components such as proteoglycans (Bordoni and Morabito, 2018; Brumitt and Cuddeford, 2015; Wu et al., 2017; Docheva et al., 2015). It is important to note that these ECM components are laid down in a random order (Docheva et al., 2015). Other characteristics of the reparative stage are an increase in cellularity and an increase in absorption of large amounts of water.

(3) Remodeling: Around 6–8 weeks after injury, tendons begin remodeling and can take anywhere from 1 to 2 years to complete (Docheva et al., 2015). This stage can actually be broken up into two substages: consolidation and maturation.
   Consolidation: Consolidation begins around 6–8 weeks after injury, and during this phase, there is a decrease in both cellularity and matrix production. The type III collagen is replaced with type I collagen, which makes the site of injury more fibrous. The once unorganized orientation of ECM now becomes organized as collagen fibers are oriented along the longitudinal axis of the tendon. The ECM organization, along with collagen replacement, begins to restore the tendon stiffness and tensile strength (Brumitt and Cuddeford, 2015; Wu et al., 2017; Docheva et al., 2015).
   Maturation: Maturation begins around 10 weeks after injury. During maturation, the cross-linking between collagen fibrils increases and there is the formation of tendon-like tissue (Brumitt and Cuddeford, 2015; Wu et al., 2017; Docheva et al., 2015).

8.20 Mechanisms of healing

There are two overlapping mechanisms for tendon healing: extrinsic and intrinsic. It is believed that these two phases act cooperatively to promote tendon healing (Docheva et al., 2015). Extrinsic healing starts first, with fibroblasts and inflammatory cells from the tendon periphery invading the site of injury and synthesizing the initial collagen matrix and cellular adhesions. These, in turn, help kick-start the healing process. At this point, intrinsic healing becomes active with the activation of local stem cells from the endotenon. These cells then migrate to the site of injury and begin reorganizing the ECM and support the newly formed vascular network (Wu et al., 2017; Docheva et al., 2015; Sun et al., 2015).

Tendon healing is a complex process that requires a large degree of synchronization and organization. The driving force behind this is mediated by a number of molecules, including inflammatory cytokines (i.e., IL-6 and IL-1β) (Wu et al., 2017; Docheva et al., 2015; Voleti et al., 2012). In later tendon healing, a number of growth factors drive the process (i.e., basic fibroblast growth factor [bFGF], bone morphogenic proteins [BMPs], transforming growth factor beta [TGF-β], insulin-like growth factor 1 [IGF-1], platelet-derived growth factor [PDGF], and vascular endothelial growth factor [VEGF]) (Apostolakos et al., 2014; Wu et al., 2017; Docheva et al., 2015; Voleti et al., 2012). During this coordination and synchronization of the healing process, tendon cells are also involved through the synthesis of enzymes, which degrade tendon matrix and help with remodeling (Apostolakos et al., 2014; Wu et al., 2017; Docheva et al., 2015). More on these signaling molecules will be discussed later in the chapter.

The healed tendon, when compared with the native uninjured tendon, has reduced integration of collagen fibers and has a higher ratio of type III to type I collagen. Type III collagen has a smaller diameter compared with type I collagen; this leads the tendon to thicken and stiffen in an attempt to overcome the difference in mechanical strength. As a result, the healed tendon usually does not regain the same mechanical integrity or functional activity as an uninjured tendon (Wu et al., 2017; Docheva et al., 2015).

There is an additional complication when looking at healing of the tendon enthesis. The tendon-bone healing occurs through the mechanism described above; however, this scar tissue does not reestablish the native tendon-bone zones of the enthesis. Thus, the resultant enthesis lack the original tendon-bone insert and is left much weaker (Apostolakos et al., 2014).

There are certain factors that the tendon also needs to properly heal, including, but not limited to, motion, tension at the repair site, adequate nutrition, vascular perfusion, and minimal gap formation at the repair site. These cannot always be achieved in a natural healing process, and even if they are, there is still a risk of weakness and lack of function as previously discussed, which is why surgery might be indicated (Lilly and Messer, 2006).

8.21 Surgical intervention

There are three types of tendon repair: primary, delayed primary, or secondary. Primary repair often occurs within the first 24 h of injury, and surgery is undertaken to fix the injury. Delayed primary repair occurs within a few days of the injury. Secondary repairs may occur 2—5 weeks or longer after the injury. Within the different types of surgery, there are a number of techniques that may be used to repair tendons. There are different indications for each approach, but that is beyond the scope of this chapter. Surgical treatment includes an end-to-end repair, a transosseous tendon repair with bone tunnels, a suture anchor tendon repair, or reconstructed from autograft or allograft tissue (Hsu and Siwiec, 2018; Tang, 2005; Pope and Plexousakis, 2018).

When treating a tendon injury surgically, you must keep in mind the original characteristics of the uninjured tendon, and they should be restored or mimicked for optimal postinjury function. For example, an increase in tendon laxity could allow abnormal and extreme movement of the joint, leading to injury of surrounding tissue (Duenwald et al., 2009). An increased in tendon compliance would result in an inability to generate force from muscle contraction (Brumitt and Cuddeford, 2015). Insufficient laxity could result in stiffness and limited movement, causing damage under normal force and movement conditions (Duenwald et al., 2009).

8.22 Tendon regeneration

Because of the relative poor outcomes of conservative or surgical interventions, the rapidly expanding field of regenerative medicine has been used in conjunction with other healing interventions in an attempt to improve outcomes (Docheva et al., 2015; Wilkins and Bisson, 2012). Experimentation in several niches of tissue regeneration has yielded promising results regarding their eventual clinical implementation. These treatment modalities include growth factor delivery, stem cell therapy, grafts utilizing biomimetic scaffolds, gene therapy, application of mechanical forces, and administration of sound and electromagnetic waves to the affected tissue (Morais et al., 2015; Nixon et al., 2012; Mehta and Mass, 2005; Zhang et al., 2017; Pesqueira et al., 2018). Therapies aim to either improve the mechanical durability of the tendon or augment its healing potential. The remainder of this chapter summarizes the current protocols, existing commercial products, and emerging concepts pertaining to these techniques.

8.23 Utilization of growth factors in tendon healing

Growth factors are small signaling peptides released from platelets, macrophages, and polymorphonuclear cells at the injury site, which stimulate cell growth and differentiation, connective tissue synthesis, chemotaxis, and inflammation. In regard to tendon healing, they play a key role in orchestrating the stepwise healing process of the tendon. These molecules act on the injured tissue by binding to a cell surface receptor and initiating a signaling cascade that influences DNA transcriptional expression or quantity. This signal eventually leads to increased collagen synthesis, cellularity, vascularization and tissue volume at the site of injury to facilitate repair (Nixon et al., 2012).

Several studies have demonstrated that overexpression of certain growth factors at specific time intervals improves the outcome of tendon injury (Docheva et al., 2015; Mehta and Mass, 2005; Longo et al., 2011; Chen et al., 2012; Sayegh et al., 2015). Currently, two distinct delivery methods are in use. Each methodology provides advantages and drawbacks and should be utilized on a case-by-case basis. Direct application of growth factor through local injection is minimally invasive and simple, yet it leads to overflow loss of growth factor outside the affected tissue and only acts on the tissue for a short time interval. This may be a concern given that tendon repair occurs over several months or even years. Conversely, application of a growth factor–impregnated suturing or scaffold leads to a much longer duration of action, yet it is far more invasive. It is unclear at this point which delivery method is more efficacious and may depend on the specific growth factor that is delivered (Docheva et al., 2015). Animal studies have demonstrated positive results using both techniques. A comparative analysis utilizing both techniques in a controlled study has yet to be published (Longo et al., 2011).

Current publications have analyzed the efficacy of a wide variety of growth factors. The following growth factors have demonstrated positive effects on the tendon regeneration process: TGF-β, connective tissue growth factor (CTGF/CCN2), BMP family (12, 13, 14), bFGF/FGF-2, IGF-1, PDGF, VEGF, and others such as adiponectin, cartilage-derived morphogenetic protein (CDMP-1/CDMP-2), IL-10, recombinant human growth differentiation factor (rhGDF), recombinant human osteogenic protein-1 (rhOP-1), and kartogenin (Docheva et al., 2015; Zhang et al., 2018; Im, 2018). A brief introduction of the growth factor and any pertinent in vitro or in vivo studies on tendon regeneration will be included in the following sections. Furthermore, the advantages and drawbacks associated with the use of platelet-rich plasma (PRP) will be discussed in detail given its current clinical intrigue.

8.23.1 Transforming growth factor beta

TGF-β is a growth factor secreted by cells present at the wound site involved in healing and ECM reformation. It also plays a role in tendon cell mitogenesis and migration. Degranulating platelets, epithelial cells, fibroblasts, and immune cells all have the capability of releasing this factor (Longo et al., 2011). It is important to note that there are three isoforms of TGF-β. While TGF-β1 is expressed in adult wound healing, TGF-β3 is expressed primarily in fetal tendon development. TGF-β1 secretion in adult tissue not only upregulates the synthesis of type I and type III collagen resulting in rapid wound repair but also contributes to the formation of scar tissue and fibrosis. This ultimately decreases the range of motion and functionality at the conclusion of the healing process. Fetal wounds do not result in fibrosis and have been shown to lack the secretion of TGF-β1 (Sayegh et al., 2015). This phenomenon has scientists particularly interested in developing strategies to control its expression in vivo. Several studies have found a statistically significant increase in range of motion postoperatively in rabbits that were treated with a TGF-β–neutralizing antibody (Heisterbach et al., 2012; Klein et al., 2002; Chang et al., 2000). Additionally, administration of TGF-β1 paired with suppression of TGF-β3 led to tendons with compromised mechanical function despite their larger cross-sectional area. These data suggest that a combination of TGF-β1 suppression and TGF-β3 exogenous administration may result in better tendon healing postoperatively. No data defining proper dosage of these isoforms have been published to date (Wu et al., 2017).

8.23.2 CTGF/CCN2

CTGF, also known as CCN2, is involved in the differentiation of bone marrow stromal cells (BMSCs) into fibroblasts. Fibroblasts are a pivotal component of all wound repairs due to their synthesis of collagen and ECM components (Longo et al., 2011). Studies have demonstrated that BMSCs in the presence of CTGF have an increased prevalence of fibroblastic surface markers, increased production of type I collagen and tenascin C (a glycosaminoglycan present early in tissue development), and decreased surface markers for other BMSC lineages such as adipocytes and chondrocytes (Zhang et al., 2018). A multitude of animal studies, mainly performed in rats, have demonstrated that delivery of CTGF to tissue contributes to the structural integrity of developing tendon (Wurgler-Hauri et al., 2007; Chen et al., 2008). However, it is important to note there is conflicting data on whether CTGF alone is sufficient to commit BMSCs to the fibroblastic lineage. In one study, delivery of CTGF led to increased mineralization of the periodontal ligament, which is indicative of the osteoblastic lineage. In response to this discovery, these researchers administered TGF-β1 in combination with CTGF, which resulted in enhanced expression fibroblastic genes. These data suggest that CTGF is sufficient to commit BMSCs to either an osteoblastic or fibroblastic lineage depending on the environment, but other factors are required to isolate fibroblastic differentiation (Wurgler-Hauri et al., 2007). Development of a reproducible protocol for BMSC differentiation to fibroblasts would have a significant impact on the field of tissue engineering and tendon regeneration but to date has not been established.

8.23.3 Bone morphogenic protein family

BMPs are part of the TGF-β superfamily and play a role in bone, fibrocartilage, ligament, and tendon embryogenesis and repair. The tenogenic members of the BMP family include BMP-12, -13, and -14, which are also known as GDF (growth and differentiation factor) -7, -6, and -5, respectively. These factors are involved in transcriptional regulation, ECM organization, and tenocyte proliferation in tendon repair and formation (Docheva et al., 2015). BMP-14 has been shown to increase the prevalence of tendon progenitor cells at the wound site while improving tensile strength and speed of healing in rats. The same results were obtained whether the growth factor was administered directly or through impregnated suture. However, introduction of BMP-14 to tendon also resulted in increased chondroid regions within the tissue (Bolt et al., 2007). BMP-12 administration increased the load-to-failure ratio of the bone–tendon interface in sheep (Sayegh et al., 2015). Taking both of these studies into consideration, it is clear that targeting specific loci within the tendon and optimizing the dosages of BMP growth factors is important in future clinical use. Although introduction of this growth factor has had promising results, more data are necessary to develop its true therapeutic potential.

8.23.4 bFGF/FGF-2

bFGF is an important growth factor for the stimulation of angiogenesis and mesenchymal stem cell (MSC) differentiation. This growth factor is elevated early in the process of tendon healing and stimulates fibroblastic activity (Sayegh et al., 2015). Studies have shown that administration of bFGF facilitates the proliferation of MSCs in a dose-dependent manner and that integrin expression was increased. Additionally, collagen fibril diameter, density, and biomechanical capabilities improved (Tang et al., 2008). These data have recently been brought into question by a study that claimed that although cell proliferation and type III collagen were indeed increased, the mechanical properties remained the same. Further data are required to verify its effects in vivo (Longo et al., 2011).

8.23.4.1 Insulin-like growth factor 1

IGF-1 is an anabolic growth factor that decreases swelling, induces cell proliferation, upregulates collagen synthesis, and stimulates synthesis of the ECM. It is primarily expressed in the inflammatory and remodeling stages of wound healing and seems to be produced in an imbalanced manner in specific tendon types and locations (Docheva et al., 2015). Data indicate that it functions synergistically with PDGF to prompt tenocyte chemotaxis and mitosis. Animal studies have shown that rats treated with IGF-1 recovered faster and had more functionality in their tendons when compared with controls. IGF-1 has also been used to improve histological scores of biomimetic scaffold grafts in animal trials (Costa et al., 2006).

8.23.4.2 Platelet-derived growth factor

PDGF has demonstrated beneficial effects in tendon healing by stimulating cell proliferation and matrix synthesis but has also been implicated in the induction of inflammatory cytokine expression leading to fibrosis. With this drawback in mind, scientists developed a controlled release delivery system consisting of fibrin and heparin that decreased inflammatory effects while improving cell density, proliferation, and collagen synthesis (Sayegh et al., 2015). PDGF has shown to be a potent stimulator of tendon growth indicators but needs to be released in the proper doses and at the proper time. PDGF receptors are present for approximately 6 months after tendon injury, which indicates its role throughout the entirety of the tendon healing process (Longo et al., 2011). Further research is required to determine which doses are most efficient at specific time intervals in the healing process.

8.23.4.3 VEGF

VEGF promotes angiogenesis and increases the permeability of capillaries through the stimulation of nitric oxide synthesis. Blood supply is an important factor in the healing of almost any wound, and tendon injuries are not exempt. VEGF is only present in injured or developing tendons and is primarily found in the proliferative and remodeling phases of tendon healing (Sayegh et al., 2015). Administration of VEGF led to significantly improved tensile strength 2 weeks postoperatively, yet by 4 weeks, no significant difference was present (Petersen et al., 2003). These data implicate that VEGF may accelerate the initial healing process but play little role in the final result. VEGF is also highly prevalent in ruptured Achilles tendons, suggesting it may have a role in tendon pathology (Wu et al., 2017). Although some positive outcomes have been obtained, less encouraging results have come from VEGF experimentation.

8.23.5 Other growth factors of interest

Although the preceding growth factors have been the subjects of several studies, other growth factors have implicated in tendon healing processes. Adiponectin, which is an adipocyte-secreted hormone, increases cell sensitivity to insulin and has demonstrated the ability to promote tendon progenitor cell differentiation into tenocytes, as well as tenocyte proliferation in vitro. Surprisingly, scientists also found no increase in adipogenic, osteogenic, or chondrogenic gene expression (Zhang et al., 2018).

CDMPs are similar to the BMP family. CDMP-2 direct application increased the mechanical strength and stiffness of animal tendons 2 weeks postoperatively and presented with a more organized healing matrix under histological analysis (Longo et al., 2011).

IL-10 is an antiinflammatory cytokine released by immunomodulation cells. The decrease in inflammation has been shown to slow or even inhibit the formation of scar tissue in animal fetuses (Ricchetti et al., 2008). No adult studies have been performed to date to analyze the effects of IL-10 on adult tendon scar formation. IL-10 inclusion in growth factor cocktails allows for the inclusion of other factors that exacerbate the immune response. Determining the proper dosage in these cocktails is a key in its inclusion in future therapies (Longo et al., 2011).

rhGDF-5 has been used to coat sutures for tendon repair. The groups that received these augmented sutures had a higher ultimate tensile load, greater stiffness, and high tendon hypertrophy early in the tendon healing process. However, no significant difference was seen between the control sutures and the rhGDF-5 sutures after 6 weeks (Dines et al., 2007).

rhOP-1 has been shown to increase proliferation of tendon cells and synthesis of ECM components in vitro (Yamada et al., 2008). No in vivo studies have been performed using rhOP-1 to date.

Kartogenin is a small molecule that is most often utilized in the differentiation of MSCs into chondrocytes. A recent publication has investigated its use in promoting bone—tendon junction regeneration and has found encouraging results (Im, 2018). Although kartogenin would have to be localized the bone—tendon interface to avoid chondroid deposits, its clinical use could be of use once this technology becomes available.

8.23.6 Autologous growth factor sources, platelet-rich plasma

Although delivery of individual growth factors has led to characterizations of their roles and functions, the vast majority of current research concerns the use of autologous PRP in tendon repair. PRP is essentially blood plasma with abnormally high levels of platelets. Other autologous sources of growth factor similar to PRP have been defined with slight alterations. These permutations include platelet rich in growth factor, platelet-rich fibrin matrix, and plasma-leukocyte membrane (Mehta and Mass, 2005).These platelets secrete growth factors causing PRP to possess growth factor concentrations far above physiological levels. Although the mechanism of action for autologous growth factor sources remains unknown, some animal models have shown that PRP improves multiple facets of tendon healing (Docheva et al., 2015). These include collagen synthesis, deposition and fiber orientation, cell proliferation, biochemical and biomechanical properties, and increased expression of PDGF and TGF-β. However, because of the lack of a consistent and reproducible preparation method, results have been all over the spectrum. Some studies have concluded that PRP administration has a no effect, while others even have found a statistically significant negative effect on tendon healing (Docheva et al., 2015; Lin et al., 2018; Schnabel et al., 2007; Schepull et al., 2011).

Unlike purified growth factors, autologous PRP is already in use clinically for tendon injuries. PRP has been found to have a positive effect on anterior cruciate ligament (ACL) repair postoperatively with a smaller defect size under MRI. In contrast to these findings, several studies have shown that PRP has no effect on the healing of ruptured Achilles tendons or epicondylar tendinopathy. Moreover, the safety of PRP has been called into question by some researchers as a possible carcinogen (Taylor et al., 2011). These conflicting results make it extremely difficult to draw strong conclusions about its efficacy. There is a desperate need for standardization of autologous growth factor preparation, patient selection, and outcome measurements to determine the efficacy of these therapies.

8.24 Stem cell—based approaches to tendon healing

Stem cells are a class of undifferentiated cell that are derived either from an embryo or an adult somatic line. These cells possess the ability to further differentiate into more specialized cell lines when exposed to the correct growth factors and microenvironment. Stem cells are often classified by their potency (i.e., totipotent, pluripotent, multipotent) and vary in their potential specializations depending on how differentiated they are at a given time interval. Adult somatic stem cells can be harvested from multiple tissue types for eventual therapeutic use. These sites include but are not limited to the bone marrow, adipose tissue, tendon tissue, periosteum, dermis, and peripheral blood (Chen et al., 2013).

MSCs are anabolic, immunomodulatory, and multipotent cells that have the ability to differentiate into tenocytes, which have made them the main focus of any cell-based tendon regeneration endeavor to date. Once isolated from other cell types using fluorescent or magnetic cell sorting, MSCs can be expanded in vitro to reach therapeutic levels. These cells are then injected into the injured tendon area where they replenish or supplement the reparative cells already present in the tissue (Nixon et al., 2012).

When compared with growth factor and tissue scaffold research, very few trials have been performed concerning stem cells. This area of interest is controversial, time consuming, and expensive but has had promising results in the literature. Stem cell therapies are often combined with other tissue engineering approaches such as the aforesaid growth factors of scaffolds. The remainder of this section provides a brief overview of the main stem cells used in the literature and their therapeutic results. Additionally, this section concludes with a brief discussion on the ethical and financial limitations surrounding the use of stem cells in current and future clinical trials.

8.24.1 Bone marrow−derived mesenchymal stem cells

BMSCs, or bone marrow−derived MSCs, are ideal cells for tendon regeneration techniques because of their nonimmunogenic and even immunosuppressive characteristics (Ahmad et al., 2012). The major disadvantage of BMSCs is the difficulty and pain associated with bone marrow harvesting and their uncontrollable phenotypes in vivo. Furthermore, even with a successful harvest BMSCs usually only represent 0.001%−0.01% of the total cell population and require culture expansion to reach therapeutic numbers (Sayegh et al., 2015).

BMSCs have been utilized in the presence of BMP-12, BMP-14, or tenogenic transcription factor cDNA (scleraxis and Smad8) with moderate success (Haddad-Weber et al., 2010; Hoffmann et al., 2006). In vitro studies found that the BMSCs adopted the tenocyte phenotype (Tan et al., 2012). In vivo studies have demonstrated that there is an improvement in histological and mechanical properties in the first 6 weeks, but no studies have studied their long-term effects (Docheva et al., 2015). In addition to their ability to differentiate into tenocytes, scientists have found alternative uses for this cell line. Several studies have found the immunomodulatory effect of BMSCs to be beneficial to tendon healing even in the absence of their differentiation. Moreover, this cell line has been suggested to stimulate resident progenitor cells through its trophic and stimulatory functions, contributing to tendon regeneration in a less direct role (Sayegh et al., 2015).

One clinical trial utilized BMSCs to assist in complete rotator cuff tears. After 12 months, the experimental group experienced improved tendon regeneration and integrity (Ellera Gomes et al., 2012). Although there is a paucity of data to support this trial, these results are encouraging for their future implementation into clinical therapies.

8.24.2 Adipose-derived mesenchymal stem cells

Adipose-derived mesenchymal stem cells (ADSCs) are obtained from adipose tissue and are of particular interest due to their relative abundance and ease of harvest when compared with bone marrow aspirate. These stem cells are very similar to BMSCs but are less efficient at differentiating into osteogenic and chondrogenic pathways. Studies have shown that ADSCs are proficient at expressing tendon-related gene markers when treated with IGF-1, TGF-β, or BMP-14 (Sayegh et al., 2015).

Studies that compared the ability of ADSCs to BMSCs have been performed, and they were found to have comparable scaffold adherence, collagen fiber organization, and proliferation potential (Sayegh et al., 2015; Kryger et al., 2007). Nevertheless, there is little data as to the efficacy of ADSCs in tendon regeneration to date. Caution should be taken in their use due to their natural proclivity to differentiation into adipose tissue rather than tenocytes. More data will need to be obtained to determine their potential.

8.24.3 Embryonic stem cells

Embryonic stem cells (ESCs) are obtained from the inner cell mass of blastocysts and are totipotent in nature (Chen et al., 2013). ESCs are ethically controversial and have teratogenic potential due to their ability to spontaneously differentiate. These two factors have been a major deterrent to their use in clinical therapies. Early stage fetal tissue ESC-like cells and stepwise differentiation of ESCs into MSCs are two techniques scientists have used to greatly lower the teratogenic potential of these cells. Improving the safety of their use could be incredibly advantageous to clinical therapies. This is due to their unlimited mitotic potential and immortality, no apparent rejection by immune systems, and their ability to be frozen and stored for future use. All of these characteristics make them superior to the BMSCs and ADSCs described previously (Docheva et al., 2015; Sayegh et al., 2015).

In animal studies, the aforementioned benefits of ESCs have been verified by results of comparative studies. Undifferentiated ESCs were injected into tendon lesions and compared with the longevity of MSCs. After 90 days, ESCs were overwhelmingly present while MSCs only existed at 5% their original application (Guest et al., 2010). ESCs have also demonstrated the ability to improve tendon healing outcomes in a superior manner to MSCs in comparative studies (Watts et al., 2011). These results suggest that ESCs, when applied safely, can be incredibly beneficial to the healing process. More data are required before human trials.

8.24.4 Induced pluripotent stem cells

Induced pluripotent stem cells (iPSCs) are adult somatic cells that are genetically reprogrammed to become stem cells through viral gene delivery. The major benefit of these cells is that they possess similar characteristics to the aforementioned ESCs, yet they lack the ethical and moral dilemma of destroying embryonic tissue. However, they also share the teratogenic potential of ESCs. Although there have been positive results in the literature, iPSCs have the tendency to form teratomas in most animal models when solely administered (Xu et al., 2013). Selecting for a certain lineage differentiation through gene priming techniques may be a possible solution to this shortcoming. More research is required to create an iPSC line that is nonteratogenic and safe for use in clinical trials.

8.24.5 Tendon-derived stem cells

Tendon-derived stem cells (TDSCs) are self-renewing multipotent stem cells that are present in mature adult tendons in small quantities. The main benefit of TDSCs over the common MSC is that these cells physiologically belong in the tendon. Therefore, they should be more familiar with the ECM environment and have a higher likelihood of survival and correct differentiation. Studies have shown that TDSCs are more likely to form tendon-like structures in vivo than BMSCs, which formed bony masses (Lui and Wong, 2012). Animal models have yielded promising results in the majority of trials. These cells are still in the preclinical stage but are an exciting new niche of study for scientists interested in tendon regeneration (Ahmad et al., 2012).

Although these cells are intriguing, they do not lack their share of shortcomings. Just as with MSCs, TDSCs have demonstrated the tendency to produce ectopic ossifications when used as a therapy in animal models (Sayegh et al., 2015). They are also very difficult to purify and expand due to insufficient knowledge about tenocyte differentiation pathways and tend to undergo phenotypic drift during expansion in culture. Finally, allogeneic cells may cause an immune response, and autologous cells have the potential to cause comorbidities to the patient donor site (Sayegh et al., 2015). A great deal of research and insight is required to prepare these cells for eventual clinical use.

8.25 The role of biologic and synthetic scaffolds in tendon healing

The current practice of autologous and allogeneic tendon transfer, while moderately effective, has significant shortcomings. Autologous tendon transfer often causes donor site morbidity, and allogeneic transfer can result in graft rejection and an elevated immune response. The development of biologic and synthetic scaffolds that provide a template for growth and short-term biomechanical stability to the injured tendon provides a novel alternative to the two conventional methods (Sayegh et al., 2015). To be effective, these scaffolds must be a biocompatible construct with high surface area that supports cell attachment and growth. In regard to tendon scaffolds, they must promote tenocyte differentiation and mimic the native architecture of the tendon.

The literature describes three distinct scaffold forms: collagen-based constructs, tissue-based constructs, and synthetically engineered constructs (Huang et al., 2006). Each scaffold form comes with its own set of benefits and shortcomings. This section will discuss the current constructs under investigation and the relative findings involved.

8.25.1 Collagen-based constructs

The majority of collagen-based scaffold research has involved type I collagen gels and composites. These gels are composed of randomly oriented collagen fibrils interspersed with a stem cell line. Multiple studies have found that tendons treated with collagen gel implants exhibited higher strength and tensility than controls. They also found that the gel to cell ratio should mimic the ratio of the tendon being grafted to facilitate de novo synthesis of tendon matrices and remodeling of the degradable graft (Docheva et al., 2015).

Collagen-based grafts have also utilized nanotechnology to create constructs that ordered collagen fibers longitudinally rather than randomly. The point of this construct was to more accurately mimic the fibril structure of the native tendon. The alignment of collagen fibers induced stem cell differentiation into morphology similar to tenocytes, promoted upregulation of tenocyte-related genes, and resulted in tendons that were larger and stiffer than controls (Kishore et al., 2012; Alfredo Uquillas et al., 2012). These aligned nanofibers mimicked the microenvironment of native tendon and stimulated the implanted cells to differentiate following the tenogenic lineage. Although these results are impressive, the major limiting factor of these grafts is the dimensions. Scientists have yet to make a nanotechnology forged collagen graft of suitable size for human tendon trials (Docheva et al., 2015).

Other endogenous proteins have been utilized for scaffold construction including fibrin, hyaluronic acid, and elastin. These studies have not yielded results as promising as the collagen-based constructs but demonstrated a performance superior to nontreated controls (Longo et al., 2011). They will not be discussed in further detail here because of their lack of literature or current interest.

8.25.2 Tissue-based constructs

The use of xenografts is a possible solution to the dimensional problem presented by the nanotech collagen grafts. Tissue samples from other species that match the structure, tensility, and physiology of human tendon have been in the literature for quite some time. Porcine small intestine submucosal grafts are FDA approved and have been utilized with mixed outcomes in several studies investigating rotator cuff and Achilles healing. The intestinal submucosa was found to deteriorate within 4 weeks, which is an opportune time to make room for the remodeled tissue and cells. However, a few studies stated the grafts caused a noninfectious effusion at the graft site and that there were no long-term benefits to the structural integrity of the experimental group (Huang et al., 2006; Schlegel et al., 2006; Dejardin et al., 2001).

Researchers have also found interest in using decellularized tendons to more perfectly replicate the microenvironment of the tissue. These tendons are subsequently seeded with BMSCs that assume a phenotype similar to tenocytes. This is further evidence that simulating the microenvironment of the native tissue has a major effect of the phenotypic fate of graft cells. Although this is promising, scientists are having trouble establishing proper cell populations in the deeper sections of the graft and have found that the decellularization process affects the structural integrity of the graft (Longo et al., 2012). Further innovation is required for decellularized tissue to be a viable option.

The dermis of humans, cows, and pigs has a vascular network and collagenous structure that is ideal for grafts. Several studies have investigated their efficacy in rotator cuff and Achilles repair. All dermis donor specimen have positive results in the literature, but human allograft dermis seems to be the gold standard in this category. More than a few studies have found early return to physical activity, improved tensile strength, reduced pain, and improved mobility for dermis graft recipients (Docheva et al., 2015; Lee, 2008). One investigation noted the increased immune response at the donor site when utilizing xenograft material in mice (Valentin et al., 2006). This may be the cause of the improved outcomes of allograft dermis recipients. More data are required to make a more definitive claim about the advantages and disadvantages of each dermal graft.

8.25.3 Synthetically engineered constructs

Numerous synthetically engineered constructs have been used in the attempt to produce scaffolds. Although the list is extensive, a few have received more attention in the literature. These include Synthasome X-repair (poly-L-lactide), Biomerix Rotator Cuff Repair Patch (polycarbonate polyurethane urea), polyglycolic acid (PGA), chitosan-based hyaluronan hybrid, synthetic oligo(poly(ethylene glycol)fumarate) (OPF)-based biomaterials, and most recently, magnetically responsive constructs (Longo et al., 2011). These materials have demonstrated success due to their specifically tailored architecture and their degradation characteristics, which can be modified based on the polymers involved. The finding of each polymer will be briefly discussed.

Synthasome X-Repair and Biomerix Rotator Cuff Repair Patch have both demonstrated biocompatibility superior to the porcine small intestine submucosa patch described previously. Patients who have received this scaffold experienced improved biomechanical function of their infraspinatus tendon (Sayegh et al., 2015). Several studies have shown that these scaffolds allow for BMSC attachment and proliferation, a critical characteristic for optimization of therapies (Derwin et al., 2009; Santoni et al., 2010).

PGA has also been shown to be biocompatible with tendon cells. One study involving a PGA scaffold demonstrated that mechanical forces play a role in the differentiation and strength of tendon tissue. The PGA fibers were seeded with tenocytes under mechanical strain, which resulted in an in vitro tendon with superior strength and maturity compared with controls (Cao et al., 2006). Although this is promising, scientists are concerned about the carbon dioxide produced by PGA as it degrades. This can lower the pH of the donor site resulting in cell and tissue necrosis (Longo et al., 2011).

Chitosan-based hyaluronan hybrid scaffold and synthetic OPF-based biomaterials were both tested with fibroblasts to promote tendon regeneration. The OPF materials demonstrated biocompatibility and fibroblast clustering but demonstrated no significant increase in tendon integrity. The chitosan hybrid resulted in enhanced type I collagen production and improved tensile strength 4−12 weeks postoperatively (Longo et al., 2011; Brink et al., 2009). Both scaffolds require further data to draw strong conclusions.

The use of magnetically responsive biomaterials is cutting-edge research that derives its rationale from data that demonstrate the importance of mechanical stimulation in cell differentiation pathways. Recently, scientists created magnetic constructs that were introduced into a magnetic field once seeded with stem cells. These cells were stressed in a uniform direction similar to the strain placed on the cells in vivo. Preliminary data demonstrate that this technique does influence the tenocyte differentiation pathway (Pesqueira et al., 2018). More data are required to determine its reproducibility.

8.26 The role of gene transfer in tendon healing

Gene therapy is a treatment protocol that involves delivering genes that encode for a protein product to the injured tissue, rather than the protein itself. This gene is then incorporated by the cells in the injured tissue, which proceed to produce the protein product internally. This method has many benefits over direct or augmented suture application of growth factor because the exposure to the factor is no longer limited by the survival of the protein but rather by the survival of the cells. Moreover, if the native cells are synthesizing the protein, these products go through the proper post-translational processing and are less likely to elicit an immune response. Gene therapy is also useful because of its ability to transfer genes for intracellular proteins (Nixon et al., 2012). In theory, this technique would be superior to direct injection or impregnated sutures. Unfortunately, it is highly experimental, and no human studies have employed gene transfer for tendon repair. This may be because of the bad publicity received due to a death from gene therapy in 1999 (Docheva et al., 2015).

Genes can be delivered by two distinct vectors, either viral or nonviral. Viral vectors are used to transfer cDNA to target cells. Viruses are perfect vehicles for the gene of choice because of their natural ability to transfer their gene product to target cells. To prevent virulence, the viruses are attenuated before loading. These vectors can then theoretically be used to infect target cells without the risk of adverse effects (Nixon et al., 2012).

The three viral vectors used primarily in the literature are adenovirus (AV), adeno-associated virus (AAV), and hemagglutinating virus of Japan (HVJ). All three viral vehicles have experienced success in vitro and in vivo, but they are not without shortcomings (Docheva et al., 2015). Although AV was the most effective vehicle in the short term, it stimulated a significant immune response. This response will prevent redosing of the gene if not enough cells are transduced. AAV transfection resulted in increased expression of gene product and elevated collagen synthesis, but transfection was far less efficient than AV. HVJ was not as effective as AV but still elicited a considerable immune response. Immunomodulatory elements have been considered as a future element of viral vector studies (Evans, 2014).

Nonviral vectors usually consist of polynucleotide strands that are transferred into the cell via electroporation. The efficiency of transfer is often lower than viral vectors, but they often elicit a smaller immune response. Additionally, to transfect the cells with electroporation, the cells must be removed from their native location to receive the gene product. The most common nonviral vectors include antisense oligonucleotides, plasmids, and polydeoxyribonucleotides (PDRNs). Antisense oligonucleotides are designed to knock down the expression of their target gene. Scientists effectively reduced the expression of type V collagen to increase the diameter of type I collagen molecules in tenocytes. While this had no therapeutic bearing, the proof of concept is an exciting advancement (Shimomura et al., 2003). Plasmids have also been transfected via electroporation successfully, which yielded mixed results. Multiple scientists have attempted to use a liposome carrier for plasmids, which elicited a violent immune response. Discovery of a biocompatible carrier is important for further consideration of this option (Longo et al., 2011). PDRNs are incredibly potent molecules that stimulate the A2A cell receptor. Although the majority of research regarding PDRNs is in the skin, one study has tested their ability to assist tendon healing, yielding positive results. The experimental group exhibited greater ECM growth, cell growth, cell migration, and reduced inflammation when compared with controls (Veronesi et al., 2017). This extremely promising result warrants verification to determine reproducibility.

8.27 Future of tendon regeneration

As discussed above, there are a number of advantages and drawbacks to each of the regenerative techniques covered in this chapter. Some techniques are still in need of additional research and data to accurately assess its role in the future of the field. However, the role of regenerative medicine in tendon healing is vast and shows signs of having a key role alongside the use of surgical intervention to facilitate complete tendon healing. There have been a number of studies, as discussed above, which have shown improved outcomes from the use of regenerative medicine in the setting of tendon injury. With the field being relatively young, we can expect to see a growing pool of data and clinical studies using a wide variety of regenerative techniques to help facilitate tendon healing.

References

Alfredo Uquillas, J., Kishore, V., Akkus, O., 2012. Genipin crosslinking elevates the strength of electrochemically aligned collagen to the level of tendons. J. Mech. Behav. Biomed. Mater. 15, 176–189.

Ahmad, Z., et al., 2012. Exploring the application of stem cells in tendon repair and regeneration. Arthroscopy 28 (7), 1018–1029.

Angeline, M.E., Rodeo, S.A., 2012. Biologics in the management of rotator cuff surgery. Clin. Sports Med. 31 (4), 645–663.

Apostolakos, J., et al., 2014. The enthesis: a review of the tendon-to-bone insertion. Muscles Ligaments Tendons J. 4 (3), 333–342.

Benjamin, M., et al., 2002. The skeletal attachment of tendons—tendon "entheses". Comp. Biochem. Physiol. Mol. Integr. Physiol. 133 (4), 931–945.

Benjamin, M., et al., 2006. Where tendons and ligaments meet bone: attachment sites ('entheses') in relation to exercise and/or mechanical load. J. Anat. 208 (4), 471–490.

Benjamin, M., Ralphs, J.R., 2001. Entheses—the bony attachments of tendons and ligaments. Ital. J. Anat. Embryol. 106 (2 Suppl. 1), 151–157.

Benjamin, M., Kaiser, E., Milz, S., 2008. Structure-function relationships in tendons: a review. J. Anat. 212 (3), 211–228.

Bolt, P., et al., 2007. BMP-14 gene therapy increases tendon tensile strength in a rat model of Achilles tendon injury. J. Bone Joint Surg. Am. 89 (6), 1315–1320.

Bordoni, B., Morabito, B., 2018. Anatomy, tendons. In: StatPearls. StatPearls Publishing, StatPearls Publishing LLC, Treasure Island (FL).

Brink, K.S., Yang, P.J., Temenoff, J.S., 2009. Degradative properties and cytocompatibility of a mixed-mode hydrogel containing oligo[poly(ethylene glycol)fumarate] and poly(ethylene glycol)dithiol. Acta Biomater. 5 (2), 570−579.

Brockis, J.G., 1953. The blood supply of the flexor and extensor tendons of the fingers in man. J. Bone Joint Surg. Br. 35-b (1), 131−138.

Brumitt, J., Cuddeford, T., 2015. Current concepts of muscle and tendon adaptation to strength and conditioning. Int. J. Sports Phys. Ther. 10 (6), 748−759.

Buchanan, C.I., Marsh, R.L., 2002. Effects of exercise on the biomechanical, biochemical and structural properties of tendons. Comp. Biochem. Physiol. Mol. Integr. Physiol. 133 (4), 1101−1107.

Cao, D., et al., 2006. In vitro tendon engineering with avian tenocytes and polyglycolic acids: a preliminary report. Tissue Eng. 12 (5), 1369−1377.

Carroll, C.C., et al., 2008. Influence of aging on the in vivo properties of human patellar tendon. J. Appl. Physiol. (1985) 105 (6), 1907−1915.

Chang, J., et al., 2000. Studies in flexor tendon wound healing: neutralizing antibody to TGF-beta1 increases postoperative range of motion. Plast. Reconstr. Surg. 105 (1), 148−155.

Chaplin, D.M., 1973. The vascular anatomy within normal tendons, divided tendons, free tendon grafts and pedicle tendon grafts in rabbits. A microradioangiographic study. J. Bone Joint Surg. Br. 55 (2), 369−389.

Chen, C.H., et al., 2008. Tendon healing in vivo: gene expression and production of multiple growth factors in early tendon healing period. J. Hand Surg. Am. 33 (10), 1834−1842.

Chen, L., et al., 2012. Synergy of tendon stem cells and platelet-rich plasma in tendon healing. J. Orthop. Res. 30 (6), 991−997.

Chen, H.S., et al., 2013. Stem cell therapy for tendon injury. Cell Transplant. 22 (4), 677−684.

Christensen, B., et al., 2008. Effects of long-term immobilization and recovery on human triceps surae and collagen turnover in the Achilles tendon in patients with healing ankle fracture. J. Appl. Physiol. (1985) 105 (2), 420−426.

Costa, M.A., et al., 2006. Tissue engineering of flexor tendons: optimization of tenocyte proliferation using growth factor supplementation. Tissue Eng. 12 (7), 1937−1943.

Cutts, A., Alexander, R.M., Ker, R.F., 1991. Ratios of cross-sectional areas of muscles and their tendons in a healthy human forearm. J. Anat. 176, 133−137.

D'Agostino, M.A., 2010. Chapter 9 − enthesitis. In: Wakefield, R.J., D'Agostino, M.A. (Eds.), Essential Applications of Musculoskeletal Ultrasound in Rheumatology. W.B. Saunders, Philadelphia, pp. 103−109.

Dejardin, L.M., et al., 2001. Tissue-engineered rotator cuff tendon using porcine small intestine submucosa. Histologic and mechanical evaluation in dogs. Am. J. Sports Med. 29 (2), 175−184.

Derwin, K.A., et al., 2009. Rotator cuff repair augmentation in a canine model with use of a woven poly-L-lactide device. J. Bone Joint Surg. Am. 91 (5), 1159−1171.

Deymier, A.C., et al., 2017. Micro-mechanical properties of the tendon-to-bone attachment. Acta Biomater. 56, 25−35.

Dines, J.S., et al., 2007. The effect of growth differentiation factor-5-coated sutures on tendon repair in a rat model. J. Shoulder Elb. Surg. 16 (5 Suppl), S215−S221.

Docheva, D., et al., 2015. Biologics for tendon repair. Adv. Drug Deliv. Rev. 84, 222−239.

Duenwald, S.E., Vanderby, R., Lakes, R.S., 2009. Viscoelastic relaxation and recovery of tendon. Ann. Biomed. Eng. 37 (6), 1131−1140.

Ellera Gomes, J.L., et al., 2012. Conventional rotator cuff repair complemented by the aid of mononuclear autologous stem cells. Knee Surg. Sport. Traumatol. Arthrosc. 20 (2), 373−377.

Ensey, J.S., et al., 2009. Response of tibialis anterior tendon to a chronic exposure of stretch-shortening cycles: age effects. Biomed. Eng. Online 8, 12.

Evans, C., 2014. Using genes to facilitate the endogenous repair and regeneration of orthopaedic tissues. Int. Orthop. 38 (9), 1761−1769.

Fenwick, S.A., Hazleman, B.L., Riley, G.P., 2002. The vasculature and its role in the damaged and healing tendon. Arthritis Res. 4 (4), 252−260.

Guest, D.J., Smith, M.R., Allen, W.R., 2010. Equine embryonic stem-like cells and mesenchymal stromal cells have different survival rates and migration patterns following their injection into damaged superficial digital flexor tendon. Equine Vet. J. 42 (7), 636−642.

Haddad-Weber, M., et al., 2010. BMP12 and BMP13 gene transfer induce ligamentogenic differentiation in mesenchymal progenitor and anterior cruciate ligament cells. Cytotherapy 12 (4), 505−513.

Heisterbach, P.E., et al., 2012. Effect of BMP-12, TGF-beta1 and autologous conditioned serum on growth factor expression in Achilles tendon healing. Knee Surg. Sport. Traumatol. Arthrosc. 20 (10), 1907−1914.

Hoffmann, A., et al., 2006. Neotendon formation induced by manipulation of the Smad8 signalling pathway in mesenchymal stem cells. J. Clin. Investig. 116 (4), 940−952.

Hsu, H., Siwiec, R.M., 2018. Patellar tendon rupture. In: StatPearls. StatPearls Publishing LLC, Treasure Island (FL).

Huang, D., Balian, G., Chhabra, A.B., 2006. Tendon tissue engineering and gene transfer: the future of surgical treatment. J. Hand Surg. Am. 31 (5), 693−704.

Im, G.I., 2018. Application of kartogenin for musculoskeletal regeneration. J. Biomed. Mater. Res. A 106 (4), 1141−1148.

Johnson, G.A., et al., 1994. Tensile and viscoelastic properties of human patellar tendon. J. Orthop. Res. 12 (6), 796−803.

Jozsa, L., et al., 1991. Three-dimensional ultrastructure of human tendons. Acta Anat. (Basel) 142 (4), 306−312.

Kannus, P., 2000. Structure of the tendon connective tissue. Scand. J. Med. Sci. Sports 10 (6), 312−320.

Kelc, R., et al., 2013. The Physiology of Sports Injuries and Repair Processes, pp. 43−86.

Kirkendall, D.T., Garrett, W.E., 1997. Function and biomechanics of tendons. Scand. J. Med. Sci. Sports 7 (2), 62−66.

Kishore, V., et al., 2012. Tenogenic differentiation of human MSCs induced by the topography of electrochemically aligned collagen threads. Biomaterials 33 (7), 2137−2144.

Kjaer, M., 2004. Role of extracellular matrix in adaptation of tendon and skeletal muscle to mechanical loading. Physiol. Rev. 84 (2), 649−698.

Klein, M.B., et al., 2002. Flexor tendon healing in vitro: effects of TGF-beta on tendon cell collagen production. J. Hand Surg. Am. 27 (4), 615−620.

Kohler, J., et al., 2013. Uncovering the cellular and molecular changes in tendon stem/progenitor cells attributed to tendon aging and degeneration. Aging Cell 12 (6), 988−999.

Kryger, G.S., et al., 2007. A comparison of tenocytes and mesenchymal stem cells for use in flexor tendon tissue engineering. J. Hand Surg. Am. 32 (5), 597−605.

Lee, D.K., 2008. A preliminary study on the effects of acellular tissue graft augmentation in acute Achilles tendon ruptures. J. Foot Ankle Surg. 47 (1), 8−12.

Lilly, S.I., Messer, T.M., 2006. Complications after treatment of flexor tendon injuries. J. Am. Acad. Orthop. Surg. 14 (7), 387−396.

Lin, J., et al., 2018. Cell-material interactions in tendon tissue engineering. Acta Biomater. 70, 1−11.

Longo, U.G., et al., 2011. Tissue engineered biological augmentation for tendon healing: a systematic review. Br. Med. Bull. 98, 31—59.
Longo, U.G., et al., 2012. Scaffolds in tendon tissue engineering. Stem Cells Int. 2012, 517165.
Lu, H.H., Thomopoulos, S., 2013. Functional attachment of soft tissues to bone: development, healing, and tissue engineering. Annu. Rev. Biomed. Eng. 15, 201—226.
Lui, P.P., Wong, O.T., 2012. Tendon stem cells: experimental and clinical perspectives in tendon and tendon-bone junction repair. Muscles Ligaments Tendons J. 2 (3), 163—168.
Maganaris, C.N., Paul, J.P., 1999. In vivo human tendon mechanical properties. J. Physiol. 521 (Pt 1), 307—313.
McCrum, C., et al., 2018. Alterations in leg extensor muscle-tendon unit biomechanical properties with ageing and mechanical loading. Front. Physiol. 9, 150.
McGonagle, D., Benjamin, M., 2015. 123 — Enthesopathies. In: Hochberg, M.C., et al. (Eds.), Rheumatology, sixth ed., pp. 1014—1020 Philadelphia.
McNeilly, C.M., et al., 1996. Tendon cells in vivo form a three dimensional network of cell processes linked by gap junctions. J. Anat. 189 (Pt 3), 593—600.
Mehta, V., Mass, D., 2005. The use of growth factors on tendon injuries. J. Hand Ther. 18 (2), 87—92 quiz 93.
Morais, D.S., et al., 2015. Current approaches and future trends to promote tendon repair. Ann. Biomed. Eng. 43 (9), 2025—2035.
Nixon, A.J., Watts, A.E., Schnabel, L.V., 2012. Cell- and gene-based approaches to tendon regeneration. J. Shoulder Elb. Surg. 21 (2), 278—294.
O'Brien, M., 2005. Anatomy of tendons. In: Maffulli, N., Renström, P., Leadbetter, W.B. (Eds.), Tendon Injuries: Basic Science and Clinical Medicine. Springer London, London, pp. 3—13.
Peacock Jr., E.E., 1959. A study of the circulation in normal tendons and healing grafts. Ann. Surg. 149 (3), 415—428.
Pesqueira, T., Costa-Almeida, R., Gomes, M.E., 2018. Magnetotherapy: the quest for tendon regeneration. J. Cell. Physiol. 233 (10), 6395—6405.
Petersen, W., et al., 2003. The angiogenic peptide vascular endothelial growth factor (VEGF) is expressed during the remodeling of free tendon grafts in sheep. Arch. Orthop. Trauma Surg. 123 (4), 168—174.
Pope, J.D., Plexousakis, M.P., 2018. Quadriceps tendon rupture. In: StatPearls. StatPearls Publishing LLC, Treasure Island (FL).
Ralphs, J.R., et al., 1998. Regional differences in cell shape and gap junction expression in rat Achilles tendon: relation to fibrocartilage differentiation. J. Anat. 193 (Pt 2), 215—222.
Rathbun, J.B., Macnab, I., 1970. The microvascular pattern of the rotator cuff. J. Bone Joint Surg. Br. 52 (3), 540—553.
Ricchetti, E.T., et al., 2008. Effect of interleukin-10 overexpression on the properties of healing tendon in a murine patellar tendon model. J. Hand Surg. Am. 33 (10), 1843—1852.
Santoni, B.G., et al., 2010. Biomechanical analysis of an ovine rotator cuff repair via porous patch augmentation in a chronic rupture model. Am. J. Sports Med. 38 (4), 679—686.
Sayegh, E.T., et al., 2015. Recent scientific advances towards the development of tendon healing strategies. Curr. Tissue Eng. 4 (2), 128—143.
Schepull, T., et al., 2011. Autologous platelets have no effect on the healing of human achilles tendon ruptures: a randomized single-blind study. Am. J. Sports Med. 39 (1), 38—47.
Schlegel, T.F., et al., 2006. The effects of augmentation with Swine small intestine submucosa on tendon healing under tension: histologic and mechanical evaluations in sheep. Am. J. Sports Med. 34 (2), 275—280.

Schnabel, L.V., et al., 2007. Platelet rich plasma (PRP) enhances anabolic gene expression patterns in flexor digitorum superficialis tendons. J. Orthop. Res. 25 (2), 230−240.

Screen, H.R., 2008. Investigating load relaxation mechanics in tendon. J. Mech. Behav. Biomed. Mater. 1 (1), 51−58.

Shimomura, T., et al., 2003. Antisense oligonucleotides reduce synthesis of procollagen alpha1 (V) chain in human patellar tendon fibroblasts: potential application in healing ligaments and tendons. Connect. Tissue Res. 44 (3−4), 167−172.

Sun, H.B., et al., 2015. Biology and mechano-response of tendon cells: progress overview and perspectives. J. Orthop. Res. 33 (6), 785−792.

Svensson, R.B., et al., 2016. Effect of aging and exercise on the tendon. J. Appl. Physiol. (1985) 121 (6), 1237−1246.

Tan, S.L., et al., 2012. Effect of growth differentiation factor 5 on the proliferation and tenogenic differentiation potential of human mesenchymal stem cells in vitro. Cells Tissues Organs 196 (4), 325−338.

Tang, J.B., 2005. Clinical outcomes associated with flexor tendon repair. Hand Clin. 21 (2), 199−210.

Tang, J.B., et al., 2008. Adeno-associated virus-2-mediated bFGF gene transfer to digital flexor tendons significantly increases healing strength. an in vivo study. J. Bone Joint Surg. Am. 90 (5), 1078−1089.

Tardioli, A., Malliaras, P., Maffulli, N., 2012. Immediate and short-term effects of exercise on tendon structure: biochemical, biomechanical and imaging responses. Br. Med. Bull. 103 (1), 169−202.

Taylor, D.W., et al., 2011. A systematic review of the use of platelet-rich plasma in sports medicine as a new treatment for tendon and ligament injuries. Clin. J. Sport Med. 21 (4), 344−352.

Tempfer, H., Traweger, A., 2015. Tendon vasculature in health and disease. Front. Physiol. 6, 330.

Valentin, J.E., et al., 2006. Extracellular matrix bioscaffolds for orthopaedic applications. A comparative histologic study. J. Bone Joint Surg. Am. 88 (12), 2673−2686.

Veronesi, F., et al., 2017. Polydeoxyribonucleotides (PDRNs) from skin to musculoskeletal tissue regeneration via adenosine A2A receptor involvement. J. Cell. Physiol. 232 (9), 2299−2307.

Voleti, P.B., Buckley, M.R., Soslowsky, L.J., 2012. Tendon healing: repair and regeneration. Annu. Rev. Biomed. Eng. 14, 47−71.

Wang, J.H.C., Guo, Q., Li, B., 2012. Tendon biomechanics and mechanobiology − a mini-review of basic concepts and recent advancements. J. Hand Ther. 25 (2), 133−141.

Watts, A.E., et al., 2011. Fetal derived embryonic-like stem cells improve healing in a large animal flexor tendonitis model. Stem Cell Res. Ther. 2 (1), 4.

Wilkins, R., Bisson, L.J., 2012. Operative versus nonoperative management of acute Achilles tendon ruptures: a quantitative systematic review of randomized controlled trials. Am. J. Sports Med. 40 (9), 2154−2160.

Wren, T.A., Beaupre, G.S., Carter, D.R., 2000. Tendon and ligament adaptation to exercise, immobilization, and remobilization. J. Rehabil. Res. Dev. 37 (2), 217−224.

Wu, F., Nerlich, M., Docheva, D., 2017. Tendon injuries: basic science and new repair proposals. EFORT Open Rev. 2 (7), 332−342.

Wurgler-Hauri, C.C., et al., 2007. Temporal expression of 8 growth factors in tendon-to-bone healing in a rat supraspinatus model. J. Shoulder Elb. Surg. 16 (5 Suppl), S198−S203.

Xu, W., et al., 2013. Human iPSC-derived neural crest stem cells promote tendon repair in a rat patellar tendon window defect model. Tissue Eng. A 19 (21−22), 2439−2451.

Yamada, M., et al., 2008. Effect of osteogenic protein-1 on the matrix metabolism of bovine tendon cells. J. Orthop. Res. 26 (1), 42–48.

Zhang, N., et al., 2017. Ultrasound as a stimulus for musculoskeletal disorders. J. Orthop. Translat. 9, 52–59.

Zhang, Y.J., et al., 2018. Concise review: stem cell fate guided by bioactive molecules for tendon regeneration. Stem Cells Transl. Med. 7 (5), 404–414.

Zhang, J., Wang, J.H., 2010. Mechanobiological response of tendon stem cells: implications of tendon homeostasis and pathogenesis of tendinopathy. J. Orthop. Res. 28 (5), 639–643.

Zhou, Z., et al., 2010. Tendon-derived stem/progenitor cell aging: defective self-renewal and altered fate. Aging Cell 9 (5), 911–915.

Integration of dental implants: molecular interplay and microbial transit at tissue—material interface

9

Smitha Chenicheri[1,2], Remya Komeri[3]
[1]Microbiology Division, Biogenix Research Center for Molecular Biology and Applied Sciences, Thiruvananthapuram, Kerala, India; [2]Department of Microbiology, PMS College of Dental Science and Research, Thiruvananthapuram, Kerala, India; [3]Regional Cancer Center, Division of Biopharmaceuticals and Nanomedicine, Thiruvananthapuram, Kerala, India

Chapter outline

9.1 Evolution of the concept of biointegration of dental implants 222
9.2 Mechanisms of biointegration of dental implants 223
9.3 Establishing biological gingival seal 223
9.4 Early inflammatory phase 224
9.5 Neovascularization at peri-implant zone 225
9.6 Osteoconduction 225
    9.6.1 *De novo* bone formation, bone remodeling, and osseointegration 226
9.7 Soft tissue healing and biointegration 227
9.8 Cell signaling and integration of dental implants 228
9.9 Genetic networks in osseointegration 229
9.10 Microbial interplay in osseointegration of dental implants 232
9.11 Interface biofilms: a unique pulpit for microbial homing 234
9.12 Implant failure and enhancement of biointegration 236
9.13 ECM disorganization 237
9.14 Microbial versus host cell signaling at the interface 237
9.15 Conclusions 239
References 240

Implant dentistry has evolved for more than a century as a proven treatment to replace the missing teeth in edentulous patients. Recently, the global burden of disease study reports that oral disorders accounts for 16.5 million years lived with disability in which with one-third was contributed by edentulism (Mokdad et al., 2016). Tyrovolas et al. highlighted the prevalence of edentulism-associated health issues in younger generation, especially in regard to the psychological concerns such as depression (Tyrovolas et al., 2016). Because oral health has a direct impact on the general health of people, WHO proposed "*8020 Movement*" that recommends the healthy retention of 20 teeth at the age of 80 (Advanced Ceramics for Dentistry, 2014). However, partial to complete edentulism is still prevalent in elder population as a final marker of disturbed oral health. Fixed or removable dental prosthesis is available to rehabilitate the oral health in edentulous patients. The factors such as interjaw relationship, interforaminal space, oral hygiene, cost-effectiveness, and comfort of patients have been considered while selecting the fixed or removable dental prosthesis.

Fixed implant-supported dental prosthesis is often preferred over conventional removable prosthesis due to the increased stability, retention, esthetic appearance, comfort, and psychological advantages (López et al., 2016). The fixed implants can be endosteal (placed into alveolar bone of mandible or maxilla and pierce only one cortical bone; e.g., blade implant, Ramus Frame implant, Root Form implant), subperiosteal (has a substructure and superstructure in which the custom cast frame is placed directly beneath the periosteum), and transosteal (that crosses both cortical plates) based on the location of implantation (Ananth et al., 2015). Depending on the bone volume of patients and the clinical conditions of surrounding soft tissue, clinicians recommend either fixed full arch implants or hybrid implants. Fixed full arch implants are supported with screw or cement, prescribed in the case of sufficient bone volume and interarch space, whereas, in soft tissue deficiencies, hybrid implants with metal framework and complete denture components are preferred (López et al., 2016). Four to six fixtures or screws are required to fix the full arch prosthesis in edentulous patients. Selection of suitable materials and design of appropriate shape/size for these screws or fixtures are significant to rehabilitate the oral health in patients.

9.1 Evolution of the concept of biointegration of dental implants

On investigating the historical background of dental implant materials, a modern era was evolved after 1925, in which the clinicians began to think about the biological performance of implant materials. Because most dental biomaterials are associated with the regenerative therapy of tooth, the biointegration dealing in this chapter is mainly attributed to osseointegration. Initially, gold, platinum, ivory, tantalum, and cobalt alloys were employed as dental implant materials. In 1937, Venable et al. analyzed the interaction of these metal implant materials with bone and tissue fluids and reported the corrosion of implant materials in tissue fluids due to galvanic reactions (Ananth et al., 2015). Two years later, Strock proposed Vitallium, a cast alloy composed of cobalt, chromium, and molybdenum, as inert and stable implant material for a dental prosthesis. In 1952, Branemark designed a threaded implant design of pure titanium and evaluated the biological and functional implications as an endosteal implant (Jayesh and Dhinakarsamy, 2015). He observed the growth of bone tissue around the thin spaces of implanted titanium chambers. Natural tooth attaches to the bone via fibrous periodontal ligament, whereas, direct attachment of bone is observed with the implant surface. According to American Academy of Implant Dentistry, osseointegration is the contact established between the normal and remodeled bone with the implant surface without the interposition of nonbone or connective tissue. Later, in 2012, osseointegration was redefined as "a time-dependent healing process whereby clinically asymptomatic rigid fixation of alloplastic materials is achieved and maintained in bone during functional loading" (Zarb and Koka, 2012). This suggests that the process of osseointegration involves interplay between cellular, extracellular, and molecular mechanism in the vicinity of peri-implant zone.

9.2 Mechanisms of biointegration of dental implants

The success rate of dental implants is usually appreciable; however, still some cases are reported with complications followed by implant removal. Implant failure is influenced by general patient health status, smoking habits, oral hygiene, implant characteristics, premature loading implant location, and clinician experience. Implant failure is a multifactorial process influenced by overheating and trauma during surgery, poor bone quantity and quality, immediate loading, and lack of primary stability, which may further complicate late events such as peri-implantitis, occlusal trauma, and marginal bone loss (Levin, 2010). The implant failure is associated with continuous pain, impaired mobility, and loss of esthetic outcome that may necessitate implant replacement surgery as an extra burden to patients. Herein, a thorough knowledge on biointegration of dental implants is essential to engineer implants to ensure long-term stability and performance. The healing and remodeling events followed by implantation of dental prosthesis involve both soft and hard tissues. The intermittent relation of soft tissue healing with the osseo-remodeling influences the maintenance and survival of dental implant in the oral cavity.

9.3 Establishing biological gingival seal

Dental implants, irrespective of the position of fixation, consist of a coronal portion supported by a post that passes through the oral mucosa to reach the oral cavity. The biochemical events associated with integration of dental biomaterials initiate in oral mucosa, composing of gingival epithelium, lamina propria, and submucosal layer which are in constant contact with the implant surface. Immediately postimplantation, the gingival epithelium migrates toward the implant surface. In normal teeth, junctional epithelium attracts the gingival soft tissue with tooth surface, having an external basal lamina toward the connective tissue of gingiva and internal basal lamina (IBL) against the enamel. The IBL is rich in laminin-332, a cell adhesion protein that binds the junctional epithelium over the tooth surface to establish gingival seal. However, in peri-implant zone, the IBL composed of adhesive laminin binds only the lower part of the peri-implant epithelium (PIE) with the implant surface (Larjava et al., 2011). The hemi-desmosomes associated with the gingival epithelial cells bind the cell adhesion proteins in basal lamina and facilitate tight contact of the newly formed epithelial layer to form a gingival seal. The PIE constitutes a biological seal that protect the dynamic interface from oral toxins and bacterial infections (Jayesh and Dhinakarsamy, 2015). In normal teeth, the collagen fibers align in a perpendicular orientation to the tooth surface and form bundles that attach the tooth gingival epithelium, whereas in peri-implant zone, collagen fibers align parallel to the implant surface and form a ring that bind gingival epithelium (Silva et al., 2014). The direction of gingival fibers in peri-implant mucosa differs from the natural tooth structure, which increases the chances of bacterial penetration through epithelium toward the peri-implant crevicular fluid. Therefore, proper oral hygiene is recommended for patients to avoid the risk of peri-implant mucositis after surgical implantation.

9.4 Early inflammatory phase

Recent reports underlie the significance of inflammatory responses in the biointegration of dental implants in the oral cavity. Soon after the implantation of a foreign body, the host immune system comprising the immune cells and complements detect the threat and initiate the inflammatory reactions. The phagocytic neutrophils, macrophages, and monocytes are recruited to the site of trauma to remove the cell/tissue debris. Trindade et al. studied the role of immune activation in osseointegration around the titanium implant. The gene expression analysis at 10−28 days interval postimplantation revealed the presence of neutrophils, group 2 innate lymphoid cells, and activation of M2 macrophages that implies the initiation of type-2 inflammatory response around the titanium implant (Trindade et al., 2018). They have observed the presence of neutrophils even after the inflammatory phase, supported by the population of macrophages that inhibit the apoptotic death of neutrophils in a concern to promote vascularization under hypoxic conditions.

The surface topography of implants influences M1/M2 polarization of macrophages. Ma et al. demonstrated that monocytes differentiate into M1 phenotype on the surface of the nanostructured TiO_2 implant with tube size of 20 nm, whereas to M2 phenotype when tube size was reduced to 5 nm. The cytokines released from M1 (IL-1β, IL-6, IL-12, IL-23, tumor necrosis factor alpha (TNF-α), interferon gamma (IFN-γ), monocyte chemo-attractant protein 1 (MCP-1), macrophage inflammatory protein 1 beta (MIP-1β), inducible nitric oxide synthase) and M2 macrophages (IL-1Ra, IL-10, arginase-1, vascular endothelial growth factor A or VEGF-a, platelet-derived growth factor-BB or PDGF-BB, transforming growth factor-β or TGF-β) have significant role with respect to the implant stabilization. TGF-β and PDGF-BB promote the migration proliferation and differentiation of bone marrow—derived mesenchymal stem cells (bMSC) to osteoblasts in the peri-implant zone. In spite of inflammatory responses, the macrophages can also differentiate into osteoclasts under the influence of receptor activator of NF-κB ligand (RANKL) and osteoprotegerin, and monocyte colony stimulating factor, and the cytokines released from bMSC that balances the bone tissue homeostasis (Ma et al., 2018).

9.5 Neovascularization at peri-implant zone

The peri-implant neovascularization has been considered to be detrimental in deciding the extent of osteogenesis and osseointegration. The surgical implantation results in local hemorrhage and ischemia, and the implant surface interacts with blood cells and proteins that initiate a cascade of biochemical events. The initial phase includes the interaction of platelets with the fibrinogen adsorbed on the implant surface via GPIIb/IIIa integrin (Davies, 2003). The adhered platelets release PDGF, TGF-β, and vasoactive factors such as serotonin and histamine, which function as chemotactic factors for fibroblasts, neutrophils, smooth muscle cells, osteogenic cells, and bone marrow—derived cells. The blood coagulation and fibrin clot formation cease the hemorrhage, however, restrict the cell migration as well. To remove the blood clot, the neutrophils and macrophages infiltrate to the site and activate the phagocytosis of the fibrin clot. The cytokines released from platelets and macrophages signal the angiogenesis from postcapillary venules toward the peri-implant zone, guided by the hypoxic gradient toward the center of the wound. The endothelial cells invade the subendothelial basement membrane and proliferate to form capillary sprouts which grows longer and anastomose to loops creating a functional vasculature with blood flow. Khosravi et al. reported neo-angiogenesis and maturation into the functional vascular network within 42 days after introducing nanotopographically complex (TiNT) and machined-surface (TiMA) Ti cranial implants. The establishment of vascular network in peri-implant zone facilitates direct delivery of osteogenic precursors, improved osteogenesis, and faster osseointegration (Khosravi et al., 2018).

9.6 Osteoconduction

The phase of osteogenesis and bone formation is inseparably overlapped with the early stages of wound healing that begins within the first week following the surgery. According to the current understanding, bone matrix undergoes mineralization that restricts their growth on demand. Moreover, the bone-forming osteoblast cells differentiate into nonsecretory osteocytes which has minimal role in further bone formation. This necessitates the migration of osteogenic cells to the implant surface for new bone formation at the peri-implant site of the endosseous implant. For example, the titanium implant inserted into the rabbit bone exhibited the presence of mesenchymal stem cells from bone marrow at the site of implantation within 3 days of implantation. These MSCs differentiate to bone-forming osteoblast cells, which are polarized secretory cells which actively synthesize new bone over the old bone (distant osteogenesis) or directly over the implant surface (contact osteogenesis) (Osborn, 1980). However, these biological responses vary with respect to the nature of implant. The biotolerant implants facilitate distance osteogenesis and fibrous connective tissue formation on integration, whereas bioinert type is characterized with contact osteogenesis and de novo bone formation. Bioactive implants allow new bone formation and create a chemical bond with the bone. The migration of osteoblast toward the implant surface termed osteoconduction is influenced by the local microenvironment of the peri-implant zone. The fibrin network in the healing peri-implant zone, generated by the blood coagulation event, interacts with the implant surface and serves as a temporary matrix for the tractional migration of osteoblast cells toward the implant surface. Retention of fibrin matrix on the implant surface is influenced by the implant surface design, material composition, and wetting of implants, which in turn determines the extent of contact osteogenesis and de novo bone formation (Davies, 2003).

9.6.1 De novo *bone formation, bone remodeling, and osseointegration*

The de novo bone formation following maturation through bone remodeling comprises the establishment phase of osseointegration. The dynamic nature of bone tissue contributes to the maintenance phase of osseointegration that balances the bone formation and bone resorption by the interplay of cellular and molecular events (Venkataraman et al., 2015). Once the osteogenic cells are recruited to the implant surface, they secrete a collagen-free matrix enriched with osteopontin, bone sialoprotein, and proteoglycans that cement the new bone over the implant. The new bone formed over the implant is characteristically similar to woven spongy bone with small medullary space containing blood capillaries, osteocytic lacunae, and numerous osteoblasts and osteoclasts. After a time period of 3 months, the spongy bone transforms into the calcified compact bone with empty osteocytic lacunae, rarely observed osteoclasts and fewer capillaries between the bone—implant interface (Mello et al., 2016). Berglundh et al. studied the early healing process and alveolar bone formation around a titanium solid screw implant with a SLA (sandblasted large grit and acid etched) surface configuration in twenty Labrador dogs (Berglundh et al., 2003). The staggered screw implant was installed by replacing the mandibular premolars both in left and right side. After 2 h of installation, a fibrin coagulum with blood cells was identified in histological analysis. Coagulum was replaced with new tissue after 4 days of installation, and characteristic occurrence of fibroblast like cells was identified around the SLA surface (Fig. 9.1a). In 1 week of healing, sprouting vascular structure, collagen fibrils and newly formed woven bone was observed in the healing chamber. The formation of large areas of bone towards the apical region of implant associated with remodeling of adjacent parent bone was obvious after 2 weeks (Fig. 9.1b). After 4 weeks, healing chamber was filled with mineralized parallel-fibered and lamellar bone with well developed primary spongiosa rich in vascular structure (Fig. 9.1c).

9.7 Soft tissue healing and biointegration

The healthy soft tissue in the peri-implant zone is detrimental for the survival of dental implants. After establishing the peri-implant biological seal within 4 days of implantation, the surrounding tissue provides the protection of peri-implant zone from bacterial invasion. The further healing process continues with the colonization of leukocytes at the coronal portion and collagen-producing fibroblast at the apical portion of the peri-implant zone. In natural tooth, the collagen fibers are aligned perpendicular connecting the tooth cementum to the alveolar bone, which serves as a barrier to bacterial invasion and epithelial down growth. Because the implant surface is devoid of cementum, the fibers orient themselves in a parallel fashion, which compromise the integrity of the seal. Lack of an intact connective tissue around the implant leads to the infiltration of B cells and plasma cells that mediate chronic periodontitis or peri-implantitis by periodontal pathogens such as *Staphylococcus aureus* (Wang

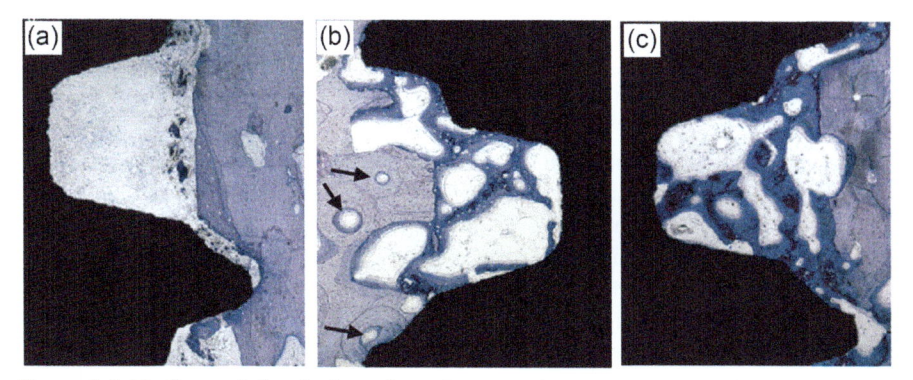

Figure 9.1 Healing and alveolar bone formation around a titanium based solid screw dental implant in Dogs. Histological analysis after 4 days (a), 2 week (b) and 4 weeks (c) after implantation. Arrows in (b) indicate the remodeling in adjacent parent bone. Figures taken from Berglundh et al., 2003.

et al., 2016). Attempts to facilitate the more-perpendicular like fiber orientation in peri-implant zone appreciates better management of soft tissue health following the surgical implantation.

9.8 Cell signaling and integration of dental implants

Well-regulated interplay of extracellular matrix (ECM) proteins, cell phenotypes, and mechanical signals ensures the proper integration and long-term survival of dental implants. The initial credit goes to the proper sensing of the peri-implant microenvironment and timely initiation of specific cell signaling pathways. Recent findings resolve the intricacies associated with the biomolecular cell signaling mechanism in the peri-implant zone to explain the potential of possible therapeutic targets.

Hypoxia signaling in the peri-implant zone is dependent on the nature of implant materials placed in the oral cavity. The TiO_2 implants continue to be oxidized which depletes the oxygen availability in the peri-implant zone and induces a hypoxic microenvironment. Moreover, the initial phase of inflammation also contributes to the generation of reactive free radicals. The hypoxic signals suppress the prolyl hydroxylase domain enzymes responsible for the ubiquitination of hypoxia-inducible factor-α (HIF-1α), which leads to the accumulation in the cytoplasm, translocation to nucleus, and activation of hypoxia-regulated genes. HIF mediates the coupling of angiogenesis and osteogenesis through the transcriptional activation of VEGF and SDF-1 (stromal cell—derived factor-1) during bone and soft tissue healing at peri-implant zone (Drager et al., 2015). HIF-1α upregulates the expression of Sox-9, which induces the chondrogenic differentiation of bMSC (Khan et al., 2010). The hypoxic or xenobiotic microenvironment at the peri-implant zone activates the HIF/ARNTL (aryl hydrocarbon receptor nuclear translocator-like) pathway. ARNTL is a PAS (per-ARNT-sim) domain containing molecular sensor which dimerizes with neuronal PAS domain-containing protein 2 (NPAS2) or circadian locomotor output cycles kaput (CLOCK) and regulates the transcription of genes essential for peri-implant bone formation (Nishimura, 2013).

The canonical and noncanonical Wnt signaling regulates the healing in the bone—implant interface from the initial phase itself. The Wnt1, 3a, 7, 8, and 10b are involved in canonical and Wnt4, 5a, and 11 in the noncanonical pathway, respectively. The canonical Wnt signaling plays a significant role in regulating the proliferation and differentiation of osteoblasts and drives the bone formation. The Wnt molecules at the zone of injury bind the Frizzled or LRP5/6 complex and stabilize the β-catenin in the nucleus through the mediators such as dishevelled, Axin, and Frat 1 protein. The β-catenin is associated with adherence junction including cadherin, enabling cell—cell interaction and cell migration essential for the healing process. Moreover, β-catenin pathway activates the nuclear factors such as RunX2, rich in residues of glutamine and alanine, which activate osteocalcin (OC) and Col1A1 genes and promote the mineralization of mesenchymal stem cells and osteoblast. The noncanonical pathway facilitates the polarization of fibroblasts to synthesize a collagenous matrix at the peri-implant zone. The noncanonical pathway may be Fz dependant (planar cell polarity orPCP) pathway or calcium dependent (Wnt/Ca^{++} pathway). PCP signaling activated by Wnt5a facilitates vascular remodeling at the injury site. Wnt10b control bone formation under the influence of mechanical load (Romanos, 2016). Mouraret et al. have studied the integration of dental implants in Axin2$^{LacZ/LacZ}$ mice with elevated Wnt signaling and reported less fibrous encapsulation, enhanced osseointegration that reduced the chances of implant failure (Mouraret et al., 2014).

TGF-β signaling is composed of multifunctional polypeptide growth factors having potent local effect on modulating ECM formation in wound repair. It also contributes to the osteoblastogenesis and bone formation via the expression of bone morphogenetic protein (BMP), which phosphorylates R-Smads and forms complexes with co-Smads. The Smad further activates Runx2 transcriptional factor that induces osteoblastic differentiation of MSC. Cornelini et al. reported the higher TGF-β expression around the failing dental implants, associated with the stimulation of endothelial cells and fibroblast cells for angiogenesis and wound repair, respectively. In addition, it is a proinflammatory cytokine with chemotactic activity over neutrophils, mast cells, and lymphocytes, which contributes to the inflammatory phase following implantation (Cornelini et al., 2003).

9.9 Genetic networks in osseointegration

The long-term stability and efficient clinical performance of dental implants relay on the sustained immunological equilibrium, foreign body equilibrium, and bone formation—resorption equilibrium following the implantation. The tightly synchronized genetic network downstream to the cell signaling pathways decides the fate of implants in the oral cavity. The genes associated with the osseointegration are tabulated as shown in the Table 9.1.

Master switch genes: The ultimate action of extracellular effector molecules like BMP, TGF-β, VEGF, IGF, Wnt-B-catenin pathway, and other cytokine modulators associated with bone formation and remodeling after implantation relays on the expression of Runx2 gene, which is considered as the master switch gene in bone formation and osseointegration. Byers et al. reported better bone formation and mineralization of polymeric scaffolds incorporated with Runx2 gene—transfected bone marrow stromal cells (Byers et al., 2004). RUNX2 is a master osteoblast—specific Runt-related heterodimeric transcription factor, also named Cbfa1 or AML3, which regulates the osteoblast commitment and osteogenic differentiation of MSCs. The RUNX2 protein enhances the expression of collagen type I (Col1a1), bone alkaline phosphatase (bALP), and OC proteins essential for bone formation. RUNX2 also

Table 9.1 Genetic network in osseointegration.

Days after implantation	Upregulated genes	Functions	Significance
Day 3/4	TNF-a, IL-6, IL-2	Proinflammatory cytokines	Initiation of the inflammatory phase
	CCL18, CXCL10, CXCL14	Chemokines	Migration of inflammatory cells
	Tollip, IL9, IL22	Antiinflammatory cytokines	Inhibit inflammatory responses
Day 7	HOX, Sp3	MSC genes	Osteogenesis
	Runx2, Osx, OCN, OPN, BMP6, BSP	Transcription factors	Osteogenesis
Day 14	BDNF (brain-derived neurotrophic factor, NTF3 (neurotrophin 3)	Neurotrophic factors	Neurogenesis
	ANXA, EPAS1	Vascular endothelial growth factor signalling	Angiogenesis
	BMP4, BMP2K	Bone morphogenetic proteins	Osteogenesis
	MMPs (matrix metalloproteinases), TIMPs (tissue inhibitor metallopeptidase)	Extracellular matrix remodeling	Bone remodeling
	CTSK, ACP5	Cathepsin K	Osteoclast activity
	PLODs (procollagen lysyl hydroxylase), LOX (lysyl oxidase), PCOLCE (procollagen C-endopeptidase enhancer)	Extracellular matrix remodeling	Extracellular matrix organization
	Coll (collagen), OC, ON (noncollagen proteins), ALP (alkaline phosphatase)	Extracellular matrix proteins	Extracellular matrix deposition

influences the expression of bone sialoprotein and osteopontin proteins and assists bone formation. BMP also function through the suppression of smurf1 (Smad ubiquitin regulatory factor 1) protein involved in RUNX2 degradation. The significant role of Runx2 gene in the mineralization of bone implants opens new avenues for improving osseointegration after implant placement.

Circadian-regulated genes in biointegration: The HIF/ARNTL pathways favor the expression of period (Per) and cryptochrome (Cry) genes coding for molecular clock proteins involved in circadian rhythm. These genes are involved in a positive feedback loop over the ARNTL pathway and provide homeostatic control over the bone remodeling by regulating the osteogenesis and osteoclastogenesis. PER protein induces the differentiation of recruited MSC to osteoblasts, whereas CRY insures the osteoclastogenesis. Kushibiki et al. reported the role of low-level laser radiations (at wavelength of 405 nm) on downregulation of CRY1 protein, which enhances the osteogenesis by MSC. They proposed laser radiations as a molecular switch to control the bone formation at the implanted site (Kushibiki and Awazu, 2008). CLOCK/NPAS2 pathway is also influenced by implant placement that directly enhances the transcription of Col2 and Col10 genes coding for collagen protein essential for chondrogenesis. Type 10 collagen increases the tensile strength of interfacial collagen in the peri-implant zone. The possible signaling pathways and genetic network associated with healing process and osseointegration of dental implant is represented in figure 9.2.

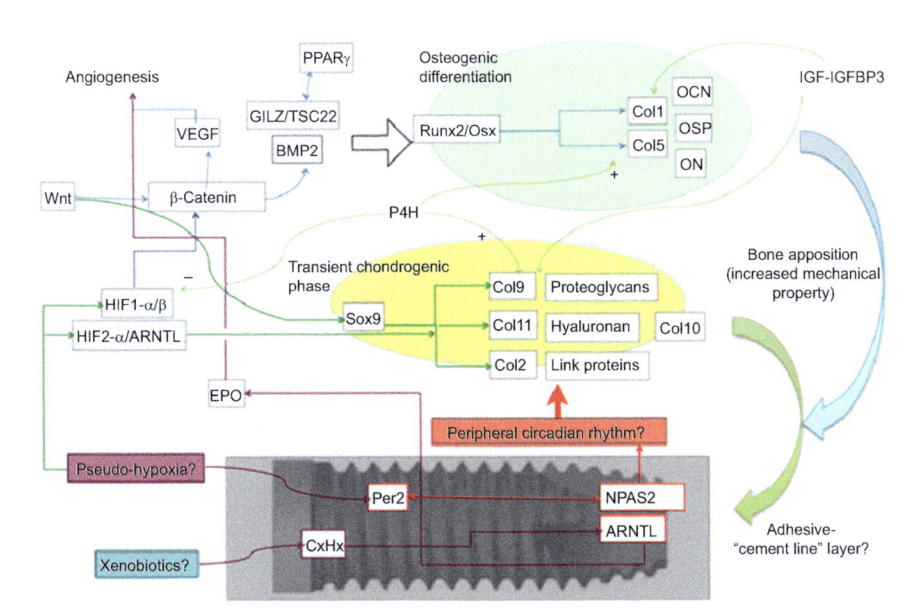

Figure 9.2 Genetic network in osseointegration. Figure courtesy to Nishimura et al., 2013.

9.10 Microbial interplay in osseointegration of dental implants

Role of microbial colonization is very crucial in the success of a dental implant. There are well-established experimental evidences showing the shift in composition of microbial flora, both quantitatively and qualitatively pre- and postimplantation of a dental implant. Plaque microflora on the implant surface induces peri-implant invasive diseases (PIIDs) such as peri-implant mucositis and peri-implantitis. Hence, the extent of oral hygiene and antimicrobial therapy improves the longevity and success of implant therapy (Mombelli and Lang, 1998).

On the basis of etiopathogenesis, an implant failure can be classified as infectious or noninfectious. The duration of implant failure is categorized as early or late failure. Broadly, the early implant failures results from failure to establish osseointegration and are governed by factors such as bacterial infection, surgical trauma, and premature loading of implant. In addition, prosthetic rehabilitation leads to late implant failure which is due to impaired osseointegration. Inflammatory lesions that develop around implants are being described as peri-implant disease, peri-implant mucositis, and peri-implantitis. In general, peri-implant diseases describe the inflammatory reactions in the tissues surrounding an implant, where peri-implant mucositis is a reversible inflammation in the mucosa at the implant zone with minimal loss of supporting bone and is comparable to gingivitis around natural teeth. Peri-implantitis presents inflammation and loss of supporting bone around the natural teeth, analogous to periodontitis (Marrone et al., 2013). Hence, a complete and infection-free establishment of bone–implant integration has become a persistent challenge in oral rehabilitation.

The implant surface and adjacent gingival tissues are colonized by oral microbiota within few minutes after the implantation procedure. Within 10 days, the microbiota mimics the periodontal microbiomes. Studies show the differences in nature and type of microorganisms colonizing the stable and failing implants. A stable and healthy implant is predominantly colonized by gram-positive anaerobic cocci and rods such as *Streptococcus thermophilus, Streptococcus oralis, Streptococcus sanguinis, Streptococcus infantis, Actinomyces naeslundii, Actinomyces gerencseriae, Actinomyces meyeri, Actinomyces oris, Streptococcus mitis,* and *Peptostreptococcus* sp, *Neisseria* sp., *Gemella haemolysans,* and *Rothia* sp. Subgingival plaque flora, such as *Haemophilus* sp. and *Veillonella parvula*, accounts for more than 50% of the bacteria. *Fusobacteria* sp. and black-pigmenting gram-negative rods are also occasionally present. These are similar to that of a healthy gingival sulcus, where periodontal pathogens and spirochetes are present in very low or undetectable level (Buddula, 2013). The presence of periodontopathogenic species in healthy peri-implant sulcus without any signs of inflammation suggests that the host's response to microbial population is very crucial in PIIDs.

Peri-implant microbiota is established soon after implantation, which remains stable and unaltered with the course of time. The higher proportions of *Porphyromonas gingivalis, Prevotella intermedia, Tannerella forsythia, Aggregatibacter actinomycetemcomitans, Fusobacterium* sp, and spirochetes are found around failing implants. Implants failing due to overload and trauma demonstrated the microbial flora of healthy periodontium and those due to infection presented periodontopathic flora. Periodontal diseases lead to partial alveolar bone loss, and the pathogens remain lodged in the periodontal pockets, which colonize the implants even after successful osseointegration leading to implant failure. These microbes also adhere to the teeth, crowns, and implants, directly influence the peri-implant microbiota to promote the plaque development. Hence, the periodontal health status of a patient is noteworthy before an implant procedure (Kumar et al., 2017).

Owing to resemblance in the environmental conditions of implant crevices and subgingival sulcus, the microbial population of the periodontium is also found to colonize a stable dental implant. Peri-implant crevicular fluid in the vicinity of the implants provides required nutrition for microbial survival in the region. Hence, the nature, volume, and content of the fluid play a crucial role in survival and performance of dental implants. For instance, different types of nutrients such as amino acids and peptides, anaerobic conditions around the implant, and neutral pH of the fluid favor the growth of periodontal microbes. The implant crevices present anaerobic condition at the interphase between the implant material and mucosal epithelium, encouraging the colonization of periodontal microbes. Hence, the predominant flora of a healthy implant is grampositive anaerobic cocci and bacilli. The falling implants have microbial flora similar to those seen in periodontal infections, which can pave way for inflammatory pathways. Peri-implant mucositis shows a shift in subgingival microbial flora from supragingival population. Therefore, there is a decrease in *Streptococcal* and *Actinomyces* population with an increase in the proportion of orange complexes such as *Fusobacterium nucleatum, P. intermedia*, and *Eubacterium* sp. Persistent inflammation triggers the penetration of infection resulting in peri-implantitis. This condition is marked by the presence of pathogenic bacteria from orange and red complexes including *P. gingivalis, T. forsythia, Prevotella nigrescens, Prevotella oris,* and *F. nucleatum.* Apart from the usual periodontal pathogens, other bacteria inhabitants include *Pseudoramibacter alactolyticus, Eubacterium* sp., *Veillonella* sp., *Synergistete* sp., *Enterobacteriaceae, Candida* sp., *Filifactor alocis, Dialister invisus, Mitsuokella* spp., *Peptococcus* sp., *Clostridiales* sp., *Catonella morbid, Chloroflexi* sp., *Tenericutes* sp., *Porphyromonas* sp., *P. nigrescens*, and *P. oris* (Pokrowiecki et al., 2017). Interestingly, several in vitro studies cite the presence and affinity of *S. aureus* and *Candida albicans* for titanium surfaces. *S. aureus* and coliforms such as *Escherichia coli* and *Pseudomonas aeruginosa* detected on implant surfaces as cited in few literatures may represent cross infection as they are rarely encountered in oral infections.

Microbial surface components recognizing adhesive matrix molecules (MSCRAMMS) expressed on the surface of *S. aureus* interact with host proteins such as fibrinogen, fibronectin, and collagen subtypes for interaction with the implant surface and ECM deposition on the implant material (Dhir, 2013). The peptidoglycan layer in gram-positive bacteria such as *S. aureus* and *Streptococcus* sp. imparts partial hydrophobic character to these bacteria which enables their adhesion to implant surfaces.

9.11 Interface biofilms: a unique pulpit for microbial homing

Microbial biofilms play a crucial role in peri-implant tissue inflammation and contribute to about 65% of diseases. This can progress to alveolar bone loss associated with the implant threads. The inherent physiological factors such as increased blood circulation, collagen fibers, periosteal vascular plexus, gingival fiber orientation, and minimal fibroblast activation contribute to majority of implant failure. Microcrack between the prosthetic restorations, implant bodies, and connectors may enhance microbial seepage and deeper penetration of acids, enzymes, microbial toxin, or metabolic products affecting periodontal tissue contributing to late failure of implant material (Tallarico et al., 2017). In addition, the microbial biofilm formation also depends on morphology, roughness, surface hydrophilicity or hydrophobicity and chemistry, thermal and mechanical stability, and abutment material of the implant material, all of which greatly influence the survival of implant (Tesmer et al., 2009) (Singh et al., 2012).

Biofilm formation on dental implant follows similar pattern of microbial colonization as in normal biofilms on the tooth surfaces and is occupied by the flora of adjacent tooth and oral environments such as gingiva, buccal mucosa, and saliva. Following implantation, the diffusion of proteins and other biomolecules from extracellular fluid facilitates the formation of a provisional matrix on the surface of implant. This temporary matrix drives the formation of salivary acquired pellicle (rich in proteins and salivary glycoproteins); however, it interferes with albumin adsorption. The pellicle formation begins within 30 min of exposure in oral cavity and is featured by low albumin absorption. The microbial adherence to salivary pellicle is facilitated by the adhesin molecules such as lectins, which are cell surface molecules present on the bacterial cell surface. The initial colonization is complexed by further intercellular bacterial adhesion, secretion, and deposition of ECM polysaccharide including levans and dextrans. These molecules promote the microbial adhesion and coaggregation on to the implant surfaces. Surface wettability influences the protein adsorption and interaction with the implant surface. Highly hydrophobic surface enhances protein adsorption, and hydrophilic surface reduces the hydrophobic interaction between proteins and implant surface leading to lower adsorption affinity.

Bone cells are attracted to hydrophilic surface, favoring osteoconduction and osseointegration. On the other hand, bacteria possess biomolecules such as lipopolysaccharide (LPS) in their cell wall, which renders bacterial cell more hydrophilic and are attracted to hydrophilic implant surfaces (Strevett and Chen, 2003). This accounts for colonization of gram-negative bacteria because they possess LPS in their outer membrane. The early colonizers are predominantly the gram-positive anaerobic cocci, rods, *Streptococcus* sp. and *Actinomyces* sp. *F. nucleatum* acts as the bridging bacteria, which coaggregates between early colonizers and late colonizers forming the complex mature plaque. The periodontal pathogens such as *P. gingivalis* and *P. intermedia* are the late colonizers. Red and orange complex species such as *P. gingivalis*, *T. forsythia*, *Treponema denticola*, *F. nucleatum,* and *P. intermedia*, respectively, colonizes the implant in peri-implantitis. Fully edentulous mouth implants are heavily colonized by black pigmented bacteroids, motile bacilli, and coccoid bacteria, whereas partially edentulous mouths harbor *P. gingivalis* and *P. intermedia* on the implant surface (Blankenship and Mitchell, 2006).

The implant abutment junction of a two-piece implant is highly vulnerable to biofilm-related infection due to microleakages. The major microbes associated include *A. actinomycetemcomitans*, *P. gingivalis*, *T. forsythia*, *T. denticola*, *P. intermedia*, *Parvimonas micra*, *F. nucleatum*, *Campylobacter rectus*, *Eikenella corrodens*, *C. albicans*, and *Enterococcus faecalis* (Canullo et al., 2016). These interactions are strongly influenced by surface roughness, chemical composition, surface properties, including wettability, surface tension, and surface free energy. To cite, the rough surface promotes biofilm formation as the microbes are entrapped and protected within the micropits of the implants. In addition, the biofilm formation increases with surface roughness Ra> 0.2 μm leading to peri-implant failure (Bollen et al., 1996; de Avila et al., 2014). Physiochemical surface modification of dental implants enhances osseointegration and thereby improves success rates of implant therapy. Such modifications enhance interactions with biological fluids and cells and accelerate peri-implant bone healing as well as improve osseointegration at sites that lack sufficient quantity or quality of bone. Methods such as grit-blasting, acid etching, and combinations are commonly used to enhance implant surface microroughness, which also reduces microbial adhesion. Drug-releasing implants in dentistry offers promising antimicrobial activity and biocompatibility (Han et al., 2016).

Pure titanium and Ti—Al—V alloy are the gold standard in implant dentistry, but ceramic materials with use of zirconium dioxide and other innovative metallic alloys are also gaining considerable interest in implantology. Various methods of surface modifications of Ti implant explore improved performances of Ti implants such as increased osseointegration, protection from body fluid—driven chemical corrosion, and reduced bacterial adhesion. Plasma treatments of Ti implant imparts antibacterial properties and are also used for cleaning and sterilization of dental implant. Plasma-polymerized polyterpenol films hinder microbial adhesion and proliferation. Inorganic functional coating of Ti implants also reduces bacterial adhesion. Calcium phosphate (CaP)—based alloys such as hydroxyapatite (HA), calcium phosphate cements (CPC), titanium oxide (TiO_2), nitride coatings, and incorporation of Ag nanoparticles on TiO_2 coatings incorporate antibacterial properties to dental implants. Other techniques such as magnetron sputtering, metal-organic chemical vapor deposition, and low-pressure chemical vapor deposition for titanium oxide deposition prevent bacterial adhesion. Zn-doped TiO_2 and Cu- or Ag-doped TiN reduces bacterial adhesion and ZrN and ZrCN coatings by cathodic arc deposition possess antibacterial properties, which improve dental implant qualities. Oxides of Zr, Si, and Zn are also used as alternatives to TiO$_2$ implants that reduce bacterial adhesion, improve biocompatibility, and are corrosion resistant (Inoue and Matsuzaka, 2015).

9.12 Implant failure and enhancement of biointegration

Implant failure is multifactorial in nature. Accelerating the process of osseointegration, improvement of material biocompatibility and prevention of bacterial adhesion to the implant are the most challenging factors in implant therapy.

The deleterious effects of anaerobic plaque bacteria are one of the major causes of implant failure. The presence of periodontopathogens such as *P. gingivalis, P. intermedia, T. forsythia,* and *A. actinomycetemcomitans* are directly related to the clinical parameters such as sulcular bleeding and plaque index. Oral hygiene and health status of preiodontium are deciding factors for the success of implant therapy. Commonly, implant placement procedures are adjunct with antimicrobial therapy and improve the clinical status of peri-implantitis patients. Delay in implant surgery post tooth extraction reduced peri-implantitis rates, and exact preoperative and postoperative antibiotic regiments along with improved dental hygiene enhance the success rates of implant therapy.

Multilevel approaches are being adopted for the management of peri-implantitis and implant failure. Prevention of pathogenic microbial community and biofilm development is of prime importance. Biofilm ingress can be prevented at different levels like prevention of microleakage at IAJ. Modification of implant and implant abutment designs readily reduce microbial seepage at IAJ (Taiyeb-Ali et al., 2009). The presence of biofilm with protective extracellular slime matrix restricts the action of antimicrobial agents and host immune responses, hence mechanical methods are mostly preferred (Saini, 2011). Contaminated implant surfaces are debrided by using various methods such as mechanical, photodynamics/lasers, and surgical treatments. Laser-assisted antimicrobial photodynamic therapy significantly reduces microbial load on infected implants and promotes rapid inactivation of microbial cell by liberation of oxygen species, which destructs the cell membrane. Laser therapy efficiently reduces the levels of clinical markers of inflammation in peri-implantitis such as bleeding on probing and clinical attachment loss without risking the integrity of implant or alveolar bone (Romeo et al., 2016; Alshehri, 2016).

Treatment of biofilm-related infection is another strategy to prevent implant failure. Antimicrobial therapy with local or systemic antibiotics was proven to be beneficial. Systemic local drugs including minocycline and tetracyclines have shown promising clinical outcomes by decreasing the levels of the *P. gingivalis*, *T. forsythia*, and *A. actinomycetemcomitans*. Oral antimicrobial rinses and application of antiplaque agents at two-stage surgeries with 0.2% chlorhexidine prevent microbial colonization and maintain oral hygiene thereby increasing success rate of implant therapy (Salvi et al., 2007). Orthopedic implants employ antibiotic, antiseptics, and antimicrobial peptides with antimicrobial properties as coatings to expand the antimicrobial microbial property. Pure metal implants such as iron, titanium, nickel, and silver exhibit inherent bacteriostatic activity.

9.13 ECM disorganization

The efficient osseointegration is required for peri-implant endosseous healing. The initial cell adhesion by physiochemical interactions is followed by a second level of focal adhesion. During osseointegration, the fibroblast growth factor (FGF) activates fibroblasts to secrete ECM proteins such as collagen (type I, III), chondroitin sulfate, fibronectin, vitronectin, and proteoglycans. Focal adhesions provide a vehicle for cross talk between the ECM and host cell. The ECM creates a physical microenvironment necessary for the cell to survive and to function, for cell anchorage, and acts as a tissue scaffold for cell communication, migration, proliferation, and differentiation. A multifunctional cellular structure consisting of a complex network of transplasma membrane integrins and cytoplasmic proteins links ECM to the cytoskeleton. ECM ligands that interact with the extracellular domain of integrins include bronectin, vitronectin, and collagen. The integrins act as mechanoreceptors, activating intracellular signal transduction pathways, which generate biochemical cellular responses (Smeets et al., 2016). The components, biomechanics, and structures mimicking ECM are highly important in osseointegration. Hence, ECM proteins coated onto the dental implant surface enhances cellular proliferation and differentiation. Degradation of ECM is caused by the acute and chronic inflammation induced by peri-implantitis pathogens, with the activation of matrix metalloproteinases (MMPs), especially MMP-1-3-8-13, and inflammatory cytokines (tumor necrosis factor alpha [TNF-α], interleukin-1 [IL-1], IL-6). ECM plays an important role in the adhesion of *P. gingivalis* to the implant surface (Mahmoud et al., 2012). Studies show that high levels of cytokines and chemical mediators of inflammation in periodontitis and peri-implantitis lead to the formation of autoantibodies toward ECM constituents, which may mediate bacterial adhesion to the implant surface and promote biofilm formation (Papi et al., 2017).

9.14 Microbial versus host cell signaling at the interface

Peri-implantitis and peri-implant mucositis are the most significant peri-implant infective diseases which are manifested by inflammatory lesions with microbial etiopathogenesis. These manifestations are due to deposition of biofilms on the implant surface similar to that of gingivitis and periodontitis. Shift of symbiotic balance in periodontopathogenic bacterial population alters the host immune response posing potential risk for infection. Ribulose biphosphate carboxylase, succinyl-CoA:3-ketoacid-coenzyme A transferase, and DNA-directed RNA polymerase subunit-beta are important bacterial markers used in chronic periodontitis and peri-implantitis (Baliban et al., 2012). FadA adhesin of *F. nucleatum* and AdpB of *Prevotella* sp. are significant virulence factors, which contribute for bacterial adhesion, lymphocyte aggregation, epithelial cell invasion, and coaggregation of primary and secondary colonizers. Msp, cfpA, and dentilysin of *T. denticola* and karilysin and prtH and bspA of *T. forsythia* are the virulence factors (Zhu et al., 2013; Ksiazek et al., 2015). These bacterial markers play important roles in synergistic polymicrobial biofilm formation along with *P. gingivalis* and coaggregation with other bacteria, destruction of host epithelial cell, and also contribute to antibiotic drug resistance. These bacterial components induce local deregulation of cytokines, upregulation of IL-8, induce proteolytic inactivation complement system, favor alveolar bone loss, and promote the degradation of immunoglobulins.

Bacterial metabolites in the peri-implant sulcus attract PMN migration through connective tissues. This leads to vasodilation in the peri-implant epithelium (PIE), which initiates inflammatory reactions. Proinflammatory cytokines such as IL-1β, IL-8a, and cathepsins are released to the implant zone. The increased levels of bacterial products including exotoxins, LPS and enzymes in peri-implant region hurdle tissue permeability and degeneration of fibroblasts population owing to extensive disruption of the ECM. In addition, ECM disorganization results in vasodilation and vasoproliferation leading to increased PMN infiltration, cytokine release and costimulation of macrophages, T cells, B cells, and/or dendritic cells. As the infection progress, there is further proliferation of epithelium into collagen-depleted areas of ECM leading to pocket deepening. The pH drop of peri-implant crevicular fluid (PICF) paves way to dissolution and corrosion of the implant surface. The eroded ions are phagocytosed by macrophages which further releases proinflammatory cytokines especially IL-8b and pave ways for bone osteolysis. Infiltration of PMNs produces enzymes, ROS and stimulates fibroblasts to release MMP8, thereby aggravates the inflammatory responses.

Compared to healthy gingival mucosa, peri-implant mucosa presents lesser Langerhans cells, which are the major APCs and higher number of interstitial cells, whereas peri-implantitis is characterized by even higher proportions of neutrophils, macrophages, and T- and B-cells. PICF contains higher levels of IL-β, IL-10, IL-12, IL-8, IL-6, and TNF-α than gingival crevicular fluid (Ujiie et al., 2012).

Gingipains (HRgpA, RgpB, and Kgp) derived from *P. gingivalis* play a pivotal role in the pathogenesis and sustained colonization in peri-implant mucositis and peri-implantitis. These are arginine- and lysine-specific proteases contributing toward bacterial virulence and tissue damage. Gingipains stimulate the expression of MMPs in fibroblasts and activate ECM degradation. Gingipains challenge the host-defense systems by degrading cytokines, components of the complement system, and several receptors, including macrophage CD14, T-cell CD4, and CD8. Furthermore, gingipains degrade CD14 from the surface of monocytes, which is a major receptor for bacterial LPS, rendering hyporesponsivity of monocytes to bacterial LPS. LPS present in gram-negative periodontal pathogens upregulates the gene expression of IL-1α, IL-1β, TNF-α, and IL-6 via the TLR4 signaling and the downstream activation and assembly of NLRP3 inflammasome. IL-6 enhances RANKL expression and stimulates differentiation and activation of osteoclasts. Presence of LPS in the proximity to the bone supports the survival of mononuclear osteoclasts through NF-kB activation and induces bone resorption and osteoclast formation. Studies suggested that IL-6 along with other cytokines leads to bone loss and progressive failure of dental implants. Natural or synthetic inhibitors of gingipains and other bacterial products are being investigated for prophylactic measures in implant therapy, however warrants further research.

The interaction between invading bacteria and host immune system is very complex in a polymicrobial environment such as peri-implant sulcus. It is proposed that soon after the implantation procedure, there is a *"race to the surface"* theory, which states that the relative speed/competition of bacteria and osteogenic cells to attach the implant surface determines the fate of the implant (Gristina, 1987). Accordingly, if bacteria colonize the implant surface before the attachment of osteogenic cells, it will result in prosthetic infection and failure of the implant. In case of a chronic inflammation or peri-implantitis, endogenous activators of innate immunity and cytokines provide booster signals, which leads to injurious immune responses. The inflammatory cytokines, mainly tumor necrosis factor (TNF-α) and IL-6, induce factors such as receptor activator of nuclear factor kappa-B ligand (RANKL) for the production of osteoclastogenic factors. Extensive formation of osteoclasts leads to bone destruction before the neo-osteogenesis at the implant interface (Takayanagi, 2007). Microbial products which act as pathogen-associated molecular patterns (PAMPs) also activate the host innate immune responses. The PAMPs were previously known as bacterial endotoxins like lipoteichoic acid, peptidoglycan, and LPS. The PAMPs are recognized by pattern recognition receptors, including toll-like receptors (TLRs). TLR4 serves as a receptor for bacterial LPS and PAMPs from gram-positive bacteria activating TLR2, which elicit proinflammatory protective immune response against microbial invasion. TLR4 also serves as receptor for several kinds of DAMPs (damage-associated molecular patterns) such as HMGB1 (high mobility group box-1), S100A8/S100A9 proteins, syndecans, and heparan sulfate, which elicits sterile inflammation. The additive effects of PAMPs and DAMPs cause the severity of inflammation by classical inflammatory pathways and activation of inflammasome-driven inflammation (Vogl et al., 2007).

Moreover, the local effects of PAMPs from biofilm lead to increased corrosion of biomaterial surfaces and lead to loosening of the implant even without infection. There are evidences that contaminating PAMPs impair the osseointegration of the implants which in turn facilitates aseptic loosening of the implant. A large body of evidences suggests that alarmins also lead to aseptic loosening either along with or in the absence of PAMPs, by activating TLRs; however, the elucidation of exact mechanism and interconnection with bacterial PAMPs needs to be explored (Greenfield, 2014).

9.15 Conclusions

The fate of dental implant therapies depends on successful integration of the implant with the host tissue. Biointegration depends on the establishment and integration of molecular communication between the implant surface and the host tissue in contact. Oral cavity being a natural polymicrobial environment is indispensable from the accumulation of biofilm. Hence, microbial association and biofilm formation with the implant surface is an ever-posing challenge in the success of implant therapy. Immune hyperactivation and subsequent ECM disorganization owing to the microbial products also hurdles the biointegration of dental materials to host tissues (usually bone). Advancements in the understanding of molecular pathology of dental diseases and the recent improvements in biomaterials science upgrade the biocompatibility and the durability of dental implants.

The next-generation dental implant biology focuses on smart/intelligent biomaterials which senses the pathological signals and elicits respective in situ responses which would solve most of the hurdles associated with implant failure.

References

Advanced Ceramics for Dentistry, 2014. Elsevier. https://doi.org/10.1016/C2011-0-07169-7.
Alshehri, F.A., 2016. The role of lasers in the treatment of peri-implant diseases: a review. Saudi Dent. J. 28, 103–108. https://doi.org/10.1016/j.sdentj.2015.12.005.
Ananth, H., Kundapur, V., Mohammed, H.S., Anand, M., Amarnath, G.S., Mankar, S., 2015. Rev. Biomat. Dental Implantol. 11, 8.
Baliban, R.C., Sakellari, D., Li, Z., DiMaggio, P.A., Garcia, B.A., Floudas, C.A., 2012. Novel protein identification methods for biomarker discovery via a proteomic analysis of periodontally healthy and diseased gingival crevicular fluid samples. J. Clin. Periodontol. 39, 203–212. https://doi.org/10.1111/j.1600-051X.2011.01805.x.
Berglundh, T., Abrahamsson, I., Lang, N.P., Lindhe, J., 2003. De novo alveolar bone formation adjacent to endosseous implants. Clin. Oral Implants Res. 14, 251–262. https://doi.org/10.1034/j.1600-0501.2003.00972.x.
Blankenship, J.R., Mitchell, A.P., 2006. How to build a biofilm: a fungal perspective. Curr. Opin. Microbiol., Growth and Dev. 9, 588–594. https://doi.org/10.1016/j.mib.2006.10.003.
Bollen, C.M., Papaioanno, W., Van Eldere, J., Schepers, E., Quirynen, M., van Steenberghe, D., 1996. The influence of abutment surface roughness on plaque accumulation and peri-implant mucositis. Clin. Oral Implants Res. 7, 201–211.
Buddula, A., 2013. Bacteria and dental implants: a review. J. Dent. Implant. 3, 58. https://doi.org/10.4103/0974-6781.111698.
Byers, B.A., Guldberg, R.E., García, A.J., 2004. Synergy between genetic and tissue engineering: Runx2 overexpression and in vitro construct development enhance in vivo mineralization. Tissue Eng. 10, 1757–1766. https://doi.org/10.1089/ten.2004.10.1757.
Canullo, L., Peñarrocha-Oltra, D., Covani, U., Botticelli, D., Serino, G., Penarrocha, M., 2016. Clinical and microbiological findings in patients with peri-implantitis: a cross-sectional study. Clin. Oral Implant. Res. 27, 376–382. https://doi.org/10.1111/clr.12557.
Cornelini, R., Rubini, C., Fioroni, M., Favero, G.A., Strocchi, R., Piattelli, A., 2003. Transforming growth factor-beta 1 expression in the peri-implant soft tissues of healthy and failing dental implants. J. Periodontol. 74, 446–450. https://doi.org/10.1902/jop.2003.74.4.446.
Davies, J.E., 2003. Understanding peri-implant endosseous healing. J. Dent. Educ. 67, 932–949.
de Avila, E.D., de Molon, R.S., Vergani, C.E., de Assis Mollo, J., Salih, V., 2014. The relationship between biofilm and physical-chemical properties of implant abutment materials for successful dental implants. Materials 7, 3651–3662. https://doi.org/10.3390/ma7053651.
Dhir, S., 2013. Biofilm and dental implant: the microbial link. J. Indian Soc. Periodontol. 17, 5–11. https://doi.org/10.4103/0972-124X.107466.
Drager, J., Harvey, E.J., Barralet, J., 2015. Hypoxia signalling manipulation for bone regeneration. Expert Rev. Mol. Med. 17, e6. https://doi.org/10.1017/erm.2015.4.

Greenfield, E.M., 2014. Do genetic susceptibility, Toll-like receptors, and pathogen-associated molecular patterns modulate the effects of wear? Clin. Orthop. Relat. Res. 472, 3709−3717. https://doi.org/10.1007/s11999-014-3786-4.

Gristina, A.G., 1987. Biomaterial-centered infection: microbial adhesion versus tissue integration. Science 237, 1588−1595.

Han, A., Tsoi, J.K.H., Rodrigues, F.P., Leprince, J.G., Palin, W.M., 2016. Bacterial adhesion mechanisms on dental implant surfaces and the influencing factors. Intl. J. Adhesion and Adhesives 69, 58−71. https://doi.org/10.1016/j.ijadhadh.2016.03.022.

Inoue, T., Matsuzaka, K., 2015. Surface modification of dental implant improves implant−tissue interface. In: Sasaki, K., Suzuki, O., Takahashi, N. (Eds.), Interface Oral Health Science 2014. Springer Japan, pp. 33−44.

Jayesh, R.S., Dhinakarsamy, V., 2015. Osseointegration. J. Pharm. BioAllied Sci. 7, S226−S229. https://doi.org/10.4103/0975-7406.155917.

Khan, W.S., Adesida, A.B., Tew, S.R., Lowe, E.T., Hardingham, T.E., 2010. Bone marrow-derived mesenchymal stem cells express the pericyte marker 3G5 in culture and show enhanced chondrogenesis in hypoxic conditions. J. Orthop. Res. 28, 834−840. https://doi.org/10.1002/jor.21043.

Khosravi, N., Maeda, A., DaCosta, R.S., Davies, J.E., 2018. Nanosurfaces modulate the mechanism of peri-implant endosseous healing by regulating neovascular morphogenesis. Commun. Biol. 1, 72. https://doi.org/10.1038/s42003-018-0074-y.

Ksiazek, M., Mizgalska, D., Eick, S., Thøgersen, I.B., Enghild, J.J., Potempa, J., 2015. KLIKK proteases of Tannerella forsythia: putative virulence factors with a unique domain structure. Front. Microbiol. 6. https://doi.org/10.3389/fmicb.2015.00312.

Kumar, A., Kawadkar, A., Mathew, D., Hegde, S., Shankar, R.K., 2017. Comparison of clinical and microbiological status of osseointegrated dental implant with natural tooth. J. Dent. Implant. 7, 46. https://doi.org/10.4103/jdi.jdi_3_17.

Kushibiki, T., Awazu, K., 2008. Controlling osteogenesis and adipogenesis of mesenchymal stromal cells by regulating a circadian clock protein with laser irradiation. Int. J. Med. Sci. 5, 319−326.

Larjava, H., Koivisto, L., Häkkinen, L., Heino, J., 2011. Epithelial integrins with special reference to oral epithelia. J. Dent. Res. 90, 1367−1376. https://doi.org/10.1177/0022034511402207.

Levin, L., 2010. Dealing with dental implant failures. Refuat Hapeh Vehashinayim 27, 6−12, 73.

López, C.S., Saka, C.H., Rada, G., Valenzuela, D.D., 2016. Impact of fixed implant supported prostheses in edentulous patients: protocol for a systematic review. BMJ Open 6, e009288. https://doi.org/10.1136/bmjopen-2015-009288.

Ma, Q.-L., Fang, L., Jiang, N., Zhang, L., Wang, Y., Zhang, Y.-M., Chen, L.-H., 2018. Bone mesenchymal stem cell secretion of sRANKL/OPG/M-CSF in response to macrophage-mediated inflammatory response influences osteogenesis on nanostructured Ti surfaces. Biomaterials 154, 234−247. https://doi.org/10.1016/j.biomaterials.2017.11.003.

Mahmoud, H., Williams, D.W., Hannigan, A., Lynch, C.D., 2012. Influence of extracellular matrix proteins in enhancing bacterial adhesion to titanium surfaces. J. Biomed. Mater. Res. B Appl. Biomater. 100, 1319−1327. https://doi.org/10.1002/jbm.b.32698.

Marrone, A., Lasserre, J., Bercy, P., Brecx, M.C., 2013. Prevalence and risk factors for peri-implant disease in Belgian adults. Clin. Oral Implant. Res. 24, 934−940. https://doi.org/10.1111/j.1600-0501.2012.02476.x.

Mello, A.S. da S., dos Santos, P.L., Marquesi, A., Queiroz, T.P., Margonar, R., de Souza Faloni, A.P., 2016. Some aspects of bone remodeling around dental implants. Revista Clínica de Periodoncia, Implantología y Rehabilitación Oral. https://doi.org/10.1016/j.piro.2015.12.001.

Mokdad, A.H., Forouzanfar, M.H., Daoud, F., Mokdad, A.A., El Bcheraoui, C., Moradi-Lakeh, M., Kyu, H.H., Barber, R.M., Wagner, J., Cercy, K., Kravitz, H., Coggeshall, M., Chew, A., O'Rourke, K.F., Steiner, C., Tuffaha, M., Charara, R., Al-Ghamdi, E.A., Adi, Y., Afifi, R.A., Alahmadi, H., AlBuhairan, F., Allen, N., AlMazroa, M., Al-Nehmi, A.A., AlRayess, Z., Arora, M., Azzopardi, P., Barroso, C., Basulaiman, M., Bhutta, Z.A., Bonell, C., Breinbauer, C., Degenhardt, L., Denno, D., Fang, J., Fatusi, A., Feigl, A.B., Kakuma, R., Karam, N., Kennedy, E., Khoja, T.A.M., Maalouf, F., Obermeyer, C.M., Mattoo, A., McGovern, T., Memish, Z.A., Mensah, G.A., Patel, V., Petroni, S., Reavley, N., Zertuche, D.R., Saeedi, M., Santelli, J., Sawyer, S.M., Ssewamala, F., Taiwo, K., Tantawy, M., Viner, R.M., Waldfogel, J., Zuñiga, M.P., Naghavi, M., Wang, H., Vos, T., Lopez, A.D., Al Rabeeah, A.A., Patton, G.C., Murray, C.J.L., 2016. Global burden of diseases, injuries, and risk factors for young people's health during 1990-2013: a systematic analysis for the Global Burden of Disease Study 2013. Lancet 387, 2383−2401. https://doi.org/10.1016/S0140-6736(16)00648-6.

Mombelli, A., Lang, N.P., 1998. The diagnosis and treatment of peri-implantitis. Periodontol 17, 63−76.

Mouraret, S., Hunter, D.J., Bardet, C., Popelut, A., Brunski, J.B., Chaussain, C., Bouchard, P., Helms, J.A., 2014. Improving oral implant osseointegration in a murine model via Wnt signal amplification. J. Clin. Periodontol. 41, 172−180. https://doi.org/10.1111/jcpe.12187.

Nishimura, I., 2013. Genetic networks in osseointegration. J. Dent. Res. 92, 109S−118S. https://doi.org/10.1177/0022034513504928.

Osborn, J.E., 1980. Dynamic aspect of the implant-bone interface. Dental Implants : Mat. Syst. 111−123.

Papi, P., Di Carlo, S., Rosella, D., De Angelis, F., Capogreco, M., Pompa, G., 2017. Peri-implantitis and extracellular matrix antibodies: a case-control study. Eur. J. Dermatol. 11, 340−344. https://doi.org/10.4103/ejd.ejd_28_17.

Pokrowiecki, R., Mielczarek, A., Zareba, T., Tyski, S., 2017. Oral microbiome and peri-implant diseases: where are we now? Ther. Clin. Risk Manag. 13, 1529−1542. https://doi.org/10.2147/TCRM.S139795.

Romanos, G.E., 2016. Biomolecular cell-signaling mechanisms and dental implants: a review on the regulatory molecular biologic patterns under functional and immediate loading. Int. J. Oral Maxillofac. Implant. 31, 939−951. https://doi.org/10.11607/jomi.4384.

Romeo, U., Nardi, G.M., Libotte, F., Sabatini, S., Palaia, G., Grassi, F.R., 2016. The antimicrobial photodynamic therapy in the treatment of peri-implantitis. Int J Dent. 7692387. https://doi.org/10.1155/2016/7692387.

Saini, R., 2011. Oral biofilm and dental implants: a brief. Natl. J. Maxillofac. Surg. 2, 228−229. https://doi.org/10.4103/0975-5950.94490.

Salvi, G.E., Persson, G.R., Heitz-Mayfield, L.J.A., Frei, M., Lang, N.P., 2007. Adjunctive local antibiotic therapy in the treatment of peri-implantitis II: clinical and radiographic outcomes. Clin. Oral Implant. Res. 18, 281−285. https://doi.org/10.1111/j.1600-0501.2007.01377.x.

Silva, E., Félix, S., Rodriguez-Archilla, A., Oliveira, P., Martins dos Santos, J., 2014. Revisiting peri-implant soft tissue - histopathological study of the peri-implant soft tissue. Int. J. Clin. Exp. Pathol. 7, 611−618.

Singh, A.V., Vyas, V., Salve, T.S., Cortelli, D., Dellasega, D., Podestà, A., Milani, P., Gade, W.N., 2012. Biofilm formation on nanostructured titanium oxide surfaces and a micro/nanofabrication-based preventive strategy using colloidal lithography. Biofabrication 4, 025001. https://doi.org/10.1088/1758-5082/4/2/025001.

Smeets, R., Stadlinger, B., Schwarz, F., Beck-Broichsitter, B., Jung, O., Precht, C., Kloss, F., Gröbe, A., Heiland, M., Ebker, T., 2016. Impact of dental implant surface modifications on osseointegration [WWW document]. BioMed Res. Int. 1−16. https://doi.org/10.1155/2016/6285620. Article ID 6285620.

Strevett, K.A., Chen, G., 2003. Microbial surface thermodynamics and applications. Res. Microbiol. 154, 329−335. https://doi.org/10.1016/S0923-2508(03)00038-X.

Taiyeb-Ali, T.B., Toh, C.G., Siar, C.H., Seiz, D., Ong, S.T., 2009. Influence of abutment design on clinical status of peri-implant tissues. Implant Dent. 18, 438−446. https://doi.org/10.1097/ID.0b013e3181ad8e7a.

Takayanagi, H., 2007. Osteoimmunology: shared mechanisms and crosstalk between the immune and bone systems. Nat. Rev. Immunol. 7, 292−304. https://doi.org/10.1038/nri2062.

Tallarico, M., Canullo, L., Caneva, M., Özcan, M., 2017. Microbial colonization at the implant-abutment interface and its possible influence on periimplantitis: a systematic review and meta-analysis. J Prosthodont Res 61, 233−241. https://doi.org/10.1016/j.jpor.2017.03.001.

Tesmer, M., Wallet, S., Koutouzis, T., Lundgren, T., 2009. Bacterial colonization of the dental implant fixture-abutment interface: an in vitro study. J. Periodontol. 80, 1991−1997. https://doi.org/10.1902/jop.2009.090178.

Trindade, R., Albrektsson, T., Galli, S., Prgomet, Z., Tengvall, P., Wennerberg, A., 2018. Osseointegration and foreign body reaction: titanium implants activate the immune system and suppress bone resorption during the first 4 weeks after implantation. Clin. Implant Dent. Relat. Res. 20, 82−91. https://doi.org/10.1111/cid.12578.

Tyrovolas, S., Koyanagi, A., Panagiotakos, D.B., Haro, J.M., Kassebaum, N.J., Chrepa, V., Kotsakis, G.A., 2016. Population prevalence of edentulism and its association with depression and self-rated health. Sci. Rep. 6, 37083. https://doi.org/10.1038/srep37083.

Ujiie, Y., Todescan, R., Davies, J.E., 2012. Peri-implant crestal bone loss: a putative mechanism [WWW document]. Int. J. Dentistry 1−14. https://doi.org/10.1155/2012/742439. Article ID 742439.

Venkataraman, N., Bansal, S., Bansal, P., Narayan, S., 2015. Dynamics of bone graft healing around implants. J. Intl. Clin. Dental Res. Org. 7, 40. https://doi.org/10.4103/2231-0754.172930.

Vogl, T., Tenbrock, K., Ludwig, S., Leukert, N., Ehrhardt, C., van Zoelen, M.A.D., Nacken, W., Foell, D., van der Poll, T., Sorg, C., Roth, J., 2007. Mrp8 and Mrp14 are endogenous activators of Toll-like receptor 4, promoting lethal, endotoxin-induced shock. Nat. Med. 13, 1042−1049. https://doi.org/10.1038/nm1638.

Wang, Y., Zhang, Y., Miron, R.J., 2016. Health, maintenance, and recovery of soft tissues around implants. Clin. Implant Dent. Relat. Res. 18, 618−634. https://doi.org/10.1111/cid.12343.

Zarb, G.A., Koka, S., 2012. Osseointegration: promise and platitudes. Int. J. Prosthodont. (IJP) 25, 11−12.

Zhu, Y., Dashper, S.G., Chen, Y.-Y., Crawford, S., Slakeski, N., Reynolds, E.C., 2013. Porphyromonas gingivalis and Treponema denticola synergistic polymicrobial biofilm development. PLoS One 8, e71727. https://doi.org/10.1371/journal.pone.0071727.

Biointegration of bone graft substiutes from osteointegration to osteotranduction

10

F.B. Fernandez[1], Suresh S. Babu[1], Manoj Komath[1], Harikrishna Varma[1,2]
[1]Division of Bioceramics, Department of Biomaterial Sciences and Technology, Biomedical Technology Wing, Sree Chitra Tirunal Institute for Medical Sciences and Technology, Trivandrum, Kerala, India; [2]Head, Biomedical Technology Wing, Sree Chitra Tirunal Institute for Medical Sciences and Technology, Trivandrum, Kerala, India

Chapter outline

10.1 Introduction 245
10.2 Bone, the hard tissue 246
10.3 Bone grafts 246
10.4 Synthetic bone graft substitutes 247
    10.4.1 Sintered calcium phosphate ceramics 247
    10.4.2 Bioglass and calcium phosilicates 249
    10.4.3 Composites and coatings 249
    10.4.4 Bioactive self-setting cements 250
10.5 Biointegration of synthetic bone graft substitutes 251
    10.5.1 Integration of sintered ceramics (osteointegration) 252
    10.5.2 Integration of bioglass (transforming integration) 252
    10.5.3 Integration of composites and coatings 254
    10.5.4 Integration of bioactive self-setting cements 255
10.6 Conclusion 257
References 257

10.1 Introduction

Biointegration refers to the complete fusion between synthetic biomaterials implanted in the body with their biological counterparts. The concept of biointegration becomes most relevant in the case of bioceramics used for bone graft substitute applications, as they are designed to integrate bone and take part in bone remodeling.

    Over the years, different classes of bioceramics have been generated and commercialized as surgical grafts for orthopedics, spine, maxillofacial, and periodontal applications. The earliest synthetic bone grafts were based on calcium phosphate ceramics. It was followed by the generation based on bioactive glassy composition such as bioglass and calcium phosphosilicates. Composites and coatings of these materials were designed for specific applications. Bioactive self-setting cements were the next generation in bone graft substitutes. Extensive literature on the design, development, testing, and biological response of these classes of materials is available. This chapter reviews the characteristics of the different classes of materials (sintered calcium phosphate ceramics, bioglass and calcium phosphosilicates, composites and coatings, and bioactive self-setting cements) with reference to biological response. The property of biointegration is viewed here as the ability of bone to bond and integrate with the host site over the healing period. Better understanding will be obtained if the interface between the implanted material and the host bone is closely observed. Certain examples of histological analysis of the implantation in animals are given to substantiate the approach.

10.2 Bone, the hard tissue

Bones not only constitute the framework of skeleton but also provide protection to vital organs such as the brain, heart, and lungs. The skeleton, along with muscles, articulates body movements precisely and ensures mobility via joints to help the accomplishment of physical tasks. Damage to the bones or joints leads to considerable disability and disruption of daily activities.

Bones mainly consist of collagen fibers impregnated with nanosized inorganic mineral "hydroxyapatite" (HA) woven in a hierarchial structure that provides mechanical strength required for the function intended. This basic physical structure is augmented by cells (osteoblasts, osteoclasts, and osteocytes), blood vessels, and nerves intercalated into it. The activity of the cells makes it resorb and rebuild, through the process called "remodeling." Thus, bone structure is considered as "hard tissue".

Damage or loss of hard tissue content is dealt with primarily by mobilization of local resources. The repair process mirrors many of the embryonic development steps as resolution of the injury is undertaken. Repair in bone generally tends to restore the area to preinjury cellular composition, structure, and function. In cases of comorbidities or extraneous factors, there is a higher risk of nonhealing. Stabilization followed by supportive therapy is used in most classes of fractures with the application of bone growth stimulators or growth factors decided on a case-to-case basis (Einhorn and Gerstenfeld, 2015). The loss of considerable amount of bone, which may lead to a critical loss of healing, is faced in many situations. Graft materials, in the form of autograft, allograft, xenograft, or synthetic materials, play a key role in the healing process here.

10.3 Bone grafts

An osseous graft harvested from an anatomic site within an individual and transplanted to another site within the same individual is autografting (Roberts TT and Rosenbaum, 2012). These grafts possess osteoconduction, osteoinduction, and osteogenic properties and are held as the gold standard of graft materials. Their use is limited by only the complications linked to harvesting process, related complications, and limited volume of material available (Khan et al., 2005).

Cancellous autografts are the most commonly used for autologous bone grafting with few osteoblasts, but with the survival of abundant mesenchymal stem cells, they ensure the maintenance of the osteogenic potential. Proteins on the graft are well preserved, leading to excellent osteoinduction when appropriately treated (Bhatt and Rozental, 2012). Cortical autografts are applied for their mechanical integrity and structural processes with a time line of several years for complete integration within the host system (Chiarello et al., 2013).

Allogeneic bone grafts refer to tissues that are harvested from an individual and transplanted to a genetically different individual within the same species (Bone grafts, 2019). Allografts are the best alternative to autografts and have been applied in diverse complex clinical scenarios for resolution. They are amenable to processing and customization and are thus available in multiple formats (Bhatt and Rozental, 2012). Compared with other graft systems, allografts are at times highly immunogenic and demonstrate a commensurately higher failure rate, which is believed to be due to activation of the major histocompatibility complex antigens (Stevenson and Horowitz, 1992). Based on availability, prior pathogen testing, and related issues, application of fresh allografts is limited, and preserved modified, processed allografts are preferred in clinical practices (Urist, 1980).

10.4 Synthetic bone graft substitutes

Materials science, from 1970s onwards, has been heavily borrowed from nature to derive bioceramic materials, which could be safely used inside human body for functional applications. The early use of bioceramics was as bioinert support structures in joint prostheses. Materials such as alumina and zirconia having mechanical strength and dimensional stability were used for the purpose. Bioactive ceramics were developed later as synthetic substitutes for bone grafts. These materials included HA, bioactive glass, glass ceramics, and resorbable calcium phosphates, which possessed bioactivity along with biocompatibility and osteoconductivity. Already, these products have made notable clinical impact in various specialties related to skeletal repair with large number of brands in the market.

The different material classes used as synthetic bone substitutes are described below.

10.4.1 Sintered calcium phosphate ceramics

The first-generation bone graft substitutes were calcium phosphates, which are intuitive from the identification of the mineral part of bone. Calcium phosphates with the Ca/P ratio in the range of 1.5—1.67 were found to be sinterable to ceramic form with in vivo stability, bioresorbability, and osteoconductivity. Key functional parameters of the CaP ceramics are dependent on their Ca/P ratios, crystal structure, and nature of porosity (Wang and Yeung, 2017). CaP ceramics can be processed into various formats thereby increasing their application spectrum. It is bioabsorbabale with osteoconductive nature and hence has received prime interest and is used widely in clinical studies (Schwartz and Bordei, 2005) (Scheer and Adolfsson, 2009) (Oonishi et al., 1997).

HA ($Ca_{10}(PO_4)_6(OH)_2$) attracted major clinical interest among the CaP ceramics, being the basic mineral of bone. Jarcho 1976 (Azenha et al., 2015) and Aoki 1977 independently demonstrated that ($Ca_{10}(PO_4)_6(OH)_2$) a sintered HA also bonds to living bone. HA ceramics had mechanical properties better than cancellous bone and showed excellent osteointegration and osteoconduction (Bhatt and Rozental, 2012). HA is resistant to compressive loads (Zwingenberger et al., 2012) but brittle and weak to tension and shear (Kitsugi et al., 1987). Relatively high Ca/P ratio and its crystalline nature delay the resorption rate of Hap, which is a process carried out by giant cells and macrophages (Eggli et al., 1988). Pore size and connectivity of the ceramic play major role in ensuring homing of cells, differentiation and proliferation of osteoprogenitor cells, revascularization, and subsequent replacement by new bone at the defect site. Porosity in the range of 300−500 um had been proved to be optimum (Di Luca et al., 2016). The state of the art in this area is the development of nanocrystalline apatites, which has lower sintering temperature and increase the resorption rates (Kattimani et al., 2016).

Tricalcium phosphate (TCP), specifically the β-form, has the Ca/P ratio of 1.5, which causes an accelerated degradation and absorption in vivo. Majority of β-TCP will be resorbed by cellular breakdown action of phagocytes within 6−24 months. Because of the inherent thermodynamical instability at physiological pH, the implanted β-TCP is converted partially to HA in vivo and retained for longer time periods (Daculsi et al., 1989). It is found effective for the closure of trauma or benign tumor defects but is not preferred as a substitute for load-bearing tricortical graft (Azenha et al., 2015).

This property renders it effective for the closure of trauma or benign tumor defects but is not preferred as a graft (Finkemeier, 2002). Enhanced angiogenesis has been investigated in the case of application of TCP for bone defects (Thomas and Puleo, 2009), and comparison of various HA against β-TCP indicated its ability to promote proliferation and angiogenesis in human umbilical vein endothelial cells (Fernandez de Grado et al., 2018). In mice intramuscular implantation, there is an increase in density of microvessels (Fernandez de Grado et al., 2018). On comparing the resorption of porous cylinders of HA and β-TCP in animals, they were found to resorb with volume reduction of 5.4% and 85.4%, respectively, under identical conditions (Eggli et al., 1988). To balance the resorption rate, bioactivity, and mechanical properties, combination of HA and β-TCP ceramics was developed to take advantages of their standalone properties (Bohner, 2000). Certain specific ratios (such as 70%HA−40% TCP) of these biphasic calcium phosphates (BCP) became very popular in clinical use.

10.4.2 Bioglass and calcium phosphosilicates

A new generation of bone substitute materials originated when Hench, in the 1970s, proved glasses made from silica, calcium, and phosphate mixtures possess bone bonding properties (Hench and Wilson, 1984). These materials, commercially known as "bioglass," had no analogues in nature but possessed higher bioactivity compared with calcium phosphate ceramics. A typical bioglass composition contains silicon dioxide (SiO_2), sodium oxide (Na_2O), calcium oxide (CaO), and phosphorus pentoxide (P_2O_5). It is amorphous in structure and has the capacity to form apatitic layers over the surface from body fluids. The bioactivity and resorbability will depend on the ratio of SiO_2-CaO-P_2O_5. There is lack of mechanical strength, and the resorption rate is high. It has been mainly applied in the reconstruction of facial defects (Azenha et al., 2015) and in the delivery of biological factors or active moieties (Azenha et al., 2015) (Hench, 2006). The material because of the alkaline dissolution end products has demonstrated antimicrobial properties under laboratory conditions (Zhang et al., 2010; Allan et al., 2001). Concerns about the high silica content and brittleness of bioglass compounds have lead to further innovation in the same line.

In 1981, Kokubo et al. (1982) developed a glass-ceramic containing 38 wt% of crystalline oxyfluoroapatite ($Ca_{10}(PO_4)_6(OH)_2$) and 34 wt% of β-wollastonite ($CaSiO_3$) by sintering and crystallization of a glass powder compact having the composition (in wt%) 4.6, MgO; 44.7, CaO; 34.0, SiO_2; 16.2, P_2O_5; and 0.5, CaF_2 (Raggatt and Partridge, 2010). The glass-ceramic product was named as A-W after the crystalline phases, which ensures its direct bonding to bone. The newly developed glass bonded to bone faster than sintered HA but slower than bioglass. With the bending strength of the material higher than that of human bone at 215 MPa, Kokubo et al. determined that the material could withstand over 10 years of continuous loading at a bending stress of 65 MPa (Kitsugi et al., 1987). It has led to the use of the material on a large scale in intervertebral disks, iliac crests with over 60,000 patients receiving the material.

One significant advantage of bioglass is that it can bond with soft tissue also, which gave it an edge in applications such as middle ear implant. However, the fast degradation of bioglass may lead to a higher pH value and accumulated ions in the microenvironment, which did not favor cell activity thereby, jeopardizing the bone ingrowth (Jones, 2013). John et al. developed a novel bioactive glass coated porous HA bioceramic, which combines the mechanical strength of sintered ceramic with higher bioactivity at lower doses of silica components (John et al., 2008). The developed triphasic material was denoted as HA coated with silica and was further evaluated at multiple levels (Sandeep et al., 2006) and also as a cell carrier for tissue engineering applications (Nair et al., 2008).

10.4.3 Composites and coatings

Composite materials are another class of synthetic grafts, which refer to particles of bioactive ceramics or glasses dispersed in a biocompatible polymer matrix. These could be fabricated to custom shape and having tuned mechanical properties corresponding to the natural bone at the repair site. The bioactive particles exposed on the surface provide osteoconductivity and form stable tissue interface (Bonfield, 1993). Composites help in constructing large single pieces of grafts (like the cranial bone), which is not possible with the brittle bioceramics and glasses. Polymer matrices used are polymethyl methacrylate, polyethylene (PE) and hydroxyethyl methacrylate. These composites are generally nonresorbable, whereas composites made using polymers such as poly-L-lactate are biodegradable and undergo slow resorption. Composites, combined with modern fabrication techniques such as additive manufacture, are of immense help in orbital floor reconstruction, maxillofacial correction, etc. Several new polymer−ceramic composites (polymers such as polylactates, polyglyconates, polyhydroxybuterates, and methacrylate and their derivatives, reinforced with HA and alumina) are developed in the recent years for custom applications (Bonfield, 1993). Coatings of bioactive materials on metallic implants also have gained popularity in skeletal repair, particularly for joint prosthesis in orthopedics and single tooth implants in dentistry. Despite the bone bonding ability, bioceramics and bioactive glasses are considered unfit for load-bearing applications because of brittleness and low machinability, whereas metallic implants do not possess bioactivity and tends to form a fibrous interface with the host bone and undergo loosening in due course. The longevity of metallic implant could be enhanced by providing coatings of calcium phosphate ceramics or bioactive glass on their surface (Shepperd and Apthorp, 2005). Various techniques are adopted to make coatings of bioactive materials on metallic implant surface, such as plasma spraying, sputtering, ion beam deposition, pulsed laser deposition, sol−gel deposition, electrochemical coating, and biomimetic processing (Lacefield, 1998).Out of these, only plasma spraying is proved to be commercially feasible. In this technique, the ceramic powder is fed to a stream of gas plasma at high temperatures and directed toward the substrate (implant). The partially molten powder gets adhered and spread on the surface. It is possible to coat large areas at high rates, with this technique (de Groot et al., 1987). HA plasma spray coatings are proved to induce bone contact to the implant and to improve the implant fixation (Sandén et al., 2002).

However, achieving adhesion of the coating with substrate (titanium implant) surface is a big challenge. Alteration of chemical phases during the spray process and poor control of thickness and surface morphology are other concerns.

10.4.4 Bioactive self-setting cements

This class of materials could be viewed as third-generation bone substitutes, after Ca-P ceramics and bioglass. They constitute the heterogeneous composition including one or more solid-dispersed phase and liquid solvent, which when mixed, develops dough/ paste consistency. The physicochemical reaction in the mix leads to setting and subsequent hardening at definite time period and develops mechanical properties to meet the clinical requirement for bone defect repair (Barinov and Komlev, 2011). These compositions have the unique combination of mouldability, osteoconductivity, and biodegradability. The inorganic self-setting bone cement systems are classified in to two groups—calcium phosphate cements (CPCs) and calcium sulfate cements (CSCs).

CPCs are generally designed as "apatitic cements," which essentially contain the Ca/P ratio of 1.67 corresponding to HA, the bone mineral. In 1986, Brown and Chow developed poorly crystalline precipitated HA through the setting reaction of equimolar mixtures of tetracalcium phosphate (TTCP) and dicalcium phosphate anhydrous/dicalcium phosphate dihydrate (DCPA/DCPD) with phosphate solution at pH of 7.5−8.0 (LC, 2009). The setting time to 5 min was tried with neutral phosphate salt solution. Brushite (or dicalcium phosphate)-based bone cement formulations were also developed in the class of CPCs (Bohner, 2000). These cements were found highly useful in orthopedics and dentistry (Komath and Varma, 2004).

CSCs are based on the hydration reaction of calcium sulfate hemihydrate to dihydrate form (gypsum) in a slightly exothermic reaction. Although this forms a cost-effective and simple cement, its use as a bone graft material was very sparse till the past decade. The reasons may be limited osteoconductivity and fast resorption (Larsson and Hannink, 2011), which creates an acidic microenvironment in the surrounding tissues and leaves the implanted site mechanically unstable (Thomas and Puleo, 2009). Although a few products are available in the market, there is a lack of adequate evidence for clinical use as a bone filler or bone graft (Fernandez de Grado et al., 2018).

However, there is a revival of interest in gypsum bone cements in recent years, owing to the plausibility of local drug delivery in pathological, metabolic, and oncological challenges of the bone. CPC is less useful in this aspect as the acid−base reaction during setting is likely to decompose the drug molecules. The setting mechanism of CSC involves only rehydration and recrystallization, and hence, any water-soluble drugs can safely be incorporated. The use of CSCs as drug delivery medium for osteoporosis and osteomyelitis has already been demonstrated (Thomas and Puleo, 2009). Currently, CaS material suitable for drug loading, in putty form as well as preformed beads, is commercially available for the treatment of bone diseases (Hughes et al., 2015) (Fernandez de Grado et al., 2018). A modified form of CSCs has been developed, which is proved to have bioactivity and in vivo stability (Sandhya et al., 2017).

10.5 Biointegration of synthetic bone graft substitutes

Bone graft substitutes are used in lieu of graft material wherein there is a lack of native autograft or due to other considerations. Availability of synthetic substitutes provides an opportunity for tailoring the capabilities of the material to the need at the graft site ensuring an optimal outcome. As various classes of materials have been developed, their interaction with the host system is crucial in graft success. The bonding of sintered HA to bone can be considered as a true case of biointegration (Baino, 2017).

10.5.1 Integration of sintered ceramics (osteointegration)

Branemark defined osseointegration as *"A direct connection between living bone and a load-carrying endosseous implant at the light microscopic level."* (Brånemark and Chien, 2005). The desired effect in application of synthetic graft materials is achieving osteointegration. The ability of material to stimulate and sustain bony ingrowth also promotes anchorage and long-term implant integration. Osteointegration can be in principle encouraged by the creation of a continuum between graft and native bone. This microenvironment will ensure the retention and proliferation of cells committed to the osteogenic lineage and normal bone homeostasis (Raggatt and Partridge, 2010). The wide use of sintered ceramics is due to their chemical bonding with bone, which in long term translates to biointegration. The volume of bone formed in contact with ceramic interfaces of apatite tends to be higher than those seen adjacent to titanium in earlier stages of bonding (Biointegration, 2019). A series of studies by Kjeld Soballe et al. (Søballe, 1993) (Søballe et al., 1992) validated its use also as a coating that would promote osseointegration with materials that would not natively support such properties. The generation of a chemical bond with the substrate material by the host milieu without an intervening fibrous tissue layer is achieved by the use of sintered ceramics. This layer can be visualized as a clear cut boundary as set down by Branemark et al.

Fig. 10.1 shows a perception of the "osteointegration concept" through histological pictures of sintered HA implanted in animal bone. HA porous granules have been used in rabbit tibia, and the healing was observed through histopathology at time periods 12, 26, and 52 weeks. The sections were stained with Stevenels' blue followed by counterstaining in van Gieson's picrofuchsin. At 12 weeks, the implanted area contained mostly the ceramic granule, with osteoclast cell in the pores and crevices (A1 and A2). By 26 weeks, a part of the ceramic got resorbed, and the histology showed the new bone apposing the remaining ceramic (B1 and B2). Considerable part of the ceramic remained even after 52 days (C1 and C2), and signs of healing were evident in the bone part. The haversian systems have been formed in the apposing bone indicating complete healing. Yet, there remained a clear boundary between the granule parts and the healed bone.

10.5.2 Integration of bioglass (transforming integration)

Bioglass, a series of specially designed silica-based glasses, consists of a three-dimensional silicon oxide network modified by incorporation of oxides of sodium, calcium, and phosphorus. The composition of glass based on the incorporation of said components varied its properties. It was first applied in 1969 by Hench et al. and formed the true second generation of biomaterials (Hench, 2006). Bioglass provided for the rapid assimilation of material to bone via the formation of interfacial bonding of tissue to the graft material. Johnson et al. using PerioGlas in a pilot study were able to demonstrate the peripheral osteoid formation followed by bone deposition within the defect from the surgical margin toward the bone. PerioGlas particles were found interconnected by developed areas of osteoid as well as surrounded by new trabecular structures (Johnson et al., 1997). Alkali-free bioglasses have been

Biointegration of bone graft substiutes from osteointegration to osteotranduction 253

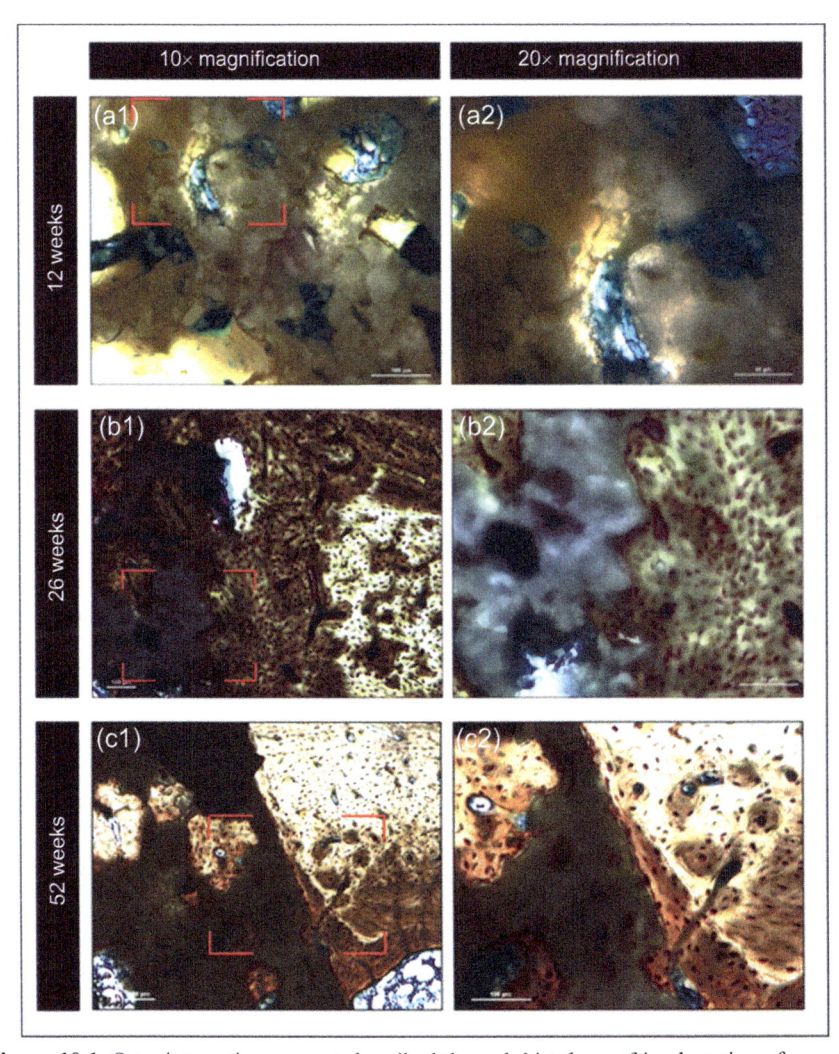

Figure 10.1 Osteointegration concept described through histology of implantation of hydroxyapatite (HA) in animal bone. Sintered HA porous granules have been implanted in rabbit tibia, and the healing was observed through histopathology at time periods 12, 26, and 52 weeks. The sections were stained with Stevenels' blue followed by counterstaining in van Gieson's picrofuchsin. The pictures in pairs A, B, and C show the histology at 12, 26, and 52 weeks, respectively. In each, the sets with suffix 1 and 2 represent magnifications 10× and 20×, with area marked in set 1 is shown magnified in set 2.

developed, and animal models are developed in comparison with the standard bioglasses. Femoral bone defect in sheep was used and developed material demonstrated biocompatibility and osteoconductive nature with a slower resorption than standard bioglass (Cortez et al., 2017). Preparation and demonstration of phytic acid—derived bioglasses for use as injectable in the case of vertebral compression fractures are also possible avenues (Zhu et al., 2017). Integration in the bioglass milieu is fluid with a quick interaction leading to anchorage followed by rapid solution- and cell-mediated resolution.

10.5.3 Integration of composites and coatings

Coatings of apatite have been applied via multiple processes to increase compatibility of implant materials as well as provide enhanced anchorage with bone bonding. Fixation is promoted by the closer apposition as well as bony ingrowth into coated surfaces. Investigations in a rabbit model of implants coated by plasma spray technique gave interesting insights. The coatings were proven to be stable with dense bone apposition to the coated areas (Darimont et al., 2002). The coating in some cases of rat studies when doped with rare elements contributed to rapid cell adhesion and integration (Husak et al., 2018). The use of coated titanium plates in the tibial proximal metaphyses of rabbits also pointed to the bone bonding and deposition of bone at implant−bone interfaces. The bone material was deposited directly on to the coated surfaces without a fibrous tissue interface (Yan et al., 1997). The metal coating system may therein play a key role in the generation of bioactive surfaces and long-term implant success. Ceramic systems have been utilized in composites with polymer-based models to increase bone bonding nature of the material. Materials containing magnetic Fe_3O_4 nanoparticles with mesoporous bioactive glass/polycaprolactone (Fe_3O_4/MBG/PCL) composite scaffolds of 60% porosity have demonstrated enhanced osteogenic activity, drug delivery capacities, and potential multifunctionality (Zhang et al., 2014). In the case of polycaprolactone-based systems, HA addition was indicated for better biological properties, whereas TCPs enhanced the mechanical properties. Wettability of the scaffolds was not changed with ceramic addition (Huang et al., 2018). Viable bone is a composite of inorganic materials as well as organic components and protein structures. Natural bone is endowed with high strength, high fracture toughness, and also deformability with a low elastic modulus. To generate an appropriate substitute, investigators are prompted to generate composites that can combine ceramic bioactives with polymers that will beneficially affect the mechanical properties. Mimicking of mineral component and microstructure of natural bone development of nano-HA−polymer composite scaffolds with high porosity and well-controlled pore architectures are prepared (Wei and Ma, 2004); they provide an excellent substrate for cell attachment and migration in bone tissue. A PE−Hap composite was attempted in the 1980s by a group who were able to achieve 45% ceramic content without diluting properties of the polymer matrix (Bonfield, 1993). Polyetheretherketone (PEEK) is a semicrystalline thermoplastic with excellent mechanical properties, high temperature durability, and good chemical, fatigue resistance. HA−PEEK composites have been widely reported. PEEK has made considerable inroads into its application in biomedical domain (Giebaly et al., 2016) in both orthopedic and dental applications (Najeeb et al., 2016). Melt mixing and injection molding have been applied to generate PEEK with upto 40% HA content. This has been in line with findings that indicate bioactivity of HA/PEEK composite increases with increasing HA volume fraction in the composite (Yu et al., 2005). Histological studies have indicated that with high HA loading, osteoblastic activities were seen in the formation of osteoid and osteocytes within lamellar bone in developing mature bone at longer implantation periods (Bakar et al., 2003). To prevent the significant decrease in osteoblast population postaddition of HA nanoparticles (Xu et al., 2009) to culture and negate time-dependent toxicological effects of HA nanoparticles on pulmonary surfactants (Fan et al., 2011), they are employed in the role of addition agents in polymer-based composites. Electrophoretic codeposition of PEEK−HA composite coatings has been developed for improving bioactivity and adhesion strength of the coatings (Baştan et al., 2018).

10.5.4 Integration of bioactive self-setting cements

Bioactive self-setting cements contain elements conducive to bone bonding as well as replacement in tandem with native bone development. In this case, calcium phosphate bone cements are also designated as osteotransductive. Their transformation into new bone postimplantation is well established (Driessens et al., 1998). Calcium-deficient HA agents can be rendered more osteotransductive by the addition of anhydrous dicalcium phosphate or calcium carbonate to the cement powder. These are mainly aimed at irregular defects, wherein the margins are not well defined and the user-friendly nature of the material allows the operator to adapt the same to custom defects. Higher compressive strength than human trabecular bone is observed for most cases in excess of 10 MPa. Incorporation of transforming growth factor beta 1 in calcium phosphate cements is shown to stimulate its osteotransductive nature in calvarial bone defects (Blom et al., 2001). For efficient bone regeneration procedures around endosseous implants, CPCs with recombinant growth factors may prove to be appropriate in ensuring early osseointegration and implant use. The use of extracellular matrix collagen−modified cements also indicates rapid assimilation with DCP-rich cements supplemented with collagen indicating highest levels of osteogenesis (Blom et al., 2001). Cements alone or augmented by soft tissue or ligamentous tissues play a key role in the regeneration of designated tissues (Wen et al., 2009).

Fig. 10.2 shows a perception of the "osteotransduction concept" through histological pictures of bioactive bone cement implanted in animal bone. The bone cement composition "BioCaS" has been used in rabbit tibia, and the healing was observed through histopathology at time periods 12, 26, and 52 weeks. The sections were stained with Stevenels' blue followed by counterstaining in van Gieson's picrofuchsin. By 12 weeks, the implanted area was seen covered with new bone, and cement particles were engulfed. Osteoclast cells are present in the crevices (D1 and D2). The 26 weeks histology showed that most of the cement part was resorbed and replaced by maturing bone (E1 and E2). In 52 weeks (F1 and F2), complete healing was observed with the haversian systems formed at the site. At any stage of healing, there were no boundaries between the cement and the healed bone.

On comparing the histological results of the bone defect healing with BioCaS cement through 52 weeks, the phenomenon of progressive material resorption and simultaneous new bone formation is evident This property is called "osteotransductivity," the ideal requirement for a bone graft substitute. Compared with ceramic granules

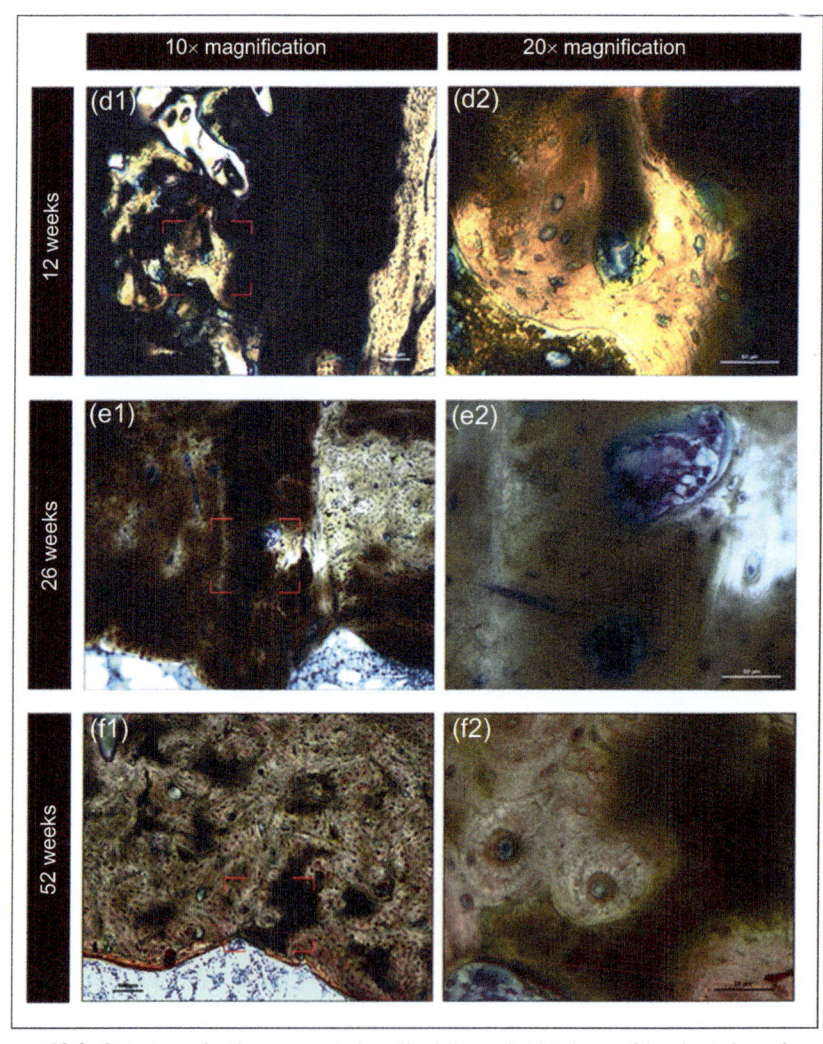

Figure 10.2 Osteotransduction concept described through histology of implantation of a bioactive bone cement in animal bone. The cement has been implanted in rabbit tibia, and the healing was observed through histopathology at time periods 12, 26, and 52 weeks. The sections were stained with Stevenels' blue followed by counterstaining in van Gieson's picrofuchsin. The pictures in pairs D, E, and F show the histology at 12, 26, and 52 weeks, respectively. In each, the sets with suffix 1 and 2 represent magnifications 10× and 20×, with area marked in set 1 is shown magnified in set 2.

(as in Fig. 10.1), BioCaS comprises of uniform micron-sized particles entangled together during cement setting. The interparticle boundaries gradually become weaker during resorption and give way to newly growing bone. The particulate structure of BioCaS offers an enormously large surface area for osteoblasts to act on and remodel the defect. Such a healing process will provide better strength to the repaired site, as the new bone replaces the material and the remodeling progresses fast.

10.6 Conclusion

Synthetic bone graft substitutes have been classified as sintered calcium phosphate ceramics, bioglass and calcium phosphosilicates, composites and coatings, and bioactive self-setting cements. The biointegration aspects of each of these were discussed in detail. The sintered ceramics show typical osteointegration, wherein clear cut boundary can be observed between the surface of the material and the remodeling host bone. Even in a period of 52 weeks, very little resorption of the ceramic could be noticed. In the case of bioglass category, a "transforming" kind of integration could be observed, which shows changing boundary with time. The new bone will replace the material slowly, as and when the resorption of the material occurs. Integration of composites will be in a mixed fashion with multiple interfaces, corresponding to nonbioactive polymer surface and bioactive filler part. Bioactive cements show a special characteristic of *osteotransduction* without any boundary between the material and the new bone during the healing period. The new bone constantly replaces the material and leads to complete healing of the site.

References

Allan, I., Newman, H., Wilson, M., June 2001. Antibacterial activity of particulate bioglass against supra- and subgingival bacteria. Biomaterials 22 (12), 1683–1687.

Azenha, M.R., de Lacerda, S.A., Marão, H.F., Filho, O.P., Filho, O.M., September 1, 2015. Evaluation of crystallized biosilicate in the reconstruction of calvarial defects. J. Maxillofac. Oral. Surg 14 (3), 659–665.

Baino, F., 2017. 7 – ceramics for bone replacement: commercial products and clinical use [Internet]. In: Palmero, P., Cambier, F., De Barra, E. (Eds.), Advances in Ceramic Biomaterials. Woodhead Publishing, pp. 249–278. Available from: http://www.sciencedirect.com/science/article/pii/B9780081008812000075.

Bakar, M.S.A., Cheang, P., Khor, K.A., 2003. Tensile properties and microstructural analysis of spheroidized hydroxyapatite-poly (etheretherketone) biocomposites. Mater. Sci. Eng. A 1–2 (345), 55–63.

Barinov, S.M., Komlev, V.S., December 1, 2011. Calcium phosphate bone cements. Inorg. Mater. 47 (13), 1470–1485.

Baştan, F.E., Atiq Ur Rehman, M., Avcu, Y.Y., Avcu, E., Üstel, F., Boccaccini, A.R., May 3, 2018. Electrophoretic co-deposition of PEEK-hydroxyapatite composite coatings for biomedical applications. Colloids Surf. B Biointerfaces 169, 176–182.

Bhatt, R.A., Rozental, T.D., November 1, 2012. Bone graft substitutes. Hand Clin 28 (4), 457–468.

Biointegration – An Overview | ScienceDirect Topics. [Internet]. [cited 2019 Mar 31]. Available from: https://www.sciencedirect.com/topics/engineering/biointegration.

Blom, E.J., Klein-Nulend, J., Yin, L., van Waas, M.A., Burger, E.H., December 2001. Transforming growth factor-beta1 incorporated in calcium phosphate cement stimulates osteotransductivity in rat calvarial bone defects. Clin. Oral Implant. Res. 12 (6), 609–616.

Bohner, M., December 2000. Calcium orthophosphates in medicine: from ceramics to calcium phosphate cements. Injury 31 (Suppl. 4), 37–47.

Bonfield, W., 1993. Design of bioactive ceramic-polymer composites. In: An Introduction to Bioceramics. Advanced Series in Ceramics, Vol. 1. World Scientific, pp. 299−303 [Internet]. Available from: https://www.worldscientific.com/doi/abs/10.1142/9789814317351_0016.

Brånemark, P.-I., Chien, S., 2005. The Osseointegration Book: From Calvarium to Calcaneus. Quintessence Publishing Company, p. 25.

Chiarello, E., Cadossi, M., Tedesco, G., Capra, P., Calamelli, C., Shehu, A., et al., 2013 Oct. Autograft, allograft and bone substitutes in reconstructive orthopedic surgery. Aging Clin Exp Res 25 (Suppl 1), S101−103.

Cortez, P.P., Brito, A.F., Kapoor, S., Correia, A.F., Atayde, L.M., Dias-Pereira, P., et al., 2017. The in vivo performance of an alkali-free bioactive glass for bone grafting, FastOs® BG, assessed with an ovine model. J. Biomed. Mater. Res. B Appl. Biomater. 105 (1), 30−38.

Daculsi, G., Legeros, R.Z., Nery, E., Lynch, K., Kerebel, B., 1989. Transformation of biphasic calcium phosphate ceramics in vivo: ultrastructural and physicochemical characterization. J. Biomed. Mater. Res. 23 (8), 883−894.

Darimont, G.L., Cloots, R., Heinen, E., Seidel, L., Legrand, R., June 2002. In vivo behaviour of hydroxyapatite coatings on titanium implants: a quantitative study in the rabbit. Biomaterials 23 (12), 2569−2575.

de Groot, K., Geesink, R., Klein, C.P., Serekian, P., December 1987. Plasma sprayed coatings of hydroxylapatite. J. Biomed. Mater. Res. 21 (12), 1375−1381.

Di Luca, A., Ostrowska, B., Lorenzo-Moldero, I., Lepedda, A., Swieszkowski, W., Van Blitterswijk, C., et al., March 10, 2016. Gradients in pore size enhance the osteogenic differentiation of human mesenchymal stromal cells in three-dimensional scaffolds. Sci. Rep. 6, 22898.

Driessens, F.C., Planell, J.A., Boltong, M.G., Khairoun, I., Ginebra, M.P., 1998. Osteotransductive bone cements. Proc. Inst. Mech. Eng 212 (6), 427−435.

Eggli, P.S., Müller, W., Schenk, R.K., July 1988. Porous hydroxyapatite and tricalcium phosphate cylinders with two different pore size ranges implanted in the cancellous bone of rabbits. A comparative histomorphometric and histologic study of bony ingrowth and implant substitution. Clin. Orthop. (232), 127−138.

Eggli, P.S., Müller, W., Schenk, R.K., July 1988. Porous hydroxyapatite and tricalcium phosphate cylinders with two different pore size ranges implanted in the cancellous bone of rabbits. A comparative histomorphometric and histologic study of bony ingrowth and implant substitution. Clin. Orthop. (232), 127−138.

Einhorn, T.A., Gerstenfeld, L.C., January 2015. Fracture healing: mechanisms and interventions. Nat. Rev. Rheumatol. 11 (1), 45−54.

Fan, Q., Wang, Y.E., Zhao, X., Loo, J.S.C., Zuo, Y.Y., August 23, 2011. Adverse biophysical effects of hydroxyapatite nanoparticles on natural pulmonary surfactant. ACS Nano 5 (8), 6410−6416.

Fernandez de Grado, G., Keller, L., Idoux-Gillet, Y., Wagner, Q., Musset, A.-M., Benkirane-Jessel, N., et al., June 4, 2018. Bone substitutes: a review of their characteristics, clinical use, and perspectives for large bone defects management. J. Tissue Eng 9 [Internet]. Available from: https://www.ncbi.nlm.nih.gov/pmc/articles/PMC5990883/ https://doi.org/10.1177/2041731418776819.

Finkemeier, C.G., March 2002. Bone-grafting and bone-graft substitutes. JBJS 84 (3), 454.

Giebaly, D.E., Twaij, H., Ibrahim, M., Haddad, F.S., September 29, 2016. Cementless hip implants: an expanding choice. Hip. Int. J. Clin. Exp. Res. Hip. Pathol. Ther 26 (5), 413−423.

Hench, L.L., November 2006. The story of bioglass. J. Mater. Sci. Mater. Med. 17 (11), 967−978.
Hench, L.L., Wilson, J., November 9, 1984. Surface-active biomaterials. Science 226 (4675), 630−636.
Huang, B., Caetano, G., Vyas, C., Blaker, J.J., Diver, C., Bártolo, P., January 14, 2018. Polymer-ceramic composite scaffolds: the effect of hydroxyapatite and β-tri-calcium phosphate. Materials 11 (1) [Internet]. Available from: https://www.ncbi.nlm.nih.gov/pmc/articles/PMC5793627/.
Hughes, E., Yanni, T., Jamshidi, P., Grover, L.M., February 1, 2015. Inorganic cements for biomedical application: calcium phosphate, calcium sulphate and calcium silicate. Adv. Appl. Ceram 114 (2), 65−76.
Husak, Y., Solodovnyk, O., Yanovska, A., Kozik, Y., Liubchak, I., Ivchenko, V., et al., November 2018. Degradation and in vivo response of hydroxyapatite-coated Mg alloy. Coatings 8 (11), 375.
John, A., Varma, H.K., Vijayan, S., Bernhardt, A., Lode, A., Vogel, A., et al., November 2008. Vitroinvestigations of bone remodeling on a transparent hydroxyapatite ceramic. Biomed. Mater. 4 (1), 015007.
Johnson, M.W., Sullivan, S.M., Rohrer, M., Collier, M., December 1997. Regeneration of peri-implant infrabony defects using PerioGlas: a pilot study in rabbits. Int. J. Oral Maxillofac. Implant. 12 (6), 835−839.
Jones, J.R., January 2013. Review of bioactive glass: from Hench to hybrids. Acta Biomater 9 (1), 4457−4486.
Kattimani, V.S., Kondaka, S., Lingamaneni, K.P., January 1, 2016. Hydroxyapatite—past, present, and future in bone regeneration. Bone Tissue Regen Insights 7, S36138. BTRI.
Khan, S.N., Cammisa, F.P., Sandhu, H.S., Diwan, A.D., Girardi, F.P., Lane, J.M., 2005 Feb. The biology of bone grafting. J Am Acad Orthop Surg 13 (1), 77−86.
Kitsugi, T., Yamamuro, T., Nakamura, T., Kakutani, Y., Hayashi, T., Ito, S., et al., April 1987. Aging test and dynamic fatigue test of apatite-wollastonite-containing glass ceramics and dense hydroxyapatite. J. Biomed. Mater. Res. 21 (4), 467−484.
Komath, M., Varma, H.K., September 2004. Fully injectable calcium phosphate cement—a promise to dentistry. Indian J. Dent. Res. Off. Publ. Indian Soc. Dent. Res 15 (3), 89−95.
Lacefield, W.R., 1998. Current status of ceramic coatings for dental implants. Implant Dent 7 (4), 315−322.
Larsson, S., Hannink, G., September 2011. Injectable bone-graft substitutes: current products, their characteristics and indications, and new developments. Injury 42 (Suppl. 2), S30−S34.
LC, C., January 2009. Next generation calcium phosphate-based biomaterials. Dent. Mater. J. 28 (1), 1−10.
Nair, M.B., Suresh Babu, S., Varma, H.K., John, A., January 1, 2008. A triphasic ceramic-coated porous hydroxyapatite for tissue engineering application. Acta Biomater 4 (1), 173−181.
Najeeb, S., Zafar, M.S., Khurshid, Z., Siddiqui, F., January 2016. Applications of poly-etheretherketone (PEEK) in oral implantology and prosthodontics. J. Prosthodont. Res 60 (1), 12−19.
Oonishi, H., Iwaki, Y., Kin, N., Kushitani, S., Murata, N., Wakitani, S., et al., January 1, 1997. Hydroxyapatite in revision of total hip replacements with massive acetabular defects. J. Bone Joint Surg. Br 79-B (1), 87−92.
Raggatt, L.J., Partridge, N.C., August 13, 2010. Cellular and molecular mechanisms of bone remodeling. J. Biol. Chem. 285 (33), 25103−25108.

Roberts, T.T., Rosenbaum, A.J., 2012 Dec. Bone grafts, bone substitutes and orthobiologics: the bridge between basic science and clinical advancements in fracture healing. Organogenesis 8 (4), 114−124.

Sandén, B., Olerud, C., Petrén-Mallmin, M., Larsson, S., April 2002. Hydroxyapatite coating improves fixation of pedicle screws. A clinical study. J. Bone Joint Surg. Br 84 (3), 387−391.

Sandeep, G., Varma, H.K., Kumary, T.V., Babu, S.S., John, A., 2006. Characterization of novel bioactive glass coated hydroxyapatite granules in correlation with in vitro and in vivo studies. Trends in Biomaterials and Artificial Organs 19 (2), 99−107.

Sandhya, S., Mohanan, P.V., Sabareeswaran, A., Varma, H.K., Komath, M., 2017 03. Preclinical safety and efficacy evaluation of "BioCaS" bioactive calcium sulfate bone cement. Biomed. Mater. Bristol. Engl 12 (1), 015022.

Scheer, J.H., Adolfsson, L.E., March 1, 2009. Tricalcium phosphate bone substitute in corrective osteotomy of the distal radius. Injury 40 (3), 262−267.

Schwartz, C., Bordei, R., August 1, 2005. Biphasic phospho-calcium ceramics used as bone substitutes are efficient in the management of severe acetabular bone loss in revision total hip arthroplasties. Eur. J. Orthop. Surg. Traumatol. 15 (3), 191−196.

Shepperd, J.A.N., Apthorp, H.S., 2005. A contemporary snapshot of the use of hydroxyapatite coating in orthopaedic surgery. J. Bone Joint Surg. Br 87 (8), 1046−1049.

Søballe, K., January 1993. Hydroxyapatite ceramic coating for bone implant fixation: mechanical and histological studies in dogs. Acta Orthop. Scand. 64 (Suppl. 255), 1−58.

Søballe, K., Hansen, E.S., B-Rasmussen H, Jørgensen, P.H., Bünger, C., 1992. Tissue ingrowth into titanium and hydroxyapatite-coated implants during stable and unstable mechanical conditions. J. Orthop. Res. 10 (2), 285−299.

Stevenson, S., Horowitz, M., July 1992. The response to bone allografts. JBJS 74 (6), 939.

Thomas, M.V., Puleo, D.A., February 2009. Calcium sulfate: properties and clinical applications. J. Biomed. Mater. Res. B Appl. Biomater. 88 (2), 597−610.

Urist, M.R., 1980. Fundamental and Clinical Bone Physiology. Lippincott, p. 440.

Wang, W., Yeung, K.W.K., December 1, 2017. Bone grafts and biomaterials substitutes for bone defect repair: a review. Bioact. Mater 2 (4), 224−247.

Wei, G., Ma, P.X., August 2004. Structure and properties of nano-hydroxyapatite/polymer composite scaffolds for bone tissue engineering. Biomaterials 25 (19), 4749−4757.

Wen, C.-Y., Qin, L., Lee, K.-M., Chan, K.-M., May 2009. The use of brushite calcium phosphate cement for enhancement of bone-tendon integration in an anterior cruciate ligament reconstruction rabbit model. J. Biomed. Mater. Res. B Appl. Biomater. 89 (2), 466−474.

Xu, J.L., Khor, K.A., Sui, J.J., Zhang, J.H., Chen, W.N., October 2009. Protein expression profiles in osteoblasts in response to differentially shaped hydroxyapatite nanoparticles. Biomaterials 30 (29), 5385−5391.

Yan, W.-Q., Nakamura, T., Kawanabe, K., Nishigochi, S., Oka, M., Kokubo, T., September 1, 1997. Apatite layer-coated titanium for use as bone bonding implants. Biomaterials 18 (17), 1185−1190.

Yu, S., Hariram, K.P., Kumar, R., Cheang, P., Aik, K.K., May 2005. In vitro apatite formation and its growth kinetics on hydroxyapatite/polyetheretherketone biocomposites. Biomaterials 26 (15), 2343−2352.

Zhang, D., Leppäranta, O., Munukka, E., Ylänen, H., Viljanen, M.K., Eerola, E., et al., May 2010. Antibacterial effects and dissolution behavior of six bioactive glasses. J. Biomed. Mater. Res. A 93 (2), 475−483.

Zhang, J., Zhao, S., Zhu, M., Zhu, Y., Zhang, Y., Liu, Z., et al., October 15, 2014. 3D-printed magnetic Fe_3O_4/MBG/PCL composite scaffolds with multifunctionality of bone regeneration, local anticancer drug delivery and hyperthermia. J. Mater. Chem. B 2 (43), 7583−7595.

Zhu, T., Ren, H., Li, A., Liu, B., Cui, C., Dong, Y., et al., June 15, 2017. Novel bioactive glass based injectable bone cement with improved osteoinductivity and its in vivo evaluation. Sci. Rep. 7 (1), 3622.

Zwingenberger, S., Nich, C., Valladares, R.D., Yao, Z., Stiehler, M., Goodman, S.B., August 1, 2012. Recommendations and considerations for the use of biologics in orthopedic surgery. BioDrugs 26 (4), 245−256.

Stem cell−based therapeutic approaches toward corneal regeneration

11

Balu Venugopal, Bernadette K. Madathil, Anil Kumar P.R.
Division of Tissue Culture, Department of Applied Biology, Biomedical Technology Wing, Sree Chitra Tirunal Institute for Medical Sciences and Technology, Thiruvananthapuram, Kerala, India

Chapter outline

- **11.1 Introduction** 264
    - 11.1.1 Cornea and corneal layers 264
    - 11.1.2 Corneal epithelial homeostasis 266
    - 11.1.3 Limbal epithelial stem cells and its characteristics 266
    - 11.1.4 Corneal stem cell niche 267
    - 11.1.5 Limbal stem cell markers 267
    - 11.1.6 Limbal deficiency conditions 268
- **11.2 Corneal blindness and current therapies** 268
    - 11.2.1 Amniotic membrane transplantation 270
    - 11.2.2 Autologous conjunctival limbal transplant 270
    - 11.2.3 Allogeneic limbal stem cell transplant 271
    - 11.2.4 Cultivated limbal epithelial transplantation 271
    - 11.2.5 Simple limbal epithelial transplantation 272
    - 11.2.6 Corneal stromal stem cells 272
- **11.3 Other cell-based approaches—nonlimbal sources** 273
    - 11.3.1 Cultivated oral mucosal epithelial transplantation 273
    - 11.3.2 Mesenchymal stem cells 274
    - 11.3.3 Hair follicular stem cells 275
    - 11.3.4 Dental pulp stem cells 276
    - 11.3.5 Skin epidermal stem cells 277
    - 11.3.6 Human embryonic stem cells 277
    - 11.3.7 Induced pluripotent stem cells 277
- **11.4 Biomaterials in corneal reconstruction** 278
    - 11.4.1 Biological materials for corneal regeneration 278
        - 11.4.1.1 Collagen 278
        - 11.4.1.2 Amniotic membrane 279
        - 11.4.1.3 Silk 280
        - 11.4.1.4 Gelatin 281
    - 11.4.2 Synthetic biomaterials 282
        - 11.4.2.1 Polyvinyl alcohol 282
        - 11.4.2.2 Poly(2-hydroxyethyl methacrylate) 282

　　　　11.4.2.3　Polyethylene glycol diacrylate　283
　　　　11.4.2.4　Poly(lactide-co-glycolide)　283
　　　　11.4.2.5　Thermoresponsive polymers　283
　　　　11.4.2.6　Poly N-isopropylacrylamide-co-glycidylmethacrylate　285
11.5　Translational and clinical perspective　286
References　286

11.1　Introduction

The eyes have a predominant position among the sensory organs as they enable the complex mechanism of vision in all living organisms. The structure and organization of eyes differs with organisms. The human eye is made up of multiple layers and is organized into anterior and posterior segments. Anterior region collects or perceives light, while the posterior section helps to convert the collected light in to images enabling vision. Structurally, the eye can be divided further into three parts: the cornea and sclera; the iris, choroid, and ciliary body; and the retina. Light enters the eye through the outermost clear front "window" of the eyeball called cornea. The cornea allows passage of light to the iris and then to the lens. The light rays from the lens is focused on to the retina where the incident light rays are converted to electric impulses by the light receptors—the rod cells and cone cells. These electric impulses transmitted to the vision processing region of the brain are converted to an image. Any interference or condition in the eye that prevents the formation of a proper image can be considered visual impairment.

A recent report in the Lancet on the global data of visual impairments highlighted that 253 million people across the globe are affected by some degree of vision loss and recorded 36 million blind people (Bourne et al., 2017). India accounts for more than 8 million blind people, which is more than 30% of the world's blind population. According to the World Health Organization, corneal surface diseases are the major cause of blindness globally (5.1%) along with cataract, glaucoma, and age-related macular degeneration. Corneal surface disorders are considered as the second largest cause of blindness in India next to cataract, and above 6.8 million people suffer from visual impairment due to the damage of the surface of the eye (Gupta et al., 2013; Vashist et al., 2017). Visual loss by any reason will have profound effects on the patient, based on their psychological, financial, and social situations.

11.1.1　Cornea and corneal layers

The cornea, as mentioned, is the transparent dome-shaped outermost part of the eye. It helps to remove harmful ultraviolet (UV) radiations, protect inner eye from foreign objects, external stimuli entry, and also contributes to the refractive power of eye (Karring et al., 2004). Cornea has to be structurally and functionally intact for clear and perfect vision. In humans, the cornea measures approximately 12 mm vertically and 11 mm horizontally. The cornea is comparatively thick at the periphery and reduces towards the centre (Prospero Ponce, et al., 2009). It is avascular and hence is not nourished with blood vessels; instead, it receives nutrient supply via the aqueous humour and tears (Sweeney et al., 1998). The cornea meets its oxygen requirements directly from atmospheric oxygen that gets dissolved in the tears, which then diffuses to the cornea (Holden and Mertz, 1984).

The incoming light passes the cornea and gets focused on to the retina, so it is important that the cornea must remain clear and exhibit normal shape and curvature for the incident light to pass through. The cornea is structurally organized into five different layers: the corneal epithelium, Bowman's layer, stroma, Descemet's membrane, and the corneal endothelium, which help in maintaining corneal transparency for proper vision (DelMonte and Kim, 2011). Even small blood vessels, cloudiness, and opacity can interfere with the proper focusing of incoming light. This scarring or cloudiness causes minor irritations and can result in blindness. Corneal epithelium being the outermost layer of the cornea is prone to various kinds of damage including thermal or chemical injuries.

The corneal epithelium is approximately 50 µm in thickness and forms a barrier against entry of foreign objects. It consists of nonkeratinized stratified squamous epithelial cells held together by tight junctions (Klyce, 1972) organized in 4—5 layers. The outer layers comprise of terminally differentiated epithelial cells. Beneath this layer lie the suprabasal cells followed by the basal or stem-like cells (Agrawal and Tsai, 2003). The basal cells are attached to the underlying basement membrane, and they mediate cell migration during epithelial injury (Pajoohesh-Ganji and Stepp, 2005). The basement membrane acts as an anchorage for the epithelial cells to grow and organize.

The Bowman's layer is 8—14 µm in thickness and consists of irregularly arranged collagen fibers. This cellular layer separates the outermost epithelial layer and underlying stroma, protecting these underlying layers from injury and harmful stimuli. The corneal stroma is vascular and is the thickest layer of the cornea comprising mainly of water, regularly organized collagen fibrils, and stromal keratocytes. These fibrils are organized in a regular lattice-like pattern that accounts for corneal transparency, strength, elasticity, and form (Meek and Knupp, 2015). Stromal keratocytes help in the synthesis of collagen and proteoglycans. The Descemet's membrane is a rapidly regenerating acellular layer composed of collagen type IV and VIII (Pajoohesh-Ganji and Stepp, 2005). It is a thin but sturdy layer that acts as a protective barrier against injuries and serves as a basement membrane for the corneal endothelium. The corneal endothelium is the innermost layer of the cornea with a single layer of cuboidal cells. The endothelium helps in maintaining fluid balance by regulating the flow of fluid across the stroma (Waring et al., 1982).

This chapter deals with corneal surface disorders and the application of stem cells toward identifying various therapeutic strategies in rectifying corneal damages. This review will focus on the corneal epithelium, its function, associated disease conditions, and stem cell—based approaches adopted for treating ocular surface disorders.

11.1.2 Corneal epithelial homeostasis

The terminally differentiated epithelial cells in the superficial layers of the cornea are continuously sloughed off from the surface and replenished at regular intervals by proliferating cells from the basal layer (Yazdanpanah et al., 2017). These proliferating cell populations reside in the lower layers of the corneal epithelium in the corneoscleral junction called the limbus, and these stem cells are known as limbal stem cells (Dua and Azuara-Blanco, 2000). They undergo mitotic division in the limbal region and become postmitotic when they migrate into the suprabasal layers where they differentiate into squamous epithelial cells (Kruse, 1994). This vertical migration of limbal stem cells due to the proliferative pressure of dividing cells in the basal layers helps in self-renewal and replacement of the corneal epithelial cells (Lavker et al., 1991). The vertical movement is followed by horizontal migration from peripheral cornea to central cornea under physiological conditions (Dua and Forrester, 1990). The dogma in corneal homeostasis states that the total epithelial cell mass remains constant, i.e., the amount of cells lost by sloughing is equal to the amount of cells replaced by mitotic division and migration of basal stem cells (Sharma and Coles, 1989).

According to the XYZ hypothesis of corneal epithelial maintenance by Thoft and Friend (1983), the limbus is a reservoir of stem cells that undergo asymmetric division forming daughter stem cells and a transient-amplifying cell (TAC) (Thoft and Friend, 1983). TACs migrates vertically forming terminally differentiated superficial squamous cells, while the stem cells replenish the stem cell pool. Thus, the hypothesis states that vertically migrating basal cells (X) and centripetally migrating cells (Y) equally compensate the sloughing superficial cells from the surface (Z). However, clinical indications for the presence of stem cells in the central cornea apart from the limbal region are not explained by the XYZ hypothesis (Majo et al., 2008).

11.1.3 Limbal epithelial stem cells and its characteristics

Initial evidences for the presence of stem cells in the basal layer of limbus were elucidated from immunostaining studies with cytokeratin (CK) 3, the characteristic marker of differentiated corneal epithelium. The absence of CK3 staining in the limbal basal region was an indication that the basal layer consists of more progenitor cells and stem cells (Schermer et al., 1986). The presence of limbal stem cells was confirmed in the limbal region by the identification of label-retaining, slow-cycling cells (Cotsarelis et al., 1989). Furthermore, these cells also displayed a greater tendency to form colonies (Pellegrini et al., 1999). It was hypothesized that these slow-cycling cells in the limbal basal layer move upwards forming TAC, which later migrate centripetally forming terminally differentiated corneal epithelial cells. Limbal cells are cuboidal in shape with a high nucleus-to-cytoplasm ratio and have a diameter of around 10 µm. The cytoplasm appears smooth and sparse with less organelles and intracellular junctions (German et al., 2006). These cells exist in an undifferentiated state and have a high multipotent differentiation potential when encountered with appropriate cellular signals.

11.1.4 Corneal stem cell niche

A stem cell niche is an anatomical location where stem cells reside. A niche provides physical protection along with modulating a variety of chemical signals that help maintain the stemness of the cells in the stem cell pool. Various experimental studies have discussed the existence of different hypothetical niches to house limbal stem cells in the peripheral cornea. These include the palisades of Vogt, limbal epithelial crypts (LECs), and focal stromal projections (Yoon et al., 2014). These niche structures help facilitate adhesion, provide growth factors and nutrients, and act as a barrier to external stimuli such as applied forces, UV rays, and oxidative damage (Bessou-Touya et al., 1998; Echevarria and Di Girolamo, 2011). These structures are identified by rete folds, melanocyte expression, nerve endings, and underlying blood vessels (Ordonez and Di Girolamo, 2012). The limbal niche has a unique anatomical structure that consists of an array of radially oriented fibrovascular ridges called the palisades of Vogt (Goldberg and Bron, 1982). There is a cluster of cells from the posterior end of the palisades of Vogt that extends toward the underlying stroma called the LEC or limbal crypt (LC) (Dua et al., 2005). These clusters of cells were shown to have a high concentration of stem cells and are expected to be a limbal niche. The presence of CK14, ABCG2, and P63 was added as evidence to suggest that corneal stem cells reside in LCs.

Maintaining stem cell properties within the limbal niche is an interplay with various signaling pathways. Limbal niche—associated signaling pathways include the sonic hedgehog, Wnt/β-catenin, TGF-β, and Notch signaling pathway (Li et al., 2007). Self-renewal and cell cycle entry of these stem cells are modulated by the Wnt and Notch signaling pathways, while the Wnt/β-catenin signaling regulates morphogenesis in limbal stem cells (Mukhopadhyay et al., 2006; Kulkarni et al., 2010).

In the limbal niche, the limbal stem cells undergo asymmetric division giving rise to a daughter stem cell and one TAC. The TACs migrate out of the limbal niche maturing to corneal epithelial cells, while the daughter stem cells help replenish the stem cell pool (Sun and Lavker, 2004).

11.1.5 Limbal stem cell markers

As discussed, various cell markers were used by researchers to identify limbal stem cells. A definite marker towards determining the limbal stem cell population is still debated. Limbal stem cells within the niche are an undifferentiated cell population, which was previously determined or elucidated by the negative expression of differentiation markers such as CK3 (Schermer et al., 1986). The set of stem cell—associated marker currently used by research groups to identify limbal stem cells includes p63, integrin's, ABCG2, alpha enolase, vimentin, and keratin 19.

p63 is a transcription factor in the p53 family and is present in abundance in proliferating keratinocytes including cornea. This protein was found localized to the nucleus and plays a key role in regulating morphogenesis and differentiation of these cells. Limbal stem cells express p63, while TACs weekly express this marker and are absent in differentiated epithelial cells (Pellegrini et al., 2001). Alpha 9 (α9) integrin and beta 1 (β1) integrin are found localized to limbal basal layer. α9 integrin is a marker for

TAC but presence of α9 integrins in limbal stem cells is not well elucidated. β1 integrin stains both limbal and corneal basal cells but is absent in the superficial layers (Stepp et al., 1995; Gomes et al., 2010).The ATP-binding cassette subfamily G, member 2 (ABCG2) is considered as a universal stem cell marker and is also expressed by limbal basal stem cells. ABCG2 is not expressed by the limbal suprabasal cells and corneal epithelial cells (Gomes et al., 2010). Vimentin and CK19 are two markers that are coexpressed in limbal basal stem cells (Kasper et al., 1992). The cells expressing stem cell markers showed label-retaining properties when pulsed with nucleotide analogues, indicating their stem cell origin, but were not useful to differentiate between limbal stem cells and TACs (Zieske, 1994). CK3, CK 12, and Connexion 43 are the markers that are absent in limbal basal cells but expressed by corneal epithelial cells (Schermer et al., 1986; Kurpakus et al., 1990; Dong et al., 1994).

11.1.6 Limbal deficiency conditions

Any disruption to the corneal epithelial homeostasis can, with time, affect proper vision. Such disruption occurs when there is damage to the limbal niche, preventing proliferation and migration of the limbal stem cells to the peripheral corneal surface. This affects renewal of the corneal epithelium leading to vision loss (Dua et al., 2000). This loss of limbal stem cells is called limbal stem cell deficiency (LSCD) and prevents the eye from repairing itself (Gupta et al., 2013). It is usually caused by thermal or chemical burns or by inherited disease conditions such as ocular cicatricial pemphigoid (OCP), Stevens–Johnson syndrome. LSCD can compromise ocular surface integrity resulting in scarring, opacification, and extensive damage to the corneal surface, the clear window of the eye. A compromised epithelium will result in conjunctival over—growth on the corneal surface causing blurred vision and even blindness.

LSCD is manifested mainly by blurred vision, sensitivity to light, tear discharge, swelling, vascularization, and conjunctival growth on the corneal surface to chronic inflammatory symptoms and increased opacification (Dua et al., 2000; Sejpal et al., 2013). LSCD is termed partial, where a portion of the limbus is damaged while the remaining region is intact and functional. When the damage affects the entire limbal ring, the clinical condition is termed as complete LSCD. Identifying the extent of damage is crucial in determining treatment methodologies as this varies from topical medication to surgical transplantation procedures.

11.2 Corneal blindness and current therapies

Corneal blindness is considered as an avoidable and treatable condition (Vashist et al., 2017). Treatment options are decided by determining the extent of limbal damage. This can vary from eye drops for topical medication to corneal epithelial transplantation procedures in worst case scenarios. The extent of damage in the limbus is determined by the limbal clock hour damage criteria set by Dua et al., which considers the limbus as a 12-h clock; it takes into the consideration the extent of limbal and conjunctival involvement in the damage and grades the damage from I to VI. As per the current classification, in general, grade IV to VI are considered for transplantation procedures (Dua et al., 2001). Transplanting the patient's own cells from the good eye (autologous transplantation) or from a donor tissue (allogeneic transplantation) allows replenishing of the lost stem cells (Ghezzi et al., 2015), thereby allowing corneal epithelial restoration. Transplantation can be performed directly using limbal tissue or by expanding the required number of cells ex vivo in a laboratory. In case of bilateral limbal damages, i.e., the limbus of both eyes are damaged, allogenic sources from donors or cadavers remain the gold standard therapy (Tan et al., 1996). Such allogeneic transplantation faces issues of less number of suitable donor corneas, graft rejection, recurrent infections, and long-term immunosuppression treatments (de Araujo and Gomes, 2015). Therefore, researchers have identified or experimented alternative treatment strategies. Corneal surface reconstruction using stem cells has become a promising approach over corneal grafts. Stem cells are harvested from various tissues and are expanded in laboratory conditions for transplantation (Pellegrini et al., 1997). Provision for ex vivo expansion of stem cells, ease of availability, and access to stem cells help address issues on critical shortage of reliable ocular donor tissue. This review outlines various stem cell–based approaches and other tissue-engineering strategies toward addressing ocular surface reconstruction. Deficiency of limbal stem cells can lead to ocular pain, decreased vision, and photophobia. Fig. 11.1 provides an overall classification of current treatment options.

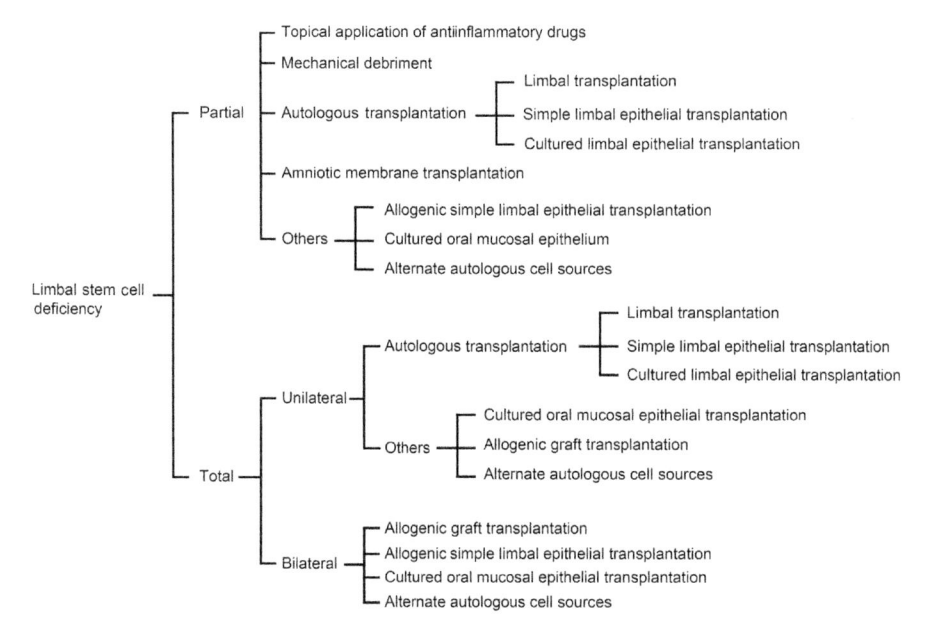

Figure 11.1 Classification of various treatment modalities practiced for the reconstruction of the ocular surface in different clinical conditions of LSCD.

11.2.1 Amniotic membrane transplantation

Amniotic membrane (AM), the innermost layer of the placenta, is widely used in treating ocular surface damages. Use of AM as a therapeutic tool was initially reported in early 1900's from its use in skin transplantation (Davis 1910) and was also demonstrated in ocular surface damages with chemical burn (de RÖ, 1940). The therapeutic effect of AM can be attributed to the cytokines and growth factors harbored in the epithelium and stroma (Koob et al., 2013). To date, AM has been successfully used for the treatment of Stevens–Johnson syndrome, chemical or thermal burns, Ocular Cicatricial Pemphigoid OCP, and LSCD conditions (Rahman et al., 2009). AM promotes ocular surface reconstruction in various means. It is used as a patch over the eye as a biological bandage (Gruss and Jirsch, 1978), where the AM supports the underlying tissue to perform their functions and also mobilize host epithelial cells (Zhang et al., 2015). AM is also used as a substrate during stem cell transplantation in ocular surface therapies (Du et al., 2003). Sangwan et al. demonstrated the use of AM membrane in partial LSCD conditions with reported success rate of 60%–70% depending on the conjunctival involvement in the corneal damage (Sangwan et al., 2004).

However, there are several factors that limit the use of the AM including availability of raw material, consistency in processing of membrane for therapeutic purpose, cold storage conditions in long-term storage, and risk of cryptic infections. The biological variability of AM limits development of a standardized, reproducible "off-the-shelf" product without batch-to-batch variations (Rahman et al., 2009; Malhotra and Jain, 2014). This indicates the need of a standardized, reproducible AM alternative but, unfortunately, to date no available substrate was shown to fully recapitulate the ability of the AM to support cell growth and stem cell survival.

11.2.2 Autologous conjunctival limbal transplant

Conjunctival limbal autograft (CLAU) is a well-documented and performed procedure for ocular surface damages. This is performed routinely in case of unilateral, partial, or total limbal damages (Fernandes et al., 2004). In this procedure, a conjunctival limbal graft from the contralateral eye is transplanted to the damaged corneal surface. This was first reported by Jose Barraquer in 1964 and later in 1977 by Richard Thoft for unilateral ocular surface injuries (Amescua et al., 2014). Kenyon and Tseng in 1989 were the first to report the successful outcome of CLAU procedure in clinical setting with 21 patients and later got wide spread acceptance (He and Yiu, 2014). This procedure involves dissection of a 4–5 mm conjunctival tissue spanning the limbal region along with a little of the underlying stromal region from the fellow eye and is transplanted to the damaged corneal surface. Two strips of around 120° are dissected. The recipient eye undergoes a 360° peritomy and removal of the fibrovascular pannus, and then the graft is either sutured or glued using fibrin glue (Fernandes et al., 2004; Bakhtiari and Djalilian, 2010; He and Yiu, 2014). The main disadvantage of this technique is the significant loss of limbal region from the healthy eye, leading to donor site morbidity (Ebrahimi et al., 2009), but being an autograft, there are no issues associated with immune rejection in CLAU procedure. However, a 40%–50% of limbal resection is not expected to cause donor site morbidity. Also attempts to treat unilateral LSCD with smaller graft dimensions (>120°) were comparatively unsuccessful compared with corneal surface reconstruction using larger donor grafts (Liang et al., 2009). But, in case of bilateral limbal damage conditions, CLAU is impossible due to the absence of a contralateral healthy limbus.

11.2.3 Allogeneic limbal stem cell transplant

In case of bilateral limbal damage, where both limbus are damaged beyond self-repair, allogenic transplantation remains a clinically viable therapeutic approach. This can either be a living-related conjunctival limbal allograft (Lr-CLAL) or a cadaveric keratolimbal allograft (KLAL). Both these approaches use foreign limbal tissue for transplantation, and so they pose the risk of immune rejection with poor long-term survival of transplanted graft when compared with autograft transplantation. The clinical procedure is similar to CLAU except in the use of donor tissue. Lr-CLAL is preferred over KLAL as some degree of compatibility can be identified through HLA typing of potential donors. Apart from Lr-CLAL where a small piece of tissue is dissected from the donor cornea surface, KLAL allows excision of 360° limbal ring with a small portion of scleral tissue and peripheral cornea. The KLAL procedure eliminates issue of donor site morbidity but has the disadvantage of immune suppression and immune rejection. Severe bilateral LSCD is associated with hyperactivity of the host immune system, and this also accelerates the failure rate of allograft transplantation despite the immune suppressive therapy (He and Yiu, 2014).

Lr-CLAL allows dissection of a small piece of healthy graft tissue from the donor, while KLAL provides a 360° limbus for treating the damaged cornea, but Lr-CLAL has reported better results compared with KLAL (Liang et al., 2009; Bakhtiari and Djalilian, 2010). Holland et al. introduced a combined technique that maximizes the benefits of both the procedures called the "cincinnati procedure." The keratolimbal tissue is used to fill the gaps and voids after grafting the conjunctival tissue in the damaged eye (Holland and Schwartz, 2004).

11.2.4 Cultivated limbal epithelial transplantation

Risk of immune rejection associated with allografts leads to modifications in the existing methods of corneal reconstruction therapies. This led to the attempt to expand the limbal epithelial stem cells ex vivo, thereby enabling generation of the required number of stem cells for corneal surface reconstruction. The scope of using ex vivo expanded limbal stem cells was exploited by Pellegrini et al., and also demonstrated the transplantation of ex vivo expanded autologous limbal stem cells in LSCD conditions caused by unilateral alkaline injury (Pellegrini et al., 1997). Following this, Li et al. demonstrated the use of ex vivo expanded autologous stem cells using AM as a carrier in treating LSCD conditions due to chemical burn (Tsai et al., 2000). Cultivated limbal epithelial transplantation (CLET) was able to address the issue of donor site morbidity in autologous therapies where a very small 1–2 mm tissue was dissected from the healthy fellow eye of the patient. Later, these small tissue pieces were used to establish limbal stem cells using definite culture systems in controlled culture conditions in a laboratory setup. This will allow the stem cells to migrate out of the explant and expanded in an optimized culture conditions. Based on the components of the culture system, it can be either a "xenobiotic culture system" or a "xenobiotic-free culture system." Xenobiotic culture system contains some animal-derived components within them, which promote cell growth and maintenance, but their presence raises safety concerns including immune rejection when used for transplantation purpose (He and Yiu, 2014). Serum, being the major xenogenic component in any culture system, was successfully replaced by using autologous human serum or human blood—derived serum with improved clinical outcomes (Chakraborty et al., 2012; Shahdadfar et al., 2012). The expansion or growth of limbal stem cells also depends on the type of matrix on which the cells are grown. This includes AM and fibrin glue, which have already been clinically proven. CLET also avoids use of immunosuppression as it can use autologous tissue without donor site morbidity in case of partial LSCD, and CLET is also practiced from allogeneic donors and cadavers (Behaegel et al., 2017).

11.2.5 Simple limbal epithelial transplantation

Simple limbal epithelial transplantation (SLET) was introduced as a modified surgical technique addressing the limitations associated with the previous techniques in corneal surface therapies. This procedure was able to remove the complications that may arise due to the ex vivo expansion of stem cells including interaction with xenogenic components. This was demonstrated by Sangwan et al., and they reported successful reconstruction in six patients who underwent SLET for unilateral or bilateral limbal damage (Sangwan et al., 2012). Further long-term follow-up study reported 76% of successfully regenerated corneal surface (Basu et al., 2016).This technique utilizes a small 2×2 mm donor limbal tissue that is dissected into smaller pieces. After removal of fibrovascular tissue from the damaged cornea, these small limbal explants are placed over the damaged corneal surface and covered with an AM. The AM helps to modulate inflammation and provides a supportive microenvironment for the stem cells to expand in vivo and migrate out of the limbal transplants (He and Yiu, 2014). The AM with the limbal transplants is held to the corneal surface using fibrin glue. Various modifications have been reported including use of AMs to sandwich limbal stem cells (Amescua et al., 2014). SELT is a procedure that is less complicated to perform and is highly reproducible. Long-term results from SLET were found satisfactory. SLET involves less surgical procedural steps, avoiding requirement of a special laboratory setup, addresses issues with donor site morbidity, and has no requirement for immune suppression (if in case of unilateral LSCD).

11.2.6 Corneal stromal stem cells

The quest for more stem cell niches in the cornea due to donor shortage and immune rejection led to the discovery of another major cell source within the corneal stroma. Hopkinson's group identified a small group of stem cells within the corneal stroma with striking similarities to mesenchymal stem cells (MSCs) (Branch et al., 2012). These cells called corneal stromal stem cells (CSSCs) displayed high proliferation potential along with maintaining stemness. Similar to MSCs, they exhibited colony-forming potential, multilineage differentiation potential, asymmetric division, and also expressed the most common MSC markers CD105, CD73, CD90, and ABCG2 (Pinnamaneni and Funderburgh, 2012). These cells also fulfilled the criteria for MSC as provided by the International Society of Cellular Therapy and was shown to have epithelial differentiation potential when induced (Hashmani et al., 2013). CSSCs are believed either to be of neural crest origin or of bone marrow—derived cells (Sosnova et al., 2005). CSSCs are shown to control neutrophil infiltration and thereby modulate formation of stromal scar tissue (Hertsenberg et al., 2017). The identification of this novel population of stem cells with mesenchymal properties within the corneal stromal niche may open up avenues in the field of corneal surface therapy, but more clinically proven outcomes are still awaited.

11.3 Other cell-based approaches—nonlimbal sources

In partial and unilateral conditions, limbal stem cells are still considered the gold standard. In case of bilateral deficiencies, other cells sources have to be explored to address issues with graft rejection and immune suppression. To overcome these problems, alternative autologous sources with potential for ocular surface reconstruction have been explored (Fig. 11.2).

11.3.1 Cultivated oral mucosal epithelial transplantation

There were concerns regarding treating bilateral LSCD conditions, where autologous limbal stem cells were not available and long-term failure of allogeneic grafts lead researchers to look at other available cell sources. Oral mucosal epithelial cells were one of the first attempted nonocular cell source for treatment of corneal surface damages. The use of oral mucosal epithelial cells for ocular epithelial surface reconstruction was first demonstrated by Nakamura and Kinoshita with experiments in rabbit models (Nakamura et al., 2003; Nakamura and Kinoshita, 2003). This technique addressed the issue of using allogenic cells by identifying an alternative cell source from a nonlimbal region. These cells were demonstrated to reconstruct a regular stratified epithelium with quick cellular turnover, minimum keratinization, and exhibited a low risk of immunologic reaction (Dobrowolski et al., 2015). As practiced in CLET, a healthy oral mucosal biopsy was dissected from inner cheek, cut into small pieces, and was expanded in vitro on denuded AM till confluence, and this was then transplanted to the damaged corneal surface and secured with suture (Nakamura and Kinoshita, 2003). Long-term follow-up after COMET surgery was reported to have satisfactory clinical results with good visual outcomes (Satake et al., 2011; Sotozono et al., 2013). COMET addressed the problem of use of allogenic tissues and was shown to be a better replacement for KLAL or allogenic SLET, both requiring long-term

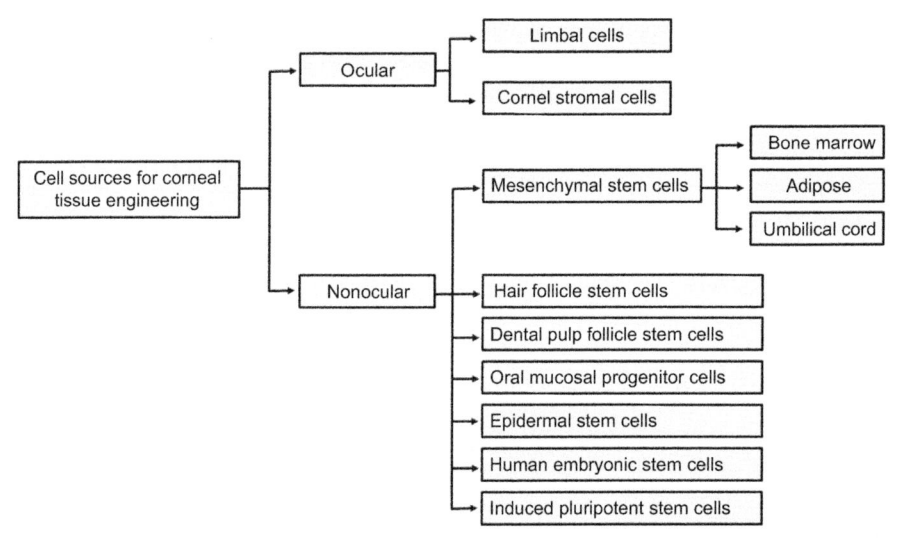

Figure 11.2 Classification of various stem cell sources used for corneal tissue engineering. The stem cells from ocular tissue are mainly from the limbus. Nonocular tissues are alternative stem cell sources in the body that offers autologous cells.

immune suppression therapy. This was one of the initial alternative stem cell therapy approaches shown successfully in clinical application.

11.3.2 Mesenchymal stem cells

MSCs comprise another promising group of adult stem cells with proven applications in regenerative medicine. Use of MSC for corneal surface reconstruction has also been reported with ample evidence to prove the role of MSCs in corneal wound repair (Zhang et al., 2015).

MSCs are a self-renewing population of adult stem cells that can differentiate into multiple lineages including the bone, cartilage, fat, muscles, neurons, cardiac muscle cells, and epithelial cells (Jackson et al., 2007). They have the advantage of being isolated or extracted from various anatomical locations such as the bone marrow, adipose tissue, Wharton's jelly, dental pulp, peripheral blood, cord blood (Law and Chaudhuri, 2013), and limbal stroma of the human eye (Hashmani et al., 2013). Their minimal expression of major histocompatibility class II and other costimulatory CD markers under normal conditions (Gao et al., 2016) along with antiinflammatory and angiogenic effects makes MSCs an ideal candidate for use in corneal regenerative medicine therapies.

MSCs are now widely attempted in corneal research (Li and Zhao, 2014). They can either be administered directly to the corneal surface (Yao et al., 2012) or by substrate carriers such as AM (Ma et al., 2006) or fibrin gels (Gu et al., 2009). MSCs were also reported to transdifferentiate to corneal epithelial cells expressing CK3/12, a corneal epithelial differentiation marker, under a suitable differentiation environment.

Gu et al. demonstrated that the bone marrow MSCs can differentiate into corneal epithelial–like cells in vivo and in vitro (Gu et al., 2009). Either differentiated or undifferentiated MSC is said to be beneficial, as both have been shown to play a role in corneal wound healing (Zhang et al., 2015). All these evidences show that MSCs are an ideal alternative autologous cell source to limbal stem cells for ocular surface regeneration. Similar to bone marrow MSCs, adipose tissue–derived MSCs are also equally considered as a potential source of stem cells for corneal wound healing. The differentiation of stem cells to corneal epithelial–like cells has more understanding now. One of the transcription factors that have been identified to take part in the differentiation of limbal stem cells is paired box 6 (PAX 6). When PAX6 is introduced into adipose-derived MSCs from Sprague Dawley rats, the cells transdifferentiated into corneal epithelial–like cells in 21 days (Sun et al., 2018). Subsequent transplantation of EGFP-labeled PAX6$^+$-transfected ADMSCs to rabbit LSCD cornea showed characteristics of corneal epithelial cells in vivo. Similarly, human ADMSCs transplanted to rabbit LSCD models showed improved corneal and limbal epithelial phenotypes (Galindo et al., 2017). The potential of ADMSC in ocular surface repair as a novel treatment is also expected to get clinical success in future.

The porous connective tissue that surrounds umbilical cord vessels harbors another group of MSCs called Wharton's jelly MSCs (WJMSCs), which are of extraembryonic and partly embryonic origin (De Miguel et al., 2009). Another advantage of WJMSCs is the unique expression of tumor suppressor genes, secretion of hematopoietic cytokines, and that they do not form teratomas (Fong et al., 2011). The differentiation of WJMSC to corneal epithelial cells in vitro was first demonstrated by Garzon et al. on three-dimensional fibrin-agarose scaffolds (Garzon et al., 2014). The differentiated WJMSCs showed differentiation toward epithelium and stroma by expressing markers for epithelium and extracellular matrix (ECM) components. Because of the potential of MSCs to differentiate to corneal and stromal component, lot of research has been done to differentiate and use MSCs as an alternative source of cells for corneal surface reconstruction.

11.3.3 Hair follicular stem cells

The hair follicles (HFs) are part of the pilosebaceous unit located within the skin and are easily accessible. They harbor stem cells from diverse developmental origins that continuously self-renew, differentiate, regulate hair growth, and contribute to skin homeostasis. The main reservoirs of the stem cells are the bulge area (BG), the dermal papillae (DP), and the dermal shaft (DS). The BG cells arise from the ectoderm, while the DP and DS cells are derived from the mesoderm (Schneider et al., 2009). The BG stem cells are surrounded by cell populations with different developmental origins, for example, follicular epithelium ectoderm, melanocyte stem cell–neural crest, and DP/DS-mesoderm. These cells act in cohort to coordinate and exchange signals to generate a pigmented hair shaft (Mistriotis and Andreadis, 2013). The BG stem cells are characterized by KRT15^{high}/CD200$^+$/CD34$^-$/CD271$^-$ profile for cell surface marker. The DP/DS stem cells are positive for CD90, CD44, CD49b, CD105, and CD73, which is similar to the cell surface profile characteristic of MSCs (Mistriotis and Andreadis, 2013).

Stem cells have increased proliferation and broad differentiation capacity; hence, they can be used for tissue-engineering applications. Hair follicle stem cells (HF-SCs) are highly proliferative in vitro and multipotent and thus have potential in the field of regenerative medicine. Blazejewska et al., 2009, had reported on the transdifferentiation of murine vibrissae HF stem cells into corneal epithelial-like cells. The stem cells were isolated from the BG area and grown on different ECM substrates (type IV collagen, fibronectin, laminin-1, laminin-5) with different conditioned medium obtained from central and peripheral corneal fibroblasts, limbal stromal fibroblasts, and 3T3 fibroblasts. Their results showed that laminin-5, a major component of the corneolimbal basement membrane zone, and conditioned medium from limbal stromal fibroblasts, facilitated the HF-SCs to adhere rapidly and form regularly arranged stratified cell sheets (Blazejewska et al., 2009). The group later used a transgenic mice K12(rtTA/rtTA)/tetO-cre/ROSA(mTmG), in which HFSCs change color, from red to green, once they differentiate to corneal epithelial cells and express the corneal epithelial-specific differentiation marker Krt12. The HF-SCs isolated from the transgenic mice were expanded in vitro and later transplanted to a wild-type mice with LSCD. The ocular surface was found to be reconstructed in nearly 80% of the LSCD mice. These results support the therapeutic potential of HF-SCs in corneal tissue engineering (Meyer-Blazejewska et al., 2011). Yang and his coworkers could transdifferentiate BG stem cells into corneal epithelial-like cells in conditioned medium containing corneal limbus soluble factors. They have also explained that the transdifferentiation depends on the upregulation of the transcription factors PAX6 and downregulation of β-catenin and Lef-1 (Yang et al., 2009). Much research still needs to be done in the field of HF-SCs and their potential use in regenerative medicine and tissue engineering with an ultimate aim of translating these findings to clinical therapeutic applications.

11.3.4 Dental pulp stem cells

Human dental pulp harbours a group of stem cells capable of multilineage differentiation. These cells are reported to express both MSCs and embryonic stem cell markers (Haagdorens et al., 2016). It was also reported that these group of stem cells are positive for P63, β1-integrin, and ABCG2, indicating their similarity with limbal epithelial stem cells. When these cells were transplanted into damaged ocular surface, it resulted in reconstruction of ocular surface with gradual improvement in corneal transparency with time (Monteiro et al., 2009). Another study reported reorganization of corneal epithelium after transplantation of dental pulp stem cells onto mild and severe burn models of the corneal surface. Complete reorganization was reported in mild damage models, while loose corneal epithelium was observed in severe damage models (Gomes et al., 2010). Dental pulp stem cells are relatively easy to access, but need for extraction of the tooth is a disadvantage.

11.3.5 Skin epidermal stem cells

Skin epidermal stem cells have also been evaluated for their potential to regenerate the damaged corneal surface. Epidermal stem cells are usually transplanted along with other supporting matrices such as AM, fibrin, etc. (Sehic et al., 2015). Yang et al. had developed a goat LSCD model and have transplanted epidermal stem cells from goat ear skin to the goat eyes using a denuded AM. Post transplantation, the corneal epithelium was reformed; visual acuity was regained with time and was found to express CK3, CK12, and PAX6. In the follow-up period, the corneal surface remained transparent without serious complications (Yang et al., 2008). In another set of studies, skin epithelial stem cells were transduced with PAX6, a transcription factor for corneal epithelial differentiation. This was proposed to help the differentiation of the cells into limbal stem cell–like cells, and on transplantation to corneal rabbit injury models, these reprogrammed cells were able to reform the corneal epithelial surface (Ouyang et al., 2014). The transplantation of these stem cells in animal model showed good success rates. The ease of access of skin epidermal stem cells is another major advantage of these stem cells.

11.3.6 Human embryonic stem cells

Embryonic stem cells have potential in the field of tissue engineering due to their increased plasticity and pluripotent nature. These cells are derived from the inner cell mass of the human embryo. It has also been demonstrated that embryonic stem cells have the potential to differentiate into corneal epithelial–like cells (Haagdorens et al., 2016). Embryonic cells are cultured on either collagen- or gelatin-coated surfaces and are usually transferred as carrier free cell sheets onto the damaged corneal surface. In a study from Zhu et al., human embryonic stem cells cultured on acelluar porcine cornea were differentiated to limbal epithelial–like cells. When transplanted in rabbit LSCD models, the cell sheet helped in corneal epithelial reconstruction forming CK3 positive cells. Basal layer of cells was shown to be positive for P63 and ABCG2 (Zhu et al., 2013).The restored epithelium exhibited high levels of expression of CD44 and E-cadherin, which are important in corneal epithelial wound healing (Homma et al., 2004). Although much research studies using embryonic stem cells have been conducted in the field of regenerative medicine, ethical concerns, ease of access, and chances of tumorogenesis and immunogenicity (Sehic et al., 2015) are still major issues of concern.

11.3.7 Induced pluripotent stem cells

Induced pluripotent stem cells (IPSCs) are derived from differentiated adult stem cells by reprogramming them back into stem cells through the forced induction of reprogrammed genes. IPSCs have also been generated from human corneal epithelium and human dermal fibroblast; these cells have been evaluated for their potential to form corneal epithelial cells. Corneal epithelium derived IPSC were more efficient in generating limbal epithelial cells in comparison with IPSCs from dermal fibroblast. The efficiency of differentiation is comparatively less, and heterogeneity is more when using IPSCs for corneal epithelial differentiation (Hayashi et al., 2012; Sareen et al., 2014). Another approach is to differentiate IPSCs into a homogenous population of p63 positive cells and then differentiate this homogenous population into corneal epithelial—like cells (Mikhailova et al., 2014).

11.4 Biomaterials in corneal reconstruction

Limited availability of healthy donor corneas remains a major setback to the surgical treatment of corneal dystrophies. The tissue-engineering approach for corneal regeneration circumvents this problem via the in vitro expansion of stem cells onto biomaterials. The biomaterials to be used as substrates or scaffolds must have suitable mechanical properties, be transparent, and biocompatible. They may be either synthetic or natural polymers (Chen et al., 2018). Different synthetic and natural polymers have been used as biomaterial scaffolds for corneal tissue engineering and regeneration. The major difficulty in generating a corneal equivalent in vitro is that it should exhibit strength and transparency similar to those of native tissue. As collagen is the most critical component in cornea in determining the transparency, collagen-based corneal alternatives have been extensively studied. An interpenetrating substitute from porcine collagen and phospholipid was fabricated by Liu et al. The authors also showed that porcine atelocollagen can be replaced with recombinant human collagen (Liu et al., 2009). Another report by the same group showed that transplantation of collagen substitutes helped in regeneration of corneal cells, nerves, and optical clarity (McLaughlin et al., 2010). The two main characteristics other than biocompatibility required in scaffolds for corneal tissue engineering is the high transparency with suitable refractive power and the ability to withstand the tension of intraocular pressure and movements of eye (Ruberti and Zieske, 2008). Much research is being done to design biomimetic matrices and scaffolds for corneal tissue engineering.

11.4.1 Biological materials for corneal regeneration

11.4.1.1 Collagen

Collagen forms the major constituent of the ECM in many tissues including the cornea. Although the main function of collagen is structural, it does play an active role in wound repair and healing. In vivo collagen exists as a triple helix, and 28 different forms of collagen in vertebrates have been identified (Shoulders et al., 2009). Collagen is biodegradable, biocompatible, easily available, and versatile; thus, it is a widely used natural polymer in tissue engineering. Collagen-based biomaterials are obtained either as a decellularized matrix that retains the original tissue shape or as extracted and purified collagen solution that can be polymerized or cross-linked to form various collagen-based scaffolds (Rémi Parenteau-Bareil et al., 2010). These scaffolds may be films, sheets, disks, or hydrogels. Cross-linking of collagen is done by chemical, physical, or biological methods. The frequently used chemicals for cross-linking are aldehydes such as formaldehyde and glutaraldehyde, carbodiimide (EDC), and isocyanates (hexamethylene diisocyanate). Plant extracts, e.g., genipin, oleuropein; photoreactive agents, e.g., riboflavin; and carbohydrates, e.g., glucose have been used to a lesser extent. Physical cross-linking is mediated by UV or dehydrothermal treatment. Biological cross-linking using enzymes such as transglutaminase can enhance the mechanical toughness of collagen (Rémi Parenteau-Bareil et al., 2010).

Collagen is the main component of the corneal stroma. In corneal tissue engineering, collagen is mainly used as a cell growth matrix for various corneal cell types. Liu et al. (2006) have reported on the use of cross-linked collagen patches for promoting corneal cell and nerve regeneration in rabbit and porcine models (Liu et al., 2006). The use of recombinant human collagen for generation of collagen scaffolds for corneal regeneration has also been reported (Dravida et al., 2008). Collagen vitrigel has also been used as a substrate for corneal epithelial expansion in vitro and has shown promising results in LSCD rabbit models (Chae et al., 2015). Ascorbic acid has also been used to stimulate stromal fibroblasts in vitro to secrete their own ECM and collagen. These can be stacked to form a thick stroma on which the corneal epithelium can be seeded to form functional bioengineered corneal constructs (Proulx et al., 2010). Electrospun type 1 collagen scaffolds have also been used to culture rabbit corneal fibroblasts (Phu et al., 2011). Wu Z et al., (2018) have used a polyvinyl acetate (PVA)/collagen (PVA-COL) scaffold and have shown them to be efficient scaffolds for growth of human keratocytes and human corneal epithelial cells (hiCECs) (Wu et al., 2018). The electrospinning technique can be used to generate aligned fibers that better mimic the stromal architecture.

11.4.1.2 Amniotic membrane

Human amniotic membrane (HAM) contains a large number of biologically active factors that promote healing while suppressing inflammation, which advocates the use of AMs as wound dressings. In surgical ophthalmology, AM has been successfully used for the treatment of Stevens—Johnson syndrome, chemical or thermal burns, and unilateral LSCD conditions (Rahman et al., 2009). HAM has been used either as bandages to promote wound healing or has been used as substrates for in vitro culture of corneal epithelial cells and their subsequent application in vivo in the damaged cornea.

Battle and Perdomo, in 1993, used AM preserved in 95% ethyl alcohol as a substitute for conjunctival membranes in fornix reconstruction and in the treatment of recurrent pterygium and alkali burns. Tseng and Kim, in 1995, performed AM transplantations in rabbits for ocular surface reconstruction (Utheim et al., 2018). Tsai et al. reported on the use of HAM as a substrate for the expansion and culture of limbal epithelial stem cells (Tsai et al., 2000). The use of bioengineered grafts developed using expanding limbal epithelial stem cells ex vivo on AM became a new method for treating LSCD (Grueterich et al., 2003). Sangwan et al. have reported on the use of HAM as a substrate for the culture of limbal stem cells and the subsequent use of these constructs for the treatment of LSCD; the success rates of these constructs depended on the extent of involvement of the conjunctiva (Fatima et al., 2006).To increase the transparency of HAM constructs, Zhang et al. (2016) have developed ultrathin AM (UAM) by using collagenase IV to remove most of the stroma and bring down the stromal thickness to 30 μm (Zhang et al., 2016). These UAM were demonstrated to be efficient scaffolds for the culture of rabbit limbal epithelial cells. A recent study (Joubert et al., 2017) reported on the improved wound healing properties of amniotic membrane incorporated with retinoic acid used to treat experiment mice models with chemically induced corneal burns.

HAM can be used either intact along with the amniotic epithelial cells or in the denude state where the epithelial cells are removed. Stringent processing and sterilization techniques need to be followed to maintain the consistency of the membrane for preservation. HAM is known to reduce scarring, inflammation, and to provide antifibrotic effects; however, this membrane is associated with some drawbacks, including risks of contamination and transmission of infectious diseases and biologic variability between donor tissues (Niknejad et al., 2008; Kong and Mi, 2016).

11.4.1.3 Silk

The use of silk in the field of wound healing and regenerative medicine is well documented. Silk is produced from silkworms, spider, mites, and beetles; however, the most common sources of silk for tissue-engineering application are silkworm and spiders. The structural arrangement of proteins, mechanical properties, and ease of processing of the silk fiber varies across species (DeBari and Abbott, 2018). Although spider silk does not contains sericin, the extraction process is more time consuming and expensive than extraction of silk from the silkworms particularly *Bombyx mori* (Romer and Scheibel, 2008). Degumming is the process of removal of sericin. The resultant fibroin fibers can be woven or processed to form scaffolds, hydrogels, and films. These films can be modified with cell-signaling molecules to form biologically functional matrices (Lawrence et al., 2009). Alternatively these surfaces may also be patterned by lithographic techniques; such patterning is found to effect cell adhesion, alignment, and migration (Fitton et al., 1998).

The optical clarity, mechanical strength, and controlled degradability make silk a suitable material for corneal tissue engineering. Lawrence et al., 2009 have reported on the use of optically transparent patterned silk films for the culture of rabbit and human corneal cell cultures. These cultured films were also stacked to form 3D corneal constructs. The cells were found to proliferate and be viable over a period of 10 days. Silk fibroins can also be electropsun into mats or membranes. Biazar et al., 2015, have reported on the growth of human corneal limbal epithelial cells on these membranes. Immunohistochemical and real-time PCR analysis revealed no change in the expression profile of the cells grown on the mat as compared with those grown on HAM (Biazar et al., 2015). Biofunctionalization of silk films by conjugation with Arg-Gly-Asp is the expansion (RGD) peptides has been shown to improve the attachment, proliferation, and alignment of human corneal fibroblasts. These cells were also found to express genes for collagen-I, collagen-V, decorin, and biglycan (Gil et al., 2010). Wang L et al. have also reported on the biocompatibility of RGD-functionalized silk films when implanted into the corneal stroma of rabbits. These films maintained their integrity and transparency over a 180-day period without inducing any neovascularization or immunogenic response (Wang et al., 2015). Transparent silk hydrogels have been developed by treating silk fibroin solution with organic solvents. These gels when seeded with hiCECs have shown promising results (Mitropoulos et al., 2015). Recently, Gosselin EA et al. (2018) have applied both the techniques of patterning and biofunctionalization with collagen type IV and fibronectin to form silk films that were used to coculture human corneal epithelial and stromal cells. When the cells reached nearly 80% confluency, the films were stacked to form a 3D corneal construct (Gosselin et al., 2018).

11.4.1.4 Gelatin

The ease of availability, low cost, and biocompatibility make gelatin an ideal biomaterial for tissue-engineering applications. Gelatin is obtained through the hydrolysis of collagen. Gelatin hydrogels are prepared by dehydrothermal or chemical cross-linking (Zhi et al., 2018).

De la Mata et al. have demonstrated the suitability of chitosan (CS)–gelatin biopolymers scaffolds for the culture of human corneal epithelial and human limbal epithelial stem cells. The cells showed the expression for corneal (K3, K12) and limbal epithelial stem cell (K15) markers (Mata et al., 2013). Collagen-gelatin-hyaluronic acid films were generated by cross-linking using 1-ethyl-3-(3-dimethyl amino propel)-carbodiimide (EDC) and N-hydroxysuccinimide (NHS). These films also supported the growth of hiCECs (Liu et al., 2013). Porous gelatin hydrogels, prepared by chemical cross-linking using glutaraldehyde, have been seeded with keratocytes, and the matrix is found to be efficient in promoting cell migration and facilitates transfer of oxygen and nutrients. These seeded scaffolds when used to reconstruct a damaged corneal stroma in a rabbit model, no inflammatory response or increase in ocular pressure was noted in these rabbits (Mimura et al., 2011). Nanoscale modification of porous gelatin materials with chondroitin sulfate using carbodiimide chemistry was used to construct corneal stromal equivalents and was shown to support corneal keratocyte growth (Lai, 2013) Electrospinning is a versatile technique through which nanofibrous scaffolds with increased mechanical strength can be fabricated. In the technique, a high voltage is applied between a syringe and a collector to draw nano- or microscale fibers from the material dispensed by the syringe (Ortega et al., 2013a,b). Gelatin has been electrospun to form nanofibrous scaffolds for corneal tissue-engineering application, but they are reported to have low mechanical properties (Tonsomboon and Oyen, 2013; Tonsomboon et al., 2013). To increase mechanical properties blended (PHBV)/gelatine and gelatin/PLLA have been electrospun. Yan et al. have shown that alignment of the electrospun fibers increases the tensile strength and break strength of the scaffold. These electrospun scaffolds were efficient substrates for in vitro culture of corneal epithelial cells (Yan et al., 2012).The PHBV/gelatine nanofibrous scaffolds have been reported to promote human limbal stem cell growth and successfully formed a 3D corneal epithelium that was viable for 2 weeks and the expression profile of these cells were same as those grown on HAM which served as the control (Baradaran-Rafii et al., 2015).

The semisynthetic gelatine methacrylate (GelMA) is being increasingly used in tissue engineering due to its superior mechanical properties, tunable surface characteristics, and high biocompatibility. Kilic Bektas and Hasirci (2018) have reported on the viability of human keratocytes loaded into GelMA hydrogels. These cells were found to synthesize collagen type I and V, decorin, and biglycan, confirming their functionality in vitro. Rizwan et al. (2017) have cross-linked GelMA to form GelMA+, the surface of which was patterned and proved to be an efficient substrate for in vitro culture of human corneal endothelial cells. GelMA currently finds wide application as an ink for 3D printing of cells (Rizwan et al., 2017).

11.4.2 Synthetic biomaterials

Synthetic biomaterials are also studied extensively in corneal tissue engineering. Synthetic polymers are used in ocular therapy as a substitute for cornea or as a substrate for corneal cells. Zhi et al. has reviewed the various types of synthetic polymers that are related to corneal tissue engineering (Zhi et al., 2018). Some of the popular biomaterials are detailed in the following sections.

11.4.2.1 Polyvinyl alcohol

Polyvinyl alcohol is used for ophthalmic applications, such as preparing contact lenses and eye drops, based on its high degree of swelling and biocompatibility (Hyon et al., 1994).It has been also been proposed as a biomaterial scaffold for corneal tissue engineering. Sayed et al. proposed a copolymer of synthetic polymer PVA and the natural polymer CS as an alternative to AM for human limbal epithelial cell delivery (Seyed and Vijayaraghavan, 2018). They cross-linked a blend of PVA−CS mixture using with EDC and 2 NHS and studied the cytocompatibility in vitro. In another study, Wu et al. reported aligned nanofibrous electrospun scaffold fabricated using PVA−collagen composite for application in tissue-engineered cornea (Wu et al., 2018).

11.4.2.2 Poly(2-hydroxyethyl methacrylate)

As early as 1949, PMMA has been proposed as a biomaterial for intraocular lens and is now widely used across the world. However, being hydrophilic, it has been modified and used as a substrate for cell growth for use as corneal grafts. Zainuddin et al. modified the surface of PHEMA hydrogel into a cell-adhesive surface by atom transfer radical polymerization (ATRP) grafting (Zainuddin et al., 2008). Human corneal epithelial cells adhered and spread on the surface modified PMMA. In another study, Xiang et al. proposed a structurally and mechanically effective artificial cornea by modifying the surface of PMMA into a T-style design of a keratoprosthesis (Xiang et al., 2015). In vivo implantation further confirmed the suitability of T-style keratoprosthesis.

11.4.2.3 Polyethylene glycol diacrylate

Polyethylene glycol diacrylate (PEGDA) is a polymer that has been studied for potential use as an alternative to human corneal limbus. Ortega developed an ex vivo model as an initial step toward developing an implantable limbus (Ortega et al., 2013a,b) using microstereolithograpy to study corneal stem cell function within a niche environment and ultimately to develop an implantable limbus for future clinical use. Microstereolithography was used for the fabrication of PEGDA-based rings on a macroscopic (1.2 cm) scale containing unique microfeatures (pockets), which were then modified with fibronectin to promote cell adhesion (Yanez-Soto et al., 2013).

11.4.2.4 Poly(lactide-co-glycolide)

Synthetic polymers are used in clinical applications either alone or combined with biological polymers because of the advantages of their mechanical properties over natural materials. Poly(lactide-*co*-glycolide) (PLGA) is a material identified for clinical application and approved by regulatory agencies such as the Food and Drugs Administration of the United States. Electrospun nanofibrous scaffolds of PLGA have been used as cell carrier for corneal tissue engineering. It has been reported that limbal epithelial cells seeded on both sides of nanofibrous mat form multilayered structures and are found to degrade within 2 weeks. Hence, cell-seeded PLGA scaffolds are suggested as an alternate to AM, the commonly used cell carrier in corneal tissue engineering (Deshpande et al., 2010). PLGA has also been shown to be a good substrate for corneal endothelial cell adhesion. When coated with collagen, PLGA sheets enhanced endothelial cell adhesion with characteristic morphology, proliferation, and gene expression (Kim et al., 2017).

11.4.2.5 Thermoresponsive polymers

The development of smart polymers and their subsequent use in biomedical applications have made a significant impact in the field of drug delivery, tissue engineering, and cell-based therapeutics. Smart polymers or stimuli responsive polymers change their properties in response to variations internal or external stimuli such as pH, enzymes, electricity, temperature, light, magnetic field, and ultrasound. Hence, to a certain degree, these polymers can be controlled (Ashammakhi and Kaarela, 2017). To date, the majority of the published reports are on the use of pH and temperature responsive smart materials. Pluronics and poly(N-isopropylacrylamide) (NIPAAm) are most commonly used temperature/thermoresponsive polymer, while poly(2-propylacrylic acid) and CS/polyethyleneimine blend are pH-responsive polymers (Khan and Tanaka, 2017).

Poly(N-isopropylacrylamide)

Thermoresponsive polymers and their derivatives exhibit reversible hydrophilic—hydrophobic transition and are widely used as cell culture substrates. Among them, NIPAAm, with a lower critical solution temperature (LCST $\sim$ 32°C), is the mostly studied system. When the temperature is above the LCST, the polymer surface is hydrophobic and facilitates cell adhesion. At temperatures below the LCST, the surface becomes hydrophilic, and the polymer chains extend out and facilitate cell detachment from the substrate without the use of enzymes or harmful chemicals. The cell—cell and cell—ECM interactions are maintained enabling detachment of intact cell sheets (Dai et al., 2016).

NIPAAm can be grafted onto tissue culture polystyrene dishes (TCPS) or glass by various techniques such as electron beam (EB)—induced polymerization, ATRP, reversible addition-fragmentation chain transfer polymerization, block copolymer coating via spin coating, photopolymerization though visible light using photoinitiator, and UV irradiation—induced cross-linking (Nagase et al., 2018). The thickness of a grafted thermoresponsive polymer layer plays a role in cell adhesion and detachment and how efficient a cell sheet can be obtained. In fact, cell adhesion and detachment were better modulated when the polymer coating was thin at approximately 15.5 nm. The type of substrate on which grafting was done also played an indirect role. In fact, relatively thick PIPAAm layer on TCPS compared with that on glass substrates can dehydrate and adhere cells (Akiyama et al., 2004; Fukumori et al., 2010).

The use of thermoresponsive polymers facilitate the generation of intact cell sheets without any adherent carrier or scaffolding biomaterial and finds much application in corneal tissue engineering. The absence of carriers overrides the problems of transparency, immune rejection, and degradation of the biomaterial. The pioneering works (Nishida et al., 2004) demonstrated the use of PNiPAAm-grafted cell culture surfaces to generate compact mulitlayered rabbit limbal epithelial cell sheets. The cells were expanded in vitro on the thermoresponsive substrate, and intact cell sheets were harvested after 2 weeks by reducing the temperature to 20°C. Using a poly(vinylidene difluoride) membrane, these cell sheets attached to an exposed rabbit corneal stroma within minutes without suturing, and the corneal epithelium was successfully reconstructed. Hayashi et al. (2017) reported on the use of human pluripotent stem cells to derive hiCECs, which were subsequently expanded on the thermoresponsive substrate UpCell (Hayashi et al., 2017). The cells were cultured to confluency and were permitted to undergo stratification. Intact cell sheets were harvested by lowering the temperature to 20°C. Patterned thermoresponsive culture dishes have also been used for generation of cocultured cell sheets containing several cell types or for generation of aligned cell sheets, where cells are guided to align with a specific orientation. Micropatterning can be done through various techniques such as photolithography, photomasking, and microcontact printing (Elloumi-Hannachi et al., 2010). Nara et al. (2015) had utilized the direct-write assembly (DWA) to develop microperiodic parallel patterns of silk-Poly(N-isopropylacrylamide) (polyNIPA) and gelatin-polyNIPA (Nara et al., 2015). This patterned surface was used as a substrate for the culture of corneal keratocytes. Narrowing down of the pattern dimension increases orientation of the cells; however, efficiency of retrieval of intact cell sheets was low.

Cell sheet technology facilities the harvest of corneal epithelial cells sheets without the need for any protease treatment. Stable transplantation of these epithelial cell sheets without the need of sutures is an added advantage of the technique. The thermally detached sheets are fragile, and the use of compatible delivery methods is important for the surgical manipulation of sheet grafts.

11.4.2.6 Poly N-isopropylacrylamide-co-glycidylmethacrylate

NIPAAm can be copolymerized with glycidylmethacrylate to form poly N-isopropylacrylamide-co-glycidylmethacrylate (NGMA). Our laboratory has reported on the synthesis of NGMA using free radical polymerization. The copolymer so formed has an LCST of 31°C. Thermoresponsive NGMA substrates are generated by the method of solution casting, wherein the polymer is coated onto TCPS dishes, oven dried overnight at 60°C, washed with ice cold distilled water to remove unbound polymer, and air-dried. The culture dishes were sterilized using ethylene oxide. These NGMA-coated dishes were used for the in vitro expansion and generation of rabbit limbal epithelial cell sheets (Nithya et al., 2011). These cell sheets were harvested by simply lowering the temperature to below the LCST of NGMA. The cell—cell and cell—ECM junctions were intact as evidenced by positive staining of ZO-1. The feasibility of using the NGMA-coated dishes as thermoresponsive substrates for the generation of intact rabbit corneal endothelial cells was also reported (Madathil et al., 2014). The harvested endothelial cell sheets exhibited intact morphology and characteristic cobble stone morphology during scanning electron microscopic evaluation of the retrieved sheets. The positive expression for the markers aquaporin 1, collagen IV, Na+-K+ ATPase, and Fetal Liver Kinase 1 (FLK-1) by reverse transcription polymerase chain reaction analysis confirms the presence of functionally active endothelial cells in the cell sheet.

A foremost aim of corneal tissue engineering is to develop a full thickness construct that would mimic the natural cornea. The construct would ideally include the prime three corneal cell types i.e., the corneal epithelial, stromal, and endothelial cells. Typically, tissue-engineering strategies employ a "top-down" approach, in which cells are seeded on a polymeric scaffold on which they populate and produce their own appropriate ECM with the help of perfusion, growth factors, and/or mechanical stimulation. Recently in engineering corneal tissues, the "bottom-up" approach is being promoted, which focuses on developing modular microtissues with repeated functional units that have specific microarchitectural features (Connon, 2015). As an initial attempt, our laboratory had reported on the use of thermoresponsive NGMA substrates to develop a sequentially layered corneal construct. Rabbit corneal endothelial cells were cultured onto the NGMA dishes and allowed to grow to confluency, following which they were seeded with rabbit stromal cells to obtain a coculture. As endothelial cells are contact inhibited, they remain as a monolayer on which the stromal cells are allowed to grow to form multilayers. The corneal epithelial cell sheet retrieved from another NGMA dish served as the final layer over the endothelial—stromal bilayered construct. The layered corneal construct was harvested from the NGMA dish by lowering the temperature to below the LCST of NGMA (Ramesh et al., 2014). The use of micropatterned thermoresponsive substrates facilitates controlled cell orientation (Nagase et al., 2018). This technique would have potential for development of engineered corneal stroma with aligned cells that mimic the microarchitecture of the native cornea.

NGMA has epoxy rings that can be opened for the incorporation of biomolecules and would aid in the development of functionalized thermoresponsive substrates. This is one of the key research areas of the laboratory.

11.5 Translational and clinical perspective

The above sections review the status on research and development of stem cell—based corneal surface reconstruction methods experimented and performed in laboratory and clinical settings. While highlighting the different substrates and various cell types that are promising for translation of proposed methods to corneal surface reconstruction in clinics, it is important to look through a clinician's perspective. This substantiates the need for the involvement of clinician in the development of tissue-engineered products and procedures. While proposing tissue engineering strategies, it is necessary to develop a technology that can be immediately applied without much preoperative procedure, easy to perform minimally invasive surgical requirements, minimum postoperative medical care, and above all without compromising the efficient recovery of the damaged cornea. SLET was an important development in the field that is successfully practiced, but treatment using allogenic systems is yet to be tailored to clinicians and patients requirements. The potential health risks associated with ex vivo cultivated limbal stem cell transplantation are the process and the transplantation techniques. The process involves cultivation of cells that uses allogenic and/or xenogenic components that pose risks of infection or toxicity. Variation in the protocols used by different research groups also needs uniformity. Xenogenic-free culture methods in a Good Manufacturing Practice facility would be the mandatory requirement that has to be implemented at early stage globally (Behaegel et al., 2017). Donor shortage or availability of a healthy donor cornea is one of the major issues faced by clinicians. The stem cells from limbus, conjunctivolimbal graft, and keratolimbal grafts have improved the vision restoration. There is a major drive to identify other autologous nonocular cell sources to address this issue. Alternate stem cell sources have shown the required functionality of the graft and hence have got a clinical relevance in ocular surface regenerative therapy. This will offer clinicians with more therapeutic opportunities to avoid issues related to immune rejection or requirement of immune suppression. Tissue engineering—based methods are very instrumental in this field of corneal surface reconstruction by improving and devising therapies viable to translation for clinical settings.

References

Agrawal, V.B., Tsai, R.J., 2003. Corneal epithelial wound healing. Indian J. Ophthalmol. 51 (1), 5−15.

Akiyama, Y., Kikuchi, A., et al., 2004. Ultrathin poly(N-isopropylacrylamide) grafted layer on polystyrene surfaces for cell adhesion/detachment control. Langmuir 20 (13), 5506−5511.

Amescua, G., Atallah, M., et al., 2014. Modified simple limbal epithelial transplantation using cryopreserved amniotic membrane for unilateral limbal stem cell deficiency. Am. J. Ophthalmol. 158, 469−475.

Ashammakhi, N., Kaarela, O., 2017. Stimuli-responsive biomaterials: next wave. J. Craniofac. Surg. 28 (7), 1647−1648.

Bakhtiari, P., Djalilian, A., 2010. Update on limbal stem cell transplantation. Middle East Afr. J. Ophthalmol. 17 (1), 9−14.
Baradaran-Rafii, A., Biazar, E., et al., 2015. Cellular response of limbal stem cells on PHBV/gelatin nanofibrous scaffold for ocular epithelial regeneration. Int. J. Polym. Mater. Polym. Biomater 64 (17), 879−887.
Basu, S., Sureka, S.P., et al., 2016. Simple limbal epithelial transplantation: long-term clinical outcomes in 125 cases of unilateral chronic ocular surface burns. Ophthalmology 123 (5), 1000−1010.
Behaegel, J., Ni Dhubhghaill, S., et al., 2017. Safety of cultivated limbal epithelial stem cell transplantation for human corneal regeneration. Stem Cell. Int. 2017, 6978253.
Bessou-Touya, S., Picardo, M., et al., 1998. Chimeric human epidermal reconstructs to study the role of melanocytes and keratinocytes in pigmentation and photoprotection. J. Investig. Dermatol. 111 (6), 1103−1108.
Biazar, E., Baradaran-Rafii, A., et al., 2015. Oriented nanofibrous silk as a natural scaffold for ocular epithelial regeneration. J. Biomater. Sci. Polym. Ed. 26 (16), 1139−1151.
Blazejewska, E.A., Schlotzer-Schrehardt, U., et al., 2009. Corneal limbal microenvironment can induce transdifferentiation of hair follicle stem cells into corneal epithelial-like cells. Stem Cell. 27 (3), 642−652.
Bourne, R.R.A., Flaxman, S.R., et al., 2017. Magnitude, temporal trends, and projections of the global prevalence of blindness and distance and near vision impairment: a systematic review and meta-analysis. Lancet Glob. Health 5 (9), e888−e897.
Branch, M.J., Hashmani, K., et al., 2012. Mesenchymal stem cells in the human corneal limbal stroma. Invest. Ophthalmol. Vis. Sci 53 (9), 5109−5116.
Chae, J.J., Ambrose, W.M., et al., 2015. Regeneration of corneal epithelium utilizing a collagen vitrigel membrane in rabbit models for corneal stromal wound and limbal stem cell deficiency. Acta Ophthalmol 93 (1), e57−66.
Chakraborty, A., Dutta, J., et al., 2012. Effect of cord blood serum on ex vivo human limbal epithelial cell culture. J. Ocul. Biol. Dis. Infor 5 (3−4), 77−82.
Chen, Z., You, J., Liu, X., Cooper, S., Hodge, C., Sutton, G., Crook, J.M., Wallace, G.G., 2018 Mar 6. Biomaterials for corneal bioengineering. Biomed Mater 13 (3), 032002. https://doi.org/10.1088/1748-605X/aa92d2.
Connon, C.J., 2015. Approaches to corneal tissue engineering: top-down or bottom-up? Procedia Eng 110, 15−20.
Cotsarelis, G., Cheng, S.Z., et al., 1989. Existence of slow-cycling limbal epithelial basal cells that can be preferentially stimulated to proliferate: implications on epithelial stem cells. Cell 57 (2), 201−209.
Dai, Z., Shu, Y., et al., 2016. Effects of culture substrate made of poly(N-isopropylacrylamide-co-acrylic acid) microgels on osteogenic differentiation of mesenchymal stem cells. Molecules 21 (9).
Davis, J., 1910. Skin transplantation with a review of 550 cases at the Johns Hopkins Hospital. Johns Hopkins Hospital Rep 15, 307−310.
de Araujo, A.L., Gomes, J.A., 2015. Corneal stem cells and tissue engineering: current advances and future perspectives. World J. Stem Cell. 7 (5), 806−814.
De Miguel, M.P., Arnalich Montiel, F., et al., 2009. Epiblast-derived stem cells in embryonic and adult tissues. Int. J. Dev. Biol. 53 (8−10), 1529−1540.
de RÖ, T.A., 1940. Plastic repair of conjunctival defects with fetal membranes. Arch. Ophthalmol. 23 (3), 522−525.
DeBari, M.K., Abbott, R.D., 2018. Microscopic considerations for optimizing silk biomaterials. Wiley Interdiscip. Rev. Nanomed. Nanobiotechnol e1534.

DelMonte, D.W., Kim, T., 2011. Anatomy and physiology of the cornea. J. Cataract Refract. Surg. 37 (3), 588−598.

Deshpande, P., McKean, R., et al., 2010. Using poly(lactide-co-glycolide) electrospun scaffolds to deliver cultured epithelial cells to the cornea. Regen. Med. 5 (3), 395−401.

Dobrowolski, D., Orzechowska-Wylegala, B., et al., 2015. Cultivated oral mucosa epithelium in ocular surface reconstruction in aniridia patients. BioMed Res. Int 2015, 7.

Dong, Y., Roos, M., et al., 1994. Differential expression of two gap junction proteins in corneal epithelium. Eur. J. Cell Biol. 64 (1), 95−100.

Dravida, S., Gaddipati, S., et al., 2008. A biomimetic scaffold for culturing limbal stem cells: a promising alternative for clinical transplantation. J. Tissue Eng. Regenerat. Med. 2 (5), 263−271.

Du, Y., Chen, J., et al., 2003. Functional reconstruction of rabbit corneal epithelium by human limbal cells cultured on amniotic membrane. Mol. Vis. 9, 635−643.

Dua, H.S., Azuara-Blanco, A., 2000. Limbal stem cells of the corneal epithelium. Surv. Ophthalmol. 44 (5), 415−425.

Dua, H.S., Forrester, J.V., 1990. The corneoscleral limbus in human corneal epithelial wound healing. Am. J. Ophthalmol. 110 (6), 646−656.

Dua, H.S., King, A.J., et al., 2001. A new classification of ocular surface burns. Br. J. Ophthalmol. 85 (11), 1379−1383.

Dua, H.S., Saini, J.S., et al., 2000. Limbal stem cell deficiency: concept, aetiology, clinical presentation, diagnosis and management. Indian J. Ophthalmol. 48 (2), 83−92.

Dua, H.S., Shanmuganathan, V.A., et al., 2005. Limbal epithelial crypts: a novel anatomical structure and a putative limbal stem cell niche. Br. J. Ophthalmol. 89 (5), 529−532.

Ebrahimi, M., Taghi-Abadi, E., et al., 2009. Limbal stem cells in review. J. Ophthalmic. Vis. Res. 4 (1), 40−58.

Echevarria, T.J., Di Girolamo, N., 2011. Tissue-regenerating, vision-restoring corneal epithelial stem cells. Stem Cell Rev 7 (2), 256−268.

Elloumi-Hannachi, I., Yamato, M., et al., 2010. Cell sheet engineering: a unique nanotechnology for scaffold-free tissue reconstruction with clinical applications in regenerative medicine. J. Intern. Med. 267 (1), 54−70.

Fatima, A., Sangwan, V.S., et al., 2006. Technique of cultivating limbal derived corneal epithelium on human amniotic membrane for clinical transplantation. J. Postgrad. Med. 52 (4), 257−261.

Fernandes, M., Sangwan, V.S., et al., 2004. Limbal stem cell transplantation. Indian J. Ophthalmol. 52 (1), 5−22.

Fitton, J.H., Dalton, B.A., et al., 1998. Surface topography can interfere with epithelial tissue migration. J. Biomed. Mater. Res. 42 (2), 245−257.

Fong, C.Y., Chak, L.L., et al., 2011. Human Wharton's jelly stem cells have unique transcriptome profiles compared to human embryonic stem cells and other mesenchymal stem cells. Stem Cell Rev 7 (1), 1−16.

Fukumori, K., Akiyama, Y., et al., 2010. Characterization of ultra-thin temperature-responsive polymer layer and its polymer thickness dependency on cell attachment/detachment properties. Macromol. Biosci. 10 (10), 1117−1129.

Galindo, S., Herreras, J.M., et al., 2017. Therapeutic effect of human adipose tissue-derived mesenchymal stem cells in experimental corneal failure due to limbal stem cell niche damage. Stem Cells 35 (10), 2160−2174.

Gao, F., Chiu, S.M., et al., 2016. Mesenchymal stem cells and immunomodulation: current status and future prospects. Cell Death Dis 21 (7), 327.

Garzon, I., Martin-Piedra, M.A., et al., 2014. Generation of a biomimetic human artificial cornea model using Wharton's jelly mesenchymal stem cells. Invest. Ophthalmol. Vis. Sci 55 (7), 4073−4083.

German, M.J., Pollock, H.M., et al., 2006. Characterization of putative stem cell populations in the cornea using synchrotron infrared microspectroscopy. Invest. Ophthalmol. Vis. Sci 47 (6), 2417−2421.

Ghezzi, C.E., Rnjak-Kovacina, J., et al., 2015. Corneal tissue engineering: recent advances and future perspectives. Tissue Eng. B Rev. 21 (3), 278−287.

Gil, E.S., Mandal, B.B., et al., 2010. Helicoidal multi-lamellar features of RGD-functionalized silk biomaterials for corneal tissue engineering. Biomaterials 31 (34), 8953−8963.

Goldberg, M.F., Bron, A.J., 1982. Limbal palisades of Vogt. Trans. Am. Ophthalmol. Soc. 80, 155−171.

Gomes, J.A., Geraldes Monteiro, B., et al., 2010. Corneal reconstruction with tissue-engineered cell sheets composed of human immature dental pulp stem cells. Invest. Ophthalmol. Vis. Sci 51 (3), 1408−1414.

Gosselin, E.A., Torregrosa, T., et al., 2018. Multi-layered silk film coculture system for human corneal epithelial and stromal stem cells. J. Tissue Eng. Regenerat. Med. 12 (1), 285−295.

Grueterich, M., Espana, E.M., et al., 2003. Ex vivo expansion of limbal epithelial stem cells: amniotic membrane serving as a stem cell niche. Surv. Ophthalmol. 48 (6), 631−646.

Gruss, J.S., Jirsch, D.W., 1978. Human amniotic membrane: a versatile wound dressing. Can. Med. Assoc. J. 118 (10), 1237−1246.

Gu, S., Xing, C., et al., 2009. Differentiation of rabbit bone marrow mesenchymal stem cells into corneal epithelial cells in vivo and ex vivo. Mol. Vis. 15, 99−107.

Gupta, N., Tandon, R., et al., 2013. Burden of corneal blindness in India. Indian J. Community Med. 38 (4), 198−206.

Haagdorens, M., Van Acker, S.I., et al., 2016. Limbal stem cell deficiency: current treatment options and emerging therapies. Stem Cells Int 2016, 9798374−9798374.

Hashmani, K., Branch, M.J., et al., 2013. Characterization of corneal stromal stem cells with the potential for epithelial transdifferentiation. Stem Cell Res. Ther. 4 (3), 75.

Hayashi, R., Ishikawa, Y., et al., 2012. Generation of corneal epithelial cells from induced pluripotent stem cells derived from human dermal fibroblast and corneal limbal epithelium. PLoS One 7 (9), e45435.

Hayashi, R., Ishikawa, Y., et al., 2017. Coordinated generation of multiple ocular-like cell lineages and fabrication of functional corneal epithelial cell sheets from human iPS cells. Nat. Protoc. 12 (4), 683−696.

He, H., Yiu, S.C., 2014. Stem cell-based therapy for treating limbal stem cells deficiency: a review of different strategies. Saudi J. Ophthalmol 28 (3), 188−194.

Hertsenberg, A.J., Shojaati, G., et al., 2017. Corneal stromal stem cells reduce corneal scarring by mediating neutrophil infiltration after wounding. PLoS One 12 (3), e0171712.

Holden, B.A., Mertz, G.W., 1984. Critical oxygen levels to avoid corneal edema for daily and extended wear contact lenses. Invest. Ophthalmol. Vis. Sci 25 (10), 1161−1167.

Holland, E.J., Schwartz, G.S., 2004. The Paton lecture: ocular surface transplantation: 10 years' experience. Cornea 23 (5), 425−431.

Homma, R., Yoshikawa, H., et al., 2004. Induction of epithelial progenitors in vitro from mouse embryonic stem cells and application for reconstruction of damaged cornea in mice. Invest. Ophthalmol. Vis. Sci 45 (12), 4320−4326.

Hyon, S., Cha, W., et al., 1994. Poly(vinyl alcohol) hydrogels as soft contact lens material. J. Biomater. Sci. Polym. Ed. 5 (5), 397−406.

Jackson, L., Jones, D.R., et al., 2007. Adult mesenchymal stem cells: differentiation potential and therapeutic applications. J. Postgrad. Med. 53 (2), 121−127.

Joubert, R., Daniel, E., et al., 2017. Retinoic acid engineered amniotic membrane used as graft or homogenate: positive effects on corneal alkali burns. Invest. Ophthalmol. Vis. Sci 58 (9), 3513−3518.

Karring, H., Thogersen, I.B., et al., 2004. Proteomic analysis of the soluble fraction from human corneal fibroblasts with reference to ocular transparency. Mol. Cell. Proteom. 3 (7), 660−674.

Kasper, M., Stosiek, P., et al., 1992. Cytokeratin and vimentin heterogeneity in human cornea. Acta Histochem 93 (2), 371−381.

Khan, F., Tanaka, M., 2017. Designing smart biomaterials for tissue engineering. Int. J. Mol. Sci. 19 (1).

Kilic Bektas, C., Hasirci, V., 2018. Mimicking corneal stroma using keratocyte-loaded photopolymerizable methacrylated gelatin hydrogels. J. Tissue Eng. Regenerat. Med. 12 (4), e1899−e1910.

Kim, E.Y., Tripathy, N., et al., 2017. Collagen type I−PLGA film as an efficient substratum for corneal endothelial cells regeneration. J. Tissue Eng. Regenerat. Med. 11 (9), 2471−2478.

Klyce, S.D., 1972. Electrical profiles in the corneal epithelium. J. Physiol. 226 (2), 407−429.

Kong, B., Mi, S., 2016. Electrospun scaffolds for corneal tissue engineering: a review. Materials 9 (8), 614.

Koob, T.J., Rennert, R., et al., 2013. Biological properties of dehydrated human amnion/chorion composite graft: implications for chronic wound healing. Int. Wound J. 10 (5), 493−500.

Kruse, F.E., 1994. Stem cells and corneal epithelial regeneration. Eye 8 (Pt 2), 170−183.

Kulkarni, B.B., Tighe, P.J., et al., 2010. Comparative transcriptional profiling of the limbal epithelial crypt demonstrates its putative stem cell niche characteristics. BMC Genomics 11, 526.

Kurpakus, M.A., Stock, E.L., et al., 1990. Expression of the 55-kD/64-kD corneal keratins in ocular surface epithelium. Invest. Ophthalmol. Vis. Sci 31 (3), 448−456.

Lai, J.Y., 2013. Corneal stromal cell growth on gelatin/chondroitin sulfate scaffolds modified at different NHS/EDC molar ratios. Int. J. Mol. Sci. 14 (1), 2036−2055.

Lavker, R.M., Dong, G., et al., 1991. Relative proliferative rates of limbal and corneal epithelia. Implications of corneal epithelial migration, circadian rhythm, and suprabasally located DNA-synthesizing keratinocytes. Invest. Ophthalmol. Vis. Sci 32 (6), 1864−1875.

Law, S., Chaudhuri, S., 2013. Mesenchymal stem cell and regenerative medicine: regeneration versus immunomodulatory challenges. Am. J. Stem Cells 2 (1), 22−38.

Lawrence, B.D., Marchant, J.K., et al., 2009. Silk film biomaterials for cornea tissue engineering. Biomaterials 30 (7), 1299−1308.

Li, F., Zhao, S.Z., 2014. Mesenchymal stem cells: potential role in corneal wound repair and transplantation. World J. Stem Cell. 6 (3), 296−304.

Liu, Y., Gan, L., Carlsson, D.J., Fagerholm, P., Lagali, N., Watsky, M.A., Munger, R., Hodge, W.G., Priest, D., Griffith, M., 2006 May. A simple, cross-linked collagen tissue substitute for corneal implantation. Invest Ophthalmol Vis Sci 47 (5), 1869−1875.

Li, W., Hayashida, Y., et al., 2007. Niche regulation of corneal epithelial stem cells at the limbus. Cell Res 17 (1), 26−36.

Liang, L., Sheha, H., et al., 2009. Limbal stem cell transplantation: new progresses and challenges. Eye 23 (10), 1946−1953.

Liu, W., Deng, C., et al., 2009. Collagen-phosphorylcholine interpenetrating network hydrogels as corneal substitutes. Biomaterials 30 (8), 1551−1559.

Liu, Y., Ren, L., et al., 2013. Crosslinked collagen—gelatin—hyaluronic acid biomimetic film for cornea tissue engineering applications. Mater. Sci. Eng. C 33 (1), 196—201.
Ma, Y., Xu, Y., et al., 2006. Reconstruction of chemically burned rat corneal surface by bone marrow-derived human mesenchymal stem cells. Stem Cells 24 (2), 315—321.
Madathil, B.K., Kumar, P.R., et al., 2014. N-isopropylacrylamide-co-glycidylmethacrylate as a thermoresponsive substrate for corneal endothelial cell sheet engineering. Biomed. Res. Int. 2014, 450672.
Majo, F., Rochat, A., et al., 2008. Oligopotent stem cells are distributed throughout the mammalian ocular surface. Nature 456 (7219), 250—254.
Malhotra, C., Jain, A.K., 2014. Human amniotic membrane transplantation: different modalities of its use in ophthalmology. World J. Transplant. 4 (2), 111—121.
Mata, A. d. l., Nieto-Miguel, T., et al., 2013. Chitosan—gelatin biopolymers as carrier substrata for limbal epithelial stem cells. J. Mater. Sci. Mater. Med. 24, 2819—2829.
McLaughlin, C.R., Acosta, M.C., et al., 2010. Regeneration of functional nerves within full thickness collagen-phosphorylcholine corneal substitute implants in Guinea pigs. Biomaterials 31 (10), 2770—2778.
Meek, K.M., Knupp, C., 2015. Corneal structure and transparency. Prog. Retin. Eye Res. 49, 1—16.
Meyer-Blazejewska, E.A., Call, M.K., et al., 2011. From hair to cornea: toward the therapeutic use of hair follicle-derived stem cells in the treatment of limbal stem cell deficiency. Stem Cell. 29 (1), 57—66.
Mikhailova, A., Ilmarinen, T., et al., 2014. Small-molecule induction promotes corneal epithelial cell differentiation from human induced pluripotent stem cells. Stem Cell Rep 2 (2), 219—231.
Mimura, T., Chang, J.H., et al., 2011. MT1-MMP cleavage of the antiangiogenic proteoglycan decorin: role in corneal angiogenesis. Cornea 30 (Suppl. 1), S45—49.
Mistriotis, P., Andreadis, S.T., 2013. Hair follicle: a novel source of multipotent stem cells for tissue engineering and regenerative medicine. Tissue Eng. B Rev 19 (4), 265—278.
Mitropoulos, A.N., Marelli, B., et al., 2015. Transparent, nanostructured silk fibroin hydrogels with tunable mechanical properties. ACS Biomater. Sci. Eng. 1 (10), 964—970.
Monteiro, B.G., Serafim, R.C., et al., 2009. Human immature dental pulp stem cells share key characteristic features with limbal stem cells. Cell Prolif 42 (5), 587—594.
Mukhopadhyay, M., Gorivodsky, M., et al., 2006. Dkk2 plays an essential role in the corneal fate of the ocular surface epithelium. Development 133 (11), 2149—2154.
Nagase, K., Yamato, M., et al., 2018. Poly(N-isopropylacrylamide)-based thermoresponsive surfaces provide new types of biomedical applications. Biomaterials 153, 27—48.
Nakamura, T., Endo, K., et al., 2003. The successful culture and autologous transplantation of rabbit oral mucosal epithelial cells on amniotic membrane. Invest. Ophthalmol. Vis. Sci 44 (1), 106—116.
Nakamura, T., Kinoshita, S., 2003. Ocular surface reconstruction using cultivated mucosal epithelial stem cells. Cornea 22 (7 Suppl. 1), S75—80.
Nara, S., Chameettachal, S., et al., 2015. Strategies for faster detachment of corneal cell sheet using micropatterned thermoresponsive matrices. J. Mater. Chem. B 3 (20), 4155—4169.
Niknejad, H., Peirovi, H., et al., 2008. Properties of the amniotic membrane for potential use in tissue engineering. Eur. Cells Mater. 15, 88—99.
Nishida, K., Yamato, M., et al., 2004. Corneal reconstruction with tissue-engineered cell sheets composed of autologous oral mucosal epithelium. N. Engl. J. Med. 351 (12), 1187—1196.

Nithya, J., Kumar, P.R., et al., 2011. Intelligent thermoresponsive substrate from modified overhead projection sheet as a tool for construction and support of cell sheets in vitro. Tissue Eng. C Methods 17 (2), 181–191.

Ordonez, P., Di Girolamo, N., 2012. Limbal epithelial stem cells: role of the niche microenvironment. Stem Cell. 30 (2), 100–107.

Ortega, I., Deshpande, P., et al., 2013a. Development of a microfabricated artificial limbus with micropockets for cell delivery to the cornea. Biofabrication 5 (2), 1758–5082.

Ortega, I., Ryan, A.J., et al., 2013b. Combined microfabrication and electrospinning to produce 3-D architectures for corneal repair. Acta Biomater 9 (3), 5511–5520.

Ouyang, H., Xue, Y., et al., 2014. WNT7A and PAX6 define corneal epithelium homeostasis and pathogenesis. Nature 511 (7509), 358–361.

Pajoohesh-Ganji, A., Stepp, M.A., 2005. In search of markers for the stem cells of the corneal epithelium. Biol. Cell 97 (4), 265–276.

Parenteau-Bareil, R., Gauvin, R., Berthod, F., 2010. Collagen-Based Biomaterials for Tissue Engineering Applications. Materials 3, 1863–1887.

Pellegrini, G., Dellambra, E., et al., 2001. p63 identifies keratinocyte stem cells. Proc. Natl. Acad. Sci. U. S. A. 98 (6), 3156–3161.

Pellegrini, G., Golisano, O., et al., 1999. Location and clonal analysis of stem cells and their differentiated progeny in the human ocular surface. J. Cell Biol. 145 (4), 769–782.

Pellegrini, G., Traverso, C.E., et al., 1997. Long-term restoration of damaged corneal surfaces with autologous cultivated corneal epithelium. Lancet 349 (9057), 990–993.

Phu, D., Wray, L.S., et al., 2011. Effect of substrate composition and alignment on corneal cell phenotype. Tissue Eng. 17 (5–6), 799–807.

Pinnamaneni, N., Funderburgh, J.L., 2012. Concise review: stem cells in the corneal stroma. Stem Cell. 30 (6), 1059–1063.

Prospero Ponce, C.M., Rocha, K.M., et al., 2009. Central and peripheral corneal thickness measured with optical coherence tomography, Scheimpflug imaging, and ultrasound pachymetry in normal, keratoconus-suspect, and post-laser in situ keratomileusis eyes. J. Cataract Refract. Surg. 35 (6), 1055–1062.

Proulx, S., d'Arc Uwamaliya, J., et al., 2010. Reconstruction of a human cornea by the self-assembly approach of tissue engineering using the three native cell types. Mol. Vis. 16, 2192–2201.

Rahman, I., Said, D.G., et al., 2009. Amniotic membrane in ophthalmology: indications and limitations. Eye 23 (10), 1954–1961.

Ramesh, K.N., Madathil, B.K., et al., 2014. A bioengineered sequentially layered in vitro corneal construct. Int. J. Latest Res. Sci. Technol 3 (6), 155–163.

Rizwan, M., Peh, G.S.L., et al., 2017. Sequentially-crosslinked bioactive hydrogels as nanopatterned substrates with customizable stiffness and degradation for corneal tissue engineering applications. Biomaterials 120, 139–154.

Romer, L., Scheibel, T., 2008. The elaborate structure of spider silk: structure and function of a natural high performance fiber. Prion 2 (4), 154–161.

Ruberti, J.W., Zieske, J.D., 2008. Prelude to corneal tissue engineering – gaining control of collagen organization. Prog. Retin. Eye Res. 27 (5), 549–577.

Sangwan, V.S., Basu, S., et al., 2012. Simple limbal epithelial transplantation (SLET): a novel surgical technique for the treatment of unilateral limbal stem cell deficiency. Br. J. Ophthalmol. 96 (7), 931–934.

Sangwan, V.S., Matalia, H.P., et al., 2004. Amniotic membrane transplantation for reconstruction of corneal epithelial surface in cases of partial limbal stem cell deficiency. Indian J. Ophthalmol. 52 (4), 281–285.

Sareen, D., Saghizadeh, M., et al., 2014. Differentiation of human limbal-derived induced pluripotent stem cells into limbal-like epithelium. Stem Cells Transl. Med 3 (9), 1002−1012.
Satake, Y., Higa, K., et al., 2011. Long-term outcome of cultivated oral mucosal epithelial sheet transplantation in treatment of total limbal stem cell deficiency. Ophthalmology 118 (8), 1524−1530.
Schermer, A., Galvin, S., et al., 1986. Differentiation-related expression of a major 64K corneal keratin in vivo and in culture suggests limbal location of corneal epithelial stem cells. J. Cell Biol. 103 (1), 49−62.
Schneider, M.R., Schmidt-Ullrich, R., et al., 2009. The hair follicle as a dynamic miniorgan. Curr. Biol. 19, R132−R142.
Sehic, A., Utheim, Ø.A., et al., 2015. Pre-clinical cell-based therapy for limbal stem cell deficiency. J. Funct. Biomater. 6 (3), 863−888.
Sejpal, K., Bakhtiari, P., et al., 2013. Presentation, diagnosis and management of limbal stem cell deficiency. Middle East Afr. J. Ophthalmol. 20 (1), 5−10.
Seyed, M.A., Vijayaraghavan, K., 2018. Physicochemical characterization and bioactivity of an improved chitosan scaffold cross. ARRB 24 (6), 1−16.
Shoulders, M.D., Raines, R.T., 2009. Collagen structure and stability. Annu Rev Biochem 78, 929−958. https://doi.org/10.1146/annurev.biochem.77.032207.120833.
Shahdadfar, A., Haug, K., et al., 2012. Ex vivo expanded autologous limbal epithelial cells on amniotic membrane using a culture medium with human serum as single supplement. Exp. Eye Res. 97 (1), 1−9.
Sharma, A., Coles, W.H., 1989. Kinetics of corneal epithelial maintenance and graft loss. A population balance model. Invest. Ophthalmol. Vis. Sci 30 (9), 1962−1971.
Sosnova, M., Bradl, M., et al., 2005. CD34+ corneal stromal cells are bone marrow-derived and express hemopoietic stem cell markers. Stem Cells 23 (4), 507−515.
Sotozono, C., Inatomi, T., et al., 2013. Visual improvement after cultivated oral mucosal epithelial transplantation. Ophthalmology 120 (1), 193−200.
Stepp, M.A., Zhu, L., et al., 1995. Localized distribution of alpha 9 integrin in the cornea and changes in expression during corneal epithelial cell differentiation. J. Histochem. Cytochem. 43 (4), 353−362.
Sun, J., Liu, W.-H., et al., 2018. Differentiation of rat adipose-derived mesenchymal stem cells into corneal-like epithelial cells driven by PAX6. Exp. Ther. Med 15 (2), 1424−1432.
Sun, T.T., Lavker, R.M., 2004. Corneal epithelial stem cells: past, present, and future. J. Investig. Dermatol. Symp. Proc. 9 (3), 202−207.
Sweeney, D.F., Xie, R.Z., et al., 1998. Nutritional requirements of the corneal epithelium and anterior stroma: clinical findings. Invest. Ophthalmol. Vis. Sci 39 (2), 284−291.
Tan, D.T., Ficker, L.A., et al., 1996. Limbal transplantation. Ophthalmology 103 (1), 29−36.
Thoft, R.A., Friend, J., 1983. The X, Y, Z hypothesis of corneal epithelial maintenance. Invest. Ophthalmol. Vis. Sci 24 (10), 1442−1443.
Tonsomboon, K., Oyen, M.L., 2013. Composite electrospun gelatin fiber-alginate gel scaffolds for mechanically robust tissue engineered cornea. J. Mech. Behav. Biomed. Mater. 21, 185−194.
Tonsomboon, K., Strange, D.G., et al., 2013. Gelatin nanofiber-reinforced alginate gel scaffolds for corneal tissue engineering. Conf. Proc. IEEE Eng. Med. Biol. Soc 2013, 6671−6674.
Tsai, R.J., Li, L.M., et al., 2000. Reconstruction of damaged corneas by transplantation of autologous limbal epithelial cells. N. Engl. J. Med. 343 (2), 86−93.
Utheim, T.P., Aass Utheim, O., et al., 2018. Concise review: altered versus unaltered amniotic membrane as a substrate for limbal epithelial cells. Stem Cells Transl. Med 7 (5), 415−427.

Vashist, P., Senjam, S.S., et al., 2017. Definition of blindness under national programme for control of blindness: do we need to revise it? Indian J. Ophthalmol. 65 (2), 92−96.

Wang, L., Ma, R., et al., 2015. Biocompatibility of helicoidal multilamellar arginine-glycine-aspartic acid-functionalized silk biomaterials in a rabbit corneal model. J. Biomed. Mater. Res. B Appl. Biomater. 103 (1), 204−211.

Waring 3rd, G.O., Bourne, W.M., et al., 1982. The corneal endothelium. Normal and pathologic structure and function. Ophthalmology 89 (6), 531−590.

Wu, Z., Kong, B., et al., 2018. Engineering of corneal tissue through an aligned PVA/collagen composite nanofibrous electrospun scaffold. Nanomaterials 8 (2), 124.

Xiang, J., Sun, J., et al., 2015. T-style keratoprosthesis based on surface-modified poly (2-hydroxyethyl methacrylate) hydrogel for cornea repairs. Mater Sci. Eng. C Mater. Biol. Appl 50, 274−285.

Yan, J., Qiang, L., et al., 2012. Effect of fiber alignment in electrospun scaffolds on keratocytes and corneal epithelial cells behavior. J. Biomed. Mater. Res. A 100 (2), 527−535.

Yanez-Soto, B., Liliensiek, S.J., et al., 2013. Biochemically and topographically engineered poly(ethylene glycol) diacrylate hydrogels with biomimetic characteristics as substrates for human corneal epithelial cells. J. Biomed. Mater. Res. A 101 (4), 1184−1194.

Yang, K., Jiang, Z., et al., 2009. Corneal epithelial-like transdifferentiation of hair follicle stem cells is mediated by pax6 and beta-catenin/Lef-1. Cell Biol. Int. 33 (8), 861−866.

Yang, X., Moldovan, N.I., et al., 2008. Reconstruction of damaged cornea by autologous transplantation of epidermal adult stem cells. Mol. Vis. 14, 1064−1070.

Yao, L., Li, Z.-R., et al., 2012. Role of mesenchymal stem cells on cornea wound healing induced by acute alkali burn. PLoS One 7 (2) e30842−e30842.

Yazdanpanah, G., Jabbehdari, S., et al., 2017. Limbal and corneal epithelial homeostasis. Curr. Opin. Ophthalmol. 28 (4), 348−354.

Yoon, J.J., Ismail, S., et al., 2014. Limbal stem cells: central concepts of corneal epithelial homeostasis. World J. Stem Cell. 6 (4), 391−403.

Zainuddin, B.Z., Keen, I., Hill, D.J., Chirila, T.V., Harkin, D.G., 2008. PHEMA hydrogels modified through the grafting of phosphate groups by ATRP support the attachment and growth of human corneal epithelial cells. J. Biomater. Appl 23 (2), 147−168.

Zhang, L., Coulson-Thomas, V.J., et al., 2015. Mesenchymal stem cells for treating ocular surface diseases. BMC Ophthalmol 15 (Suppl. 1), 155.

Zhang, L., Zou, D., et al., 2016. An ultra-thin amniotic membrane as carrier in corneal epithelium tissue-engineering. Sci. Rep. 6, 21021−21021.

Zhi, C., Jingjing, Y., et al., 2018. Biomaterials for corneal bioengineering. Biomed. Mater. 13 (3), 032002.

Zhu, J., Zhang, K., et al., 2013. Reconstruction of functional ocular surface by acellular porcine cornea matrix scaffold and limbal stem cells derived from human embryonic stem cells. Tissue Eng. 19 (21−22), 2412−2425.

Zieske, J.D., 1994. Perpetuation of stem cells in the eye. Eye 8 (Pt 2), 163−169.

… Part Three

Drug delivery

Biocompatibility of materials and its relevance to drug delivery and tissue engineering

12

Thomas Chandy
Philips Medisize LLC, Hudson, WI, United States

Chapter outline

12.1 Biocompatibility of materials and medical applications 298
    12.1.1 Fundamental aspects of tissue response to materials 298
    12.1.2 Blood—material interactions and initiation of the inflammatory response 299
    12.1.3 Surface modifications to improve biocompatibility of materials 301
    12.1.4 Biostability of polymeric materials and biocompatibility 303
12.2 Biomaterials for controlled drug delivery 304
    12.2.1 Polymers used in drug delivery 305
    12.2.2 Modified polymers for drug delivery 306
    12.2.3 Polymer comatrix system for combination drug delivery 307
    12.2.4 Biocompatible coatings for bioactive protein delivery 307
    12.2.5 Nano versus microparticles in cancer drug delivery 309
12.3 Biomaterials for tissue engineering and regenerative medicine 310
    12.3.1 Surface-engineered biomaterials for tissue engineering 313
    12.3.2 Use of polymeric biomaterials in nanomedicine 314
12.4 Role of scaffold and the loaded drug/growth factor in the integration of extracellular matrix and cells at the interface 315
    12.4.1 Induction of angiogenesis in tissue-engineered scaffolds for bone repair: a combined gene therapy—cell transplantation approach 317
    12.4.2 Hydrogel composite materials for enhanced neurotrophin delivery in neural prostheses 318
    12.4.3 Fine-tuning Notch signaling to promote angiogenesis 318
    12.4.4 Myocardial tissue engineering via growth hormones 319
    12.4.5 Multiple factor delivery for vascular tissue engineering 320
    12.4.6 Cell sources for tissue engineering applications 320
12.5 Future outlook on combination devices with drug delivery and tissue engineering 321
References 322

12.1 Biocompatibility of materials and medical applications

The biocompatibility of a long-term implantable medical device refers to the ability of the device to perform its intended function, with the desired degree of incorporation in the host, without eliciting any undesirable local or synthetic effects. Practically speaking, the evaluation of biological response to a medical device is carried out to determine if the medical device performs as intended and presents no significant harm to the patient or user. Thus, the goal of biological response evaluation is to predict whether a biomaterial, medical device, or prostheses express any potential harm to the patient or user by evaluation conditions that simulate clinical use. In almost all situations, the practical consequence is that we select devices that irritate the host the least, through the choice of the most inert and least toxic materials and the most appropriate mechanical design.

The relationship between an implanted material and its host tissues constitutes the concept of biocompatibility, which depends on the mutual effects of the implant on the surrounding host tissues. A lack of biocompatibility can result in complications of implanted prostheses such as loosening of orthopedic joint prostheses (Abu-Amer et al., 2007), fibrotic overgrowth of mammary prostheses (Destouet et al., 1992), and thrombogenic reactions on vascular grafts (Bridges and Garcia, 2008; Gorbet and Sefton, 2004).

12.1.1 Fundamental aspects of tissue response to materials

The process of implantation of a biomaterial, prosthesis, or medical device results in injury to surrounding tissues or organs (Anderson, 1993; Singer et al., 1987). The response to injury is dependent on multiple factors including the extent of injury, the loss of basement membrane structures, blood–material interactions, provisional matrix formation, the magnitude or degree of cellular necrosis, and the extent of the inflammatory response. These events in turn may occur very early, i.e., within 2–3 weeks of the time of implantation, and may affect the subsequent perturbation of homeostatic mechanisms that lead to the cellular cascades of wound healing and the subsequent capsule development. These events are summarized in Table 12.1.

The end-stage healing response to biomaterials is generally fibrosis or fibrous encapsulation. A thin fibrous capsule is inevitable for the long-term stability of the implant, and a thick capsule shows the irritation of the tissue due to the implant. Local systemic factors may play a role in the wound healing response to biomaterials or implants. Local factors include the site (tissue or organ) of implantation, the adequacy of blood supply, and the potential for infection. Systemic factors may include nutrition, hematological and immunological derangements, glucocortical steroids, and preexisting diseases such as atherosclerosis, diabetes, and infection (Joist and Pennington, 1987; Boudot et al., 2016; Nguyen et al., 2015; Vroman, 1988).

Table 12.1 Sequence of host reactions following implantation of medical devices.

Injury
Blood—material interactions
Provisional matrix formation
Acute inflammation
Granulation tissue
Foreign body reaction
Fibrosis/fibrous capsule development

12.1.2 Blood—material interactions and initiation of the inflammatory response

Implantation of a prosthesis can activate the host's blood coagulation mechanism, trigger immunologic reactions, or introduce foreign organisms resulting in common complications such as thrombosis, formation of fibrotic hyperplasia, and bacterial infection (Schoen, 2013; Szymonowicz et al., 2013; Pandiyarajan et al., 2013). This complication is largely associated with cardiovascular devices such as intravascular stents and heart valves, which are in direct contact with blood (Wise et al., 2012). In addition to the blood—material interactions, implantation of a prosthesis in vivo triggers a temporal recruitment of inflammatory cells, such as multinucleated foreign body giant cells.

The normal host response to an implant includes trauma, inflammation, the immune system's reaction, and eventual healing or scarring. Biomaterials exhibiting a lack of biocompatibility could induce many complications, which might include long-lasting chronic inflammation or cytotoxic chemical buildup. Blood—material interactions and the inflammatory responses are intimately linked and in fact early responses to injury involved with blood and the vascular wall (Joist and Pennington, 1987; Boudot et al., 2016; Nguyen et al., 2015; Vroman, 1988; Schoen, 2013; Szymonowicz et al., 2013; Pandiyarajan et al., 2013; Wise et al., 2012; Noishiki and Chvapil, 1987). The host response to biomaterials or to an implant is demonstrated in Table 12.2. Regardless of the tissue or organ into which a biomaterial is implanted, the initial inflammatory responses, thrombi, and/or blood clots are formed because of the injury to vascular connective tissue (Table 12.2).

Thrombus formation involves activation of extrinsic and intrinsic coagulation systems, the complement systems, the fibrinolytic system, the kinin-generating system, and platelets (Anderson, 2001; Vroman. et al., 1982). The adhesion of circulating platelets at sites of injury on the vessel wall where subendothelium, collagen, is exposed to the flowing blood is an important step in the formation of a hemostatic plug. Activation of platelets by contact with biomaterials is a key event in the thromboembolic complications of prosthetic devices in contact with blood. It is known that a film of plasma protein adsorbs on biomaterials exposed to blood and that this event

Table 12.2 Host response characteristics.

Protein adsorption and desorption characteristics
Complement activation
Platelet adhesion, activation, and aggregation
Activation of intrinsic clotting cascade
Neutrophil activation
Fibroblast behavior and fibrosis
Microvascular changes
Macrophage activation, foreign body giant cell production
Osteoblast/osteoclast responses
Endothelial proliferation
Antibody production, lymphocyte behavior
Acute hypersensitivity/anaphylaxis
Delayed hypersensitivity
Genotoxicity, reproductive toxicity
Tumor formation

proceeds interaction of the surface with blood cells (Rao and Chandy, 1999; Baier et al., 1985). Platelet adhesion to biomaterial surfaces occurs in various steps including initial attachment, spreading, release of granule contents, and platelet aggregation (Liu et al., 2012; Cornelius et al., 2011; Sweet et al., 2011).

Vroman et al. (Vroman et al., 1980) extensively studied the initial phase of blood–biomaterial interactions. They have postulated that on hydrophilic surfaces (e.g., glass), fibrinogen is deposited within in seconds along with traces of high-molecular-weight kininogen (HMWK) and factor XII. The more HMWK and factor XII arrive and displace fibrinogen. Platelets adhere most, where fibrinogen remains. Thus, fibrinogen plays a key role in platelet–biomaterial attachment and subsequent thrombus formation. An early feature of activation of platelets by soluble agonists such as adenosine diphosphate or thrombin (Thr) is exposure of specific membrane glycoproteins (GPIIb-IIIa), to which fibrinogen molecule bind with high affinity (Maeguerie. et al., 1979; Tomikawa et al., 1980). It should be noted that fibrinogen dissolved in plasma does not induce platelet aggregation but rather acts as a cofactor in the process, once the platelet GPIIb-IIIa receptors are activated. Fibrinogen normally circulates in peaceful coexistence with inactivated platelets without any obvious interaction (Tomikawa et al., 1980; Brash, 2000; Andrews and Fox, 1991).

12.1.3 Surface modifications to improve biocompatibility of materials

Surface modification of biomaterials is one approach for improving their functionalities for best performance. The common purpose of surface treatment is to modify the outermost layer of a polymer by inserting some functional groups onto the surface to improve its barrier properties, wettability, sealability, its adhesion to other materials, or its interaction with a biological environment, while maintaining the desirable bulk properties of the polymer. Therefore, this section will present an overview of the consolidated routes for the surface modification of polymers through the chemical (including compounding with surface-active additives) and physical methods and will describe the new perspectives offered by polymer surface modification through biological routes for improved biocompatibility.

Various investigators (Andrews and Fox, 1991; Morois et al., 1996; Chandy et al., 1999) used physical, chemical, or biological methods to generate an inert or passive interface with blood/tissue reactions. These include albumin, collagen, laminin (LN), fibronectin (FN), heparin coatings, hirudin, antiplatelet drugs, prostaglandins, and growth factors, for normal healing and naturalization (Chandy. et al., 1999; Kito and Matsuda, 1996; Riau et al., 2017; Radtke et al., 2017). Table 12.3 gives a summary of surface modifications used for improving biocompatibility of implants. However, successful results have not yet been reported for small-caliber artificial vascular grafts. It seems modifications of these prostheses surfaces that stimulate endothelialization could reduce thrombosis, eliminate platelet deposition, resist bacterial infection, and extend graft patency (Cheng et al., 2017; Bos. et al., 1998; Ritter. et al., 1997).

Enhancement of antithrombogenicity at the luminal surface and tissue regeneration at the outer surface may lead to a vital and functional artificial graft. Surface modifications that have been used to enhance endothelial cell (EC) seeding on vascular prostheses include immobilization of FN, LN, collagen, growth factors, and peptides (Ritter et al., 1997; Sharma et al., 1987; Chandy et al., 2000). An ideal coating process should have normal healing process and naturalization. It seems an ideal healing process is as follows: as the luminal surface coating is biodegraded, while it maintains its

Table 12.3 Surface Modifications for improving biocompatibility.

1 Physical treatments	Plasma glow, corona discharge, radiation, UV, laser-induced modifications, ion beam−based process
2 Chemical modifications	Hydrogels (poly(2-hydroxy ethyl methacrylate), poly(ethylene glycol), hyaluronic acid, synthetic phosphoryl choline derivatives, silicone coatings, antibacterial coatings, self-assembled monolayers, etc.)
3 Biological modifications	Albumin coatings, gelatin, heparin, LMW heparin, hirudin, collagen IV, laminin-5, growth factors, phosphoryl choline, etc. Extracellular matrix proteins, tissue engineering

thrombus-free character, tissue formation, including endothelial regeneration process, occurs. When this coating is completely biodegraded, generated tissues are expected to replace it.

Doi and Matsuda (Doi and Matsuda, 1997) prepared microporous polyurethane grafts that were coated with photoreactive gelatin, fibroblast growth factor, and heparin and were photocured by UV irradiation. Histological evaluation of these grafts showed that the neoarterial wall of the fibroblast growth factor/heparin-impregnated graft was much thicker and was formed with ECs than that of noncoated graft. Thus, the coimmobilization of growth factor and heparin significantly accelerated neoarterial regeneration and the patency rates of small-caliber grafts. Chandy et al. (Chandy et al., 2000) prepared a series of surface coatings by modifying the argon plasma-treated polytetrafluoroethylene (PTFE) and Dacron grafts with collagen IV and LN and subsequently immobilizing bioactive molecules such as PGE_1, heparin, or phosphatidyl choline via the carbodiimide functionalities. Table 12.4 provides the information on platelet adhesion and fibrinogen adsorption to various modified PTFE and Dacron surfaces (Chandy et al., 2000). These in vitro studies showed that the fibrinogen adsorption and platelet adhesion on modified grafts were significantly reduced. This study proposed that the surface grafting of matrix components (collagen IV and LN) and subsequent immobilization of bioactive molecules (PGE1, heparin, or phosphatidyl choline) changed the surface conditioning of vascular grafts and subsequently improved their biocompatibility.

Antibacterial coatings are tried on polymers for preventing localized infection and inflammation to devices. The simplest way for producing sol—gel coatings with antibacterial properties is the incorporation of silver, both in the form of metallic nanoparticles or ions. The antibacterial activity of polyethylene (PE) and poly(vinyl chloride)

Table 12.4 Effect of immobilized biomolecules on the adhesion of platelets and fibrinogen absorption on PTFE (PTFE) and Dacron surfaces.

Surfaces	Platelet adhesion 30 min (mm^2 + SD)	Fibrinogen adsorption ($\mu g/cm^2$)
PTFE (untreated)	62.92 + 4.9	1.08 + 0.21
CL-PTFE-PGE1	21.65 + 3.3	0.41 + 0.02
CL-PTFE-heparin	15.00 + 3.5	0.26 + 0.01
CL-PTFE-PC	16.58 + 4.2	0.21 + 0.02
Dacron (untreated)	48.93 + 5.7	0.86 + 0.04
CL-Dacron-PGE1	14.05 + 3.5	0.27 + 0.01
CL-Dacron-heparin	17.50 + 3.3	0.22 + 0.02
CL-Dacron-PC	13.58 + 3.2	0.19 + 0.01

CL-Collagen- and laminin-immobilized surface.
Compiled from Chandy T., Das. G.S., Wilson. R.F., Rao G.H.R., Use of plasma glow for surface-engineering biomolecules to enhance blood compatibility of Dacron and PTFE vascular prosthesis, *Biomaterials* 21, 699—712, 2000.

films coated with silver-doped organic−inorganic hybrid coatings prepared by sol−gel processes has been reported by Marini et al (Marini et al., 2007). Structural evolution and antibacterial properties of silver-doped silica-methyl hybrid coatings have also been investigated (Riau et al., 2015, 2016; Procaccini et al., 2014). Nanocomposite hybrid coating containing silver and copper ions with or without nanoparticles of titanium dioxide was deposited onto PMMA to enhance its bacterial resistance (Slamborová et al., 2013).

12.1.4 Biostability of polymeric materials and biocompatibility

In the selection of biomaterials to be used in device design and manufacture, the first consideration should be fitness for purpose about characteristics and properties of the biomaterials, which include chemical, toxicological, physical, electrical, morphological, and mechanical properties (Wise et al., 2012). Table 12.5 presents a list of biomaterial components and characteristics that may impact the overall biological responses of the medical device. Therefore, knowledge of these components in the medical device, i.e., final product, the duration of the exposure of the device to the tissues, and the degradation products, is necessary. The range of potential biological hazards is broad and may include short-term effects, long-term effects, or specific toxic effects, which should be considered for every material and medical device (Zhao et al., 1991; Cosgriff-Hernandez et al., 2016).

Polymeric biomaterials with functional groups such as urethanes, ester, carbonates, amides, and anhydrides consisting of carbonyls bonded to heterochain elements (O, N, and S) are most vulnerable to host-induced hydrolytic processes (Simmons et al., 2004; Mishra et al., 2015; Cory, 2013). In contrast, functional groups such as hydrocarbon, alkyl, aryl, halocarbon, dimethyl siloxane, and sulfone are highly stable when subjected to hydrolytic degradation (Cory, 2013). Hydrolytic degradation is also catalyzed by ions in the extracellular fluids, enzymes secreted by phagocytic cells, and pH decrease caused by local infection (Simmons et al., 2004; Mishra et al., 2015; Cory, 2013; Liu et al., 2011; Szycher and Reed, 1992). Oxidative biodegradation by hemolytic chain reaction or heterolytic mechanism is another biochemical pathway for the unintended in vivo breakdown of polymeric implants. These reactions are initiated and propagated by oxidizing molecular species such as superoxide anion radicals and their

Table 12.5 Biomaterials and components relevant to in vivo assessment of tissue compatibility.

The material(s) of manufacture
Intended additives, process contaminants, and residues
Leachable substances
Degradation products
Other components and their interactions in the final product
The properties and characteristics of the final product

derivatives (hydrogen peroxide and hypochlorite) that are generated from activated phagocytic cells in high concentrations at the device–tissue interface (Liu et al., 2011).

Polyurethanes are used as biomaterials for a variety of applications (Zhao et al., 1991; Szycher and Reed, 1992) such as pacemaker lead insulators, catheters, total artificial heart, and heart valves. The popularity of polyurethane for biomedical applications stems from their excellent physical properties and good biocompatibility. Biomaterials to be successful as implant devices should be well accepted by the host system as well as they should not exert any adverse effect on the host. However, several studies demonstrated that the polyether soft segment is susceptible to oxidation after extended periods in vivo (Szycher and Reed, 1992; Zhao et al., 1990). Furthermore, oxidative attack on the polymeric surface might also induce shallow brittle microcracks, which are subsequently propagated to deeper and wider cracks (Cory, 2013; Liu et al., 2011; Szycher and Reed, 1992; Zhao et al., 1990; Christenson et al., 2005a,b; Wilkoff et al., 2016). It is now generally accepted that the reactive oxygen intermediates released by adherent macrophages and foreign body giant cells initiate the observed biodegradation. Because of the known susceptibility of poly(ether urethanes) (PEU) to oxidation, poly(carbonate urethanes) are currently being examined as a more biostable replacement for long-term implants (Zhao et al., 1990; Christenson et al., 2005a,b; Wilkoff et al., 2016; Christenson et al., 2005a,b; Padsalgikar et al., 2015; Chandy et al., 2009). Several in vitro and early in vivo studies showed that polyurethanes with polycarbonate soft segments are more biostable than comparable PEUs (Chaffin et al., 2013; Gallagher et al., 2017; Cozzens et al., 2010). Thus, it is inevitable to select the biomaterial(s), knowing the life of the implant, i.e., for a short-term or long-term use.

12.2 Biomaterials for controlled drug delivery

Controlled drug delivery occurs when a polymer, whether natural or synthetic, is judiciously combined with a drug or other active agent in such a way that the active agent is released from the material in a predesigned manner. The release of the active agent may be constant over a long period, it may be cyclic over a long period, or it may be triggered by the environment or other external events. In any case, the purpose behind controlling the drug delivery is to achieve more effective therapies while eliminating the potential for both under- and overdosing (Langer, 1980). Other advantages of using controlled-delivery systems can include the maintenance of drug levels within a desired range, the need for fewer administrations, optimal use of the drug in question, and increased patient compliance. While these advantages can be significant, the potential disadvantages cannot be ignored: the possible toxicity or nonbiocompatibility of the materials used, undesirable by-products of degradation, any surgery required to implant or remove the system, the chance of patient discomfort from the delivery device, and the higher cost of controlled-release systems compared with traditional pharmaceutical formulations.

In recent years, controlled drug delivery formulations and the polymers used in these systems have become much more sophisticated, with the ability to do more than simply extend the effective release period for a drug. For example, current controlled-release systems can respond to changes in the biological environment and deliver—or cease to deliver—drugs based on these changes. In addition, materials that should lead to targeted delivery systems have been developed, in which a formulation can be directed to the specific cell, tissue, or site where the drug it contains is to be delivered (Langer, 1980; Morris and Shebuski, 1995). While much of this work is still in its early stages, emerging technologies offer possibilities that scientists have only begun to explore.

12.2.1 Polymers used in drug delivery

A range of materials have been employed to control the release of drugs and other active agents. To be successfully used in controlled drug delivery formulations, a material must be chemically inert and free of leachable impurities. It must also have an appropriate physical structure, with minimal undesired aging, and be readily processable. Some of the materials that are currently being used or studied for controlled drug delivery are shown in Table 12.6. However, in recent years, additional polymers designed primarily for medical applications have entered the arena of controlled release. Many of these materials are designed to degrade within the body and are shown in Table 12.7. The mechanical properties and degradation time of a bioresorbable device can be tailored to a specific application by adjusting the molecular weight, crystallinity, and hydrophilicity of the polymer. For example, compositions with higher hydrophilic and amorphous structures and a lower molecular weight resorb faster, yet they often sacrifice mechanical strength. Conversely, higher crystallinity

Table 12.6 Polymers used for drug delivery applications.

Poly urethanes (PU)
Poly(dimethyl siloxane) or silicones (PDMS)
Poly(2-hydroxy ethyl methacrylate) (pHEMA)
Poly(N-vinyl pyrrolidone) (PVP)
Poly(methyl methacrylate) (PMMA)
Poly(vinyl alcohol) (PVA)
Poly(acrylic acid) (PAA)
Polyacrylamide (PA)
Poly(ethylene-co-vinyl acetate) (PEVA)
Poly(ethylene glycol) (PEG)
Poly(methyl methacrylic acid) (PMMA)

Table 12.7 Degradable polymers used for drug delivery.

Poly(lactic acid) (PLA)
Poly(glycolic acid) (PGA).
Poly(lactide-co-glycolide) (PLGA)
Polycaprolactone (PCL)
Polyanhydrides
Poly(ortho esters) (POEs)
Poly(propylene fumarate)
Polyphosphazenes
Chitosan
Dextran
Hyaluronic acid (HA)
Poly(alginic acid)

and molecular weight improve mechanical properties and decrease resorption rates. The greatest advantage of these degradable polymers is that they are broken down into biologically acceptable molecules that are metabolized and removed from the body via normal metabolic pathways (Chandy and Sharma, 1993; Song et al., 1997). However, biodegradable materials do produce degradation by-products that must be tolerated with little or no adverse reactions within the biological environment (Mei et al., 2018; Yang et al., 2018; Mai et al., 2018).

12.2.2 Modified polymers for drug delivery

Recurrent luminal narrowing (restenosis) because of excessive intimal hyperplasia remains a major limiting factor for the long-term success of vascular interventions (Faxon and Currier, 1995; Berk and Harris, 1995). It has been shown that to inhibit vascular smooth muscle cell proliferation, drugs must be used at a high concentration for a prolonged period. Because of the focal nature of restenosis, local delivery systems, such as polymer-based perivascular techniques and catheter-based cardiovascular techniques, have been employed (Berk and Harris, 1995; Heller, 1980; Nihant et al., 1995; Ruiz and Benoit, 1991). Various geometrical devices, such as polymeric matrices, microspheres, and circumferential wraps, have been developed for perivascular local delivery (Szycher and Reed, 1992). Polylactic acid (PLA) and copolymers of lactic and glycolic acids are well-known biodegradable and histocompatible aliphatic polymers. They are commonly used as biodegradable sutures, and they have more recently contributed to the reconstruction of deficient or injured organs and to improved galenic formulations (Ruiz and Benoit, 1991; Gref et al., 1994). During the past few years, several techniques for drug encapsulation have been developed, which use aliphatic polyesters. Studies of Bazile et al. (Bazile et al., 1995) indicated that hydrophilic coatings with poly(ethylene glycol) (PEG) can increase the blood half-life of PLA nanospheres in rats up to several hours. The main advantage of surface-coated microspheres in comparison with other long-circulating systems is their shelf stability and their ability to control the release of the encapsulated drug compound (Niu et al., 2018; Wagner et al., 2018; Salehiabar et al., 2018; Quellec et al., 1998, 1999).

Biodegradable microspheres containing Taxol were formulated with PLA−PEG polymers as a sustained drug delivery system for the control of restenosis (Das. et al., 2001). The release of the drug was sustained over 4 weeks in vitro. These studies suggest that biocompatible PEG-coated PLA microspheres provide a near zero-order, in vitro release of Taxol for therapeutic applications (Das et al., 2001). The amount of Taxol release from PLA−PEG microspheres was 14.4 mg/mg within 30 days of dissolution compared with 10.7 mg/mg for PLA−PVA (Das et al., 2001). In other words, PEG-coated microspheres release more Taxol to the site for better therapeutics.

Antiplatelet therapy potentially reduces intimal formation induced by platelet aggregation. Platelet glycoprotein receptor GPIIb/IIIa has an important role in platelet activation and remains a clinical target to inhibit thrombosis (Montalescot et al., 2001). Monoclonal antibodies against the human GPIIb/IIIa receptor provide significant inhibition of platelet aggregation when systemically infused in human therapies. To bring these agents forward toward clinical use, creative approaches to formulating, testing, and understanding the mechanisms of action of these agents in the nanoparticle context is required. Randomized, blinded trials that assess stent clinical efficacy without bias in study designs will be required to fully assess performance enhancements.

12.2.3 Polymer comatrix system for combination drug delivery

Controlled release of appropriate drugs alone and in combinations is one approach for treating coronary obstructions, balloon angioplasty, restenosis associated with thrombosis, and calcification. Chandy et al. (Chandy et al., 2001) demonstrated the possibility of encapsulating Taxol-loaded PLA microspheres within heparin−chitosan spheres to develop a prolonged release comatrix form. The comatrix system was fabricated from Taxol-loaded PLA microbeads encapsulated in chitosan beads. A controlled delivery of Taxol/heparin was achieved by coating the PLA/chitosan comatrix with PEG (Table 12.8). This study (Chandy et al., 2001) also highlights the use of a comatrix system for targeting system drug combinations, with least side effects for the therapeutic applications. This study showed the release of drug combinations having synergistic effects for prolonged periods for treating restenosis (Chandy et al., 2001).

12.2.4 Biocompatible coatings for bioactive protein delivery

Proteins and enzymes represent a growing and promising field of therapeutics and are currently administered by injection. Protein delivery from biodegradable polymer systems has been a challenging area of research because of the necessity of improving the

Table 12.8 Amount of Taxol and heparin released from poly(ethylene glycol) coated poly(lactic acid) chitosan comatrix system.

Time (days)	Amount of taxol (μg/mg bead)	Amount of heparin (units/mg bead)
1	0.83 + 0.03	13.1 + 0.37
5	1.06 + 0.12	17.7 + 0.62
15	2.17 + 0.08	23.6 + 0.05
30	2.71 + 0.02	25.8 + 1.34
40	3.82 + 0.01	26.0 + 1.4
60	4.08 + 0.13	27.1 + 0.86

delivery of newly developed macromolecular drugs and antigens (Cleland, 1997; Wang et al., 2018; Louzao et al., 2018). During the last few years, several techniques for drug encapsulation have been developed, which currently use aliphatic polyesters. Bazile et al. (Bazile et al., 1995) found that hydrophilic coatings with PEG could increase the blood half-life of PLA nanospheres in rats up to several hours. The main advantage of PEG-coated nanospheres over other long-circulating systems is their shelf stability and ability to control the release of an encapsulated compound (Gref et al., 1994; Bazile et al., 1995). Chandy et al. (Chandy et al., 2002) encapsulated human serum albumin (HSA) and Thr in PEG-coated, monodisperse, biodegradable microspheres with a mean diameter of about 10 μm. The PLA−PEG microspheres demonstrated an initial burst release followed by a constant slow release of HSA and Thr for a period of 20 days (Chandy et al., 2002).

The main advantage of PEG-coated nanospheres over other long-circulating systems is their shelf stability and ability to control the release of an encapsulated compound (Cleland, 1997; Chandy et al., 2002). PEG appears to work as a protective colloid for the emulsion droplets during the preparation. The PEG molecules adsorbed on the surface of the droplets prevented the coalescence of droplets. Therefore, it appears that the PEG coating can increase the payload of drugs and ensure better stabilization. Nagaoka et al. (Nagaoka et al., 1984) synthesized a graft copolymer of methacrylates with PEG and found the resulting polymer to be quite nonthrombogenic. PEG-grafted polymer surfaces have also been shown to reduce protein adsorption and are highly resistant to mammalian and bacterial cell adhesion (Gref et al., 1994; Quellec et al., 1999). Therefore, these new microcapsules fabricated from PLA/PEG may serve to provide controlled protein delivery and immunoprotection, whereas the outer layer of PEG may serve to enhance biocompatibility and reduce biodegradation. PEG-coated PLA microspheres show great potential for protein-based drug delivery.

Peptide drug delivery by routes other than the parenteral one has gained much attention in recent years. Many studies are currently being conducted to address other absorption sites for peptide delivery through different targeting mechanisms such as entrance via the Pyers patch or mucoadhesion (Ali and Ahmed, 2018; Chandy et al., 1998; Saffran et al., 1986). Oral delivery is the easiest method of administration and allows for a more varied load to be released; however, proteins are quickly denatured and degraded in the hostile environment of the stomach. A potential solution to these problems is the use of microencapsulation process for oral release of therapeutic agents (Gallagher et al., 2017). The protein is encapsulated in a core material that is covered by a biocompatible, semipermeable membrane. The membrane controls the diffuse release rate of the protein from the capsule to the surrounding medium while protecting the remaining encapsulated protein from biodegradation.

Alginates (Alg), chitosan, and PEG matrices had been reported potentially useful for medical and pharmaceutical applications such as artificial skin, artificial kidney, cell encapsulation, and as drug carrier for target delivery (Chandy et al., 2002; Nagaoka et al., 1984; Ali and Ahmed, 2018; Chandy et al., 1998; Saffran et al., 1986). PEG-coated chitosan/calcium Alg system was used to encapsulate albumin and hirudin for use as an oral delivery system (Chandy et al., 1998). The protein release was less in stomach pH (acidic), and these acid-treated capsules had released almost all the entrapped proteins into intestinal media (pH 7.4) within 6 h (Chandy et al., 1998). The released hirudin showed their biological activity when tested with specific coagulations assays (Chandy et al., 1998). This study proposed that chitosan-Alg microbeads may be used as a vehicle for delayed release of protein drugs. The incorporation of biocompatible PEG can protect the protein from degradation, and subsequently, their bioavailability. PEGs are used for used for improving the blood compatibility of polymers and can preserve the biological properties of proteins (Bazile et al., 1995; Chandy et al., 1998). Thus, it seems that the chitosan/PEG-Alg system is a good candidate for oral delivery of proteins and other biomolecules.

12.2.5 Nano versus microparticles in cancer drug delivery

Cancer has a physiological barriers (Chatterjee and Zhang, 2007; Aghajanzadeh et al., 2018; Nosrati et al., 2018; Li et al., 2018; Jain, 2001) such as vascular endothelial pores, heterogeneous blood supply, heterogeneous architecture, etc. For treatment to be successful, it is important to get over these barriers. Cancer represents an enormous biomedical challenge (Jain, 2001; Wu et al., 2018; Stylios et al., 2005) for drug delivery, and the treatment is dependent on the method of delivery. In the past, cancer patients were using various anticancer drugs, with radiation therapy, but these drugs and treatments were less successful and had major side effects (Brigger, 2002). Nanoparticles have attracted the attention of scientists because of their multifunctional character. Several nano- and microtechnologies, mostly nanoparticles, have been used to facilitate drug delivery in cancer (Jain, 2001; Wu et al., 2018; Stylios et al., 2005) due to their multifunctional character and fast recovery from the disease.

Several nano- and microtechnological approaches have been used to improve the delivery of chemotherapeutic agents to cancer cells with the goal of minimizing toxic effects on healthy tissues with maintaining antitumor activity (Stylios et al., 2005; Brigger, 2002; Behl et al., 2018; Portney and Ozkan, 2006). Nanoscale devices smaller than 50 nm can easily enter most cells, and those smaller than 20 nm can move out of blood vessels as they circulate through the body (Stylios et al., 2005; Brigger, 2002; Behl et al., 2018; Portney and Ozkan, 2006). Nanoparticles can serve as customizable, targeted drug delivery vehicles capable of delivering large doses of chemotherapeutic agents or therapeutic genes into malignant cells while sparing healthy cells, greatly reducing the side effects of current cancer therapies. This specific targeting of drugs to cancer cells can reduce the toxicity of drugs to normal tissue (Portney and Ozkan, 2006; Yezhelyev et al., 2006).

In medicine, nanoparticles first found use in the diagnosis of tumors in the liver and spleen using magnetic resonance tomography. In practice, colloidal gold nanoparticles are the most commonly used nanoparticles for diagnostics and tumor-specific drug delivery (Yezhelyev et al., 2006). Development of cancer receptor-specific gold nanoparticles will allow efficient targeting/optimum retention of engineered gold nanoparticles within tumors and thus provide synergistic advantages in oncology as it relates to molecular imaging and therapy. Recent studies have demonstrated the feasibility of using targeted gold nanoparticles embedded in cancer cells for CT imaging under in vitro conditions (Popovtzer, 2008). Target specificity is achieved through hybrid nanoparticles that are produced by conjugating gold nanoparticles with tumor-specific biomolecules, including monoclonal antibodies, peptides, or various receptor-specific substrates (Yezhelyev et al., 2006; Popovtzer, 2008). The use of nanoparticles as drug delivery vehicles for anticancer therapeutics has great potential to revolutionize the future of cancer therapy. The nanoparticles specifically accumulate at the tumor site, and their use of drug delivery vectors results in the localization of a greater amount of the drug load at the tumor architecture, thus improving cancer therapy and reducing the harmful nonspecific side effects of chemotherapeutics. In addition, formulation of these nanoparticles with imaging contrast agents provides a very efficient system for cancer diagnostics.

12.3 Biomaterials for tissue engineering and regenerative medicine

In recent years, much attention has been focused on the development of biomaterials for use in tissue engineering applications. The increasing shortage of tissue and organ donors has driven the development of viable tissue substitutes. Tissue engineering is a multidisciplinary field of biomedical research that merges the fields of cellular and molecular biology, cell and tissue culture techniques, and materials sciences to recreate tissue for treatments involving reconstruction or replacement. New biocompatible materials are developed constantly for use as scaffolds in tissue engineering of a wide variety of tissues and structures such as the cartilage, bone, liver, blood vessels, heart valves, myocardial tissues, and knee meniscus (Vacanti et al., 1991; Santos et al., 2018; Li et al., 2018; Zarrintaj et al., 2018; Tandon et al., 2018; Grad et al., 2003; Buma et al., 2003). Tissue engineering is defined as the creation of new tissue for the therapeutic reconstruction of the human body, by the deliberate and controlled stimulation of selected target cells through a systematic combination of molecular and mechanical signals (Vacanti et al., 1991; Grad et al., 2003).

In the late 2000s, tissue engineering strategies for tissue repair were transformed by the combination of stem cells with scaffold-based therapy to form a new integrated field called "regenerative medicine" or "regenerative engineering." This is a rapidly growing multidisciplinary field seeking stem cell therapy to aid in the regeneration process of a diseased organ or injury. In 2010, Laurencin redefined tissue engineering in terms of regenerative engineering as "the integration of materials science and tissue engineering with stem and developmental cell biology and regenerative medicine toward the regeneration of complex tissues, organs, or organ systems" (Laurencin et al., 2010). Significant developments have been reported in the past two decades in tissue engineering and regenerative medicine for the repair of damaged tissues such as the cartilage, skin, bladder, muscle, bone, and blood vessels using different biomaterials.

In general, however, depending on the tissue that is to be engineered, two main tissue engineering approaches can be distinguished. In the first and currently the most popular approach, cells are seeded onto scaffolds and cultured in vitro to form a construct that subsequently is implanted into a laboratory animal (Ramrattan. et al., 2005; Vannozzi et al., 2018). The concept is to take a suitable material that can give physical form or outline to the area of tissue to be regenerated and to invest it with those tissue components that are responsible for tissue growth and repair. The combination, often called a construct, can help the patient regenerate new functional tissue. The tissue components may be cells or biomolecules or both. Based on this analysis, a definition of tissue engineering has been produced as follows' tissue engineering is the persuasion of the body to heal itself through the delivery to the appropriate sites of molecular signals, cells, and/or supporting structures (Nikoubashman et al., 2018; Hutmacher et al., 2003).

Among various biomaterials, polymers possess significant importance due to their chemical tunability to allow scaffolds with appropriate physical, biological, and mechanical properties. Additional important factors in tissue engineering are the scaffold architecture, biodegradability, and the physical stability of scaffold to meet the complex functionalities possessed by each tissue type (Hutmacher et al., 2003). Moreover, a scaffold should have the capability to promote greater material—cell interaction and improve cell adhesion and proliferation. Also, a scaffold should allow adequate transfer of gas, nutrients, and growth factors required for cellular development. Proper exchange of nutrients and growth factors will be possible through scaffold porosity and pore interconnectivity, and hence, scaffold morphology plays a key role in tissue repair and regeneration. Another important factor to be mentioned is the degradability of the scaffold. Ideally, the degradation rate of the polymeric scaffold should be in accordance with regeneration of the tissue.

The most common type of tissue engineering product is one which involves a biodegradable polymeric support (for example, in the form of a porous scaffold) into which are introduced cells of the appropriate phenotype, along with some suitable drugs such as growth factors. This construct may be cultured in a sterile bioreactor, when the cells produce the regenerated tissue, which can then be implanted in the host (Ramrattan et al., 2005; Vannozzi et al., 2018; Nikoubashman et al., 2018; Hutmacher et al., 2003; Frenkel and Di Cesare, 2004). This is a very ambitious objective, and only limited success has been achieved so far in relatively simple areas such as the skin. Developments are well advanced in bone and cartilage regeneration, and there is much activity in experimental systems for arteries and nerves.

In an alternative approach that is gaining popularity, especially with the advent of the so-called smart scaffolds (Ramrattan et al., 2005; Vannozzi et al., 2018; Nikoubashman et al., 2018), acellular scaffolds are implanted into a laboratory animal. Rather than relying on in vitro seeding of cells as in the first approach, this alternative approach relies on in vivo in growth of surrounding tissue into the scaffold with subsequent proliferation and differentiation into the desired tissue. Especially in large three-dimensional (3D) scaffolds, with a size exceeding the critical diffusion distance, this approach will be useful when ingrowth of vascularized tissue can be induced (Frenkel and Di Cesare, 2004; Harris and Cooper, 2004; Barralet et al., 2003; Xu et al., 2004). The development of new biomaterials for tissue engineering, and the enhancement of tissue ingrowth into existing materials by means of attaching growth factors ("smart scaffolds"), creates the necessity to develop tools for assessment of tissue ingrowth rates into porous biomaterials.

A variety of materials have been considered for the scaffolds (Harris and Cooper, 2004; Barralet et al., 2003; Xu et al., 2004; He and Matsuda, 2002; Rothenburger et al., 2002; Ye, 2014), the most common being the biodegradable polyesters mentioned previously and some natural biopolymers such as collagen and a variety of polysaccharides (Table 12.9). Natural polymers include collagen, gelatin, elastin (Ela), actin, keratin, albumin, chitosan, alginic acid, chitin, cellulose, silk, and hyaluronic acid (HA) (Hegewald et al., 2015; Hansson et al., 2017; Tienen et al., 2003). Many natural polymeric materials are more easily accepted by biological systems where they can be metabolically processed through established pathways. However, natural biomaterials have some disadvantages including possible immunogenicity, structural complexity, and inferior biomechanical properties.

Biodegradable synthetic polymers offer many advantages over other materials for developing scaffolds in tissue engineering, drug delivery, and in vivo sensing

Table 12.9 The scaffold materials used for tissue engineering applications.

Synthetic biodegradable polymers (PLA, PLGA, PGA, PCL, etc.)
Synthetic nonbiodegradable polymers (as hybrids)
Natural biopolymers (proteins, polysaccharides)
Self-assembled biological structures
Tissue-derived structures (SIS, porcine pericardium, porcine dermis, amniotic membrane, etc.)
Bioactive ceramics and glass ceramics
Composites, including nanocomposites
Multilayered structures
Temperature/pH-responsive polymers (cell sheet)

PCL, polycaprolactone; *PGA*, poly(glycolic acid); *PLA*, polylactic acid; *PLGA*, poly(lactide-*co*-glycolide).

(Ramrattan et al., 2005; Ye, 2014; Hegewald et al., 2015; Hansson et al., 2017). Compared with natural polymers, synthetic polymers can easily be tailored into any form suitable for tissue engineering applications. Such materials should provide a 3D structure that not only plays a supportive role for the tissue but also interacts with cells to control their function and differentiation (Hutmacher et al., 2003; Barralet et al., 2003) PLA, polyglycolic acid, and their copolymers dominate as scaffold materials because the materials are already approved by the FDA and because the degradation rate can be manipulated. However, the materials are much weaker and much less elastic than soft biological tissue, cell adhesion is only desirable with surface modifications, and the degradation rate is very fast. Other classes of degradable polymers have been investigated, such as polycaprolactone, polyanhydrides, poly(propylene fumarate), polyphosphazenes, and so on (Xu et al., 2004; He and Matsuda, 2002; Rothenburger et al., 2002; Ye, 2014; Hegewald et al., 2015; Hansson et al., 2017; Tienen et al., 2003; Flanagan and Pandit, 2003). The recent developments have proposed that copolymers of PEG with PLA or poly(lactide-*co*-glycolide) (PLGA) and the elastomeric polyglycerol sebacate need attention to be used for soft tissue applications (Wang. et al., 2003).

12.3.1 Surface-engineered biomaterials for tissue engineering

A wide variety of natural and synthetic biodegradable polymers are currently available on the market for regenerative/tissue engineering applications. Most of them are either collagen-or polyester-based materials. Use of advanced processing techniques and numerous synthetic organic routes still allows the modification of existing polymers to alter their properties suitable for tissue/organ repair and other biomedical applications. Appropriate biocompatibility and degradability could be achieved by combining different polymers by physical blending or chemical modifications.

A subset of naturally derived scaffolds is extracellular matrix (ECM)—based biomaterials. This includes extracted components of the ECM such as collagen, dextran, fibrin, Matrigel, and HA, as well as bulk ECM obtained through decellularization procedures (Caliari et al., 2011). These natural ECM-based scaffolds typically have high biocompatibility and tissue growth. Several ECM-like materials that combine synthetic polymer with 3D collagen gels have been investigated for tissue engineering scaffolds. For example, PEG—collagen composites have been used as in vivo scaffolds for connective tissue regeneration (Tan and Saltzman, 1990), whereas FN—collagen and LN—collagen composites have been used to grow cells in vitro. Functional receptor-mediated and signal-transmitting cell adhesion on a conventional biomaterial is mediated by ECM molecules, such as FN, vitronectin, collagen, or LN. Numerous studies have highlighted the importance of ECM components for appropriate cell migration during morphogenesis of the nervous system as well as for neurite outgrowth in vitro (Tan and Saltzman, 1990; Louie et al., 1997).

The ECM proteins such as LN and FN are expressed in distinct spacial and temporal patterns in the developing nervous system. These proteins are necessary for natural healing processes of the nervous system and appear to be a critical factor in the integration and function of biomaterial-based implants in the nervous system. Chandy and

Rao (Chandy and Rao., 2002) used a spray coating technique to prepare PLA tubes. To improve the flexibility of these devices, an elastomeric polymer, poly(ethylene vinyl acetate) (PEVAc), was added to the PLA. The PLA/PEVAc tubes were further surface modified with Ela and LN via carbodiimide and glutaraldehyde treatment. This study (Louzao et al., 2018) indicated that the surface-grafted matrix components (Ela and LN) on PLA/PEVAc tubes would offer a new approach for nerve growth and tissue engineering.

The two main classes of extracellular macromolecules that form the natural ECM are proteoglycans and fibrous proteins. Tan et al. (Tan et al., 2001) suggested that proteoglycans, containing long unbranched polysaccharide side chains covalently tethered to a fibrous backbone, form 3D networks of hydrated gels in which cells can be embedded. The development of suitable 3D matrices for the maintenance of cellular viability and differentiation is critical for applications in tissue engineering and cell biology. To this end, gel matrices of different proportions of Alg/Ela/PEG were prepared and examined (Chandy et al., 2003). The composite matrix of Alg/Ela/PEG has polysaccharide structures (Alg) and fibrous proteins (Ela) in a water-soluble PEG. This structure allows the Alg/Ela/PEG gel to maintain structural integrity and can keep biological integrity of cells to grow and multiply in the scaffold. This novel composite matrix structures are resumable to natural ECM components and may have potential biological and mechanical benefits for use as a cellular scaffold.

Tissue engineering generally involves combining mammalian cells (including stem cells) with polymer materials in such a way as to create new tissues or organs. Although this area has been the focus for considerable research (Frenkel and Di Cesare, 2004; Harris and Cooper, 2004), many challenges remain. First, it will be essential to find cell sources that yield sufficient quantities of differentiated cells. Stem cells may represent an important source, provided that their differentiation, growth, and immunogenicity can be controlled. Another possibility is autologous cells, although in many cases, one may not be able to masses using grow them quickly enough and in such a way that they maintain their differentiated state. A second challenge is ensuring that cells in engineered tissues survive. One promising approach is to develop methods for vascularizing 3D cell, either microfabricated systems or the controlled release of growth factors (Xu et al., 2004; He and Matsuda, 2002). A third major issue is immune rejection. Finding ways to prevent it—for example, using somatic nuclear transfer—is clearly important but far from reduction to practice. Finally, there are practical issues, such as the shelf life of polymer cell systems as well as scale-up and production issues. Despite these many challenges, the opportunities in tissue engineering are numerous.

12.3.2 Use of polymeric biomaterials in nanomedicine

Nanomedicine is an emerging area with applications to monitor, repair, construct, and control human biological systems through utilization of devices and structures at the nanodimension. Polymeric biomaterials used in nanomedicine are generally classified as either natural or synthetic polymers. Natural polymers are raw materials that naturally occur in the biological environment; examples include chitosan, albumin, heparin, silk, and collagen (Moreno-Vega et al., 2012; Wulf et al., 2018; Hasirci, 2007). Synthetic polymers are also extensively used in nanomedicine, which include PLA, poly(glycolic acid), copolymer PLGA, polyesters, polyurethanes, and PEG. These polymers have certain advantages such as controllable chemical, structural, and mechanical properties with minimal variations between the batches of synthesized nanomaterials.

Polymeric nanoparticles have gained considerable attention in nanomedicine due to the potential for surface modification, pharmacokinetic control, suitability for targeted delivery of therapeutics (Kamaly et al., 2012), mechanical properties (Banik and Brown, 2014), and design flexibility. More specifically, the size, surface morphology, chemistry, charge, porosity, drug diffusivity, and encapsulation efficiency are properties that push polymeric nanoparticles to the forefront for nanomedicine applications (Kumari et al., 2010). In drug delivery, nanoparticles show great promise by altering the bioavailability, pharmacokinetic, and pharmacodynamic properties of drug molecules to improve therapeutic delivery; however, clinical translation has been slow with the lack of ideal and established solutions for precise targeting, cell internalization, and controlled drug solubility and release (Banik and Brown, 2014).

Multifunctional nanoparticles are the next generation of nanomedicine that will be used to customize therapy for patient-to-patient treatment and eliminate the need for several drug dosages and the potential for deleterious effects of surrounding healthy cells and tissues. The focus of nanoparticles in drug delivery links on the concept of creating single particles with the ability to perform multiple tasks—targeting/detection, imaging to monitor progression, and therapeutic/treatment functionalities (Wang and Thanou, 2010). For example, multifunctional particles with surface functionalization that allow for simultaneous drug delivery and fluorescent biological imaging at desired wavelengths were prepared and demonstrated their dual functions. Khdair et al. (Khdair et al., 2010) introduced another type of multifunctional nanoparticle system based on the natural polymer sodium alginate to counteract the multidrug resistance effect by targeting with two drugs in one particle—chemotherapeutic drug doxorubicin and methylene blue associated with photodynamic therapy.

12.4 Role of scaffold and the loaded drug/growth factor in the integration of extracellular matrix and cells at the interface

The molecular mechanisms that determine survival, differentiation, and movement in multicellular organisms are dependent on interactions with the ECM. Cells in tissues are structurally and functionally integrated with their surrounding ECM via numerous dynamic connections. On the intracellular face of these linkages, adhesion receptors tether the contractile cytoskeleton to the plasma membrane and compartmentalize cytoplasmic signaling events, while at the extracellular face, the same receptors direct the deposition of the ECM itself (Tan and Saltzman, 1990; Salvay and Shea, 2006). Such membrane-proximal functions in turn trigger distal processes, such as alterations in the direction of cell movement, regulation of cell fate construction of ECM networks, and consequent shaping of higher order tissue structure. An understanding of the molecular events that underpin ECM function would therefore help elucidate some of the key organizing principles of multicellular life.

ECM and cell−ECM interactions also contribute widely to disease. Many of the major human diseases are either caused by defects in cell−ECM coordination, are exacerbated by aberrant use of normal cell adhesive processes, or are potentially correctable by altering tissue structure or cell movement (Tan and Saltzman, 1990; Louie et al., 1997; Salvay and Shea, 2006). For example, progressive extracellular remodeling in chronic atherosclerotic, fibrotic, and neurodegenerative diseases leads to a loss of tissue integrity, altered adhesion is a defining characteristic of malignancy, and the pathogenesis of inflammatory and thrombotic diseases relies on aberrant cell aggregation and/or trafficking. The development of strategies to correct ECM dysfunction therefore has enormous promise as a route for improving treatment of many important clinical conditions.

Growth factors initiate and control a variety of cellular processes involved in tissue formation (Table 12.10). Their use in the clinic, however, has been facilitated following advances in recombinant protein technology. Growth factors, growth factor receptors, and monoclonal antibodies are currently being employed to treat clinical conditions such as obesity, cancer, and idiopathic short stature (Xu et al., 2004; He and Matsuda, 2002; Salvay and Shea, 2006), with the potential for use in wound healing and tissue regeneration. Localized delivery of tissue inductive factors from scaffolds can function to direct progenitor cell differentiation toward the desired cell fate. Although in vitro studies of tissue formation on scaffolds can be performed simply by adding growth factors to cell culture media, translation of these studies to applications with in vivo tissue formation requires the use of delivery systems that can make these factors available at the appropriate concentration and duration.

A tissue-engineered implant is a biologic−biomaterial combination in which some components of tissue have been combined with a biomaterial to create a device for the restoration or modification of tissue or organ function. Specific drugs/growth factors, released from a delivery device or from cotransplanted cells, would aid in the induction of host paraenchymal cell infiltration and improve engraftment of codelivered cells for more efficient tissue regeneration or ameliorate disease states. Growth factors are polypeptides that transmit signals to modulate cellular activities. Growth factors can either stimulate or inhibit cellular proliferation, differentiation, migration, adhesion, and gene expression (Salvay and Shea, 2006; Powell et al., 1990; Phan et al., 2016; Beck et al., 2004; Winter et al., 2006; Bonadio, 2000). There are several characteristic properties of growth factors. The characteristic properties of growth factors are described to provide a biological basis for their use in tissue-engineered devices. The principles of polymeric device development for therapeutic growth factor or drug delivery in the context of tissue engineering are proposed.

Table 12.10 Growth factors delivered to promote tissue formation.

Growth factor	MW/ kDa	Functions
Nerve growth factor (NGF)	27.0	Promotes neuron survival and extension in CNS and PNS; modulates differentiation of various neuron types in vivo and in vitro; plays role in tissue repair and fibrosis
Insulin-like growth factor 1 (IGF-1)	27.9	Mediates actions of growth hormone; increases proteoglycans and type II collagen synthesis
Insulin-like growth factor-2 (IGF-2)	35.1	Promotes myogenic differentiation of ES cells
Epidermal growth factor (EGF)	6.2	Wound healing
Fibroblast growth factor-1 (FGF-1)	17.5	Would healing; vascular repair; fibroblast mitogen
Fibroblast growth factor-2 (FGF-2)	17.3	Chondrogenesis; angiogenesis; neuronal and endothelial cell proliferation
Platelet-derived growth factor (PDGF)	22–25	Maturation of blood vessels; recruitment of SMCs to developing vasculature; wound healing; neural regeneration
Bone morphogenic protein-2 (BMP-2)	44.7	Osteogenesis; angiogenesis
Transforming growth factor β1 (TGF-β1)	25.0	Promotes chondrogenic differentiation; increases cartilage matrix synthesis and chondrocyte proliferation
Vascular endothelial growth factor (VEGF)	19–22	Angiogenesis; vasculogenesis; osteogenesis; neurotrophic factor for motor neurons; cartilage remodeling

12.4.1 Induction of angiogenesis in tissue-engineered scaffolds for bone repair: a combined gene therapy–cell transplantation approach

One of the fundamental principles underlying tissue engineering approaches is that newly formed tissue must maintain sufficient vascularization to support its growth. Efforts to induce vascular growth into tissue-engineered scaffolds have recently been dedicated to developing novel strategies to deliver specific biological factors that direct the recruitment of EC progenitors and their differentiation (Beck et al., 2004; Bonadio, 2000). The challenge, however, lies in orchestration of the cells, appropriate biological factors, and optimal factor doses. Beck et al. (Beck et al., 2004) reported a novel approach as a step forward to resolve this dilemma by combining an ex vivo gene transfer strategy and EC transplantation. The utility of this approach was evaluated using 3D PLGA-sintered microsphere scaffolds for bone tissue engineering applications. Adipose-derived stromal cells (ADSCs) were isolated and transfected with adenovirus encoding the cDNA of vascular endothelial growth factor (VEGF). They demonstrated that the combination of VEGF releasing ADSCs and ECs results in marked vascular growth within PLGA scaffolds and thereby delineates the potential of ADSCs to promote vascular growth into biomaterials (Beck et al., 2004).

Similarly, delivery of multiple growth factors to sites of bone injury was shown to dramatically enhance bone regeneration. Dual delivery of bone morphogenic protein-2 (BMP-2) and transforming growth factor-β3 (TGF-β3) from a hydrogel promoted significant bone formation by cotransplanted bone marrow stem cells within 6 weeks of implantation (Salvay and Shea, 2006). Interestingly, the synergistic activity of these two factors allowed them to be provided at a low dose, whereas supraphysiological concentrations of the individual factors resulted in negligible bone tissue formation. In addition to dual protein delivery, DNA and protein delivery can be combined to provide the necessary factors for tissue formation. Combined delivery of plasmid DNA encoding for BMP-4, VEGF protein, and human bone marrow stromal cells significantly enhanced bone formation relative to delivery of any single factor.

12.4.2 Hydrogel composite materials for enhanced neurotrophin delivery in neural prostheses

Long-term implanted electrode arrays have a significant importance in neural prosthetics for recording the electrical signals from nearby neurons and generating electrical signals to stimulate nearby tissue. Therefore, the density of neurons and their proximity to the electrode sites play an important role in the electrode performance (Winter et al., 2006). Many biodegradable and biocompatible polymers have been used as electrode coatings to minimize the acute and chronic inflammatory response and to preserve neurons near the recording sites. As PEG is nontoxic, nonimmunogenic, and inert to most biological molecules, such as proteins, block copolymers of PEG have been studied in by various groups (Gref et al., 1994; Bazile et al., 1995; Niu et al., 2018; Wagner et al., 2018; Salehiabar et al., 2018; Quellec et al., 1998, 1999; Chandy et al., 2001; Winter et al., 2006) to form hydrogels as the scaffolds that can support or stimulate neuron growth. A composite drug delivery system comprising polymeric hydrogels (i.e., PEGPLA or PEG-polycaprolactone, [PEGPCL]) and other vehicles, such as biodegradable PLGA microspheres or PCL-based electrospun nanofibrous scaffolds, was developed to provide neurotrophins at a predetermined rate for 2–3 months in vivo (Winter et al., 2006). The stable and sustained release of neurotrophins can remarkably enhance the attraction, attachment, and restoration of neurons around the chronically implanted electrodes.

12.4.3 Fine-tuning Notch signaling to promote angiogenesis

Tissue engineering scaffolds releasing proteins mimic the natural reservoir capacity of the extracellular matrix. Coupling delivery of protein with a degradable carrier capable of organizing tissue formation is a powerful approach for tissue regeneration. Promoting angiogenesis is both a critical aspect for tissue regeneration and a potential effective therapy for cardiovascular diseases. Inspired by the recent findings that systemic introduction of Notch inhibitors reduced blood flow to tumor tissues by forming excessive yet dysfunctional vasculature, various studies were conducted (Bonadio., 2000; Babensee et al., 2000) and proposed that precisely controlled partial and local Notch inhibition might enhance regional neovascularization, by altering the responsiveness of local ECs and/or progenitors to angiogenic stimuli. In vivo delivery of an appropriate combination of gamma secretase inhibitor (GSI) and VEGF from an injectable Alg hydrogel system to ischemic hind limbs led to a greater recovery of blood flow than VEGF or GSI alone; perfusion levels reached 80% of the normal level by week 4 with combined GSI and VEGF delivery (Babensee et al., 2000; Rowley et al., 1999).

Strikingly, direct intramuscular or intraperitoneal injection of GSI did not result in the same level of improvement, suggesting that the extended presence of GSI (gel delivery) is important for its activity. The optimal dose of GSI delivered from Alg hydrogels did not show any adverse effects, in contrast to systemic introduction of GSI. Altogether, these results suggested a new approach to promote angiogenesis by finetuning Notch signaling and may provide new options to treat patients with diseases, such as diabetes, which can diminish angiogenic responsiveness (Rowley et al., 1999).

12.4.4 Myocardial tissue engineering via growth hormones

The concept that growth hormone (GH) and insulin-like growth factor 1 (IGF-1), the mediator of many of the effects of GH on peripheral tissues, target the heart has recently emerged from a series of animal and human studies. Conditions of GH/IGF-1 deficiency in humans are associated with cardiac atrophy and impaired cardiac function. The hypertrophic response with enhanced cardiac performance observed in rats subjected to chronic GH/IGF-1 excess appears to be beneficial in the setting of experimental and, recently, human heart failure (Davis et al., 2006; Bleiziffer et al., 2007). It is likely that the site-specific delivery of GH or IGF-1 via polymeric scaffolds can regenerate the cardiac tissues through cell proliferation and differentiation (Rowley et al., 1999; Davis et al., 2006).

Substantial data suggest that IGF-1 is a potent cardiomyocyte growth and survival factor (Davis et al., 2006). Mice deficient in IGF-1 have increased apoptosis after myocardial infarction, whereas cardiac-specific IGF-1 over expression protects against myocyte apoptosis and ventricular dilation after infarction (Salvay and Shea, 2006). IGF-1 over expression increases cardiac stem cell number and growth, leading to an increase in myocyte turnover and function in the aging heart. After infarction, IGF-1 promotes engraftment, differentiation, and functional improvement of embryonic stem cells transplanted into myocardium. Davis et al. (Davis et al., 2006) developed a delivery system using a "biotin sandwich" approach that allows coupling of a factor to peptide nanofibers without interfering with self-assembly. Biotinylation of selfassembling peptides allowed specific and highly controlled delivery of IGF-1 to local myocardial microenvironments, leading to improved results of cell therapy. This approach allowed a greater control of the intramyocardial environment by delivering growth factors to injured myocardium. With this system, it may be possible to design the local microenvironment to improve the endogenous regenerative response, for example, by delivery of a chemoattractant to promote stem cell migration. Although much more must be learned about why mammals have inadequate cardiac regeneration, the ability to control the local myocardial microenvironment may prove critical to preventing heart failure.

12.4.5 Multiple factor delivery for vascular tissue engineering

Tissue morphogenesis and regeneration are typically driven by the concomitant action of multiple factors, which can work synergistically on the same process, or can target different barriers to regeneration. The synergistic effect of growth factors has been reported for many developmental processes including angiogenesis, where mature blood vessels form by the combined action of VEGF and platelet-derived growth factor (PDGF) to form stable vessels (Salvay and Shea, 2006; Davis et al., 2006). Although VEGF can initiate angiogenesis, PDGF promotes vessel maturation via recruitment of smooth muscle cells to the developing endothelium. PLA/polyglycolic acid scaffolds releasing both VEGF and PDGF formed a mature vascular network within and around the scaffolds.

The concentration and duration of function for tissue inductive factors at the regenerating tissue site are critical parameters involved in promoting developmental processes and the formation of mature tissues. Therapeutic angiogenesis and antiangiogenesis reveal these concepts, as immature vessels or vessels that regress over time can lead to unsuccessful or abnormal tissue formation. Sustained expression of low to medium levels of VEGF or hormone combinations are required to promote the growth of blood vessels displaying normal morphological and functional characteristics (Salvay and Shea, 2006; Powell et al., 1990; Phan et al., 2016; Beck et al., 2004; Winter et al., 2006; Bonadio, 2000). Cells transplanted that expressed low levels of VEGF avoided the formation of aberrant vessels and hemangiomas observed with cells expressing high levels of VEGF. Thus, it is assumed that multifactorial presentation of growth factors can be more effective at stimulating natural developmental processes leading to tissue formation.

12.4.6 Cell sources for tissue engineering applications

Biomaterials-based approaches to engineering soft connective tissue can potentially offer an alternative to current tissue transfer techniques that require donor site tissue. The advantages of an off-the-shelf adipose tissue replacement include availability and avoiding donor site morbidity. Advances in the field of regenerative medicine contribute to the development of more biomimetic materials that provide the foundation for tissue repair and regeneration.

One source of cells for culture is the cells from adipose tissues. The parenchymal cells of adipose tissue are adipocytes, terminally differentiated lipid-filled cells providing a key energy reserve for the rest of the body. Although they are the primary cells transplanted in fat grafting procedures, their inability to replicate, high metabolic demands, and fragile nature limit their utility in adipose tissue regeneration. As an alternative, many researchers have looked to progenitor cells that can be readily expanded and can differentiate into adipocytes. Bone marrow adult stem cells are another source to be investigated for cell sourcing for tissue engineering applications. A subset of multipotent progenitor cells was found to reside in the bone marrow niche and could be induced to differentiate down specific mesenchymal lineages (Bleiziffer et al., 2007; Pittenger et al., 1999). The bone marrow aspirate is collected from the iliac crest, followed by selection for plastic-adherent cells.

The embryonic stem cells are also another source for similar use. Embryonic stem cells are pluripotent cells that have high proliferative capacity in culture and can be expanded through far more passages compared with adult stem cells without reaching senescence. The cells are isolated from the inner cell mass of blastocysts and can be differentiated toward numerous somatic cells with varying phenotypes. However, the use of embryonic stem cells in research is limited because of various ethical issues. Recently, researchers discovered that it was possible to induce adult somatic cells to revert to an embryonic-like state by genetically reprogramming the cells to express their various transcription factors (Takahashi et al., 2007; Takahashi and Yamanaka, 2006; Yu et al., 2007). Cells expressing these factors exhibited key characteristics of embryonic stem cell pluripotency, including embryoid body and teratoma formation, and proved capable of differentiating into cells of all three germ layers. As a result, any adult somatic cell may be used as a cell source, thereby circumventing the ethical issues controlling the use of embryonic stem cells in research.

12.5 Future outlook on combination devices with drug delivery and tissue engineering

Combination products that converge medical devices with therapeutics and diagnostics are a rapidly growing segment of the health-care industry. Combination devices—those comprising drug releasing components together with functional prosthetic implants—represent a versatile, emerging clinical technology promising to provide functional improvements to implant devices in several classes (Shmulewitz and Langer, 2006; Langer, 2006). Landmark antimicrobial catheters and the drug-eluting stent have heralded the entrance and significantly, routes to FDA approval, for these devices into clinical practice. Most prominent are new combination devices representing current orthopedic and cardiovascular implants with newly added capabilities from onboard or directly associated drug delivery systems now under development. Wound coverings and implantable sensors will also benefit from this combination enhancement. Infection mitigation, a common problem with implantable devices, is a current primary focus. Ongoing progress in cell-based therapeutics, progenitor cell exploitation, growth factor delivery, and advanced formulation strategies will provide a more general and versatile basis for advanced combination device strategies (Langer, 2006; Jeong and Kohane, 2011; Sakamoto-Ozaki et al., 2012; Luo et al., 2006). These seek to improve tissue—device integration and functional tissue regeneration.

Although the field of biomedical engineering encompasses much more than combination products, many of today's most promising areas of development combine biologic, pharmaceutical, and/or medical device components. Among the most notable biomedically engineered combination products are neuromodulating devices, tissue engineering technologies, and nanomedicines (Langer, 2006; Jeong and Kohane, 2011). The drug-eluting stents significantly reduce coronary in-stent restenosis compared with bare metal stents (Schwartz et al., 2004). However, recent studies have shown that the late stent thrombosis defined as thrombosis after 30 days of postdeployment is a growing safety concern with drug-eluting stents (Schwartz et al., 2004). The drug-eluting stents did not fully reendothelialize after 40 months of implantation, whereas the bare stents can completely reendothelialize after 6–7 months (Langer, 2006; Schwartz et al., 2004). This impaired initial healing of drug-eluting stents may be caused by the polymer matrix, the drug, or a combination of two (Schwartz et al., 2004). Recent studies have indicated that the surface engineering of drug-eluting stents with extracellular components (such as collagen and LN) can enhance the healing and reendothelialization faster (Schwartz et al., 2004; Virmani et al., 2003). Thus, it seems that a combination of drug release with surface engineering stents may be a novel way of neointima formation and naturalization.

Future combination devices might best be completely redesigned de novo to deliver multiple bioactive agents over several spatial and temporal scales to enhance prosthetic device function, instead of the current "add-on" approach to existing implant device designs never originally intending to function in tandem with drug delivery systems (Jeong and Kohane, 2011; Sakamoto-Ozaki et al., 2012). As a future perspective, it is also important to improve the polymeric biomaterials with optimum properties to reduce infection and inflammation and suit the tissue interfaces. Biointegration refers to the interconnection between a biomedical device and the recipient tissue. Thus, the selection of appropriate polymeric biomaterials can only be iteratively improved with information from postimplant monitoring of devices in existing clinical use. It is possible to enhance biointegration by means of modifying surface chemistry and by drug delivery approaches.

References

Abu-Amer, Y., Darwech, I., Clohisy, J.C., 2007. Aseptic loosening of total joint replacements: mechanisms underlying osteolysis and potential therapies. Arthritis Res. Ther. 9, S6.

Aghajanzadeh, M., Zamani, M., Rashidzadeh, H., Rostamizadeh, K., Sharafi, A., Danafar, H., 2018. Amphiphilic Y shaped miktoarm star copolymer for anticancer hydrophobic and hydrophilic drugs codelivery: Synthesis, characterization, in vitro, and in vivo biocompatibility study. J Biomed Mater Res Part A 106A, 2817–2826.

Ali, A., Ahmed, S., 2018. A review on chitosan and its nanocomposites in drug delivery. Int. J. Biol. Macromol. 109, 273–286.

Anderson, J.M., 1993. Mechanisms of inflammation and infection with implanted devices. Cardiovasc. Pathol. 2, 33S–41S.

Anderson, J.M., 2001. Biological response to materials. Ann. Rev. Mater. Res. 31, 81–110.

Andrews, R.K., Fox, J.E.B., 1991. Platelet receptors in hemostasis. Curr. Opin. Cell Biol. 2, 894−903.
Babensee, J.E., McIntire, L.V., Mikos, A.G., 2000. Growth factor delivery for tissue engineering. Pharm. Res. 17, 497−504.
Baier, R.E., Depalma, V.A., Goupil, D.W., Cohen, E., 1985. Human platelets spreading on substrata of known surface chemistry. J. Biomed. Mater. Res. 19, 1157−1167.
Banik, B.L., Brown, J.L., 2014. Polymeric biomaterials in nanomedicine. In: Natural and Synthetic Biomedical Polymers. Elsevier Inc, pp. 387−394.
Barralet, J.E., Wallace, L.L., Strain, A.J., 2003. Tissue engineering of human biliary epithelial cells on polyglycolic acid/polycaprolactone scaffolds maintains long-term phenotypic stability. Tissue Eng. 9, 1037−1046.
Bazile, D., Prudhomme, C., Bassoullet, M.T., Marlard, M., Spenlenhauer, G., Veillard, M., 1995. Stealth Me−PEG−PLA nanoparticles avoid uptake by the mononuclear phagocyte system. J. Pharm. Sci. 84, 493−498.
Beck, C., Uramoto, H., Boren, J., Akyurek, L.M., 2004. Tissue-specific targeting for cardiovascular gene transfer, potential vectors and future challenges. Curr. Gene Ther. 4, 457−467.
Behl, G., Kumar, P., Sikka, M., Fitzhenry, L., Chhikara, A., 2018. PEG-coumarin nanoaggregates as π-π stacking derived small molecule lipophile containing self-assemblies for anti-tumour drug delivery. J. Biomater. Sci. Polym. Ed. 29, 360−375.
Berk, B.C., Harris, K., 1995. Restenosis after percutaneous transluminal coronary angioplasty: new therapeutic insights from pathogenic mechanisms. Adv. Intern. Med. 40, 445−501.
Bleiziffer, O., Eriksson, E., Yao, F., Horch, R.E., Kneser, U., 2007. Gene transfer strategies in tissue engineering. J. Cell Mol. Med. 11, 206−223.
Bonadio, J., 2000. Tissue engineering via local gene delivery. J. Mol. Med. 78, 303−311.
Bos, G.W., Scharenborg, N.M., Poot, A.A., Engbers, G.H.M., Terlingen, J.G.A., Beugeling, T.B., Aken, W.V., Feijen, J., 1998. Adherence and proliferation of endothelial cells on surface-immobilized albumin-heparin conjugate. Tissue Eng. 4, 267−279.
Boudot, C., Boccoz, A., Duregger, K., Kuhnla, A., 2016. A novel blood incubation system for the in-vitro assessment of interactions between platelets and biomaterial surfaces under dynamic flow conditions: the Hemocoater. J. Biomed. Mater. Res. A 104, 2430−2440.
Brash, J.L., 2000. Exploiting the current paradigm of blood-material interactions for the rational design of blood-compatible materials. J. Biomater. Sci. Polym. Ed. 11, 1135−1146.
Bridges, A.W., Garcia, A.J., 2008. Anti-inflammatory polymeric coatings for implantable biomaterials and devices. J. Diabetes Sci. Technol. 2, 984−994.
Brigger, I., Dubernet, C., Couvreur, P., 2002. Nanoparticles in cancer therapy and diagnosis. Adv. Drug Deliv. Rev. 54, 631−651.
Buma, P., Pieper, J.S., van Tienen, T., van Susante, J.L., van der Kraan, P.M., Veerkamp, J.H., van den Berg, W.B., Veth, R.P., van Kuppevelt, T.H., 2003. Cross-linked type I and type II collagenous matrices for the repair of full-thickness articular cartilage defects: a study in rabbits. Biomaterials 24, 3255−3276.
Caliari, S.R., Ramirez, M.A., Harley, B.A.C., 2011. The development of collagen-GAG scaffold- membrane composites for tendon tissue engineering. Biomaterials 32 (34), 8990−8998.
Chaffin, K.A., Wilson, C.L., Himes, A.K., Dawson, J.W., Haddad, T.D., Buckalew, A.J., Miller, J.P., Untereker, D.F., Simha, N.K., 2013. Abrasion and fatigue resistance of PDMS containing multiblock polyurethanes after accelerated water exposure at elevated temperature. Biomaterials 34, 8030−8041.

Chandy, T., Rao, G.H.R., 2002. Preparation of surface-engineered elastin/lamin nerve guide tubes of poly (lactic acid)/poly (ethylene vinyl acetate). J. Bioact. Compat Polym. 17, 183−194.

Chandy, T., Sharma, C.P., 1993. Chitosan matrix for oral sustained delivery of ampicillin. Biomaterials 14, 939−944.

Chandy, T., Mooradian, D.L., Rao, G.H.R., 1998. Chitosan/Polyethylene glycol-alginate microspheres for oral delivery of hirudin. J. Appl. Polym. Sci. 70, 2143−2153.

Chandy, T., Rao, G.H.R., Wilson, R.F., Das, G.S., 2001. Development of poly (lactic acid)/chitosan co- matrix microspheres: controlled release of Taxol-heparin for preventing restenosis. Drug Deliv. 8, 77−86.

Chandy, T., Rao, G.H.R., Wilson, R.F., Das, G.S., 2003. The development of porous Alginate/Elastin/PEG composite matrix for cardiovascular engineering. J. Biomater. Appl. 17, 287−301.

Chandy, T., Van Hee, J., Nettekoven, W., Johnson, J., 2009. Long-term in vitro stability assessment of polycarbonate urethane micro catheters: resistance to oxidation and stress cracking. J. Biomed. Mater. Res. B Appl. Biomater. 89, 314−324.

Chandy, T., Das, G.S., Wilson, R.F., Rao, G.H.R., 1999. Surface immobilized biomolecules on albumin modified porcine pericardium for preventing thrombosis and calcification. Int. J. Artif. Organs 22, 547−558.

Chandy, T., Das, G.S., Wilson, R.F., Rao, G.H.R., 2000. Use of plasma glow for surface-engineering biomolecules to enhance blood compatibility of Dacron and PTFE vascular prosthesis. Biomaterials 21, 699−712.

Chandy, T., Das, G.S., Wilson, R.F., Rao, G.H.R., 2002. Development of polylactide microspheres for protein encapsulation and delivery. J. Appl. Polym. Sci. 86, 1285−1295.

Chatterjee, D.K., Zhang, Y., 2007. Multi-functional nanoparticles for cancer therapy. Sci. Technol. Adv. Mater. 8, 131−133.

Cheng, H., Yue, K., Kazemzadeh-Narbat, M., Liu, Y., Khalilpour, A., Li, B., Zhang, Y.S., Annabi, N., Khademhosseini, A., 2017. Mussel-inspired multifunctional hydrogel coating for prevention of infections and enhanced osteogenesis. ACS Appl. Mater. Interfaces 9, 11428−11439.

Christenson, E.M., Wiggins, M.J., Anderson, M., Hiltner, A., 2005a. Surface modification of poly(ether urethane urea) with modified dehydroepiandrosterone for improved *in vivo* biostability. J. Biomed. Mater. Res. 73A, 108−115.

Christenson, E.M., Dadsetan, M., Hiltner, A., 2005b. Biostability and macrophage-mediated foreign body reaction of silicone-modified polyurethanes. J. Biomed. Mater. Res. A 74, 141−155.

Cleland, J.L., 1997. Protein delivery from biodegradable microspheres. Pharm. Biotechnol. 10, 1−43.

Cornelius, R.M., Macri, J., Brash, J.L., 2011. Interfacial interactions of apolipoprotein AI and high-density lipoprotein: overlooked phenomena in blood-material contact. J. Biomed. Mater. Res. A 99, 109−115.

Cory, A.J., 2013. Chemical and biochemical degradation of polymers intended to be biostable. In: Ratner, B.D., Hoffman, A.S., Schoen, F.J., Lemons, J.E. (Eds.), Biomaterials Science: An Introduction to Materials in Medicine, third ed. Academic Press, Elsevier, Waltham, MA, USA.

Cosgriff-Hernandez, E., Tkatchouk, E., Touchet, T., Sears, N., Kishan, A., Jenney, C., Padsalgikar, A.D., Chen, E., 2016. Comparison of clinical explants and accelerated hydrolytic aging to improve biostability assessment of silicone-based polyurethanes. J. Biomed. Mater. Res. A 104, 1805−1816.

Cozzens, D., Ojha, U., Kulkarni, P., Faust, R., Desai, S., 2010. Long term in vitro biostability of segmented polyisobutylene-based thermoplastic polyurethanes. J. Biomed. Mater. Res. A 95, 774−782.

Das, G.S., Rao, G.H.R., Wilson, R.F., Chandy, T., 2001. Controlled delivery of Taxol from poly (ethylene glycol)-coated poly (lactic acid) microspheres. J. Biomed. Mater. Res. 55, 96−103.

Davis, M., Hsieh, P.C.H., Takahashi, T., Song, Q., Zhang, S., Kamm, R.D., Grodzinsky, A.J., Anversa, P., Lee, R.T., 2006. Local myocardial insulin-like growth factor 1 (IGF-1) delivery with biotinylated peptide nanofibers improve cell therapy for myocardial infarction. Proc. Natl. Acad. Sci. 103, 8155−8160.

Destouet, J., Monsees, B., Oser, R., Nemecek, J., Young, V., Pilgram, T., 1992. Screening mammography in 350 women with breast implants: prevalence and findings of implant complications. Am. J. Roentgenol. 159, 973−978.

Doi, K., Matsuda, T., 1997. Enhanced vascularization in a microporous polyurethane graft impregnated with basic fibroblast growth factor and heparin. J. Biomed. Mater Res. 34, 361−370.

Faxon, D.P., Currier, J.W., 1995. Prevention of post PTCA restenosis. Ann. N. Y. Acad. Sci. 748, 419−428.

Flanagan, T.C., Pandit, A., 2003. Living artificial heart valve alternatives: a review. Eur. Cell Mater. 6, 28−42.

Frenkel, S.R., Di Cesare, P.E., 2004. Scaffolds for articular cartilage repair. Ann. Biomed. Eng. 32, 26−34.

Gallagher, G., Padsalgikar, A., Tkatchouk, E., Jenney, C., Iacob, C., Runt, J., 2017. Environmental stress cracking performance of polyether and PDMS-based polyurethanes in an in vitro oxidation model. J. Biomed. Mater. Res. B Appl. Biomater. 105, 1544−1558.

Gorbet, M.B., Sefton, M.V., 2004. Biomaterial-associated thrombosis: roles of coagulation factors, complement, platelets and leukocytes. Biomaterials 25, 5681−5703.

Grad, S., Kupcsik, L., Gorna, K., Gogolewski, S., Alini, M., 2003. The use of biodegradable polyurethane scaffolds for cartilage tissue engineering: potential and limitations. Biomaterials 24, 5163−5174.

Gref, R., Minamitake, Y., Peracchia, M.T., Trubetskoy, V., Torchilin, V., Langer, R., 1994. Biodegradable long-circulating nanospheres. Science 263, 1600−1603.

Hansson, A., Wenger, A., Henriksson, H.B., Li, S., Johansson, B.R., Brisby, H., 2017. The direction of human mesenchymal stem cells into the chondrogenic lineage is influenced by the features of hydrogel carriers. Tissue Cell 49, 35−44.

Harris, C.T., Cooper, L.F., 2004. Comparison of bone graft matrices for human mesenchymal stem cell-directed osteogenesis. J. Biomed. Mater. Res. 68, 747−759.

Hasirci, N., 2007. Micro and nano systems in biomedicine and drug delivery. In: Mozafari, M.R. (Ed.), Nanomaterials and Nano Systems for Biomedical Applications. Springer, The Netherlands, pp. 1−26.

He, H., Matsuda, T., 2002. Arterial replacement with compliant hierarchic hybrid vascular graft: biomechanical adaptation and failure. Tissue Eng. 8, 213−220.

Hegewald, A.A., Medved, F., Feng, D., Tsagogiorgas, C., Beierfuß, A., Schindler, G.A., Trunk, M., Kaps, C., Mern, D.S., Thomé, C., 2015. Enhancing tissue repair in annulus fibrosus defects of the intervertebral disc: analysis of a bio-integrative annulus implant in an in-vivo ovine model. J. Tissue Eng. Regenerat. Med. 9, 405−414.

Heller, J., 1980. Controlled release of biologically active compounds from bioerodible polymers. Biomaterials 1, 51−57.

Hutmacher, D.W., Ng, K.W., Kaps, C., Sittinger, M., Klaring, S., 2003. Elastic cartilage engineering using novel scaffold architectures in combination with a biomimetic cell carrier. Biomaterials 24, 4445–4457.

Jain, R.K., 2001. Delivery of molecular and cellular medicine to solid tumors. Adv. Drug Deliv. Rev. 46, 149–168.

Jeong, K.J., Kohane, D.S., 2011. Surface modification and drug delivery for biointegration. Ther. Deliv. 2, 737–752.

Joist, J.H., Pennington, D.C., 1987. Platelet reactions with artificial surfaces. Trans. Am. Soc. Artif. Intern. Organs 4, 33–341.

Kamaly, N., Xiao, Z., Valencia, P.M., Radovic-Moreno, A.F., Farokhzad, O.C., 2012. Targeted polymeric therapeutic nanoparticles: design, development and clinical translation. Chem. Soc. Rev. 41, 2971–3010. 89.

Khdair, A., Chen, D., Patil, Y., Ma, L., Dou, Q.P., Shekhar, M.P.V., et al., 2010. Nanoparticle-mediated combination chemotherapy and photodynamic therapy overcomes tumor drug resistance. J. Control. Release 141, 137–144.

Kito, H., Matsuda, T., 1996. Biocompatible coatings for luminal and outer surfaces of small-caliber artificial grafts. J. Biomed. Mater. Res. 30, 321–330.

Kumari, A., Yadav, S.K., Yadav, S.C., 2010. Biodegradable polymeric nanoparticles based drug delivery systems. Colloids Surf. B 75, 1–18.

Langer, R., 1980. Polymeric delivery system for controlled drug release. Chem. Eng. Commun. 6, 1–48.

Langer, R., 2006. Biomaterials for drug delivery and tissue engineering. MRS Bull. 31, 477–485.

Laurencin, C.T., Kumbar, S.G., Deng, M., James, R., 2010. Nano-structured scaffolds for regenerative engineering. Honorary Series in Translational Research in Biomaterials. HE Annual Meeting, Salt Lake City, Utah, USA.

Li, Y., Ye, D., Li, M., Ma, M., Gu, N., March 14, 2018. Adaptive materials based on iron oxide nanoparticles for bone regeneration. Chemphyschem 19 (16).

Li, Y., Zhang, H., Chen, Y., Ma, J., Lin, J., Zhang, Y., Fan, Z., Su, G., Xie, L., Zhu, X., Hou, Z., 2018. Integration of phospholipid-hyaluronic acid-methotrexate nanocarrier assembly and amphiphilic drug-drug conjugate for synergistic targeted delivery and combinational tumor therapy. Biomater. Sci. 6, 1818–1833.

Liu, W.F., Ma, M., Bratlie, K.M., Dang, T.T., Langer, R., Anderson, D.G., 2011. Real-time in vivo detection of biomaterial-induced reactive oxygen species. Biomaterials 32, 1796–1801.

Liu, Z., Jiao, Y., Wang, T., Zhang, Y., Xue, W., 2012. J. Control. Release 160, 14–24.

Louie, L.K., Yannas, I.V., Hsu, H.P., Spector, M., 1997. Healing of tendon defects implanted with a porous collagen-GAG matrix; histological evaluation. Tissue Eng. 3, 187–195.

Louzao, I., Koch, B., Taresco, V., Ruiz-Cantu, L., Irvine, D.J., Roberts, C.J., Tuck, C., Alexander, C., Hague, R., Wildman, R., Alexander, M.R., February 28, 2018. Identification of novel "Inks" for 3D printing using high-throughput screening: bioresorbable photocurable polymers for controlled drug delivery. ACS Appl. Mater. Interfaces 10 (8), 6841–6848.

Luo, Y., Kobler, J., Zeitels, S., Langer, R., 2006. Effects of growth factors on extracellular matrix production by vocal fold fibroblasts in 3-dimensional culture. Tissue Eng. 12, 3365–3374.

Maeguerie, G.A., Plow, E.F., Edgungton, T.S., 1979. Human platelets possess an inducible and saturable receptor specific for fibrinogen. J. Biol. Chem. 254, 5357–5363.

Mai, B.T., Fernandes, S., Balakrishnan, P.B., Pellegrino, T., 2018. Nanosystems based on magnetic nanoparticles and thermo- or pH-responsive polymers: an update and future perspectives. Acc. Chem. Res. 51, 999−1013.

Marini, M., De Niederhausern, S., Iseppi, R., Bondi, M., Sabia, C., Toselli, M., et al., 2007. Antibacterial activity of plastics coated with silver-doped organic-inorganic hybrid coatings prepared by sol-gel processes. Biomacromolecules 8 (4), 1246_1254.

Mei, Q., Luo, P., Zuo, Y., Li, J., Zou, Q., Li, Y., Jiang, D., Wang, Y., 2018. Formulation and in vitro characterization of rifampicin-loaded porous poly (ε-caprolactone) microspheres for sustained skeletal delivery. Drug Des. Dev. Ther. 12, 1533−1544.

Mishra, A., Seethamraju, K., Delaney, J., Willoughby, P., Faust, R., 2015. Long-term in vitro hydrolytic stability of thermoplastic polyurethanes. J. Biomed. Mater. Res. A 103, 3798−3806.

Montalescot, G., Barragan, P., Wittenberg, O., Ecollan, P., Elhadad, S., Villain, P., Pinton, P., 2001. Platelet glycoprotein IIb/IIIa inhibition with coronary stenting for acute myocardial infarction. N. Engl. J. Med. 344, 1895−1903.

Moreno-Vega, A.-I., Gómez-Quintero, T., Nuñez-Anita, R.-E., Acosta-Torres, L.-S., Castaño, V., 2012. Polymeric and ceramic nanoparticles in biomedical applications. J. Nanotechnol. 2012, 1−10. https://doi.org/10.1155/2012/936041.

Morois, Y., Chakfe, N., Guidoin, R., Duhamel, R.C., Roy, R., Marois, M., King, M.W., Douville, Y., 1996. An albumin-coated polyester arterial graft: in vivo assessment of biocompatibility and healing characteristics. Biomaterials 17, 3−14.

Morris, J., Shebuski, R.J., 1995. Small molecule approaches to the prevention of restenosis. Curr. Pharm. Design 1, 469−482.

Nagaoka, S., Mori, Y., Takiuchi, H., Yokota, K., Tanzawa, H., Nishiumi, S., 1984. In: Shalaby, S.W., Hoffman, A.S., Ratner, B.D., Horbett, T.A. (Eds.), Polymers, as Biomaterials. Plenum, New York, p. 361.

Nguyen, T.Y., Cipriano, A.F., Guan, R.G., Zhao, Z.Y., Liu, H., 2015. In vitro interactions of blood, platelet, and fibroblast with biodegradable magnesium-zinc-strontium alloys. J. Biomed. Mater. Res. A 103, 297−2986.

Nihant, N., Grandfils, C., Jerome, R., Teyssie, P., 1995. Microencapsulation by coacervation of poly(lactic-*co*-glycolide) IV. Effect of the processing parameters on coacervation and encapsulation. J. Control. Release 35, 117−125.

Nikoubashman, O., Heringer, S., Feher, K., Brockmann, M.A., Sellhaus, B., Dreser, A., Kurtenbach, K., Pjontek, R., Jockenhövel, S., Weis, J., Kießling, F., Gries, T., Wiesmann, M., 2018. Development of a polymer-based biodegradable neurovascular stent prototype: a preliminary in vitro and in vivo study. Macromol. Biosci. 118, e1700292−e1800014.

Niu, Y., Li, Q., Ding, Y., Dong, L., Wang, C., 2018. Engineered delivery strategies for enhanced control of growth factor activities in wound healing. Adv. Drug Deliv. Rev. S0169−409X (18), 30133−30139.

Noishiki, Y., Chvapil, M., 1987. Healing pattern of collagen-impregnated and preclotted vascular grafts in dogs. Vasc. Surg. 11, 21−401.

Nosrati, H., Abbasi, R., Charmi, J., Rakhshbahar, A., Aliakbarzadeh, F., Danafar, H., Davaran, S., 2018. Folic acid conjugated bovine serum albumin: an efficient smart and tumor targeted biomacromolecule for inhibition folate receptor positive cancer cells. Int. J. Biol. Macromol. 117, 1125−1132.

Padsalgikar, A., Cosgriff-Hernandez, E., Gallagher, G., Touchet, T., Iacob, C., Mellin, L., Norlin-Weissenrieder, A., Runt, J., 2015. Limitations of predicting in vivo biostability of multiphase polyurethane elastomers using temperature-accelerated degradation testing. J. Biomed. Mater. Res. B Appl. Biomater. 103, 159−168.

Pandiyarajan, C.K., Prucker, O., Zieger, B., Rühe, J., 2013. Influence of the molecular structure of surface-attached poly (N-alkyl acrylamide) coatings on the interaction of surfaces with proteins, cells and blood platelets. Macromol. Biosci. 13, 873−884.

Phan, V.H., Thambi, T., Duong, H.T., Lee, D.S., 2016. Poly(amino carbonate urethane)-based biodegradable, temperature and pH-sensitive injectable hydrogels for sustained human growth hormone delivery. Sci. Rep. 6, 29978.

Pittenger, M.F., et al., 1999. Multilineage potential of adult human mesenchymal stem cells. Science 284, 143−147.

Popovtzer, R., 2008. Targeted gold nanoparticles enable molecular CT imaging of cancer. Nano Lett. 8, 4593−4596.

Portney, N.G., Ozkan, M., 2006. Nano-oncology: drug delivery, imaging, and sensing. Anal. Bioanal. Chem. 384, 620−630.

Powell, E.M., Sobarzo, M.R., Saltzman, W.M., 1990. Controlled release of nerve growth factor from a polymeric implant. Brain Res. 515, 309−311.

Procaccini, R.A., Studdert, C.A., Pellice, S.A., 2014. Silver doped silica-methyl hybrid coatings. Structural evolution and antibacterial properties. Surf. Coat. Technol. 244, 92−97.

Quellec, P., Gref, R., Perrin, L., Dellacherie, E., Sommer, F., Verbavatz, J.M., Plonso, M.J., 1998. Protein encapsulation within polyethylene glycol-coated nanospheres I: physicochemical characterization. J. Biomed. Mater. Res. 42, 45−54.

Quellec, P., Gref, R., Dellacherie, E., Sommer, F., Tran, M.D., Alonso, M.J., 1999. Protein encapsulation within polyethylene glycol-coated nanospheres. II: controlled release properties. J. Biomed. Mater. Res. 47, 388−395.

Radtke, A., Topolski, A., Jedrzejewski, T., Kozak, W., Sadowska, B., Wieckowska-Szakiel, M., Szubka, M., Talik, E., Pleth Nielsen, L., Piszczek, P., 2017. The bioactivity and photocatalytic properties of titania nanotube coatings produced with the use of the low-potential anodization of Ti6Al4V alloy surface. Nanomaterials (Basel) 7 pii: E197.

Ramrattan, N., Heijkants, R.G.J.C., VanTienen, T.G., Schouten, A.J., Veth, R.P., Buma, P., 2005. Assessment of tissue in growth rates in polyurethane scaffolds for tissue engineering. Tissue Eng. 11, 1212−1223.

Rao, G.H.R., Chandy, T., 1999. Role of platelets in blood-biomaterial interactions. Bull. Mater. Sci. 3, 633−639.

Riau, A.K., Mondal, D., Yam, G.H., Setiawan, M., Liedberg, B., Venkatraman, S.S., Mehta, J.S., 2015. Surface modification of PMMA to improve adhesion to corneal substitutes in a synthetic core-skirt keratoprosthesis. ACS Appl. Mater. Interfaces 7, 21690−21702.

Riau, A.K., Mondal, D., Setiawan, M., Palaniappan, A., Yam, G.H., Liedberg, B., Venkatraman, S.S., Mehta, J.S., 2016. Functionalization of the polymeric surface with bioceramic nanoparticles via a novel, nonthermal dip coating method. ACS Appl. Mater. Interfaces 8, 35565−35577.

Riau, A.K., Venkatraman, S.S., Dohlman, C.H., Mehta, J.S., 2017. Surface modifications of the PMMA optic of a keratoprosthesis to improve biointegration. Cornea (Suppl. 1), S15−S25.

Ritter, E.F., Kim, Y.B., Reischl, H.P., Serafin, D., Rudner, A.M., Klitzman, B., 1997. Heparin coating of vascular prostheses reduce thromboemboli. Surgery 122, 888−892.

Rothenburger, M., Volker, W., Vischer, J.P., Berendes, E., Glasmacher, B., Scheld, H.H., Deiwick, M., 2002. Tissue engineering of heart valves: formation of a three-dimensional tissue using porcine heart valve cells. ASAIO J. 48, 586−594.

Rowley, J.A., Madiambayan, G., Mooney, D.J., 1999. Alginate hydrogels as synthetic extracellular matrix materials. Biomaterials 20, 45−53.

Ruiz, J.M., Benoit, J.P., 1991. In vivo peptide release from poly (DL-lactic acid-*co*-glycolic acid) copolymer 50/50 microspheres. J. Control. Release 16, 177−186.

Saffran, M., Kumar, G.S., Savariar, C., Burnham, J.C., Williams, F., Neckers, D.C., 1986. A new approach to the oral administration of insulin and other peptide drugs. Science 233, 1081−1084.

Sakamoto-Ozaki, K., Matsumoto, Y., Kanno, Z., Iida, J., Soma, K., 2012. Development of a surgical procedure for biointegration of a newly designed orthodontic onplant. Orthodontics (Chic.) 13, 216−225.

Salehiabar, M., Nosrati, H., Javani, E., Aliakbarzadeh, F., Kheiri Manjili, H., Davaran, S., Danafar, H., 2018. Production of biological nanoparticles from bovine serum albumin as controlled release carrier for curcumin delivery. Int. J. Biol. Macromol. 115, 83−89.

Salvay, D.M., Shea, L.D., 2006. Inductive tissue engineering with proteins and DNA-releasing scaffolds. Mol. Biosyst. 2, 36−48.

Santos, M., Serrano-Ducar, S., Gonzalez-Valdivieso, J., Vallejo, R., Girotti, A., Cuadrado, P., Arias, F.J., 2018. Genetically engineered elastin-based biomaterials for biomedical applications. Curr. Med. Chem. 19, 457−465.

Schoen, F.J., 2013. Introduction: biological responses to biomaterials. In: Ratner, B.D., Hoffman, A.S., Schoen, F.J., Lemons, J.E. (Eds.), Biomaterials Science: An Introduction to Materials in Medicine, third ed. Academic Press, Elsevier, Waltham, MA, USA.

Schwartz, R.S., Chronos, N.A., Virmani, R., 2004. Preclinical restenosis models and drug eluting stents. J. Am. Coll. Cardiol. 44, 1373−1385.

Sharma, C.P., Chandy, T., Sunny, M.C., 1987. Inhibition of platelet adhesion to glow discharge modified surfaces. J. Biomater. Appl. 1, 533−549.

Shmulewitz, A., Langer, R., 2006. The ascendance of combination products. Nat. Biotechnol. 24, 277−280.

Simmons, A., Hyvarinen, J., Odell, R.A., Martin, D.J., Gunatillake, P.A., Noble, K.R., Poole-Warren, L.A., 2004. Long-term in vivo biostability of poly(dimethylsiloxane)/poly (hexamethylene oxide) mixed macrodiol-based polyurethane elastomers. Biomaterials 25, 4887−4900.

Singer, I., Hutchins, G.M., Mirowski, M., Mower, M.M., Veltri, E.P., Guarnieri, T., Griffith, L.S.C., Watkins, L., Juanteguy, J., Fisher, S., Reid, P.R., Weisfeldt, M.L., 1987. Pathologic findings related to the lead system and repeated defibrillations in patients with the automatic implantable cardioverter defibrillator. J. Am. Coll. Cardiol. 10, 382−388.

Slamborová, I., Zajícová, V., Karpíšková, J., Exnar, P., Stibor, I., 2013. New type of protective hybrid and nanocomposite hybrid coatings containing silver and copper with an excellent antibacterial effect especially against MRSA. Mater. Sci. Eng. C3, 265−273.

Song, C.X., Labhasetwar, V., Murphy, H., Qu, X., Humphrey, W.R., Shebuski, R., Levy, R.J., 1997. Formulation and characterization of biodegradable nanoparticles for intravascular local drug delivery. J. Control. Release 43, 197−212.

Stylios, G.K., Giannoudis, P.V., Wan, T., 2005. Application of nanotechnologies in medical practice. Injury 365, S6−S23.

Sweet, C.R., Chatterjee, S., Xu, Z., Bisordi, K., Rosen, E.D., Alber, M., 2011. Modelling platelet-blood flow interaction using the subcellular element Langevin method. J. R. Soc. Interface 8, 1760−1771.

Szycher, M., Reed, A., 1992. Biostable polyurethane elastomers. Med. Device Technol. 3, 42−51.

Szymonowicz, M., Rybak, Z., Paluch, D., Marycz, K., Kaliński, K., Błazewicz, S., 2013. Studies of interaction between surface of pirolytic carbon and blood cells and proteins. Polim. Med. 43, 165−173.
Takahashi, K., et al., 2007. Induction of pluripotent stem cells from adult human fibroblasts by defined factors. Cell 131, 861−872.
Takahashi, K., Yamanaka, S., 2006. Induction of pluripotent stem cells from mouse embryonic and adult fibroblast cultures by defined factors. Cell 126, 663−676.
Tan, J., Saltzman, M., 1990. Influence of synthetic polymers on neutral migration in three-dimensional collagen gels. J. Biomed. Mater. Res. 46, 465−471.
Tan, W., Krishnaraj, R., Desai, T.A., 2001. Evaluation of nanostructured composite collagen-chitosan matrices for tissue engineering. Tissue Eng. 7, 203−210.
Tandon, B., Magaz, A., Balint, R., Blaker, J.J., Cartmell, S.H., 2018. Electroactive biomaterials: vehicles for controlled delivery of therapeutic agents for drug delivery and tissue regeneration. Adv. Drug Deliv. Rev. 129, 148−168.
Tienen, T.G., Heijkants, R.G., Buma, P., De Groot, J.H., Pennings, A.J., Veth, R.P., 2003. A porous polymer scaffold for meniscal lesion repair. A study in dogs. Biomaterials 24, 2541−2553.
Tomikawa, M., Iwamota, M., Olsson, P., Soderman, S., Blomback, B., 1980. On the platelet-fibrinogen interaction. Thromb. Res. 19, 869−876.
Vacanti, C.A., Langer, R., Schloo, B., Vacanti, J.P., 1991. Synthetic polymers seeded with template for new cartilage formation. Plast. Reconstr. Surg. 88, 753−762.
Vannozzi, L., Yasa, I.C., Ceylan, H., Menciassi, A., Ricotti, L., Sitti, M., April 2018. Self-folded hydrogel tubes for implantable muscular tissue Scaffolds. Macromol. Biosci. 18 (4), e1700377−e1700389.
Virmani, R., Kolodgie, A., Lafont, A., 2003. Drug eluting stents: are human and animal studies comparable? Heart 89, 133−138.
Vroman, L., 1988. The life of an artificial device in contact with blood: initial events and their elect on its funal state. Bull. N. Y. Acad. Med. 7, 64−352.
Vroman, L., Adams, A.L., Fischer, G.C., Munoz, P.C., 1980. Interaction of high molecular weight kininogen, factor XII and fibrinogen in plasma at interfaces. Blood 55, 156−168.
Vroman, L., Adams, A.I., Fischer, G.C., Munoz, P.C., Stanford, M., 1982. Proteins, plasma and blood in narrow spaces of clot promoting surfaces. Adv. Chem. 199, 266−276.
Wagner, A.M., Gran, M.P., Peppas, N.A., 2018. Designing the new generation of intelligent biocompatible carriers for protein and peptide delivery. Acta Pharm. Sin. B 8, 147−164.
Wang, M., Thanou, M., 2010. Targeting nanoparticles to cancer. Pharmacol. Res. 62, 90−99.
Wang, F., Yang, Y., Ju, X., Udenigwe, C.C., He, R., 2018. Polyelectrolyte complex nanoparticles from chitosan and acylated rapeseed cruciferin protein for curcumin delivery. J. Agric. Food Chem. 66, 2685−2693.
Wang, Y., Kim, Y.M., Langer, R., 2003. In vivo degradation characteristics of poly (glycerol sebacate). J. Biomed. Mater. Res. 66A, 192−197.
Wilkoff, B.L., Rickard, J., Tkatchouk, E., Padsalgikar, A.D., Gallagher, G., Runt, J., 2016. The biostability of cardiac lead insulation materials as assessed from long-term human implants. J. Biomed. Mater. Res. B Appl. Biomater. 104, 411−421.
Winter, J.O., Cogan, S.F., Rizzo, J.F., 2006. Neurotrophin-eluting hydrogel coatings for neural stimulating electrodes. J. Biomed. Mater. Res. B 81B, 551−563.
Wise, S.G., Waterhouse, A., Michael, P., Ng, M.K., 2012. Extracellular matrix molecules facilitating vascular biointegration. J. Funct. Biomater. 3, 569−587.
Wu, Q., Niu, M., Chen, X., Tan, L., Fu, C., Ren, X., Ren, J., Li, L., Xu, K., Zhong, H., Meng, X., 2018. Biocompatible and biodegradable zeolitic imidazolate framework/polydopamine

nanocarriers for dual stimulus triggered tumor thermo-chemotherapy. Biomaterials 162, 132–143.

Wulf, K., Teske, M., Matschegewski, C., Arbeiter, D., Bajer, D., Eickner, T., Schmitz, K.P., Grabow, N., 2018. Novel approach for a PTX/VEGF dual drug delivery system in cardiovascular applications-an innovative bulk and surface drug immobilization. Drug Deliv. Transl. Res. 8, 719–728.

Xu, C.Y., Inai, R., Kotaki, M., Ramakrishna, S., 2004. Aligned biodegradable nano fibrous structure: a potential scaffold for blood vessel engineering. Biomaterials 25, 877–890.

Yang, J., Zhai, S., Qin, H., Yan, H., Xing, D., Hu, X., 2018. NIR-controlled morphology transformation and pulsatile drug delivery based on multifunctional phototheranostic nanoparticles for photoacoustic imaging-guided photothermal-chemotherapy. Biomaterials 176, 1–12.

Ye, D., 2014. Peramo A implementing tissue engineering and regenerative medicine solutions in medical implants. Br. Med. Bull. 109, 3–18.

Yezhelyev, M.V., Gao, X., Xing, Y., Al-Hajj, A., Nie, S., O'Regan, R.M., 2006. Emerging use of nanoparticles in diagnosis and treatment of breast cancer. Lancet Oncol. 7, 657–667.

Yu, J., et al., 2007. Induced pluripotent stem cell lines derived from human somatic cells. Science 1917–1920.

Zarrintaj, P., Manouchehri, S., Ahmadi, Z., Saeb, M.R., Urbanska, A.M., Kaplan, D.L., 2018. Mozafari MAgarose-based biomaterials for tissue engineering. Carbohydr. Polym. 187, 66–84.

Zhao, Q., Agger, M., Fitzpatrick, M., Anderson, J., Hiltner, A., Stokes, K., Urbanski, P., 1990. Cellular interactions with biomaterials: *in vivo* cracking of pre-stressed Pellethane 2363-80A. J. Biomed. Mater. Res. 24, 621–637.

Zhao, Q., Topham, N., Anderson, J., Hiltner, A., Lodoen, G., Payet, C., 1991. Foreign-body giant cells and polyurethane biostability: *in vivo* correlation of cell adhesion and surface cracking. J. Biomed. Mater. Res. 25, 177–183.

Inorganic nanoparticles for targeted drug delivery

13

Willi Paul, Chandra P. Sharma
Sree Chitra Tirunal Institute for Medical Sciences and Technology, Thiruvananthapuram, Kerala, India

Chapter outline

13.1 Introduction 333
13.2 Calcium phosphate nanoparticles 337
    13.2.1 Oral insulin delivery applications 338
    13.2.2 Theranostic applications 340
    13.2.3 Gene delivery applications 344
    13.2.4 Cancer chemotherapy applications 346
    13.2.5 Tissue engineering applications 347
13.3 Gold nanoparticles 349
    13.3.1 Cancer chemotherapy applications 351
    13.3.2 Gene delivery applications 354
13.4 Iron oxide nanoparticles 355
    13.4.1 Cancer therapy applications 356
    13.4.2 Gene delivery applications 358
    13.4.3 Tissue engineering applications 359
    13.4.4 General drug delivery and targeting 361
13.5 Conclusion 362
13.6 Biointegration concept and future perspective 363
Acknowledgments 363
References 364

13.1 Introduction

Are nanostructured materials natures' gift or scientist's brainchild? It has been present in the nature for millions of years as a giant laboratory of nanoscience and engineering. Tools to visualize and examine the samples of these scales have been developed now, which helped us to understand nature's strategy and mimicking nature's bio-nanotechnology. Nanostructured materials have unique properties and capabilities that make them suitable for specific interaction with the biological system, particularly in drug delivery applications. In a biological system, nanoparticles or nanostructured materials interact with the biomolecules such as proteins and polypeptides, lipids, and other metabolites. This interaction can influence the biological reactivity of the nanomaterials. Natural systems are made of various nanosized elements that impart unique properties to the biological system. The interaction between the gecko's feet and the climbing surface is mainly because of thousands of spatula of nanodimensions on the microscopic hair or setae on its feet. Bone and cartilage are highly organized over different length scales. Bone consists of hydroxyapatite (HA) needle-like mineral crystals of 50−100 nm length and 3−10 nm width combined with collagen fibrils of around 500 nm diameter. The hierarchical structure of bone and cartilage is as shown in Fig. 13.1 (Bonzani et al., 2006). This knowledge has led to investigations and approaches for biomimicking and developing technology for nanomaterials, utilized for different applications. Several organic nanoparticles have been studied and developed successfully from polymers, dendrimers, liposomes, and micelles. Self-assembly and the presence of zwitterionic molecules with polar and nonpolar regions are the key elements in the fabrication of many organic nanoparticles. Drawbacks such as low

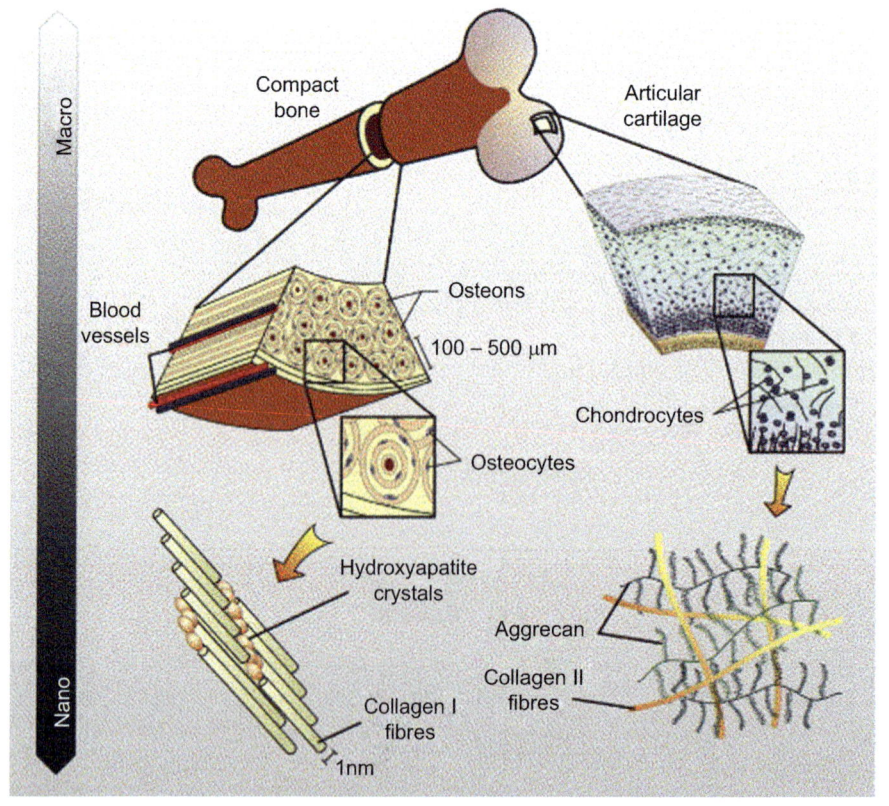

Figure 13.1 Hierarchical organization of bone over different length scales with the well-defined nanoarchitecture of the surrounding extracellular matrix.
Reprinted from Bonzani, I.C., George, J.H., Stevens, M.M., 2006. Novel materials for bone and cartilage regeneration. Curr. Opin. Chem. Biol. 10, 568−575 with permission from Elsevier.

chemical stability, drug release rate that is unsuitable to the specific application, possibility of microbial contamination, and the undesirable effects of the organic solvents used for particle preparation are some of its inherent problems. Nonuniform size distribution, lack of stability, and tendency to aggregate, toxicity, and possibility of rapid clearance by the reticuloendothelial system are also some of the drawbacks of organic nanoparticles. Inorganic nanoparticles are small particles with special and enhanced physical and chemical properties depending on the particle size. Common inorganic nanoparticles are (1) prepared from noble metals such as gold and silver; (2) magnetic nanoparticles, i.e., superparamagnetic with large magnetic moments in a magnetic field made from Ni, Co, Fe, Fe3O4, and FePt; and (3) fluorescent nanoparticles such as quantum dots and SiO_2, etc. (Zhang et al., 2014). Inorganic nanoparticles gained significant attention in preclinical development as potential diagnostic and therapeutic systems in oncology for a variety of applications, including tumor imaging, tumor drug delivery, or enhancement of radiotherapy. Inorganic nanoparticles are nontoxic, hydrophilic, biocompatible, and highly stable compared with organic materials. However, less progress has been made in the development of inorganic nanoparticles in drug delivery systems. Inorganic nanoparticles as drug or gene delivery carriers have received much attention due to their high cellular uptake capacity, nonimmunogenic response, and low toxicity. They have exhibited significantly distinct physical, chemical, and biological properties from their bulk counterparts. Electromagnetic, optical, and catalytic properties of noble metal nanoparticles, such as gold, silver, and platinum, are known to be strongly influenced by the shape and size. Biomedical applications of metal nanoparticles have been dominated by the use of nanobioconjugates that started in 1971, after the discovery of the colloidal gold labeling technique (immunogold) by Faulk and Taylor (Faulk and Taylor, 1971). Metal-based nanoconjugates are used in various biomedical applications, such as probes for electron microscopy to visualize cellular components, drug delivery (vehicle for delivering drugs, proteins, peptides, plasmids, DNAs, etc.), detection, diagnosis, and therapy (targeted and nontargeted). Drug delivery systems, designed for enhanced drug efficacy and reduced adverse effects, have evolved, accompanied by the development of novel materials. Nanotechnology is an emerging scientific area that has created a variety of intriguing inorganic nanoparticles (Murakami and Tsuchida, 2008; Zhu, 2008; Roveri et al., 2008). In recent years, research efforts worldwide have developed nanoproducts aimed at improving health care and advancing medical research (Florence, 2018; Zhang et al., 2018a; Mir et al., 2017; Rukes, 2017; Zhong, 2015a,b,c; Toth, 2014). Biomedical applications of inorganic nanotechnology are mainly suited for diagnostic techniques, nanodrugs and delivery systems, and biomedical implants (Di Martino et al., 2017; Baeza et al., 2017; Ahmad et al., 2016; Zink, 2014; Ojea-Jimenez et al., 2013; Mattoussi and Rotello, 2013; Pecorelli et al., 2010; Liong et al., 2008). Many industries are developing nanotechnology-based applications for anticancer drugs (Basha, 2018), implanted insulin pumps (Xu, 2017), and gene therapy (Chen et al., 2016). Prostheses and implants are also being developed from nanostructured materials. The global drug delivery system market was valued at approximately US$ 510 billion in 2016 and is projected to expand at compound annual growth rate (CAGR) of over 6.9% to reach approximately US$ 900 billion by 2025 (Market, 2018).

Global market for nanotechnology-enabled drug delivery systems was valued at US$ 134 billion for 2016 and forecasted to grow to more than US$ 293.1 billion by 2022, which is projected as the single largest market opportunity (Highsmith, 2014; Evers, 2017). Reformulations will be possible with the nanoenabled drug delivery systems that help in protecting the patent holders. Nanoparticles used in diagnostic imaging will also show a healthy growth.

Recent advancement in nanotechnology has led to the introduction of various other inorganic nanoparticles beside calcium phosphates as excellent drug delivery matrices. They include iron oxide nanoparticles and fullerenes. Carbon nanotubes and nanospheres have been studied as drug delivery vehicles, as their nanometer size enables them to move easily inside the body. The drug can be either inserted in the nanotube or attached to the particle surface. The advantages of inorganic nanoparticles are their very low toxicity profile, biocompatibility, and hydrophilic nature; they are not subject to microbial attack and are extremely stable. Nanotechnology-enabled drug delivery devices are expected to contribute in enabling novel pharmaceutical therapies that target the site of the disease and help in reducing the toxicity of the active ingredient and to reduce the cost in health care. Once the functionalized nanoparticles reaches near the target cells, the targeting moiety recognizes and binds to the receptors expressed by specific cells selectively. In some cases, the nanoparticles can penetrate in the cells by receptor-mediated internalization processes and release the active ingredient inside the target cells as shown in Fig. 13.2 (Degli Esposti et al., 2018).

The importance of these inorganic nanoparticles is ever increasing. It has been established that nanoparticles can be taken up by the cells by the process of endocytosis (Oh and Park, 2014). Its size and surface properties determine the biodistribution

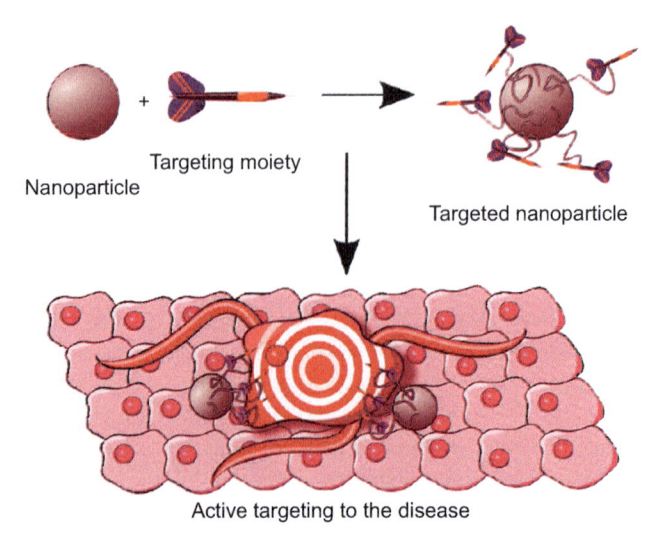

Figure 13.2 Schematic representation of the active targeting mechanism.
Reprinted from Degli Esposti, L., Carella, F., Adamiano, A., Tampieri, A., Iafisco, M., 2018. Calcium phosphate-based nanosystems for advanced targeted nanomedicine. Drug Dev. Ind. Pharm. 44, 1223−1238 with permission from Taylor & Francis.

of these nanoparticles following administration. Nanoparticles can be administered via respiratory tract, gastrointestinal tract, transdermal, and via intravenous and intramuscular routes. Through most of these routes, nanoparticles finally reach the bloodstream and the lymphatics and finally accumulate in the organs and get eliminated (Ojea-Jimenez et al., 2013). Thus, nanoparticles could be utilized to deliver nucleic acids into living cells (Ding et al., 2014). Nanoparticles now have highly advanced chemical properties (Chikkaveeraiah et al., 2018; Schmid, 2010; Caruso, 2004; Pileni, 2005), and many inorganic nanoparticles have been used as drug carriers (Liong et al., 2008; Santos et al., 2014). The inorganic nanoparticles that have been studied in delivering DNA comprise calcium phosphate, carbon nanotubes, silica, gold, iron oxide, quantum dots, strontium phosphate, magnesium phosphate, manganese phosphate, and double hydroxides (Xu et al., 2006; Murakami and Tsuchida, 2008). This chapter reviews some of the recent developments and applications of calcium phosphate nanoparticles, gold nanoparticles, and iron oxide nanoparticles in drug delivery and tissue engineering.

13.2 Calcium phosphate nanoparticles

Bioceramics are a class of advanced ceramics that are defined as ceramic products or components employed in medical and dental applications, mainly as implants and replacements. They are biocompatible and can be inert, bioactive, and degradable in physiological environments, which makes them an ideal biomaterial. Materials that are classified as bioceramics include alumina, zirconia, calcium phosphates, silica-based glasses or glass ceramics, and pyrolytic carbons. Calcium phosphates include tricalcium phosphates, HA, and tetracalcium phosphates. Calcium phosphate found in the bone is in the form of nanometer-sized needle-like crystals, with a poorly crystallized nonstoichiometric apatite phase containing other trace ions. Unlike tetracalcium phosphates and tricalcium phosphates, HA does not break down under physiological conditions. In fact, it is thermodynamically stable at physiological pH and actively takes part in bone bonding, forming strong chemical bonds with surrounding bone. This property has been exploited for rapid bone repair after major trauma or surgery. As calcium phosphates are biocompatible, resorbable, and porous, attempts have been made to utilize them as delivery systems for drugs, chemicals, and biologicals. Bajpai and coworkers initiated studies on ceramic drug delivery in the early 1980s by the introduction of aluminocalcium phosphorous oxide (ALCAP) ceramic capsules. Low cost, ease of manufacture, and biocompatibility make ceramic materials good candidates for drug delivery applications (Paul and Sharma, 2003; Thomas et al., 2015). As HA is biocompatible and is also used as a matrix for the purification of proteins, these particles could be utilized for protein and peptide drug delivery applications.

Synthetic calcium phosphates possess exceptional biocompatibility and bioactivity properties with respect to bone cells and tissues; hence, they have been widely used clinically in the form of powders, granules, dense and porous blocks, and various composites. For drug delivery, as well as for tissue engineering applications, the present trend is to develop new formulations of HA with properties closer to those of living bone, such as nanosized and monolithic structures. Nanophase calcium phosphate exhibited enhanced osteoblast functions (Webster et al., 2000, 2001; Paul and Sharma, 2007) that are very important from the implant application point of view. Nanosized calcium phosphates have been studied in drug delivery systems such as intestinal delivery of insulin (Uskokovic and Uskokovic, 2011; Paul and Sharma, 2001; Degli Esposti et al., 2018) or of other drugs such as antibiotics (Geuli et al., 2017).

13.2.1 Oral insulin delivery applications

Type I diabetes is characterized by the inefficiency of pancreatic beta cells to produce insulin. The common form of insulin therapy is by way of twice-daily subcutaneous insulin injection. Various attempts have been made to develop a noninvasive delivery system for insulin, namely via oral, buccal, transdermal delivery routes, etc., with varying levels of success (Veiseh et al., 2015). Calcium phosphates have been approved for human use in several European countries as adjuvant (Masson et al., 2017). Zinc is also being used for stabilizing insulin (long-acting insulins). Therefore, calcium phosphates, zinc phosphates, and zinc calcium phosphates seem to be suitable candidates for developing ceramic-based insulin delivery systems. Oral cavity delivery is considered to be the most desirable way of delivering drugs from a patient compliance point of view. This will be the case with insulin also, once an oral formulation has been developed. Delivered insulin in the case of microspheres follows the same pathway as naturally produced insulin by the pancreas, i.e., reaches directly to the portal circulation and to the liver, consistent with normal physiology. However, in the case of nanoparticles, the absorption of nanoparticles takes place via the Peyer's patches region, reaches the lymphatic system, bypasses the first pass metabolism (bypasses liver: degradation of insulin is significantly reduced), the particles will be degraded, and delivers the insulin in the bloodstream. The concept of oral delivery of insulin-utilizing zinc phosphate nanoparticles seems to be promising as insulin will also be stable along with zinc. BioSante Pharmaceuticals, a US-based company, had developed calcium phosphate nanoparticles that have successfully passed the first stage of toxicity studies for administration orally, into muscles, under the skin, and into the lungs by inhalation. They had been used as a vaccine adjuvant and for protein delivery. Preclinical trials of both BioOral or CAPOral and BioAir or CAPAir indicated sustained delivery of insulin with sustained control of glucose levels. However, from this point, the company discontinued the studies in 2008. Similar formulations may be used for the delivery of other proteins, for example, human growth hormone orally or to the lungs. The technology of BioSante was, however, able to significantly increase the half-life and mean resistance time of insulin in the body. Calcium phosphate particles containing insulin were synthesized in the presence of poly(ethylene glycol) (PEG) and coated with casein to obtain the calcium phosphate-PEG-insulin-casein (CAPIC) oral insulin delivery system. When tested in nonobese diabetic mice under fasting or fed conditions, it was observed that the biological activity of insulin was preserved by protecting the insulin from degradation while passing through the acidic environment of the GI tract, and it displayed a prolonged hypoglycemic effect after oral administration (Morcol et al., 2004). Oral administration of 100 U/kg CAPIC to fed-diabetic mice resulted in about 50% reduction of the initial glucose levels within the first 3 h of the treatment, as shown in Fig. 13.3. Glucose returned to the control levels within 5 h. An identical dose of insulin solution (with no calcium phosphate nanoparticles) had no significant effect on blood glucose levels.

Calcium phosphate (CaP) nanoparticles with an average particle size of 47.9 nm (D50) were synthesized, and the surface was modified by conjugating it with PEG. These modified nanoparticles had a near zero zeta potential. Protection of insulin from the gastric environment was achieved by coating the nanoparticles with a pH-sensitive polymer that would dissolve in the mildly alkaline pH environment of the intestine. The release profiles of coated nanoparticles exhibited negligible release in acidic (gastric) pH, i.e., only 2% for CaP and 6.5% for PEGylated CaP. However, a sustained release of insulin was observed at neutral (intestinal) pH for over 8 h. The conformation of the released insulin, studied using circular dichroism, was unaltered when compared with native insulin. The released insulin was also stable as studied using dynamic light scattering. The immunoreactivity of the released insulin was found to be intact. These results suggest PEGylated calcium phosphate nanoparticles as an excellent carrier system for insulin toward the development of an oral insulin delivery system (Ramachandran et al., 2009). A typical transmission electron micrograph of these nanoparticles is shown in Fig. 13.4. They are needle-shaped crystals with 20 nm width and less than 100 nm length. A calcium phosphate nanoparticle

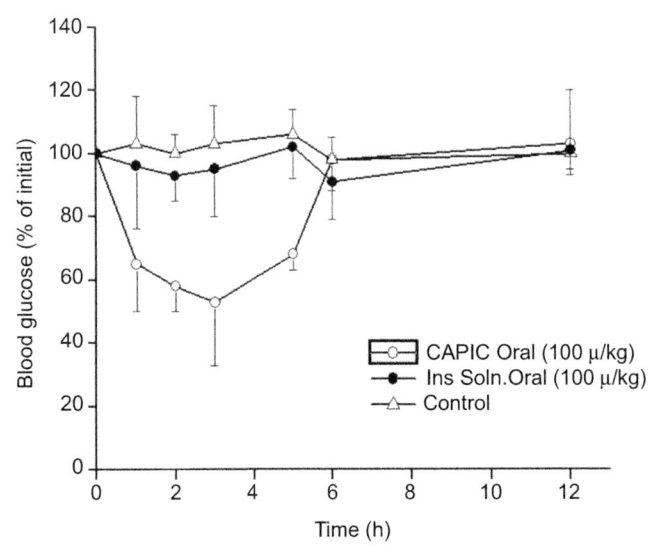

Figure 13.3 Glycemic effect of a single oral dose of calcium phosphate-PEG-insulin-casein (CAPIC) in fed diabetic mice.
Reprinted from Morcol, T., Nagappan, P., Nerenbaum, L., Mitchell, A., Bell, S.J. 2004. Calcium phosphate-PEG-insulin-casein (CAPIC) particles as oral delivery systems for insulin. Int. J. Pharm. 277, 91−97 with permission from Elsevier.

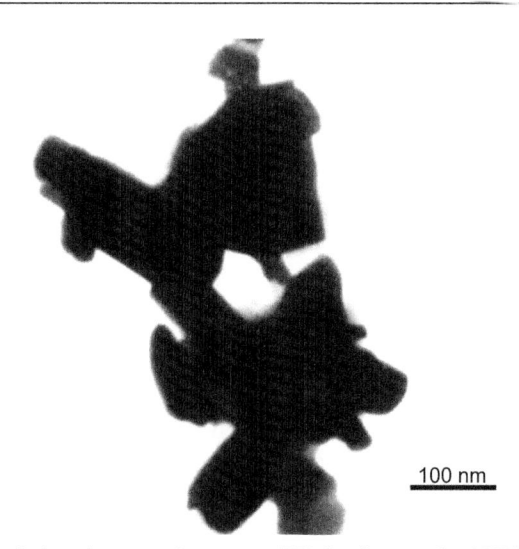

Figure 13.4 Transmission electron microscope (TEM) micrograph of PEGylated CaP nanoparticles.
Reprinted from Ramachandran, R., Paul, W., Sharma, C.P., 2009. Synthesis and characterization of PEGylated calcium phosphate nanoparticles for oral insulin delivery. J. Biomed. Mater. Res. B Appl. Biomater. 88B, 41–48 with permission from John Wiley and Sons.

formulation, where vitamin B12 grafted chitosan and alginate were coated layer by layer to form VitB12-Chi-CPNPs, exhibited paracellular and receptor-mediated uptake and an increase of insulin bioavailability in the order of fourfold (Verma et al., 2016). A sustained hypoglycemic effect of up to 12 h was observed with administration of these nanoparticles. Vitamin B12, which acted as a pH-sensitive and targeting ligand, also enhanced the oral bioavailability. Zinc calcium phosphate nanoparticles were also shown to be an excellent carrier for intestinal delivery of insulin (Paul and Sharma, 2012). Insulin resistance is a state where the glucose metabolism is affected, which increases the glucose level of an individual and is termed as Type 2 diabetes. Biomineralized insulin nanoparticles were prepared by adding $CaCl_2$ into the insulin solution, which triggers a spontaneous nucleation of calcium phosphate on acidic amino acid residue on insulin molecules (Xiao et al., 2017). These biomineralized insulin nanoparticles can bypass the cell membrane via endocytosis, release of insulin into the cytosol, and its action by triggering glucose metabolism (Fig. 13.5). This extended effect and reduction of glucose level is in contrast with the temporary effect of insulin in Type 2 diabetes. NanoHA nanoparticles were also coated with polyethylene glycol (PEG), and insulin was conjugated to the PEG coating. This decreased the degradation of HA in the gastrointestinal tract (Zhang et al., 2018b).

13.2.2 Theranostic applications

The field of nanotechnology holds incredible potential in the diagnosis and treatment of disparate diseases. Diagnosis and therapy can be effectively combined in

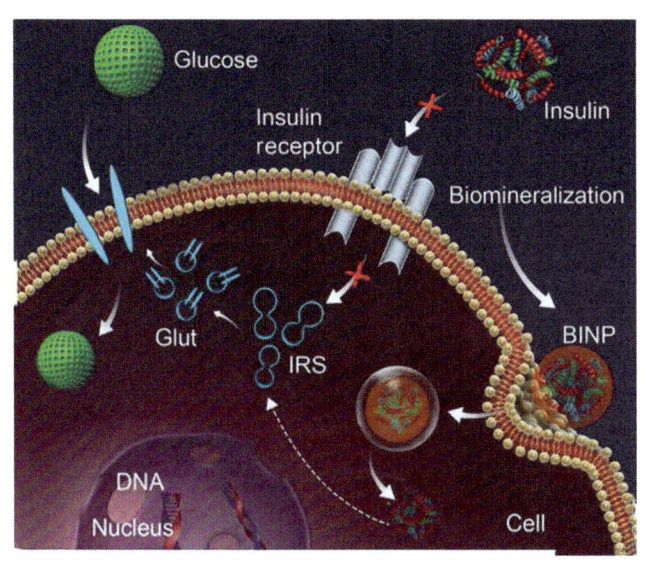

Figure 13.5 The conventional action of insulin on the cell membrane (red pathway) is inhibited by an insufficient number of receptors, resulting in type 2 diabetes; the in situ biomineralization of insulin (green pathway) delivers insulin into the intact cell through biomineralised insulin nanoparticles (BINPs), triggering an intercellular action that stimulates glucose metabolism. Reprinted from Xiao, Y., Wang, X.Y., Wang, B., Liu, X.Y., Xu, X. R., Tang, R.K., 2017. Long-term effect of biomineralized insulin nanoparticles on type 2 diabetes treatment. Theranostics 7, 4301–4312 under Creative Commons Attribution License (CC BY).

"theranostic"-based systems to interact with cancer cells to combat diseases at the molecular level and considerably improve specificity and efficacy of cancer treatment. This integration of therapeutic payloads with diagnostic agents into nanoplatforms is termed the "theranostics" (Muthu et al., 2014; Xie et al., 2010; Evangelopoulos et al., 2018; Chen et al., 2014). These theranostic nanoplatforms emphasize on diagnosis and delivery of drugs and help monitor therapeutic response and are expected to usher in a new paradigm for treatment regime (Ryu et al., 2014). In this regard, quantum dots (Ho and Leong, 2010), carbon nanotubes (Tan et al., 2011), iron oxide nanoparticles (Zheng et al., 2018), liposomes (Seleci et al., 2017), gold nanoparticles (Barbosa et al., 2014), and silica nanoparticles (Kempen et al., 2015) are considered as exceptional candidates for nanoparticle-based theranostics. Despite their promises, these systems are associated with drawbacks that include toxicity of quantum dots, nonbiodegradable nature of carbon nanotubes, and undesirable size of silica nanoplatforms to mention a few. Diverse strategies are being envisaged to address these limitations and to develop more efficient theranostic nanoplatforms.

The last few years have witnessed the use of lanthanide-doped nanoparticles as alternatives to conventional systems for biological imaging applications (Kang et al., 2014). Lanthanide-based imaging offers excellent characteristics such large Stokes shift, long lifetime, excellent quantum yield, and good stability (Ciobanu et al., 2015). In this context, it is worth noting that HA-based nanoparticles are particularly useful as nanoplatforms for effective lanthanide doping. HA, the dominant component present in natural bone mineral, possess exemplary properties such as facile functionalization, biocompatibility, bioactivity, and efficient biodegradability in biological milieu. More specifically, HA nanoparticles bequeathed with a strong tendency for ionic substitution readily replace their calcium ions leading to substitution with variant lanthanide elements. The luminous intensity of these doped HA nanoparticles depends on the crystallinity of the host material and the concentration of the doping agent added. Several research groups have reported luminous lanthanide/HA compounds incorporated with payloads such as cisplatin, Nile red (Victor et al., 2014a), and ceramide. Europium-doped apatite displayed stable red luminescence as an effective biological probe. HA doped with gadolinium and europium demonstrated significant fluorescence and provide simultaneous therapeutics and fluorescence imaging guidance. In this context, fluorescence imaging guidance in the near-infrared (NIR) region is a unique strategy for visualizing morphological details in tissue. This is ascribed to the innate low absorption of tissues in the NIR region. NIR fluorescence imaging permits deeper tissue penetration with high spatial and temporal resolution, lesser auto fluorescence, and high sensitivity. Emerging strategies like up conversion luminescence of rare earth materials address these benefits and are expected to be more viable as luminescent NIR fluorescent imaging moieties.

Stimuli-responsive nanoparticles are yet another fascinating concept, which has evolved to include combinatorial therapeutic approach for sustained delivery and codelivery of drugs. Nanoparticles equipped with different functionalities can respond to different stimuli, and the most commonly encountered are pH, temperature, and ionic strength (Yao et al., 2014). A biocompatible alginic acid−modified HA nanoplatform that responds to environmental changes in pH was studied, primarily employing neodymium as the luminescent moiety with NIR luminescent (Victor et al., 2016). Alginic acid is an anionic polysaccharide and a linear block copolymer of α-L-guluronate and β-D-mannuronate connected by 1,4-glycoside linkage. It is reported that alginic acid when used in conjunction with ceramic-like materials exhibits superior properties owing to the synergistic effects of ceramic and polymer. Stimuli-sensitive systems with alginate and calcium phosphate can thus overcome uncontrolled swelling and burst release with better controlled release of drug, facilitating its delivery via oral route. The prepared theranostic nanoplatforms exhibited minimal toxicity to the cells in vitro and demonstrated augmented drug adsorption with pH-dependent release profiles. Its effective cellular uptake was demonstrated using confocal Raman microscopy as shown in Fig. 13.6. Cellular uptake is crucial for addressing the therapeutic efficiency of these nanoplatforms. The internalization of nanoplatforms promotes site-specific delivery of drugs and helps reduce drug toxicity and augments therapeutic efficacy. Early detection and targeted chemotherapy are critical in the management of colon cancer. This HA-neodymium nanoplatform can be utilized as a theranostic system for early tumor detection, targeted tumor therapy, and monitoring of colon cancer, which can be administered via oral route. We had previously demonstrated a facile coprecipitation method for the preparation of multifunctional neodymium-doped HA−cyclodextrin nanoparticle complexes that demonstrate preferential affinity

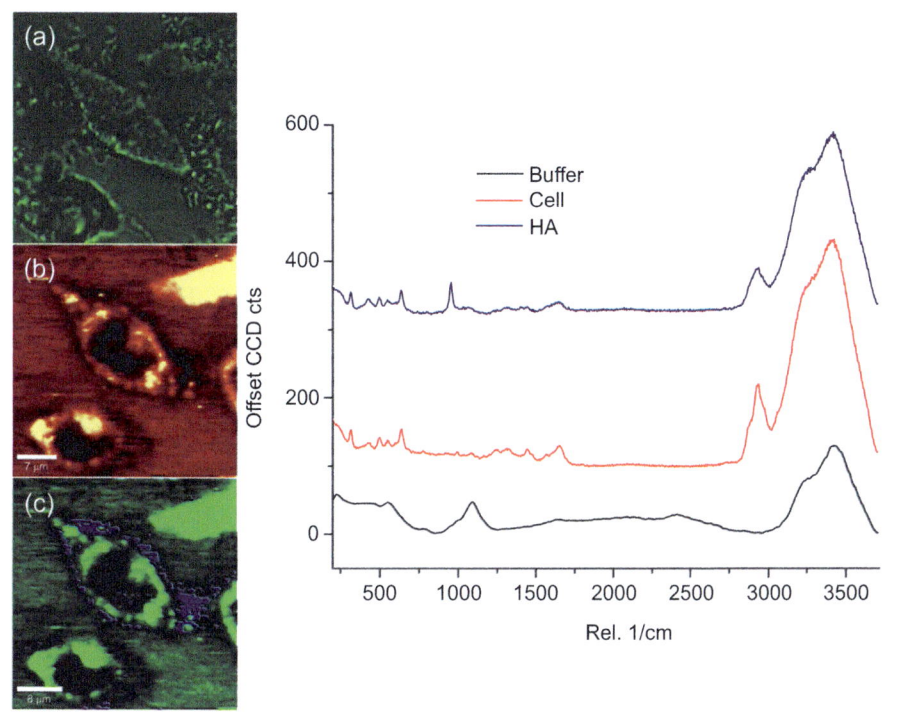

Figure 13.6 Color-coded Raman cluster map (c) of the HeLa cell treated with doped nanoparticle complex for 4 h and the corresponding spectra (d) (demixed, blue: nanoparticle complex, red: cell, black: buffer). (a) White light image and (b) the corresponding confocal Raman distribution image (2800–3100 cm^{-1}). *HA*, hydroxyapatite.
Reproduced from Victor, S.P., Paul, W., Vineeth, V.M., Komeri, R., Jayabalan, M., Sharma, C.P. 2016. Neodymium doped hydroxyapatite theranostic nanoplatforms for colon specific drug delivery applications. Colloids Surfaces B Biointerfaces 145, 539–547 with permission from Elsevier.

for albumin adsorption (Victor et al., 2014b). A novel nanoparticle based on iron-doped calcium phosphate provided dual mode magnetic resonance contrast enhancement together with thermal response suitable for ablation of solid tumors on exposure to clinically approved radiofrequency range, and power was developed for in vivo theranostic application (Ashokan et al., 2017).

Multifunctional Au@carbon/calcium phosphate core-shell nanoparticles (Au@C/CaP NPs), which act as an efficient nanoplatform for pH/NIR dual-responsive drug delivery, X-ray computed tomography imaging, and synergistic chemophotothermal cancer therapy, have been developed (Wang et al., 2017). The uptake and the intracellular release performance of the Au@C/CaP NPs were demonstrated in HeLa cells. There was a sustained release of doxorubicin, and after 4 h of uptake, these particles reached the nucleus. These nanoparticles have great potential as a theranostic nanoparticle.

13.2.3 Gene delivery applications

Gene therapy is the insertion of genes into an individual's cells and tissues to treat a disease, such as a hereditary disease in which a deleterious mutant allele is replaced with a functional one. The first approved gene therapy procedure was performed on 4-year-old Ashanthi DeSilva on September 14, 1990 (Blaese et al., 1995). Doctors removed white blood cells from the child's body, let the cells grow in the lab, inserted the missing gene into the cells, and then infused the genetically modified blood cells back into the patient's bloodstream. All viruses bind to their hosts and introduce their genetic material into the host cell as part of their replication cycle. Therefore, the scientists exploited this property and utilized viruses as vehicles to carry "good" genes into a human cell. Viral vectors have high efficiencies in delivering the gene and its expressions. However, the drawbacks, such as immunogenicity, cytotoxicity, restricted targeting, production and distribution difficulties, and high costs, led to the eventual clinical failures with viral vectors. Nonviral vectors pose certain advantages over viral vectors, with simple, large-scale production and low host immunogenicity. Several organic nanoparticles have been studied and developed successfully from cationic polymers, liposomes, and micelles. Because organic particles tend to microbial attack and were less stable, inorganic nanoparticles such as calcium phosphate got preference.

Calcium phosphate nanoparticles represented a unique class of nonviral vectors, which can serve as efficient and alternative DNA carriers for targeted delivery of genes (Maitra, 2005). It has been demonstrated that surface-modified calcium phosphate nanoparticles can be used in vivo to target genes specifically to the liver (Roy et al., 2003). Attachment of a galactose moiety onto the particle surface has increased the targetability of the particles to the liver. This surface modification makes it possible for site-specific gene delivery. Calcium phosphate nanoparticles functionalized by DNA are taken up by living cells. A clear correlation between the uptake of nanoparticles and the efficiency of transfection was found (Sokolova et al., 2007). Surface-modified calcium phosphate nanoparticles by PEGylation encapsulating p53 plasmid have been used for tumor targeted delivery (Fenske et al., 2002). The in vitro transfection studies revealed that consistent levels of gene expression could be achieved by optimizing the Ca/P ratio and the rate of mixing the calcium and phosphate precursors. The optimized forms of calcium phosphate nanoparticles were approximately 25–50 nm in size (when complexed with pDNA) and were efficient at both binding and condensing the genetic material. The differences in gene expression were not just due to a change in size of the naked calcium phosphate nanoparticles but were rather due to the combined effects of pDNA binding and condensation to the particle, which ultimately dictated the overall size of the pDNA-nanoCaP complex (Olton et al., 2007).

The benefit of calcium phosphate nanoparticles is insignificant IgE response, and importantly, they are a natural constituent of the human body. Because of these facts, CaP is well tolerated and absorbed by the body. By virtue of the potency of calcium phosphate as an adjuvant and the relative absence of side effects (He et al., 2000), calcium phosphate formulations have great potential for use in humans. Calcium

Inorganic nanoparticles for targeted drug delivery 345

ions play an important role in endosomal escape, cytosolic stability, and enhanced nuclear uptake of DNA through nuclear pore complexes. Calcium phosphate—mediated gene delivery can become more advantageous compared with other viral and nonviral carriers in the sense that the method is relatively safe (as this is in the generally recognized as safe list of FDA), is cost effective, and has high transfection efficiency.

Despite its beneficial effects such as high loading capacity, inexpensive, nontoxic, biocompatible, bioactive, easily synthesizable, and degradable in the early lysosome, calcium phosphate nanoparticles have limited success in attempts to optimize their properties for transfection comparable in efficiency to that of viral vectors, mainly because of nanoparticle aggregation. This has been overcome by addition of citrate, which in turn increased the transfection efficiency (Khan et al., 2016). Immunofluorescent image (Fig. 13.7) shows the time line of transfection events, from endocytosis of CaP-citrate carrier to the onset of enhanced green fluorescent protein expression to the complete pervasion of the transfected cells with the fluorescent signal. Calcium phosphate precipitation conditions and other compositional parameters that defined the precursor solutions affected transfection in different ways. While presence of citrate enhanced transfection when added at low concentrations, addition of polylysine extended it over longer periods of time. The synergy between these two can be beneficial for developing an effective transfection agent. In another study, a new synthesis route involving click chemistry was developed to prepare the PEGylated chelator PEG-inositol 1,3,4,5,6-pentakisphosphate that can coat and stabilize CaP nanoparticles (Huang et al., 2017). The transfection efficiency of the particles prepared

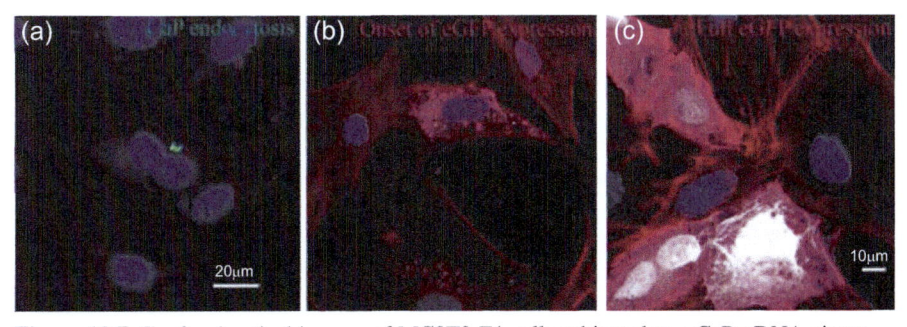

Figure 13.7 Confocal optical images of MC3T3-E1 cells subjected to a CaP-pDNA-citrate nanoparticle treatment (1 lg of pDNA per well) showing the time line of transfection events: (a) CaP particles (green) are endocytosed and are seen in the vicinity of the nucleus (blue), at the stage in which they have already delivered their plasmid payload; (b) MC3T3-E1 cells are now transfected and the expression of Enhanced green fluorescent protein (eGFP) (purple) has begun; (c) the expression of eGFP within the transfected MC3T3-E1 cells is now at its maximum, and the transfected cells glow throughout their entire volume. The images show CaP-pDNA particles stained in green, f-actin stained in red, eGFP stained in purple, and the cell nuclei stained in blue.
Reproduced from Khan, M.A., Wu, V.M., Ghosh, S., Uskokovic, V., 2016. Gene delivery using calcium phosphate nanoparticles: optimization of the transfection process and the effects of citrate and poly(L-lysine) as additives. J. Colloid Interface Sci. 471, 48–58 with permission from Elsevier.

was significantly higher, and none of the different formulations tested showed signs of cytotoxicity. A protamine sulfate−coated calcium phosphate (PS-CaP) nanoparticle has also exhibited improved stabilization of particle size and enhanced transfection efficiency (Liu et al., 2011).

13.2.4 Cancer chemotherapy applications

Hepatocellular carcinoma is the most common primary malignant tumor of the liver. When the human hepatoma cell line BEL-7402 was cultured and treated with calcium phosphate nanoparticles at various concentrations, it inhibited the growth of hepatoma cells in a dose-dependent manner (Liu et al., 2003). The clathrin-mediated endocytosis was found to be responsible for the uptake of calcium phosphate nanoparticles (Bauer et al., 2008). As the clathrin endocytic pathway keeps the particles captured in the endosomes for lysosomal digestion, the reported toxic effect of calcium phosphate nanoparticles could be caused only by cell structure damage due to accumulating calcium phosphate−filled endosomes or the toxic effect of lysosomal degraded calcium phosphate solutes in the cytoplasm.

Calcium phosphate nanoparticles have been studied for delivery of the chemotherapy agent, cisplatin. The nanoconjugate was prepared by electrostatic binding of cisplatin to the nanocalcium phosphate. The drug released from the nanoconjugate was equally effective as the free drug against the A2780cis cell line (Cheng and Kuhn, 2007). A hydrophobic cell growth inhibitor, ceramide, was successfully delivered in vitro to human vascular smooth muscle cells, via encapsulation in calcium phosphate nanoparticles. Nanoparticles encapsulating Cy3 amidite exhibited a nearly fivefold increase in fluorescence quantum yield when compared with the free dye. Thus, calcium phosphate nanoparticles can be utilized for encapsulation of imaging agents and anticancer drugs (Morgan et al., 2008). Desired levels of the therapeutic agents in the tumor sites for effective therapeutic response and minimizing damage to normal cells are the key issues in cancer therapy. A study with chitosan calcium phosphate composite nanoparticles exhibited inhibition of the growth of tumor cells and induced morphological changes typical of apoptosis (Abdel-Gawad et al., 2016). These nanoparticles showed good biocompatibility and low cytotoxicity to normal cells when injected intraperitoneally. It exhibited potent therapeutic effect on implanted solid tumor within 4 weeks of treatment and achieved complete recovery of Ehrlich ascites carcinoma after 3 months. This provided a comprehensive insight in the usage of nanoparticles of calcium phosphate in vivo as a therapy of solid tumor. In another study with calcium phosphate−polymer hybrid nanoparticle system, simultaneous delivery of paclitaxel and miRi-221/222 to their intracellular targets was achieved, leading to inhibit proliferative mechanisms of miR-221/222 and thus significantly enhancing the therapeutic efficacy of paclitaxel (Zhou et al., 2017). The intracellular delivery of miRi has been demonstrated as shown in Fig. 13.8. The white arrows also show the poor colocalization of miRi and endolysosomes, suggesting that these miRi escaped from the endolysosomes after 4 h. As NPs (miRi) enter the cells through endocytosis and localize in acidic endolysosomes, the dissolution rate of calcium phosphate binding miRi increases. Then the dissolved calcium and

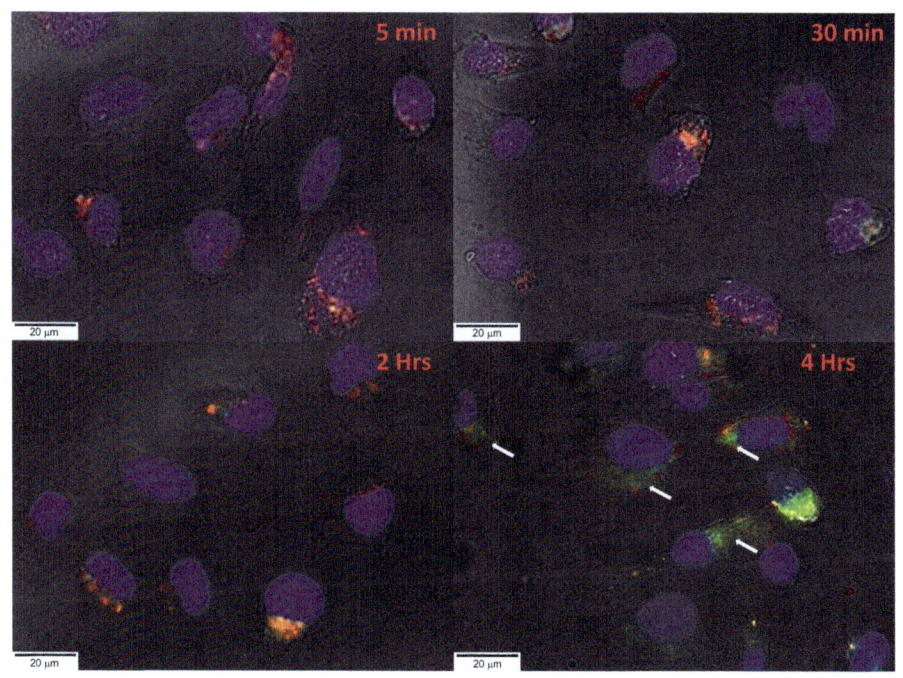

Inorganic nanoparticles for targeted drug delivery 347

Figure 13.8 Intracellular distribution of FAM-labeled miRi (green) delivered by calcium phosphate nanoparticles (NP [miRi]) over MDA-MB-231. Nuclei were stained by 4′,6-diamidino-2-phenylindole (DAPI) (blue); endolysosomes were stained by LysoTracker (red). The cells were incubated with NP (miRi) at a miRi concentration of 20 pmol/mL for 5 min, 30 min, 2 h, and 4 h at 37°C. Arrows indicate miRi in the cytoplasm.
Reprinted from Zhou, Z., Kennell, C., Lee, J.Y., Leung, Y.K., Tarapore, P., 2017. Calcium phosphate-polymer hybrid nanoparticles for enhanced triple negative breast cancer treatment via co-delivery of paclitaxel and miR-221/222 inhibitors. Nanomedicine 13, 403–410 with permission from Elsevier.

phosphate ions increase osmotic pressure and cause a swelling of the compartments. This swelling eventually ruptures the endolysosomal membranes and releases miRi into cytoplasm. Amorphous calcium phosphate nanoparticles have also been shown as a potent candidate for apoptosis induction (Pourbaghi-Masouleh and Hosseini, 2013). Codelivery of doxorubicin and paclitaxel was achieved by lipid-coated hollow calcium phosphate nanoparticles, and this codelivery system had significant synergistic in vivo antitumor activity on human lung cancer cells.

13.2.5 Tissue engineering applications

Synthetic materials could not be considered as ideal implants as the current average lifetime of an orthopedic implant, such as hip, knee, ankle, etc., is only 15 years. Conventional implants, or those implants constituted with grain size dimensions greater than one micron, could not invoke natural cellular responses to regenerate tissue that allows longer periods of successful lifetime. However, as nanophase materials can mimic the dimensions of constituent components of natural tissues, implants developed from nanophase material can be a successful alternative. Several reports on nanophase materials encourage its use for tissue engineering applications. This has been achieved by the combined effect of its ability to mimic the natural nanodimensions and also the cell responses encouraging high reactivity, which in turn helps in regenerating tissues.

NanOss bone void filler from Angstrom Medica is considered to be the first nanotechnology medical device to receive clearance by the US Food and Drug Administration (in 2005). According to the company, NanOss is an innovative structural biomaterial that is highly osteoconductive and remodels over time into human bone, with applications in the sports medicine, trauma, spine, and general orthopedics markets. It is engineered synthetic bone developed from nanocrystalline calcium phosphate and is the first material that duplicates the microstructure, composition, and performance of human bone. Utilizing nanotechnology, calcium and phosphate are manipulated at the molecular level and assembled to produce materials with unique structural and functional properties. It is prepared by precipitating nanoparticles of calcium phosphate in aqueous phase, and the resulting white powder is compressed and heated to form a dense, transparent, and nanocrystalline material. It is strong and also osteoconductive. Ostim is an injectable bone matrix in paste form, which received CE marking in 2002. It is composed of synthetic nanoparticulate HA, which is indicated for metaphyseal fractures and cysts, acetabulum reconstruction and periprosthetic fractures during hip prosthesis exchange operations, osteotomies, filling cages in spinal column surgery, combination with autogenous and allogenous spongiosa, filling in defects in children, etc.

Cell spreading is an essential function of a cell that has adhered to any surface and precedes the function of cell proliferation. Out of the bone and the ceramic material interactions that take place at the material surface, the interaction of osteoblasts is crucial in determining the tissue response at the biomaterial surface (Hunter et al., 1995). Attachment and spreading of specific bone-forming cells in cell culture has been utilized for predicting the behavior of the calcium phosphate materials in vivo (Meyer et al., 2005). The process of cell interaction on materials is highly dynamic and depends on various parameters influencing the cell responses. It is well known that the size and shape of the cell spreading area, as well as the number, size, shape, and distribution of focal adhesion plaques, are decisive for further migratory, proliferative, and differentiation behavior of anchorage-dependent cells (Bacakova et al., 2004). Cells usually do not survive if the attached cells are round and are not spreading. If the cell material contact area is significantly high, the cells tend to skip the proliferation phase and enter sooner the differentiation program. If the adhesion is intermediate, the cells are most active in migration and proliferation.

Minerals such as zinc and magnesium are known to aid bone growth, calcification, and bone density. Biphasic calcium phosphate ceramic containing zinc also promotes osteoblastic cell activity in vitro. The attachment and spreading of osteoblast-like cells onto porous ceramic materials made from nanoparticles of zinc phosphate and calcium phosphate containing zinc and magnesium has been evaluated (Paul and Sharma, 2007).

As the ceramic matrices are made from nanoparticles, this mimics the way nature itself lays down minerals. An important objective of bone tissue engineering is to develop improved scaffold materials or arrangements to control osteoblast behavior significantly affecting its response. Osteoblastic cells on HA exhibited unique attachment and subsequent behavior in vitro, which may explain why mineralized tissue formation is better on HA. Divalent cations, including Mg^{2+}, are known to be active in cell adhesion mechanisms. This investigation demonstrated that the cells were spreading well on matrix containing an optimum amount of calcium, zinc, and magnesium. It seemed that the presence of calcium and magnesium encouraged the spreading and adhesivity of osteoblast cells onto nanostructured calcium phosphate matrices. Cells were attached and spread completely on the ZnCaMgP nanomatrix, and this matrix appeared to be comparable with the control group (HA matrix), making it a promising candidate for bone tissue engineering. Attempts have been made to culture osteoblasts onto nano-HA ceramic matrix. The in vitro cultured bone may show further bone-forming capability after in vivo implantation. This tissue engineering approach is being tried on patients with skeletal problems (Ohgushi et al., 2003). However, despite numerous exciting advances in preclinical models, regulatory approval barriers, business challenges, and related intellectual property lifecycle issues have impeded clinical translation from the bench to the bedside.

The calcium phosphate nanoparticles, in the order of 100 nm, with high DNA incorporation efficiency, exhibited sustained release of DNA. The MC3T3-E1 preosteoblast cells exhibited the capacity to form bony tissue in as little as two and a half weeks when mixed with DNA nanoparticles encoding for bone morphogenic protein-2 (BMP-2) into the alginate hydrogels and injected subcutaneously in the backs of mice, showing its efficacy in bone regeneration applications. Similar osteoblast adhesion between nonfunctionalized nanocrystalline HA and cell adhesive peptide lysine-arginine-serine-arginine functionalized conventional HA was observed, demonstrating the importance of nanocrystalline particles in bone tissue engineering applications. The preparative methods, chemical interaction, biocompatibility, biodegradation, alkaline phosphatase activity, mineralization effect, mechanical properties, and delivery of nano-HA—based nanocomposites for bone tissue regeneration have been reviewed (Venkatesan and Kim, 2014; Yi et al., 2016; Kattimani et al., 2016), and these composite biomaterials proved to be promising biomaterials for bone tissue engineering.

13.3 Gold nanoparticles

In this emerging world of nanoscience, and nanotechnology with nanoparticles, gold nanoparticles, or colloid gold, the most stable metal nanoparticles, have significant importance. The behaviors of the individual particles, and size-related electronic, magnetic, and optical properties, are some of the beneficial properties of gold nanoparticles. Colloid gold was used historically for making ruby glass. The most famous example is the Lycurgus Cup that was manufactured in the 5th to 4th century BC.

It is ruby red in transmitted light and green in reflected light, due to the presence of gold colloids. The curative power of colloid gold for various diseases has been reported as early as 1618 by the philosopher and medical doctor Francisci Antonii (Antonii, 1618). Colloid gold has been used since the 1930s in modern medicine as a treatment for rheumatoid arthritis. Historically, gold nanoparticles were prepared by citrate reduction from $HAuCl_4$ to obtain particles in the range of 20 nm. This procedure is the most popular and is still being used for particle preparation. However, several other procedures are also reported by various researchers (Yeh et al., 2012) using tetraoctylammonium bromide, sodium borohydride, and various green reagents of plant origin and from polymers, using UV-induced photochemical synthesis, ultrasound-assisted synthesis, laser ablation and microbial-mediated synthesis giving particles ranging from 1 to 150 nm in size. Drug, gene, and protein delivery of gold nanoparticles have been reviewed by several investigators (Khan et al., 2014; Sengania et al., 2017).

In an attempt to deliver drug-loaded nanoparticles through the blood—brain barrier, gold nanoparticles were coated with the human serum albumin. The low surface charge and ability to absorb large amounts of creatine, and the albumin layer, may help these particles to attain their objective (Lopez-Viota et al., 2009). Gold nanoparticles, when coated with poly(gamma-glutamic acid) along with phospholipid and PEG, exhibited high stability at different pH values and long blood circulation ($t_{1/2} = 22.1$ h) on intravenous injection into mice (Prencipe et al., 2009). Reproducible and reversible phase transition and aggregation of gold nanorod—elastin-like polypeptide nanoassemblies on exposure to NIR light were observed, which find application in sensing and drug delivery (Huang et al., 2008). Surface functionalization of gold nanoparticles has been utilized for fabricating smart sensors that are capable of detecting heavy metals, glucose and specific biomolecules such as protein and DNA (Zhang, 2013), metal ions, organic molecules, proteins, nucleic acids, and microorganisms (Saha et al., 2012).

Gold-dendrimer nanoparticles exhibited high levels of uptake and selective targeting to certain organs without specific targeting moieties placed on their surfaces (Balogh et al., 2007). Gold nanoparticles were also synthesized and stabilized by new "clicked" dendrimers of generations zero to two (G[0]—G[2]) containing tri- and tetraethyleneglycol ethers (Boisselier et al., 2008). Ligand-exchanged gold quantum dots conjugated with cell-penetrating peptides are a new class of photoluminescent probes for nuclear targeting and intracellular imaging (Lin et al., 2008). It has been demonstrated that the cellular uptake of functionalized gold nanoparticles with cationic or neutral surface ligands can be readily determined using laser desorption/ionization mass spectrometry of cell lysates. The surface ligands have "mass barcodes" that allow different nanoparticles to be simultaneously identified and quantified at levels as low as 30 pmol. Subtle changes to gold nanoparticles surface functionalities can lead to measurable changes in cellular uptake propensities. It has also been demonstrated that incorporating gold nanoparticles in thermosensitive polymer microgels speeds up the response kinetics of PNIPAm, and hence enhances the sensitivity to external stimuli of PNIPAm. These microgels find potential applications for microfluidic switches or microactuators, photosensors, and various nanomedicine applications in controlled delivery and release. PEGylated gold nanoparticle conjugates, which act as water-soluble and biocompatible "cages," allow highly efficient delivery of a hydrophobic drug for photodynamic therapy. The drug delivery time required for photodynamic therapy has been greatly reduced to less than 2 h, compared with 2 days for the free drug (Cheng et al., 2008).

Gold nanoparticles can deliver recombinant proteins, nucleotides, vaccines, and multiple drug molecules and can also control release by external stimuli. One approach of drug loading in gold nanoparticles is by conjugating with drug molecules. Antibiotics can be easily conjugated via ionic or covalent bonding. Gold nanoparticles functionalized with ampicillin became potent bactericidal agents that subverted antibiotic resistance mechanisms of multiple drug-resistant bacteria (Brown et al., 2012). Conjugation of ampicillin, streptomycin, and kanamycin showed greater bactericidal activity and reduced minimal inhibitory concentration compared with their respective free forms (Bhattacharya et al., 2012; Saha et al., 2007). It also showed increased stability and activity at elevated temperatures.

13.3.1 Cancer chemotherapy applications

Cancer is the leading cause of death worldwide. For a better efficiency in chemotherapy, an improved therapeutic method to target tumor foci coupled with enhanced cytotoxicity on cancer cells and decreased side effects is needed. Gold nanoparticles—conjugated drugs have been reported to have a higher perfusion rate in targeting tumor foci, leading to reducing antitumor drug dosage for treatments and lower toxicity to normal tissues coupled with less side effects. In addition to the surface chemistry of gold nanoparticles, its physical properties were also being exploited for drug delivery applications. Gold nanoparticles can cause local increase in temperature by irradiating it with light in the range of 800–1200 nm. The potential use of gold nanoparticles in photothermal destruction of tumors has been reported by many investigators. Silica–gold nanoshells consisting of a silica dielectric core surrounded by a gold shell coated with temperature sensitive N-isopropylacrylamide-co-acrylamide hydrogel modulated drug delivery profiles for methylene blue, insulin, and lysozyme when irradiated by laser. The drug release is dependent on the molecular weight of the therapeutic molecule. Citrate-stabilized gold nanoparticles with a particle size of 30 nm were coated with anti-EGFR (epidermal growth factor receptor) to target HSC3 cancer cells (human oral squamous cell carcinoma). The use of gold nanoparticles enhanced the efficacy of photothermal therapy by 20 times.

Colloidal gold has been safely used to treat rheumatoid arthritis for 50 years and has recently been found to amplify the efficiency of Raman scattering by 14–15 orders of magnitude. Large optical enhancements can be achieved under in vivo conditions for tumor detection in live animals. PEGylated gold nanoparticles were brighter than semiconductor quantum dots with light emission in the NIR window. When conjugated to tumor-targeting ligand, single-chain variable fragment antibodies, the conjugated nanoparticles were able to target tumor biomarkers such as EGFRs on human cancer cells and in xenograft tumor models (Qian et al., 2008). PEG–gold nanoparticles have been proposed as drug carriers and diagnostic contrast agents. Nanoparticles coated with thioctic acid—anchored PEG exhibited higher colloidal stability than with monothiol-anchored PEG. Coating with high molecular weight (5000 Da) PEG was more stable. The 20-nm gold nanoparticles exhibited the lowest uptake by reticuloendothelial cells and the slowest clearance from the body; however, they showed significantly higher tumor uptake and extravasation from the tumor blood vessels and are promising potential drug delivery vehicles (Zhang et al., 2009) for tumor therapy. Selective delivery and activity of kahalalide F analogues have been reported to be improved by conjugating the peptides to gold nanoparticles, for its possible application in antitumor therapy (Hosta et al., 2009). 5-Aminolevulinic acid conjugated gold nanoparticles offered a new modality for selective and efficient destruction of tumor cells, with minimal damage to fibroblasts (Oo et al., 2008).

Tissue necrosis factor (TNF) has significant therapeutic potential for killing cancer. However, the amount of TNF that patients require has never been delivered successfully without eliciting negative side effects, such as hypotension and in some cases complete organ failure resulting in death. When TNF is coupled with colloid gold, it has been demonstrated that beneficial amounts of TNF can be delivered safely in animal models. CytImmune Sciences Inc. USA uses colloid gold particles that are typically 25 nm in size, which is small enough to pass through holes (approximately 100 nm) in the blood vessels that surround a tumor. In healthy organs, spaces between blood vessels are only 5 nm, so colloid gold particles are able to pass into a tumor but are too large to enter any organs. Once the particle passes into the tumor, the TNF is immediately available for biological activity. From the 200—300 TNF molecules around each particle, one molecule acts as the anchor, attaching to cell surface TNF receptors in and around the tumor, allowing the other TNF molecules to exert their anticancer action. It has been demonstrated that gold—dentrimer nanoparticles had high levels of targeting, and this was dependent on surface charge and size (Balogh et al., 2007). Gold nanoparticle—encapsulated alginic acid-poly[2-(diethylamino)ethyl methacrylate] monodisperse hybrid nanospheres had not only uniform size, similar surface properties, and good biocompatibility but also unique optical properties provided by the embedded gold nanoparticles. These negatively charged nanospheres were internalized by human colorectal LoVo cancer cells and hence could act as novel optical contrast reagents in tumor cell imaging by optical microscopy. Drug-loaded nanospheres exhibited similar tumor cell inhibition compared with the free drug doxorubicin (Guo et al., 2009). Functional gold nanoparticles synthesized by a ligand exchange reaction between triphenyl phosphide-stabilized precursor nanoparticles and mercaptopropionic acid were studied with the anticancer drug daunorubicin and were efficient in marking the cancer cells, which may afford potential application for the early diagnosis of the respective cancers (Song et al., 2008).

Substantial enhancement of the antiproliferative effect against K-562 leukemia cells of gold nanoparticles (4—5 nm) bearing 6-mercaptopurine-9-beta-D-ribofuranoside was achieved compared with the same drug in typically administered free form. The improvement was attributed to enhanced intracellular transport followed by the subsequent release in lysosomes. Enhanced activity and nanoparticle carriers will make possible the reduction of the overall concentration of the drug, renal clearance, and, thus, side effects. The nanoparticles with mercaptopurine also showed excellent stability over 1 year without loss of inhibitory activity (Podsiadlo et al., 2008). Improved optical coherence tomography image contrast was achieved when gold nanocages were added to tissue phantoms, along with the selective photothermal destruction of breast cancer cells in vitro (Skrabalak et al., 2007). Gold nanoparticles synthesized by employing "gellan gum" displayed greater stability to electrolyte addition and pH changes relative to the traditional citrate and borohydride-reduced nanoparticles. Anthracycline ring antibiotic doxorubicin hydrochloride loaded nanoparticles showed enhanced cytotoxic effects on human glioma cell lines LN-18 and LN-229 (Dhar et al., 2008).

Intracellular delivery of calcein was attempted by photoactivation of 100 nm gold nanoparticles and membrane-impermeable calcein. The AuNP photoactivation caused localized and reversible disruption of cell membrane, enabling calcein delivery into the cytoplasm. It was demonstrated that the optofluidic probe can be used to treat cells with single-cell precision (Doppenberg et al., 2018). The surface of bacteriophage was patterned by 6 nm gold nanoparticles, and this can hold 500 molecules of doxorubicin without interfering with the preparation process (Benjamin et al., 2018). A rapid release of the drug was demonstrated by nanosecond laser irradiation that allows highly targeted cell killing. In another approach, 23 nm gold nanoparticles were conjugated with 2,5-diphenyltetrazole and methacrylic acid by laser irradiation. This process significantly shifted the plasmon resonance of gold nanoparticles to NIR regions, which can effectively enhance the efficacy of photothermal therapy in the treatment of breast cancer (Xia et al., 2018). Another recent study demonstrated delivery of small molecule compounds in target lymphocyte populations (Yang et al., 2018). Theranostic gold nanoparticles modified with hyaluronic acid and conjugated with antiglypican-1 antibody, oridonin, gadolinium, and Cy7 dye were endocytosed into PANC-1 and BXPC-3 (overexpression GPC1) but not in 293T cells (GPC1-negative). It inhibited the viability and enhanced the apoptosis of pancreatic cancer cells in vitro. In vivo studies showed that these nanoparticles enabled multimodal imaging and targeted therapy in pancreatic tumor–xenografted mice (Qiu et al., 2018). Plant and leaf extracts conjugated gold nanoparticles exhibited antiproliferative effect on breast cancer cell lines (Jabir et al., 2019; Chahardoli et al., 2018). These Linalool-conjugated gold nanoparticles were biocompatible and could be used clinically. Multifunctional magnetic gold nanoparticles have been studied to selectively deliver doxorubicin to the tumor site in a controlled-release manner (Elbialy et al., 2018). It induced photothermal therapy by producing heat by NIR laser absorption and served as contrast agents for magnetic resonance imaging. The positive in vivo results in animal studies show promise in guided chemophotothermal synergistic therapy. Gas-filled core and shell gold nanoparticles have also been studied for photothermal treatment of tumors (Yoon et al., 2018). Gold nanoparticles can be prepared in different shapes and sizes with varied optical and photonic properties. It enabled an efficient cellular uptake along with remarkable improvements in early stage diagnosis of tumors and simultaneous drug delivery (Malik and Mukherjee, 2018; Anaya-Ruiz et al., 2018).

There are a few gold nanoparticle-based therapeutics approved by Food and Drug Administration as investigation devices (Ajnai et al., 2014; Alphandery et al., 2015).

Aurimmune is 25 nm gold nanoparticles conjugated with TNF-α and PEG. It has been shown that *Aurimmune* is less toxic than free TNF-α with same efficacy when studied on mice bearing MC38 and Lewis lung carcinoma tumors. Study shows that these nanoparticles are accumulated in tumors and not in reticuloendothelial system (RES). Clinical study carried out on patients with advanced cancer revealed its targeting potential and increased plasma half-life. *Auroshell* is another nanoparticle-based nanoformulation that has core shell architecture with a silica core and a 2 nm thin layer of gold nanoparticle conjugated with anti-HER2 antibody. These nanoparticles administered intravenously accumulated in the tumor and heated using a NIR laser. Verigene has been developed for diagnostic applications. They are gold nanoparticles of 13−20 nm in diameter that are functionalized either with oligonucelotides of antibodies that are specific to the protein of interest.

13.3.2 Gene delivery applications

Gold nanoparticles are capable of delivering large biomolecules, without restricting themselves as carriers of only small molecular drugs. Tunable size and functionality make them a useful scaffold for efficient recognition and delivery of biomolecules. They have shown success in delivery of peptides, proteins, or nucleic acids such as DNA or RNA. Complexes formed by using cationic liposomes and PEG-modified liposomes, as well as by using phosphatidylcholine liposomes, are proposed for drug and gene delivery (Kojima et al., 2008). Two different DNA oligonucleotides were loaded on two different gold nanorods via thiol conjugation. Selective releases were induced by selective melting of gold nanorods via ultrafast laser irradiation at the nanorods' longitudinal surface plasmon resonance peaks. Excitation at one wavelength could selectively melt one type of gold nanorod and selectively release one type of DNA strand. This seems potential for multiple-drug delivery strategies (Wijaya et al., 2009). Dimethyldioctadecylammonium bromide (DODAB), cationic lipid, and bilayer-coated gold nanoparticle efficiently delivered two types of plasmid DNA into human embryonic kidney cells (HEK 293). The transfection efficiency of the gold nanoparticles was about five times higher than that of DODAB (Li et al., 2008a). Stability of this complex with DNA in the presence of serum dramatically increased after coating DODAB onto the surface of the gold nanoparticles. The cytotoxicity of DODAB was also decreased (Li et al., 2008b). Colloidal gold has also been developed as vector for cellular delivery of catalytic DNA enzymes (DNAzymes), which cleave targeted messenger RNA (Tack et al., 2008). Gold nanoparticles were chemically modified with primary amine groups as intracellular delivery vehicles for therapeutic small interfering RNA (siRNA). The resultant core/shell type polyelectrolyte complexes surrounded by a protective PEG shell layer had a well-dispersed nanostructure with a hydrodynamic diameter of 96.3 ± 25.9 nm. The nanosized polyelectrolyte complexes were efficiently internalized in human prostate carcinoma cells, and thus enhanced intracellular uptake of siRNA. Furthermore, the siRNA/gold complexes significantly inhibited the expression of a target gene within the cells without showing severe cytotoxicity (Lee et al., 2008).

Multifunctional gold nanoparticles, decorated by histone motifs as histone-inspired scaffolds, exhibited sixfold enhancement in transfection efficiency and is a versatile scaffold for delivery of gene and chromatin analysis applications (Munsell et al., 2018). High gene transfection efficiency (about 85%) was achieved for lipid-coated gold nanoparticles in MCF-7 cells mediated by folic acid (FA)—based ligands (Du et al., 2017). These particles are beneficial for tumor cells that overexpress FA receptors. Plasmid DNA/c-myb conjugated with chitosan—gold nanoparticles promoted osteogenesis and inhibited osteoclastogenesis in MC-3T3 E1 cells. Titanium dental implants coated with these nanoparticles showed increased c-myb expression and upregulation of bone morphogenic proteins, osteoprotegerin, and EphB4. The c-myb delivered by these gold nanoparticles supported osseointegration of dental implant even in osteoporotic condition and supported dental implant integration and treatment in age-dependent bone destruction disease (Takanche et al., 2018). Arginine conjugated gold nanoparticles caped with starch and polyethyleneimine also improved the transfection efficiency (Dhanya et al., 2018). Noncovalently bound siRNA—gold nanoparticle core coated with a lipid layer provided efficient delivery of siRNA into human embryonic kidney cell followed by specific gene silencing (Poletaeva et al., 2018).

13.4 Iron oxide nanoparticles

Magnetic nanoparticles were initially prepared as contrast agents for magnetic resonance imaging. Because of their response to magnetic fields, these nanoparticles seem to be most promising for targeted drug delivery applications. The main disadvantage of the conventional chemotherapeutic drugs is normally their nonspecificity. The normal and healthy cells also will be attacked by the cytotoxic drug, along with the tumor cells. Drug-loaded magnetic nanoparticles can be guided toward the tumor cells by the application of a magnetic field, which can reduce the systemic distribution of the drug, thereby reducing the associated side effects (Pankhurst et al., 2003). This can also reduce the drug dosage required for the therapy and efficiently utilize the drug. Once the drug-loaded nanoparticles are concentrated at the tumor site, the drug can be released by applying stimuli such as enzymatic activity, pH change, or temperature (Alexiou et al., 2000). The use of a magnetic field for targeting cancerous tumors was proposed in the late 1970s. The magnetic nanoparticles are usually iron oxide particles and are often applied together with a suitable coating to improve their biocompatibility and functionalizability. Iron oxide nanoparticles exhibit remarkable new properties, such as superparamagnetism, high field irreversibility, high saturation field, extra anisotropy contributions, or shifted loops after field cooling. These properties are contributed by their finite size and surface effects that dominate the magnetic behavior of individual nanoparticles. These nanoparticles have a broad spectrum of applications covering industrial to medical. The nanoparticles intended for biomedical applications should have superparamagnetic behavior at room temperature and should be stable in physiological media. The stability is mainly achieved by their high charge, which avoids precipitation of the particles and settling due to gravitational forces.

They should also be nonbiodegradable in the system. Superparamagnetic nanoparticles have also been synthesized as highly monodisperse particles with hydrodynamic diameter of 1.8 nm by a templated synthesis. This film-assisted synthesis gave highly stable particles by preventing aggregation due to magnetostatic coupling (Sreeram et al., 2009).

13.4.1 Cancer therapy applications

Therapeutic application of iron oxide nanoparticles includes drug targeting and hyperthermia, where a high degree of specificity is achieved and is mainly applicable to cancer treatments. These particles allowed microparticle retention with an external magnetic field, thus reducing their clearance from the skeletal joint and releasing dexamethasone acetate for 5 days in vivo (Butoescu et al., 2009a). Iron oxide nanoparticles, when conjugated with chlorotoxin via amine-functionalized PEG silane, enabled the tumor cell−specific binding. They exhibited substantially enhanced cellular uptake and an invasion inhibition rate of approximately 98% compared with unbound chlorotoxin (Veiseh et al., 2009). In another study, iron oxide nanoparticles with maleimidyl 3-succinimidopropionate ligands were conjugated with paclitaxel molecules. These nanoparticles liberated paclitaxel molecules on exposure to phosphodiesterase. Gelatin-coated iron oxide nanoparticles demonstrated a pH-responsive drug release leading to accelerated release of doxorubicin at acidic pH compared with neutral pH with increased drug loading efficiency. Starch-coated iron oxide nanoparticles, conjugated with 5-carboxyfluorescein−labeled AGKGTPSLETTP peptide, loaded with doxorubicin, were specific to human hepatocellular carcinoma cell line and exhibited higher cytostatic effect. Nanoparticles embedded in a thermosensitive Pluronic F127 matrix on short exposure to a high-frequency magnetic field caused rapid heating and volume shrinkage of nanospheres, thereby causing instantaneous release of doxorubicin. Particles embedded in lipid also behaved similarly, releasing the drug under magnetic control. Doxorubicin-loaded superparamagnetic iron oxide nanoparticles (SPIONs) demonstrated pH-dependent release. Oleic acid−coated iron oxide and Pluronic-stabilized nanoparticles were loaded with doxorubicin and paclitaxel with a loading efficiency of 90%. These drugs in combination demonstrated highly synergistic antiproliferative activity in Michigan Cancer Foundation-7 (MCF-7) breast cancer cells. Similar oleic acid−coated nanoparticles (diameter 5−10 nm) were found highly suitable for application toward image-guided drug delivery. These 200 nm particles showed excellent stability under physiological conditions in the presence of fetal bovine serum for 7 days.

Magnetic particles could be retained in the joint, using an external magnetic field, and prolong the local release of an antiinflammatory drug. SPIONs and dexamethasone 21-acetate (DXM) were coencapsulated into biodegradable microparticles. DXM encapsulation efficacy was 90%. The microparticles were retained with an external magnet, and faster DXM release was obtained for smaller microparticles (Butoescu et al., 2008). SPIONs had an excellent biocompatibility with synoviocytes and were internalized through a phagocytic process. This could represent a suitable magnetically retainable intraarticular drug delivery system for treating joint diseases such as arthritis or osteoarthritis (Butoescu et al., 2009b).

Iron oxide magnetic nanoparticles functionalized with cisplatin demonstrated synergism between the effects of cisplatin target magnetic (MAG) nanoparticles and the application of an electromagnetic field. SPIONs stabilized by alginate exhibited magnetic targeting with an external magnetic field and did not get detained at the injection site without the magnetic field. These nanoparticles were generally considered to be biocompatible and noncytotoxic. Biodistribution, clearance, and biocompatibility of magnetic nanoparticles for in vivo biomedical applications were studied by Jain et al. by injecting them intravenously (Jain et al., 2008) to ensure their safe clinical use. The greater fraction of the injected iron was localized in the liver and spleen rather than in the brain, heart, kidney, and lung. Iron oxide nanoparticles were found to be biocompatible and did not cause long-term changes in the liver enzyme levels or induce oxidative stress. Thus, iron oxide nanoparticles can be safely used for drug delivery and imaging applications. The precise mechanisms of translocation of iron nanoparticles into targeted tissues and organs from blood circulation, as well as the underlying implications of potential harmful health effects in humans, are unknown. One of the latest studies has indicated that the exposure of iron nanoparticles induces an increase in endothelial cell permeability through reactive oxygen species (ROS) oxidative stress-modulated microtubule remodeling. This is a new understanding of the effects of nanoparticles on vascular transport of macromolecules and drugs (Apopa et al., 2009). The ability of SPION to be rapidly taken up and distributed into lymphoid tissues was demonstrated for application toward macrophage-targeted nanoformulations for diagnostic and drug therapy. Targeting of mouse xenograft tumors was performed using SPION as a model nanoparticle system. It was found that the elongated nanomaterial (nanoworm) was superior to that of spherical (nanosphere) material in tumor-targeting efficiency and blood half-life (Park et al., 2009). Incorporation of PEG as a spacer improved the targeting efficiency. Gold-coated iron oxide nanoparticle/engineered protein G hybrid systems were successfully employed as multifunctional cargo systems for the targeting, imaging, and manipulation of mitochondria. Improved specificity, extended particle retention, and increased cytotoxicity toward tumor cells were demonstrated by iron oxide nanoparticles conjugated with a chemotherapeutic agent, methotrexate, and a targeting ligand, chlorotoxin, through a PEG linker. This possesses potential for applications in cancer diagnosis and treatment.

It has been demonstrated with poly(vinyl alcohol) (PVAlc)−coated iron oxide nanoparticles that the functionalized nanoparticles displaying potential cellular uptake by human cancer cells depend both on the presence of amino groups on the coating shell of the nanoparticles and of its ratio to the amount of iron oxide. Dendrimer-coated iron oxide SPIONs exhibited selective targeting to KB cancer cells in vitro. The results were consistent between the uptake distribution quantified by flow cytometry using 6-TAMRA UV-visible fluorescence intensity and the cellular iron content determined using X-ray fluorescence microscopy.

SPIONs were functionalized with anticancer drugs 5-fluorouridine or doxorubicin via aminoPVAlc. The 5-fluorouridine−SPIONs with an optimized ester linker were taken up by human melanoma cells and proved to be efficient antitumor agents, while the doxorubicin-SPIONs linked with a Gly-Phe-Leu-Gly tetrapeptide were cleaved by lysosomal enzymes and exhibited poor uptake by human melanoma cells in culture

(Hanessian et al., 2008). Intravenously injected iron oxide nanoparticles with hydrodynamic diameters of up to 100 nm were found accumulated in gliosarcomas by magnetic targeting and were successfully quantified by MR imaging. Hence, these nanoparticles appear to be a promising vehicle for glioma-targeted drug delivery. Rhodamine-labeled Pluronic/chitosan nanocapsules encapsulating iron oxide nanoparticles were efficiently internalized by human lung carcinoma cells on exposure to an external magnetic field. Amino-functionalized SPIONs conjugated with Hepama-1, an excellent human monoclonal antibody directed against liver cancer, could markedly kill SMMC-7721 liver cancer cells and could be very useful for biomagnetically targeted radiotherapy in liver cancer treatment (Liang et al., 2007).

SPIONs coated with covalently bound bifunctional PEG were conjugated with FA. The specificity of the nanoconjugate targeting cancerous cells was demonstrated by comparative intracellular uptake of the nanoconjugate and PEG-/dextran-coated nanoparticles by human adenocarcinoma HeLa cells. Uptake of the nanoconjugate by HeLa cells after 4 h incubation was found to be 12-fold higher than that of PEG- or dextran-coated nanoparticles. A significant negative contrast enhancement was observed with magnetic resonance phantom imaging for HeLa cells over MG-63 cells, when both were cultured with the nanoconjugate. Uptake of methotrexate-immobilized iron oxide nanoparticles, conjugated via a PEG self-assembled monolayer, by glioma cells was considerably higher than that of control nanoparticles. The conjugate was highly cytotoxic to 9L cells and the particles internalized into the cellular cytoplasm, retaining its crystal structure therein for up to 144 h. Cells expressing the human folate receptor internalized a higher level of methotrexate conjugated nanoparticles than negative control cells.

13.4.2 Gene delivery applications

SPIONs were complexed with T cell—specific ligand, the CD3 single chain antibody, and were used to condense plasmid DNA into nanoparticles with an ideally small size and low cytotoxicity. There was a 16-fold enhancement in the gene transfection level in HB8521 cells, a rat T-lymphocyte line (Chen et al., 2009). Iron oxide nanoparticles were coated with carboxylated cholesterol overlaying a monolayer of phospholipid, in which Apo A1, Apo E, or synthetic amphoteric alpha-helical polypeptides were adsorbed for targeting high-density lipoprotein (HDL), low-density lipoprotein (LDL), or folate receptors, respectively. This could be utilized for in situ loading of nanoparticles into cells for MRI-monitored cell tracking or gene therapy (Glickson et al., 2009). In the presence of magnetic fields, hexanoyl chloride—modified chitosan-stabilized iron oxide nanoparticles, conjugated with viral gene (Ad/LacZ), demonstrated high transduction efficiency. The dramatic enhancement in intracellular trafficking of the adenovirus without genetically modified vesicles can lead to enhanced nuclear transfer, especially in chimeric antigen receptor cells (Bhattarai et al., 2008).

SPIONs coupled to insulin prevented endocytosis, whereas uncoated particles were internalized by the fibroblasts due to endocytosis, which resulted in disruption of the cell membrane. The derivatized nanoparticles also showed high affinity for the cell membrane. Similarly, transferrin-derivatized particles appeared to localize to the cell membrane without instigating receptor-mediated endocytosis and also induce upregulation in the cells for many genes, particularly in the area of cytoskeleton and cell signaling. This is also the case with lactoferrin- or ceruloplasmin-conjugated nanoparticles. Iron oxide nanoparticles derivatized with elastin, dextran, or albumin induced alterations in cell behavior and morphology distinct from the underivatized particles, suggesting that dermal fibroblast cell response could be directed via specifically engineered particle surfaces. However, there was no difference observed between particles of different sizes. It has been demonstrated that higher numbers of conjugated trans-activator of transcription (TAT) peptide facilitated the cellular uptake of iron oxide nanoparticles in a nonlinear fashion. Cells labeled with these optimized preparations were readily detectable by MR imaging with 100-fold sensitivity.

13.4.3 Tissue engineering applications

Gradients of bioactive signals are the key that direct the formation of structural transitions in tissues. These native features need to be reproduced for the success of bioengineered tissue constructs. One report (Li et al., 2018) uses magnetic field alignment of glycosylated SPIONs, preloaded with growth factors, to pattern biochemical gradients into a range of biomaterial systems. Gradients of bone morphogenetic protein in agarose hydrogels were used to spatially direct the osteogenesis of human mesenchymal stem cells and generate robust osteochondral tissue constructs exhibiting a clear mineral transition from the bone to cartilage. This platform technology offers great versatility and provides an exciting new opportunity for overcoming a range of interfacial tissue engineering challenges. The gradients of growth factor loaded SPIONs were generated by external magnetic field. It was demonstrated that the higher expression of osteogenic genes in the bone region compared with the cartilage region observed could spatially influence matrix composition and structure of tissue construct (Fig. 13.9).

Three-dimensional multicellular assemblies of controlled geometry were formed using external magnetic force utilizing human endothelial progenitor cells and mouse macrophages magnetically labeled with anionic citrate−coated iron oxide nanoparticles. The procedure avoided the need for substrate chemical or physical modifications. This finds applications in tissue engineering (Frasca et al., 2009). Similarly, monocrystalline iron oxide nanoparticles as an nuclear magnetic resonance (NMR) contrast agent have been studied for noninvasive monitoring of tissue-engineered constructs, optimizing construct design, and assessing therapeutic efficacy (Constantinidis et al., 2009). Noninvasive MR monitoring of tissue-engineered vascular grafts (TEVGs) in vivo, using cells labeled with iron oxide nanoparticles, has also been demonstrated. This has been achieved by labeling human aortic smooth muscle cells with ultrasmall superparamagnetic iron oxide nanoparticles. This finds application in the evaluation of in vivo TEVG performance. A method to deliver functional nanoparticles and target antibodies into cells was successfully demonstrated with gold-coated iron oxide nanoparticle/engineered protein G hybrid systems. Targeting of mouse xenograft tumors was performed using superparamagnetic iron oxide as a

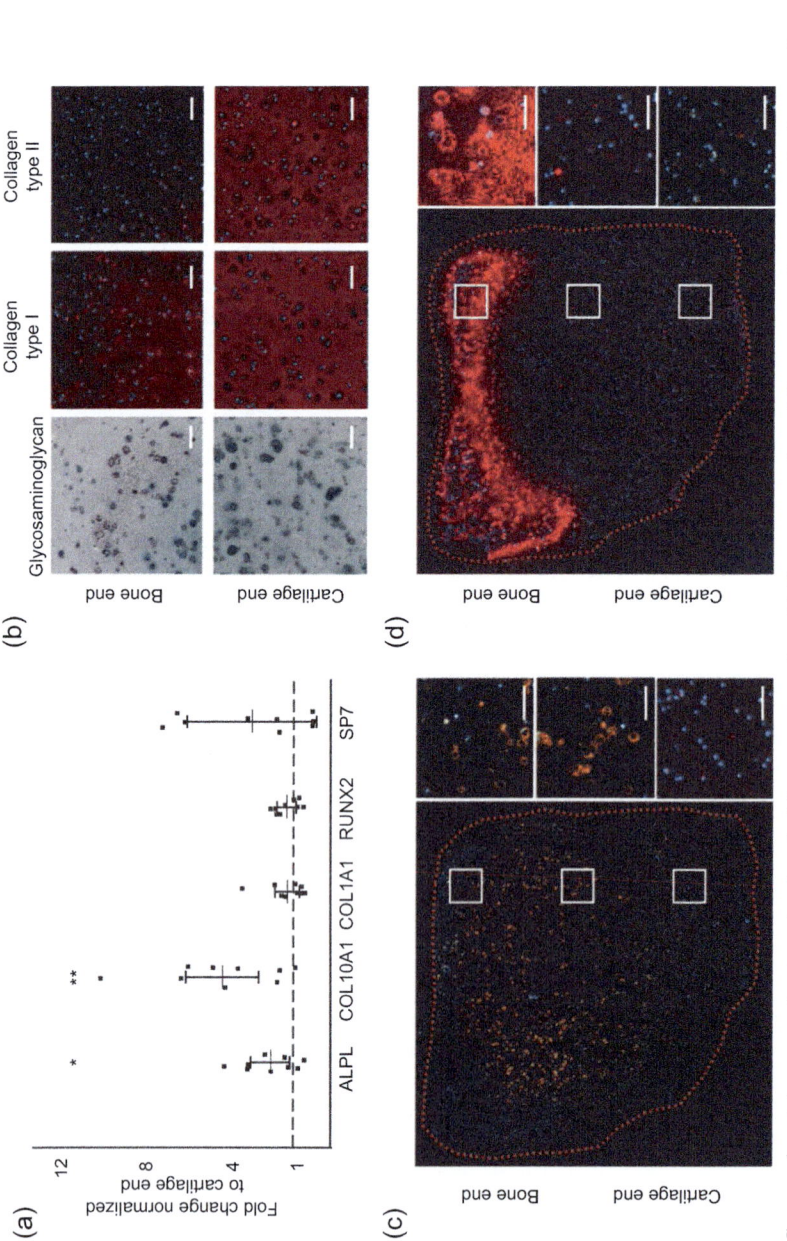

Figure 13.9 Gene and protein expression in osteochondral tissue engineering. (a) Gene expression of five osteogenic genes at the bone region of the tissue, compared with the cartilage region. Comparison of differences were made using a one sample t-test after heteroscedasticity in the data set was addressed (mean ± 95% confidence intervals, N = 3, n = 3) where $p < 0.05$ (*), $p < 0.01$ (**). (b) Histological and immunofluorescence staining of key extracellular matrix proteins present in cartilage and bone revealed deposition of sulfated glycosaminoglycans (blue) and type I and II collagen (red). Scale bars = 200 μm. (c) Immunofluorescence staining of the hypertrophic protein type X collagen (orange), and (d) the key mineralization protein osteopontin (red), which were present specifically at the bone end of the tissue. Scale bars = 200 μm.
Reprinted from Li, C., Armstrong, J.P., Pence, I.J., Kit-Anan, W., Puetzer, J.L., Correia Carreira, S., Moore, A.C., Stevens, M.M. 2018. Glycosylated superparamagnetic nanoparticle gradients for osteochondral tissue engineering. Biomaterials 176, 24–33 under Creative Commons Attribution License (CC BY).

model nanoparticle system. It was found that the elongated nanomaterial (nanoworm) was superior to that of spherical (nanosphere) material in tumor targeting efficiency and blood half-life. Incorporation of PEG as a spacer improved the targeting efficiency. Temperature-responsive magnetite/polymer nanoparticles developed from iron oxide nanoparticles and poly(ethyleneimine)-modified poly(ethylene oxide)−poly(propylene oxide)−poly(ethylene oxide) block copolymer, with an approximately 45 nm hydrodynamic diameter, underwent a sharp decrease from 45 to 25 nm, when evaluated at temperatures from 20 to 35°C. Thermoinduced self-assembly of the immobilized block copolymers occurred on the magnetite solid surfaces, which are accompanied by a conformational change from a fully extended state to a highly coiled state of the copolymer. Consequently, the copolymer shell could act as a temperature-controlled "gate" for the transit of a guest substance. The uptake and release of both hydrophobic and hydrophilic model drugs were well controlled by switching the transient opening and closing of the polymer shell at different temperatures. A sustained release of about 3 days was achieved in simulated human body conditions. In primary mouse experiments, drug-entrapped magnetic nanoparticles showed good biocompatibility and effective therapy for spinal cord damage. Such intelligent magnetic nanoparticles are attractive candidates for widespread biomedical applications, particularly in controlled drug-targeting delivery. Derivatization of the nanoparticle surface with insulin-induced alterations in cell behavior that were distinct from the underivatized nanoparticles suggests that cell response can be directed via specifically engineered particle surfaces. The uncoated particles were internalized by the fibroblasts due to endocytosis, which resulted in disruption of the cell membrane. In contrast, insulin-coated nanoparticles attached to the cell membrane, most likely to the cell-expressed surface receptors, and were not endocytosed. The presence of insulin on the surface of the nanoparticles caused an apparent increase in cell proliferation and viability.

13.4.4 General drug delivery and targeting

The high biocompatibility and versatile nature of liposomes have been combined with superparamagnetic iron oxide nanocores. The so-called magnetoliposomes find applications in enzyme immobilization for water-soluble, hydrophobic, and other applications such as MRI, hyperthermia cancer treatment, and drug delivery. Another result suggests their general applicability, for cell labeling. Iron oxide nanoparticle surfaces were coated with mannan, which induces receptor-mediated endocytosis. Mannan is a water-soluble polysaccharide having a high content of D-mannose residues, to be recognized by mannose receptors on immunate macrophages. This had excellent stability in ferrofluid and low cytotoxicity. It exhibited enhanced targeted delivery efficiency to macrophages in vitro and in vivo. These nanoparticles are suggested for their potential utility as macrophage-targeting MRI contrast agents. Iron oxide nanoparticles and (18) F-fluoride were encapsulated by hemagglutinating virus of Japan envelopes (HVJ-Es). HVJ-Es were then injected intravenously in a rat and were imaged dynamically using high-resolution PET. Magnetic force altered the biodistribution of the viral envelope to a target structure, and this could enable region-specific delivery of therapeutic vehicles noninvasively. Iron oxide nanoparticles covalently bonded to an antiprotein kinase C (PKC) α antibody. This conjugate can serve for cellular PKC localization and the inhibition of its function. The activity of the conjugate was confirmed by recognizing PKCα using the western blot method. Lactose-derivatized galactose terminal amino iron oxide nanoparticles have been developed for targeting the cell surface asialoglycoprotein receptor expressed by hepatocytes.

AminoPVAlc-SPIONs were taken up by isolated brain-derived endothelial and microglial cells at a much higher level than the other SPIONs, and no inflammatory activation of these cells was observed. Fluorescent aminoPVAlc-SPIONs derivatized with a fluorescent reporter molecule and confocal microscopy demonstrated intracellular uptake by microglial cells. Functionalized aminoPVAlc-SPIONs represent biocompatible potential vector systems for drug delivery to the brain, which may be combined with MRI detection of active lesions in neurodegenerative diseases. Nanometer-sized, dextran-coated iron oxide nanoparticles particles were tethered with N-hydroxysuccinimide folate and fluorescence isothiocyanate. Internalization of nanoparticles into targeted KB cells occurred only when nanoparticles were conjugated to folate and when the folate receptors were available and accessible on the cells.

13.5 Conclusion

Because nanotechnology focuses on the very small, it is uniquely suited to creating systems that can better deliver drugs to tiny areas within the body. Nanoenabled drug delivery also makes it possible for drugs to permeate through cell walls, which is of critical importance to the expected growth of genetic medicine over the next few years. The present review discussed the latest ongoing research and development on inorganic nanoparticles utilized as drug carriers and tissue engineering. Targetability of the drug is the key factor in enhanced efficiency and low toxicity. This can be achieved by ionic or covalent coupling of targeting ligands. However, standardization of the process of chemical ligand modification is not an easy task from the toxicity point of view and the maintenance of the stability of the active ingredient. Thus, inorganic nanoparticles require an interdisciplinary approach among different fields, including biopolymers, biopharmaceutics, materials science, and tissue engineering, to enable their potential applications in drug delivery and tissue engineering. The targetability and controlled release are usually achieved by physicochemical modification of these nanoparticles. The important aspect of inorganic nanoparticles is its high effective surface area. This further helps in conjugating multiple ligands that induce synergistic therapeutic effect. Calcium phosphate nanoparticles are particularly useful as a matrix for bone tissue engineering in addition to their utility in oral insulin delivery systems and nonviral vectors in gene delivery. Iron oxide nanoparticles are extremely effective in active targeting for anticancer therapy. Similarly, gold nanoparticles can achieve controlled delivery in response to near IR irradiation and also for theranostic applications. The toxicity and biocompatibility issues of inorganic nanoparticles have been extensively studied. Although it is generally reported to be highly biocompatible, there are also few conflicting reports on its toxicity profile. This has been mainly on its size-dependent effect and effects due to accumulation in organs. Slow biodegradation is also an important factor that has been addressed by optimization of the process parameters during nanoparticle preparation. Iron oxide nanoparticles are considered to be biocompatible; however, they have been reported to be taken up by the liver, spleen, and lymph node within 24 h. The half-life of iron oxide nanoparticles has been reported to be 22.8 days. Hence, repeated dose may cause systemic toxicity for these particles. The increased surface area of gold nanoparticles and its scattering efficiency makes it an attractive tool in imaging applications. However, a size- and dose-dependent toxicity for gold nanoparticles has also been reported. Designing the right size of nanoparticles, surface modification for cellular-specific targeting, selecting a suitable route of administration, and optimum dose can generate safe and effective inorganic nanoparticles for drug delivery, tissue engineering, sensing, and imaging applications.

13.6 Biointegration concept and future perspective

The interactions between an implant and the surrounding biological system generally are influenced by its material and surface properties. Nanomaterials interact with the proteins, cells, tissues, DNA, and other organelles of the biological system via a series of complex and dynamic physicochemical interactions that are modulated by the surface chemistry and topography, nanoscale dimension, and shape of the material. Nanotechnology plays an important role in biointegration of implants by way of coatings with nanoscale hierarchical features that significantly enhance cellular responses. It has been reported that a porosity of above 80% promotes early stage osteoinduction that signals the importance of nanofabrication of coatings, which improves biointegration and significantly elongates the implant life. These concepts are especially useful for load-bearing orthopedic implants and dental implants. Nanoparticles pose excellent uptake capability and can be functionalized with multiple ligands by various surface modifications, which makes them capable of interacting with biological entities, making them an indispensable tool in nanomedicine and as diagnostic probes or as theranostic materials. This could particularly be helpful in personalized therapy, where the disease conditions and the environmental factors (individual variations) are taken care off. We have comprehensive knowledge in the science behind the biointeraction of nanomaterials and the biological system, so that the potential of utilizing nanoparticular system in biomedicine and medical implant materials seems to be the core subject of future research.

Acknowledgments

We are grateful to Prof. Asha Kishore, Director, and Dr. H.K. Varma, Head BMT Wing of SCTIMST, for providing facilities for the completion of this work.

References

Abdel-Gawad, E.I., Hassan, A.I., Awwad, S.A., 2016. Efficiency of calcium phosphate composite nanoparticles in targeting Ehrlich carcinoma cells transplanted in mice. J. Adv. Res. 7, 143−154.

Ahmad, M.Z., Abdel-Wahab, B.A., Alam, A., Zafar, S., Ahmad, J., Ahmad, F.J., Midoux, P., Pichon, C., Akhter, S., 2016. Toxicity of inorganic nanoparticles used in targeted drug delivery and other biomedical application: an updated account on concern of biomedical nanotoxicology. J. Nanosci. Nanotechnol. 16, 7873−7897.

Ajnai, G., Chiu, A., Kan, T., Cheng, C.C., Tsai, T.H., Chang, J., 2014. Trends of gold nanoparticle-based drug delivery system in cancer therapy. J. Exp. Clin. Med. 6, 172−178.

Alexiou, C., Arnold, W., Klein, R.J., Parak, F.G., Hulin, P., Bergemann, C., Erhardt, W., Wagenpfeil, S., Lubbe, A.S., 2000. Locoregional cancer treatment with magnetic drug targeting. Cancer Res. 60, 6641−6648.

Alphandery, E., GRAND-Dewyse, P., Lefevre, R., Mandawala, C., Durand-Dubief, M., 2015. Cancer therapy using nanoformulated substances: scientific, regulatory and financial aspects. Expert Rev. Anticancer Ther. 15, 1233−1255.

Anaya-Ruiz, M., Bandala, C., Landeta, G., Martinez-Morales, P., Zumaquero-Rios, J.L., Sarracent-Perez, J., Perez-Santos, M., 2019. Nanostructured systems in advanced drug targeting for the cancer treatment: recent patents. Recent Pat. Anticancer Drug Discov. 14, 85−94.

Antonii, F., 1618. Panacea Aurea Sive Tractatus duo de ipsius Auro potabili. Bibliopolio Frobeniano, Hamburg.

Apopa, P.L., Qian, Y., Shao, R., Guo, N.L., Schwegler-Berry, D., Pacurari, M., Porter, D., Shi, X., Vallyathan, V., Castranova, V., Flynn, D.C., 2009. Iron oxide nanoparticles induce human microvascular endothelial cell permeability through reactive oxygen species production and microtubule remodeling. Part. Fibre Toxicol 6, 1.

Ashokan, A., Somasundaram, V.H., Gowd, G.S., Anna, I.M., Malarvizhi, G.L., Sridharan, B., Jobanputra, R.B., Peethambaran, R., Unni, A.K.K., Nair, S., Koyakutty, M., 2017. Biomineral nano-theranostic agent for magnetic resonance image guided, augmented radiofrequency ablation of liver tumor. Sci. Rep. 7, 14481.

Bacakova, L., Filova, E., Rypacek, F., Svorcik, V., Stary, V., 2004. Cell adhesion on artificial materials for tissue engineering. Physiol. Res. 53 (Suppl. 1), S35−S45.

Baeza, A., Ruiz-Molina, D., Vallet-Regi, M., 2017. Recent advances in porous nanoparticles for drug delivery in antitumoral applications: inorganic nanoparticles and nanoscale metalorganic frameworks. Expert Opin. Drug Deliv. 14, 783−796.

Balogh, L., Nigavekar, S.S., Nair, B.M., Lesniak, W., Zhang, C., Sung, L.Y., Kariapper, M.S., EL-Jawahri, A., Llanes, M., Bolton, B., Mamou, F., Tan, W., Hutson, A., Minc, L., Khan, M.K., 2007. Significant effect of size on the in vivo biodistribution of gold composite nanodevices in mouse tumor models. Nanomedicine 3, 281−296.

Barbosa, S., Topete, A., Alatorre-Meda, M., Villar-Alvarez, E.M., Pardo, A., Alvarez-Lorenzo, C., Concheiro, A., Taboada, P., Mosquera, V., 2014. Targeted combinatorial therapy using gold nanostars as theranostic platforms. J. Phys. Chem. C 118, 26313−26323.

Basha, M., 2018. Nanotechnology as a promising strategy for anticancer drug delivery. Curr. Drug Deliv. 15, 497−509.

Bauer, I.W., Li, S.P., Han, Y.C., Yuan, L., Yin, M.Z., 2008. Internalization of hydroxyapatite nanoparticles in liver cancer cells. J. Mater. Sci. Mater. Med. 19, 1091−1095.

Benjamin, C.E., Chen, Z., Kang, P., Wilson, B.A., Li, N., Nielsen, S.O., Qin, Z., Gassensmith, J.J., 2018. Site selective nucleation and size control of gold nanoparticle photothermal antennae on the pore structures of a virus. J. Am. Chem. Soc. 140, 17226−17233.

Bhattacharya, D., Saha, B., Mukherjee, A., Santra, C.R., Karmakar, P., 2012. Gold nanoparticles conjugated antibiotics: stability and functional evaluation. Nanosci. Nanotechnol. 2, 14−21.

Bhattarai, S.R., Kim, S.Y., Jang, K.Y., Lee, K.C., Yi, H.K., Lee, D.Y., Kim, H.Y., Hwang, P.H., 2008. N-hexanoyl chitosan-stabilized magnetic nanoparticles: enhancement of adenoviral-mediated gene expression both in vitro and in vivo. Nanomedicine 4, 146−154.

Blaese, M., Blankenstein, T., Brenner, M., COHEN-Haguenauer, O., Gansbacher, B., Russell, S., Sorrentino, B., Velu, T., 1995. Vectors in cancer therapy: how will they deliver? Cancer Gene Ther. 2, 291−297.

Boisselier, E., Diallo, A.K., Salmon, L., Ruiz, J., Astruc, D., 2008. Gold nanoparticles synthesis and stabilization via new "clicked" polyethyleneglycol dendrimers. Chem. Commun. (Camb.) 4819−4821.

Bonzani, I.C., George, J.H., Stevens, M.M., 2006. Novel materials for bone and cartilage regeneration. Curr. Opin. Chem. Biol. 10, 568−575.

Brown, A.N., Smith, K., Samuels, T.A., Lu, J.R., Obare, S.O., Scott, M.E., 2012. Nanoparticles functionalized with ampicillin destroy multiple-antibiotic-resistant isolates of *Pseudomonas aeruginosa* and *Enterobacter aerogenes* and methicillin-resistant *Staphylococcus aureus*. Appl. Environ. Microbiol. 78, 2768−2774.

Butoescu, N., Jordan, O., Burdet, P., Stadelmann, P., Petri-Fink, A., Hofmann, H., Doelker, E., 2009a. Dexamethasone-containing biodegradable superparamagnetic microparticles for intra-articular administration: physicochemical and magnetic properties, in vitro and in vivo drug release. Eur. J. Pharm. Biopharm. 72, 529−538.

Butoescu, N., Jordan, O., Petri-Fink, A., Hofmann, H., Doelker, E., 2008. Co-encapsulation of dexamethasone 21-acetate and SPIONs into biodegradable polymeric microparticles designed for intra-articular delivery. J. Microencapsul. 25, 339−350.

Butoescu, N., Seemayer, C.A., Foti, M., Jordan, O., Doelker, E., 2009b. Dexamethasone-containing PLGA superparamagnetic microparticles as carriers for the local treatment of arthritis. Biomaterials 30, 1772−1780.

Caruso, F., 2004. Colloids and Colloid Assemblies: Synthesis, Modification, Organization and Utilization of Colloid Particles. Wiley-VCH Verlag GmbH & Co. KGaA, Weinheim.

Chahardoli, A., Karimi, N., Fattahi, A., Salimikia, I., 2019. Biological applications of Phyto-synthesized gold nanoparticles using leaf extract of *Dracocephalum kotschyi*. J. Biomed. Mater. Res. A 107, 621−630.

Chen, F., Ehlerding, E.B., Cai, W., 2014. Theranostic nanoparticles. J. Nucl. Med. 55, 1919−1922.

Chen, G.H., Chen, W.J., Wu, Z., Yuan, R.X., Li, H., Gao, J.M., Shuai, X.T., 2009. MRI-visible polymeric vector bearing CD3 single chain antibody for gene delivery to T cells for immunosuppression. Biomaterials 30, 1962−1970.

Chen, J., Guo, Z., Tian, H., Chen, X., 2016. Production and clinical development of nano-particles for gene delivery. Mol. Ther. Methods Clin. Dev. 3, 16023.

Cheng, X., Kuhn, L., 2007. Chemotherapy drug delivery from calcium phosphate nanoparticles. Int. J. Nanomed. 2, 667−674.

Cheng, Y., Samia, A.C., Meyers, J.D., Panagopoulos, I., Fei, B.W., Burda, C., 2008. Highly efficient drug delivery with gold nanoparticle vectors for in vivo photodynamic therapy of cancer. J. Am. Chem. Soc. 130, 10643−10647.

Chikkaveeraiah, B.V., Malghan, S.G., Girish, K., 2018. Physicochemical properties of nanomaterials relevant to medical products. In: Pleus, R.C., Murashov, V. (Eds.), Physico-Chemical Properties of Nanomaterials. Pan Stanford, New York.

Ciobanu, C.S., Iconaru, S.L., Popa, C.L., Motelica-Heino, M., Predoi, D., 2015. Evaluation of samarium doped hydroxyapatite, ceramics for medical application: antimicrobial activity. J. Nanomater. article ID 849216, 11 pages.

Constantinidis, I., Grant, S.C., Simpson, N.E., Oca-Cossio, J.A., Sweeney, C.A., Mao, H., Blackband, S.J., Sambanis, A., 2009. Use of magnetic nanoparticles to monitor alginate-encapsulated beta TC-tet cells. Magn. Reson. Med. 61, 282−290.

Degli Esposti, L., Carella, F., Adamiano, A., Tampieri, A., Iafisco, M., 2018. Calcium phosphate-based nanosystems for advanced targeted nanomedicine. Drug Dev. Ind. Pharm. 44, 1223−1238.

Dhanya, G.R., Caroline, D.S., Rekha, M.R., Sreenivasan, K., 2018. Histidine and arginine conjugated starch-PEI and its corresponding gold nanoparticles for gene delivery. Int. J. Biol. Macromol. 120, 999−1008.

Dhar, S., Reddy, E.M., Shiras, A., Pokharkar, V., Prasad, B.L., 2008. Natural gum reduced/stabilized gold nanoparticles for drug delivery formulations. Chemistry 14, 10244−10250.

Di Martino, A., Guselnikova, O.A., Trusova, M.E., Postnikov, P.S., Sedlarik, V., 2017. Organic-inorganic hybrid nanoparticles controlled delivery system for anticancer drugs. Int. J. Pharm. 526, 380−390.

Ding, Y., Jiang, Z., Saha, K., Kim, C.S., Kim, S.T., Landis, R.F., Rotello, V.M., 2014. Gold nanoparticles for nucleic acid delivery. Mol. Ther. 22, 1075−1083.

Doppenberg, A., Meunier, M., Boutopoulos, C., 2018. A needle-like optofluidic probe enables targeted intracellular delivery by confining light-nanoparticle interaction on single cell. Nanoscale 10, 21871−21878.

Du, B., Gu, X., Han, X., Ding, G., Wang, Y., Li, D., Wang, E., Wang, J., 2017. Lipid-coated gold nanoparticles functionalized by folic acid as gene vectors for targeted gene delivery in vitro and in vivo. ChemMedChem 12, 1768−1775.

Elbialy, N.S., Fathy, M.M., AL-Wafi, R., Darwesh, R., ABDEL-Dayem, U.A., Aldhahri, M., Noorwali, A., AL-Ghamdi, A.A., 2018. Multifunctional magnetic-gold nanoparticles for efficient combined targeted drug delivery and interstitial photothermal therapy. Int. J. Pharm 554, 256−263.

Evangelopoulos, M., Parodi, A., Martinez, J.O., Tasciotti, E., 2018. Trends towards biomimicry in theranostics. Nanomaterials 8.

Evers, P., 2017. Nanotechnology in Medical Applications: The Global Market. BCC Research LLC, MA, USA.

Faulk, W.P., Taylor, G.M., 1971. An immunocolloid method for the electron microscope. Immunochemistry 8, 1081−1083.

Fenske, D.B., Maclachlan, I., Cullis, P.R., 2002. Stabilized plasmid-lipid particles: a systemic gene therapy vector. In: Phillips, M.I. (Ed.), Methods in Enzymology: Gene Therapy Methods. Elsevier BV, Amsterdam.

Florence, A.T., 2018. Nanotechnologies for site specific drug delivery: changing the narrative. Int. J. Pharm. 551, 1−7.

Frasca, G., Gazeau, F., Wilhelm, C., 2009. Formation of a three-dimensional multicellular assembly using magnetic patterning. Langmuir 25, 2348−2354.

Geuli, O., Metoki, N., Zada, T., Reches, M., Eliaz, N., Mandler, D., 2017. Synthesis, coating, and drug-release of hydroxyapatite nanoparticles loaded with antibiotics. J. Mater. Chem. B 5, 7819−7830.

Glickson, J.D., Lund-Katz, S., Zhou, R., Choi, H., Chen, I.W., Li, H., Corbin, I., Popov, A.V., Cao, W.G., Song, L.P., Qi, C.Z., Marotta, D., Nelson, D.S., Chen, J., Chance, B., Zheng, G., 2009. Lipoprotein nanoplatform for targeted delivery of diagnostic and therapeutic agents. Adv. Exp. Med. Biol. 645, 227−239.

Guo, R., Li, R.T., Li, X.L., Zhang, L.Y., Jiang, X.Q., Liu, B.R., 2009. Dual-functional alginic acid hybrid nanospheres for cell imaging and drug delivery. Small 5, 709−717.

Hanessian, S., Grzyb, J.A., Cengelli, F., Juillerat-Jeanneret, L., 2008. Synthesis of chemically functionalized superparamagnetic nanoparticles as delivery vectors for chemotherapeutic drugs. Bioorg. Med. Chem. 16, 2921−2931.

He, Q., Mitchell, A.R., Johnson, S.L., Wagner-Bartak, C., Morcol, T., Bell, S.J., 2000. Calcium phosphate nanoparticle adjuvant. Clin. Diagn. Lab. Immunol. 7, 899−903.

Highsmith, J., 2014. Nanoparticles in Biotechnology, Drug Development and Drug Delivery. BCC Research, MA, USA.

Ho, Y.P., Leong, K.W., 2010. Quantum dot-based theranostics. Nanoscale 2, 60−68.

Hosta, L., Pla-Roca, M., Arbiol, J., Lopez-Iglesias, C., Samitier, J., Cruz, L.J., Kogan, M.J., Albericio, F., 2009. Conjugation of Kahalalide F with gold nanoparticles to enhance in vitro antitumoral activity. Bioconjug. Chem. 20, 138−146.

Huang, H.C., Koria, P., Parker, S.M., Selby, L., Megeed, Z., Rege, K., 2008. Optically responsive gold nanorod-polypeptide assemblies. Langmuir 24, 14139−14144.

Huang, X., Andina, D., Ge, J., Labarre, A., Leroux, J.C., Castagner, B., 2017. Characterization of calcium phosphate nanoparticles based on a PEGylated chelator for gene delivery. ACS Appl. Mater. Interfaces 9, 10435−10445.

Hunter, A., Archer, C.W., Walker, P.S., Blunn, G.W., 1995. Attachment and proliferation of osteoblasts and fibroblasts on biomaterials for orthopedic use. Biomaterials 16, 287−295.

Jabir, M.S., Taha, A.A., Sahib, U.I., Taqi, Z.J., AL-Shammari, A.M., Salman, A.S., 2019. Novel of nano delivery system for Linalool loaded on gold nanoparticles conjugated with CALNN peptide for application in drug uptake and induction of cell death on breast cancer cell line. Mater Sci. Eng. C Mater. Biol. Appl 94, 949−964.

Jain, T.K., Reddy, M.K., Morales, M.A., Leslie-Pelecky, D.L., Labhasetwar, V., 2008. Biodistribution, clearance, and biocompatibility of iron oxide magnetic nanoparticles in rats. Mol. Pharm. 5, 316−327.

Kang, X.J., Li, C.X., Cheng, Z.Y., Ma, P.A., Hou, Z.Y., Lin, J., 2014. Lanthanide-doped hollow nanomaterials as theranostic agents. Wiley Interdiscip. Rev. Nanomed. Nanobiotechnol 6, 80−101.

Kattimani, V.S., Kondaka, S., Lingamaneni, K.P., 2016. Hydroxyapatite − past, present, and future in bone regeneration. Bone Tissue Regen. Insights 7, 9−19.

Kempen, P.J., Greasley, S., Parker, K.A., Campbell, J.L., Chang, H.Y., Jones, J.R., Sinclair, R., Gambhir, S.S., Jokerst, J.V., 2015. Theranostic mesoporous silica nanoparticles biodegrade after pro-survival drug delivery and ultrasound/magnetic resonance imaging of stem cells. Theranostics 5, 631−642.

Khan, A.K., Rashid, R., Murtaza, G., Zahra, A., 2014. Gold nanoparticles: synthesis and applications in drug delivery. Trop. J. Pharmaceut. Res. 13, 1169−1177.

Khan, M.A., Wu, V.M., Ghosh, S., Uskokovic, V., 2016. Gene delivery using calcium phosphate nanoparticles: optimization of the transfection process and the effects of citrate and poly(L-lysine) as additives. J. Colloid Interface Sci. 471, 48−58.

Kojima, C., Hirano, Y., Yuba, E., Harada, A., Kono, K., 2008. Preparation and characterization of complexes of liposomes with gold nanoparticles. Colloids Surf. B Biointerfaces 66, 246−252.

Lee, S.H., Bae, K.H., Kim, S.H., Lee, K.R., Park, T.G., 2008. Amine-functionalized gold nanoparticles as non-cytotoxic and efficient intracellular siRNA delivery carriers. Int. J. Pharm. 364, 94−101.

Li, C., Armstrong, J.P., Pence, I.J., Kit-Anan, W., Puetzer, J.L., Correia Carreira, S., Moore, A.C., Stevens, M.M., 2018. Glycosylated superparamagnetic nanoparticle gradients for osteochondral tissue engineering. Biomaterials 176, 24−33.

Li, P., Zhang, L., Ai, K., Li, D., Liu, X., Wang, E., 2008a. Coating didodecyldimethylammonium bromide onto Au nanoparticles increases the stability of its complex with DNA. J. Control Release 129, 128−134.

Li, P.C., Li, D., Zhang, L.X., Li, G.P., Wang, E.K., 2008b. Cationic lipid bilayer coated gold nanoparticles-mediated transfection of mammalian cells. Biomaterials 29, 3617−3624.

Liang, S., Wang, Y., Yu, J., Zhang, C., Xia, J., Yin, D., 2007. Surface modified superparamagnetic iron oxide nanoparticles: as a new carrier for bio-magnetically targeted therapy. J. Mater. Sci. Mater. Med. 18, 2297−2302.

Lin, S.Y., Chen, N.T., Sum, S.P., Lo, L.W., Yang, C.S., 2008. Ligand exchanged photoluminescent gold quantum dots functionalized with leading peptides for nuclear targeting and intracellular imaging. Chem. Commun. (Camb.) 4762−4764.

Liong, M., Lu, J., Kovochich, M., Xia, T., Ruehm, S.G., Nel, A.E., Tamanoi, F., Zink, J.I., 2008. Multifunctional inorganic nanoparticles for imaging, targeting, and drug delivery. ACS Nano. 2, 889−896.

Liu, Y., Wang, T., He, F., Liu, Q., Zhang, D., Xiang, S., Su, S., Zhang, J., 2011. An efficient calcium phosphate nanoparticle-based nonviral vector for gene delivery. Int. J. Nanomed. 6, 721−727.

Liu, Z.S., Tang, S.L., Ai, Z.L., 2003. Effects of hydroxyapatite nanoparticles on proliferation and apoptosis of human hepatoma BEL-7402 cells. World J. Gastroenterol. 9, 1968−1971.

Lopez-Viota, J., Mandal, S., Delgado, A.V., Toca-Herrera, J.L., Moller, M., Zanuttin, F., Balestrino, M., Krol, S., 2009. Electrophoretic characterization of gold nanoparticles functionalized with human serum albumin (HSA) and creatine. J. Colloid Interface Sci. 332, 215−223.

Maitra, A., 2005. Calcium phosphate nanoparticles: second-generation nonviral vectors in gene therapy. Expert Rev. Mol. Diagn 5, 893−905.

Malik, P., Mukherjee, T.K., 2018. Recent advances in gold and silver nanoparticle based therapies for lung and breast cancers. Int. J. Pharm. 553, 483−509.

Market, D.D.S., 2018. Drug Delivery Systems Market − Global Industry Analysis, Size, Share, Growth, Trends, and Forecast 2017 − 2025.

Masson, J.D., Thibaudon, M., Belec, L., Crepeaux, G., 2017. Calcium phosphate: a substitute for aluminum adjuvants? Expert Rev. Vaccines 16, 289−299.

Mattoussi, H., Rotello, V.M., 2013. Inorganic nanoparticles in drug delivery preface. Adv. Drug Deliv. Rev. 65, 605−606.

Meyer, U., Buchter, A., Wiesmann, H.P., Joos, U., Jones, D.B., 2005. Basic reactions of osteoblasts on structured material surfaces. Eur. Cells Mater. 9, 39−49.

Mir, M., Ishtiaq, S., Rabia, S., Khatoon, M., Zeb, A., Khan, G.M., Rehman, A.U., Din, F.U., 2017. Nanotechnology: from in vivo imaging system to controlled drug delivery. Nanoscale Res. Lett. 12.

Morcol, T., Nagappan, P., Nerenbaum, L., Mitchell, A., Bell, S.J., 2004. Calcium phosphate-PEG-insulin-casein (CAPIC) particles as oral delivery systems for insulin. Int. J. Pharm. 277, 91–97.

Morgan, T.T., Muddana, H.S., Altinoglu, E.I., Rouse, S.M., Tabakovic, A., Tabouillot, T., Russin, T.J., Shanmugavelandy, S.S., Butler, P.J., Eklund, P.C., Yun, J.K., Kester, M., Adair, J.H., 2008. Encapsulation of organic molecules in calcium phosphate nanocomposite particles for intracellular imaging and drug delivery. Nano Lett. 8, 4108–4115.

Munsell, E.V., Fang, B., Sullivan, M.O., 2018. Histone-mimetic gold nanoparticles as versatile scaffolds for gene transfer and chromatin analysis. Bioconjug. Chem. 29, 3691–3704.

Murakami, T., Tsuchida, K., 2008. Recent advances in inorganic nanoparticle-based drug delivery systems. Mini Rev. Med. Chem. 8, 175–183.

Muthu, M.S., Leong, D.T., Mei, L., Feng, S.S., 2014. Nanotheranostics – application and further development of nanomedicine strategies for advanced theranostics. Theranostics 4, 660–677.

Oh, N., Park, J.H., 2014. Endocytosis and exocytosis of nanoparticles in mammalian cells. Int. J. Nanomed. 9 (Suppl. 1), 51–63.

Ohgushi, H., Miyake, J., Tateishi, T., 2003. Mesenchymal stem cells and bioceramics: strategies to regenerate the skeleton. Novartis Found. Symp. 249, 118–127 discussion 127-32, 170-4, 239-41.

Ojea-Jimenez, I., Comenge, J., Garcia-Fernandez, L., Megson, Z.A., Casals, E., Puntes, V.F., 2013. Engineered inorganic nanoparticles for drug delivery applications. Curr. Drug Metabol. 14, 518–530.

Olton, D., Li, J., Wilson, M.E., Rogers, T., Close, J., Huang, L., Kumta, P.N., Sfeir, C., 2007. Nanostructured calcium phosphates (NanoCaPs) for non-viral gene delivery: influence of the synthesis parameters on transfection efficiency. Biomaterials 28, 1267–1279.

Oo, M.K., Yang, X., Du, H., Wang, H., 2008. 5-aminolevulinic acid-conjugated gold nanoparticles for photodynamic therapy of cancer. Nanomedicine 3, 777–786.

Pankhurst, Q.A., Connolly, J., Jones, S.K., Dobson, J., 2003. Applications of magnetic nanoparticles in biomedicine. J. Phys. D Appl. Phys. 36, R167–R181.

Park, J.H., Von Maltzahn, G., Zhang, L.L., Derfus, A.M., Simberg, D., Harris, T.J., Ruoslahti, E., Bhatia, S.N., Sailor, M.J., 2009. Systematic surface engineering of magnetic nanoworms for in vivo tumor targeting. Small 5, 694–700.

Paul, W., Sharma, C.P., 2001. Porous hydroxyapatite nanoparticles for intestinal delivery of insulin. Trends Biomater. Artif. Organs 14, 37–38.

Paul, W., Sharma, C.P., 2003. Ceramic drug delivery: a perspective. J. Biomater. Appl. 17, 253–264.

Paul, W., Sharma, C.P., 2007. Effect of calcium, zinc and magnesium on the attachment and spreading of osteoblast like cells onto ceramic matrices. J. Mater. Sci. Mater. Med. 18, 699–703.

Paul, W., Sharma, C.P., 2012. Synthesis and characterization of alginate coated zinc calcium phosphate nanoparticles for intestinal delivery of insulin. Process Biochem. 47, 882–886.

Pecorelli, T.A., Dibrell, M.M., Li, Z.X., Thomas, C.R., Zink, J.I., 2010. Multifunctional inorganic nanoparticles for imaging, targeting, and drug delivery. In: Reporters, Markers, Dyes, Nanoparticles, and Molecular Probes for Biomedical Applications II, p. 7576.

Pileni, M.P., 2005. Nanocrystals Forming Microscopic Structures. Wiley-VCH, Weinheim.

Podsiadlo, P., Sinani, V.A., Bahng, J.H., Kam, N.W.S., Lee, J., Kotov, N.A., 2008. Gold nanoparticles enhance the anti-leukemia action of a 6-mercaptopurine chemotherapeutic agent. Langmuir 24, 568–574.

Poletaeva, J., Dovydenko, I., Epanchintseva, A., Korchagina, K., Pyshnyi, D., Apartsin, E., Ryabchikova, E., Pyshnaya, I., 2018. Non-covalent associates of siRNAs and AuNPs enveloped with lipid layer and doped with amphiphilic peptide for efficient siRNA delivery. Int. J. Mol. Sci. 19.

Pourbaghi-Masouleh, M., Hosseini, V., 2013. Amorphous calcium phosphate nanoparticles could function as a novel cancer therapeutic agent by employing a suitable targeted drug delivery platform. Nanoscale Res. Lett. 8, 449−454.

Prencipe, G., Tabakman, S.M., Welsher, K., Liu, Z., Goodwin, A.P., Zhang, L., Henry, J., Dai, H.J., 2009. PEG branched polymer for functionalization of nanomaterials with ultralong blood circulation. J. Am. Chem. Soc. 131, 4783−4787.

Qian, X.M., Peng, X.H., Ansari, D.O., Yin-Goen, Q., Chen, G.Z., Shin, D.M., Yang, L., Young, A.N., Wang, M.D., Nie, S.M., 2008. In vivo tumor targeting and spectroscopic detection with surface-enhanced Raman nanoparticle tags. Nat. Biotechnol. 26, 83−90.

Qiu, W., Chen, R., Chen, X., Zhang, H., Song, L., Cui, W., Zhang, J., Ye, D., Zhang, Y., Wang, Z., 2018. Oridonin-loaded and GPC1-targeted gold nanoparticles for multimodal imaging and therapy in pancreatic cancer. Int. J. Nanomed. 13, 6809−6827.

Ramachandran, R., Paul, W., Sharma, C.P., 2009. Synthesis and characterization of PEGylated calcium phosphate nanoparticles for oral insulin delivery. J. Biomed. Mater. Res. B Appl. Biomater. 88B, 41−48.

Roveri, N., Palazzo, B., Iafisco, M., 2008. The role of biomimetism in developing nanostructured inorganic matrices for drug delivery. Expert Opin. Drug Deliv. 5, 861−877.

Roy, I., Mitra, S., Maitra, A., Mozumdar, S., 2003. Calcium phosphate nanoparticles as novel non-viral vectors for targeted gene delivery. Int. J. Pharm. 250, 25−33.

Rukes, S., 2017. Nanotechnology for drug delivery. Abstr. Pap. Am. Chem. Soc. 253.

Ryu, J.H., Lee, S., Son, S., Kim, S.H., Leary, J.F., Choi, K., Kwon, I.C., 2014. Theranostic nanoparticles for future personalized medicine. J. Control Release 190, 477−484.

Saha, B., Bhattacharya, J., Mukherjee, A., Ghosh, A.K., Santra, C.R., Dasgupta, A.K., Karmakar, P., 2007. In vitro structural and functional evaluation of gold nanoparticles conjugated antibiotics. Nanoscale Res. Lett. 2, 614−622.

Saha, K., Agasti, S.S., Kim, C., Li, X.N., Rotello, V.M., 2012. Gold nanoparticles in chemical and biological sensing. Chem. Rev. 112, 2739−2779.

Santos, H.A., Bimbo, L.M., Peltonen, L., Hirvonen, J., 2014. Inorganic nanoparticles in targeted drug delivery and imaging. In: Devarajan, P., Jain, S. (Eds.), Targeted Drug Delivery: Concepts and Design. Springer, Cham.

Schmid, G., 2010. Nanoparticles: From Theory to Application. Wiley-VCH, Weinheim.

Seleci, M., Seleci, D.A., Scheper, T., Stahl, F., 2017. Theranostic liposome-nanoparticle hybrids for drug delivery and bioimaging. Int. J. Mol. Sci. 18.

Sengania, M., Grumezescub, A.M., Rajeswaria, V.D., 2017. Recent trends and methodologies in gold nanoparticle synthesis − a prospective review on drug delivery aspect. Open Nano 2, 37−46.

Skrabalak, S.E., Au, L., Lu, X., Li, X., Xia, Y., 2007. Gold nanocages for cancer detection and treatment. Nanomedicine 2, 657−668.

Sokolova, V., Kovtun, A., Heumann, R., Epple, M., 2007. Tracking the pathway of calcium phosphate/DNA nanoparticles during cell transfection by incorporation of red-fluorescing tetramethylrhodamine isothiocyanate-bovine serum albumin into these nanoparticles. J. Biol. Inorg. Chem. 12, 174−179.

Song, M., Wang, X., Li, J., Zhang, R., Chen, B., Fu, D., 2008. Effect of surface chemistry modification of functional gold nanoparticles on the drug accumulation of cancer cells. J. Biomed. Mater. Res. A 86, 942−946.

Sreeram, K.J., Nidhin, M., Nair, B.U., 2009. Synthesis of aligned hematite nanoparticles on chitosan-alginate films. Colloids Surfaces B Biointerfaces 71, 260−267.

Tack, F., Noppe, M., Van Dijck, A., Dekeyzer, N., Van Der Leede, B.J., Bakker, A., Wouters, W., Janicot, M., Brewster, M.E., 2008. Delivery of a DNAzyme targeting c-myc to HT29 colon carcinoma cells using a gold nanoparticulate approach. Pharmazie 63, 221−225.

Takanche, J.S., Kim, J.E., Kim, J.S., Lee, M.H., Jeon, J.G., Park, I.S., Yi, H.K., 2018. Chitosan-gold nanoparticles mediated gene delivery of c-myb facilitates osseointegration of dental implants in ovariectomized rat. Artif. Cells Nanomed. Biotechnol. 1−11.

Tan, A., Yildirimer, L., Rajadas, J., DE LA Pena, H., Pastorin, G., Seifalian, A., 2011. Quantum dots and carbon nanotubes in oncology: a review on emerging theranostic applications in nanomedicine. Nanomedicine 6, 1101−1114.

Thomas, S.C., Harshita, Mishra, P.K., Talegaonkar, S., 2015. Ceramic nanoparticles: fabrication methods and applications in drug delivery. Curr. Pharmaceut. Des. 21, 6165−6188.

Toth, I., 2014. Thematic issue: nanotechnology for drug delivery applications. Curr. Drug Deliv. 11, 665, 665.

Uskokovic, V., Uskokovic, D.P., 2011. Nanosized hydroxyapatite and other calcium phosphates: chemistry of formation and application as drug and gene delivery agents. J. Biomed. Mater. Res. B Appl. Biomater. 96, 152−191.

Veiseh, O., Gunn, J.W., Kievit, F.M., Sun, C., Fang, C., Lee, J.S., Zhang, M., 2009. Inhibition of tumor-cell invasion with chlorotoxin-bound superparamagnetic nanoparticles. Small 5, 256−264.

Veisch, O., Tang, B.C., Whitehead, K.A., Anderson, D.G., Langer, R., 2015. Managing diabetes with nanomedicine: challenges and opportunities. Nat. Rev. Drug Discov. 14, 45−57.

Venkatesan, J., Kim, S.K., 2014. Nano-hydroxyapatite composite biomaterials for bone tissue engineering—a review. J. Biomed. Nanotechnol. 10, 3124−3140.

Verma, A., Sharma, S., Gupta, P.K., Singh, A., Teja, B.V., Dwivedi, P., Gupta, G.K., Trivedi, R., Mishra, P.R., 2016. Vitamin B12 functionalized layer by layer calcium phosphate nanoparticles: a mucoadhesive and pH responsive carrier for improved oral delivery of insulin. Acta Biomater. 31, 288−300.

Victor, S.P., Paul, W., Jayabalan, M., Sharma, C.P., 2014a. Cucurbituril/hydroxyapatite based nanoparticles for potential use in theranostic applications. Cryst. Eng. Comm. 16, 6929−6936.

Victor, S.P., Paul, W., Jayabalan, M., Sharma, C.P., 2014b. Supramolecular hydroxyapatite complexes as theranostic near-infrared luminescent drug carriers. Cryst. Eng. Comm. 16, 9033−9042.

Victor, S.P., Paul, W., Vineeth, V.M., Komeri, R., Jayabalan, M., Sharma, C.P., 2016. Neodymium doped hydroxyapatite theranostic nanoplatforms for colon specific drug delivery applications. Colloids Surf. B Biointerfaces 145, 539−547.

Wang, H., Zhang, M., Zhang, L., Li, S., Li, L., Li, X., Yu, M., Mou, Z., Wang, T., Wang, C., Su, Z., 2017. Near-infrared light and pH-responsive Au@carbon/calcium phosphate nanoparticles for imaging and chemo-photothermal cancer therapy of cancer cells. Dalton Trans. 46, 14746−14751.

Webster, T.J., Ergun, C., Doremus, R.H., Siegel, R.W., Bizios, R., 2000. Enhanced functions of osteoblasts on nanophase ceramics. Biomaterials 21, 1803−1810.

Webster, T.J., Siegel, R.W., Bizios, R., 2001. Nanoceramic surface roughness enhances osteoblast and osteoclast functions for improved orthopaedic/dental implant efficacy. Scripta Mater. 44, 1639−1642.

Wijaya, A., Schaffer, S.B., Pallares, I.G., Hamad-Schifferli, K., 2009. Selective release of multiple DNA oligonucleotides from gold nanorods. ACS Nano 3, 80–86.

Xia, H., Gao, Y., Yin, L., Cheng, X., Wang, A., Zhao, M., Ding, J., Shi, H., 2019. Light-triggered covalent assembly of gold nanoparticles for cancer cell photothermal therapy. Chembiochem 10, 667–671.

Xiao, Y., Wang, X.Y., Wang, B., Liu, X.Y., Xu, X.R., Tang, R.K., 2017. Long-term effect of biomineralized insulin nanoparticles on type 2 diabetes treatment. Theranostics 7, 4301–4312.

Xie, J., Lee, S., Chen, X., 2010. Nanoparticle-based theranostic agents. Adv. Drug Deliv. Rev. 62, 1064–1079.

Xu, C., 2017. Nano-Implants to Remove the Pain of Diabetes Injections [Online]. AIBN Communications, Brisbane. Available: https://aibn.uq.edu.au/article/2017/02/nano-implants-remove-pain-diabetes-injections.

Xu, Z.P., Zeng, Q.H., Lu, G.Q., Yu, A.B., 2006. Inorganic nanoparticles as carriers for efficient cellular delivery. Chem. Eng. Sci. 61, 1027–1040.

Yang, Y.S., Moynihan, K.D., Bekdemir, A., Dichwalkar, T.M., Noh, M.M., Watson, N., Melo, M., Ingram, J., Suh, H., Ploegh, H., Stellacci, F.R., Irvine, D.J., 2019. Targeting small molecule drugs to T cells with antibody-directed cell-penetrating gold nanoparticles. Biomater. Sci. 7, 113–124.

Yao, X.M., Chen, L., Chen, X.F., Zhang, Z., Zheng, H., He, C.L., Zhang, J.P., Chen, X.S., 2014. Intracellular pH-sensitive metallo-supramolecular nanogels for anticancer drug delivery. ACS Appl. Mater. Interfaces 6, 7816–7822.

Yeh, Y.C., Creran, B., Rotello, V.M., 2012. Gold nanoparticles: preparation, properties, and applications in bionanotechnology. Nanoscale 4, 1871–1880.

Yi, H., Rehman, F.U., Zhao, C.Q., Liu, B., He, N.Y., 2016. Recent advances in nano scaffolds for bone repair. Bone Res 4.

Yoon, Y.I., Pang, X., Jung, S., Zhang, G., Kong, M., Liu, G., Chen, X., 2018. Smart gold nanoparticle-stabilized ultrasound microbubbles as cancer theranostics. J. Mater. Chem. B 6, 3235–3239.

Zhang, G., Yang, Z., Lu, W., Zhang, R., Huang, Q., Tian, M., Li, L., Liang, D., Li, C., 2009. Influence of anchoring ligands and particle size on the colloidal stability and in vivo biodistribution of polyethylene glycol-coated gold nanoparticles in tumor-xenografted mice. Biomaterials 30, 1928–1936.

Zhang, G.M., 2013. Functional gold nanoparticles for sensing applications. Nanotechnol. Rev. 2, 269–288.

Zhang, J., Chen, L., Tse, W.H., Bi, R., Chen, L., 2014. Inorganic nanoparticles, engineering for biomedical applications. IEEE Nanotechnol Mag 8, 21–28.

Zhang, R.X., Li, J., Zhang, T., Amini, M.A., He, C.S., Lu, B., Ahmed, T., Lip, H., Rauth, A.M., Wu, X.Y., 2018a. Importance of integrating nanotechnology with pharmacology and physiology for innovative drug delivery and therapy – an illustration with firsthand examples. Acta Pharmacol. Sin. 39, 825–844.

Zhang, Y., Zhang, L., Ban, Q., Li, J., Li, C.H., Guan, Y.Q., 2018b. Preparation and characterization of hydroxyapatite nanoparticles carrying insulin and gallic acid for insulin oral delivery. Nanomedicine 14, 353–364.

Zheng, J.J., Ren, W.Z., Chen, T.X., Jin, Y.H., Li, A.J., Yan, K., Wu, Y.J., Wu, A.G., 2018. Recent advances in superparamagnetic iron oxide based nanoprobes as multifunctional theranostic agents for breast cancer imaging and therapy. Curr. Med. Chem. 25, 3001–3016.

Zhong, J., 2015a. Nanotechnology for drug delivery: part I. Curr. Pharmaceut. Des. 21, 3064–3065.
Zhong, J., 2015b. Nanotechnology for drug delivery: part II. Curr. Pharmaceut. Des. 21, 4129–4130.
Zhong, J., 2015c. Nanotechnology for drug delivery: part III. Curr. Pharmaceut. Des. 21, 6037, 6037.
Zhou, Z., Kennell, C., Lee, J.Y., Leung, Y.K., Tarapore, P., 2017. Calcium phosphate-polymer hybrid nanoparticles for enhanced triple negative breast cancer treatment via co-delivery of paclitaxel and miR-221/222 inhibitors. Nanomedicine 13, 403–410.
Zhu, Y.J., 2008. Inorganic Nanostructures for Drug Delivery. CRC Press, Boca Raton, FL.
Zink, J.I., 2014. Multifunctional inorganic nanoparticles controlled by nanomachines for in vitro and in vivo drug delivery. Abstr. Pap. Am. Chem. Soc. 248.

Applications of alginate biopolymer in drug delivery

14

Lisbeth Grøndahl, Gwendolyn Lawrie, A. Anitha, Aparna Shejwalkar
School of Chemistry and Molecular Biosciences, The University of Queensland, Brisbane, QLD, Australia

Chapter outline

14.1 Introduction 375
14.2 Alginate biopolymer 376
  14.2.1 Biocompatibility 377
  14.2.2 Degradation 377
  14.2.3 Chemically modified alginate 379
  14.2.4 Ionically cross-linked alginate hydrogels 381
  14.2.5 Alternative methods for hydrogel fabrication 383
14.3 Drug delivery using alginate matrices 384
  14.3.1 Use of chemically modified alginate 386
  14.3.2 Microencapsulation for transplantation 388
14.4 Concluding remarks and future directions 395
Acknowledgments 395
References 396

14.1 Introduction

Drug delivery is a core area of biotechnology, pharmacy, and materials science. The use of a drug delivery system allows tailoring of both the rate of drug delivery and the site of drug delivery. It can overcome issues of systemic drug toxicity and thus allow effective treatment of disease or assist in tissue remodeling. The investigation of synthetic polymers in the drug delivery field, such as polyethylene and silicon rubber, dates as far back as the 1960s (Desai et al., 1965; Folkman and Long, 1964). Since then, the field has moved into the use of naturally occurring polymers, which include collagen, gelatin, chitosan, alginate, and starch (Ranade and Hollinger, 2004). Each of these polymers offers different advantages, but alginates offer the greatest versatility in their ability to form ionically cross-linked hydrogels (Lee and Mooney, 2012), and chemically modified alginates greatly expand this versatility (Pawar and Edgar, 2012). Alginate has therefore been widely embraced as a desirable gel carrier candidate. To date, many types of alginate-based hydrogels are under investigation, and some are used commercially. The Food and Drug Administration granted the "generally referred as safe" material status to alginates (CFR, 2018). Moreover, major regulatory bodies such as European Medicine Agency, Therapeutic Goods Administration, and Health Canada have approved various alginate-based products. Commercially available alginate-based drug delivery products are used in various forms such as wound dressing sponges, microspheres, and injectable options with the companies 3M, Medtronic, and Coloplast as key players in the wound dressing market.

14.2 Alginate biopolymer

Alginate is typically derived from the extracellular matrix of brown algae where it acts as structural material. Major algae sources are *Laminaria hyperborea*, *Macrocystis pyrifera*, and *Ascophyllum nodosum*. Some other, less frequently used, sources for derivation of alginates are *Laminaria digitata* and *Laminaria japonica* (Smidsrød and Skåk-Bræk, 1990). Alginates are linear unbranched polymers containing β-(1 → 4)—linked D-mannuronic acid (M)—and α-(1 → 4)—linked L-guluronic acid (G) residues (Fig. 14.1). They are not random copolymers, instead they consist of blocks of similar and strictly alternating residues (i.e., MMMMM, GGGGG, and GMGMGM) (Skaugrud et al., 1999). The distribution of the M and G blocks is dependent on the type and source of algae and also on the age and the component of the algae (such as stem or leaves). For example, alginates derived from *L. hyperborea* have displayed the highest G content, while those from *L. Japonica* have been characterized by a low G content, and bacterial alginates produced from the *Pseudomonas* species are characterized by absence of G blocks. The functional properties of alginates correlate strongly with the M/G ratio. Alginate biopolymers may be formulated with a wide range of molecular weights (50–100,000 monomer units) to suit the application.

Enzymes can convert M blocks into G blocks by epimerization to achieve more control over alginate properties. The discovery of the *Azotobacter vinelandii* that encodes at least seven different mannuronan C-5 epimerases opened possibilities of overcoming limitations in applications where high G content was desired (Ertestvag et al., 1994; Svanem et al., 1999). The enzymes use different mechanisms or patterns for conversion

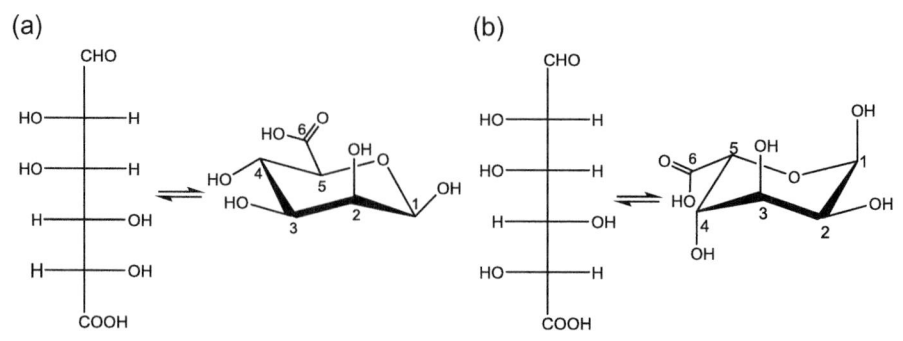

Figure 14.1 Chemical structures of the two isomeric monomers of alginate: (a) D-mannuronic acid and (b) L-guluronic acid.

of the M blocks into G, for example, AlgE1 enzymes produce long stretches of G blocks, whereas AlgE4 results in strictly alternating sequences. Using the epimerized alginate, Donati et al. made an important discovery of gelation ability of alternating MG blocks with Ca^{2+} ions, which was previously attributed to G-rich stretches of alginate (Donati et al., 2005). Some other sources of the epimerases have also been identified.

14.2.1 Biocompatibility

For the purpose of drug delivery, the biocompatibility of alginate is paramount and has been studied extensively in the literature. Raw alginates contain contaminants such as heavy metals, polyphenols, endotoxins, and pyrogens and immunogenic materials such as proteins that can induce toxic immunogenic responses in the body. These alginates stimulate monocytes to produce high levels of cytokines including interleukin-1, interleukin-6, and tumor necrosis factor (TNF) α (Soon-Shiong et al., 1991). Alginate can be purified using free-flow electrophoresis where the contaminants are separated from the alginic acids under an applied electrical field (Zimmermann et al., 1992). Alternatively, alginate can be purified by a method that involves elution and extraction of contaminants from barium cross-linked beads Klock et al. (1994). A third published method for purifying alginate is based on a multiple step procedure including the filtration and precipitation of alginate as alginic acid and subsequent extraction of proteins using organic solvents (de Vos et al., 1997). However, medical-grade alginates are now commercially available to meet the advanced needs of medical applications. DuPont is a major supplier of such medical-grade alginates. The PRONOVA sodium alginates product group (previously owned by FMC Health and Nutrition's NovaMatrix) offers ultrapure alginates and sterile alginates developed especially for pharmaceutical applications. Additionally, pharmaceutical-grade alginate brands are available, specifically PROTANAL, PROTACID, MANUCOL, and KELCOLOID. Analysis and purification of some of these pharmaceutical-grade alginates revealed that they contained high levels of endotoxins and proteins. After purification, acceptable levels of endotoxins resulted and the protein levels were reduced; however, Cl and Si contaminants were introduced (Tam et al., 2011). Introduction of contaminants can also occur during the gelation reactions, e.g., from the chemical compounds used to produce the gels (Nunamaker et al., 2007).

14.2.2 Degradation

Degradation of alginate occurs as a result of cleavage of the glycosidic bonds. This can be influenced or initiated by acidic- or alkaline-catalyzed reactions, thermal depolymerization, or by modification of the polymer such as by oxidation reactions. The M-rich alginates are more susceptible to acid and thermal degradation than the G-rich alginates (Holme et al., 2003). Thermal degradation of the polymer occurs when the temperature is above 80°C. In an acidic solution below pH 5, alginate undergoes hydrolysis at the glycosidic bond. The mechanism is shown in Fig. 14.2. The first step involves the protonation of the glycosidic oxygen. The reaction proceeds via a carbonium—oxonium ion to which a water molecule is added (Haug et al., 1963; Smidsrød et al., 1966). In strong

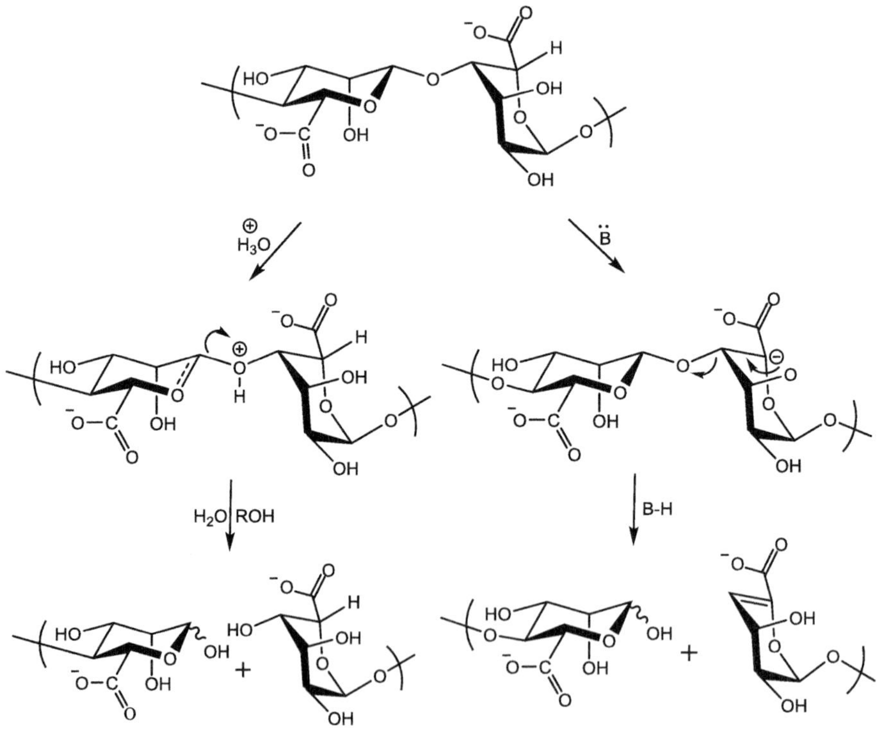

Figure 14.2 Mechanism for acid- and base-catalyzed hydrolysis.

alkaline environments above pH 10, alginate undergoes degradation through β-elimination. The mechanism is shown in Fig. 14.2. Initial proton abstraction from the C-5 position is followed by the rearrangement of the resulting intermediate leading to the formation of an unsaturated compound and an anion that is subsequently protonated. The proton abstraction from C-5 position is facilitated by the electron-withdrawing effects of the carbonyl groups from the C-6 position (Haug et al., 1963, 1967; Tsujino and Saito, 1961).

Alginates are susceptible to biodegradation in the presence of a group of enzymes called alginases present in marine algae, marine mollusks, and microorganisms (Gacesa, 1992). Alginases function efficiently at neutral pH and have been employed in past studies to determine block length and diad frequencies. The enzyme alginase is, however, absent in mammals. Thus far, a single study by Al-Shamkhani and Duncan published in 1995 reports the biodistribution profile of a chemically modified radiolabeled alginate (propylene glycol alginate-tyrosinamide labeled with ^{125}I) in male Wistar rats. The study reports that the in vivo fate of systemically administered chemically modified alginate depends on the molecular weight and that alginates with a molecular weight of 48 kg/mol or below are cleared through renal filtration within 24 h (Al-Shamkhani and Duncan, 1995). The recommendation is thus to use low molecular weights alginates unless the application is in oral delivery where the polymer is exposed to a very acidic environment (e.g., gastrointestinal) (Balakrishnan and Jayakrishnan, 2005; Bouhadir et al., 2001; Al-Shamkhani and Duncan, 1995).

14.2.3 Chemically modified alginate

Chemical modification of the alginate biopolymer is used either to change the reactivity of alginate by introducing highly reactive functional groups (e.g., aldehyde groups) or introduce chemical (e.g., phosphate or sulfate) or biochemical (e.g., amino acids) groups that can increase the biointegration and bioaffinity of alginate-based materials (Pawar and Edgar, 2012). As the alginate biopolymer itself contains both hydroxyl and carboxylic acid functional groups, it offers great versatility for chemical modification as will be illustrated below.

Oxidation of alginate using periodate was first described by Malaprade (Malaprade, 1928) and recently reviewed by Reakasame and Boccaccini (2018). It involves cleavage of the C2–C3 bond transforming the uronic acid subunits into an open chain adduct containing a dialdehyde known as alginate dialdehyde (ADA). These aldehyde groups can react spontaneously with hydroxyl groups present on the adjacent uronic acid subunits in the polymer chain to form a hemiacetal (Balakrishnan et al., 2005). Its formation can prevent complete oxidation of the biopolymer and consumption of sodium periodate. Oxidation of alginate is usually carried out in an aqueous solution at room temperature (Balakrishnan et al., 2005; Gomez et al., 2007). The degree of oxidation can be controlled, and up to 87% oxidation has been demonstrated (Balakrishnan et al., 2005). The degree of oxidation can be determined via iodometry by titrating any periodate remaining after reaction (Whistler and Wolfrom, 1962). Characterization is carried out using ^{1}H nuclear magnetic resonance (NMR) (aldehyde content) (Gomez et al., 2007) and ^{13}C NMR (to confirm the polymer scissioning that is accompanied by oxidation). Extensive scissioning of the polymer chain is reflected in the decrease in molecular weight with an increase in the amount of oxidation. Vold et al. (2006) and Smidsrød and Painter (1973) demonstrated changes in the intrinsic viscosity and chain stiffness in the oxidized alginate samples and concluded that the stiffness and the length of chains were decreased with an increase in the oxidation of alginate samples. The enhanced flexibility of the polymer chains was linked to the open-chain adducts. This open-chain adduct is much more susceptible to hydrolytic scission than unmodified alginate, and while alginates are resistant to biodegradation and are not broken down in mammals, controlled degradation of alginate can be achieved by partially oxidizing the alginate polymer. This property opened venues for use of ADA in tissue engineering applications (Bouhadir et al., 2001).

Phosphorylation of alginate can be achieved using the so-called urea phosphate method (Coleman et al., 2011) in which the reaction is done of the alginate polymer dispersed in dimethylformamide (DMF) (Mucalo et al., 1995). In this manner, phosphate groups are introduced at carbons 2 and 3 through reactions with the hydroxyl groups, and the dominant sites of phosphorylation were found from detailed NMR characterization to be G-2 and M-3. These are the equatorial groups and therefore more reactive than their axial counterparts due to reduced steric hindrance. Some degradation of the biopolymer occurs concurrently with the phosphorylation reaction, and the process achieves 2%—26% phosphorylation (per alginate subunit), which was paralleled with a decrease in molecular weight from approximately 140 to 64—38 kDa. An alternative method of phosphorylation using a $H_3PO_4/P_2O_5/Et_3PO_4$/hexanol method has since been explored (Kim et al., 2015). This yielded a degree of phosphorylation of 12.5%.

A number of different approaches to the sulfation of alginate have been reported in the literature with the use of the DCC-H_2SO_4 and $ClSO_3H$/formamide methods being most extensively studied (Arlov and Skjåk-Bræk, 2017). The DCC-H_2SO_4 method is done using the tributyl ammonium salt of alginate (TBA-Alg) dissolved in DMF. Analysis by NMR confirmed the sites of sulfation as C2 and C3, and the process resulted in a degree of sulfation of 80% with a concomitant reduction in molecular mass from 100 to 10 kDa (Freeman et al., 2008). The $ClSO_3H$/formamide method facilitates the synthesis of various degrees of substitution ranging from 5%—120% (per alginate subunit) by varying the concentration of $ClSO_3H$. A similar trend was found for sulfation of poly-M, poly-G, and poly-MG in this regard (Arlov et al., 2014). Detailed characterization by NMR revealed that sulfation occurred at C2 and C3 with no preference for either site. Importantly, the analysis also confirmed that this method of sulfation did not lead to any side reactions or side products. The $ClSO_3H$/formamide method has also been applied to sulfation of ADA producing sulfated alginates with enhanced chain flexibility (Arlov et al., 2015).

Coupling of single amino acids (e.g., cysteine) or small peptides (e.g., containing the Arginine, Glycine, Aspartate (RGD) sequence) to alginates has been studied in detail. In all cases, the amine terminal of the amino acid/peptide is reacted with the carboxylic acid group of alginate using carbodiimide chemistry (i.e., activation of the carboxylic acid group). More specifically, cysteine has been coupled to alginate with 7% substitution (of alginate subunits, 400 µmol/g) when using stoichiometric amounts of cysteine as determined photometrically (Greimel et al., 2007). Other amino acids including lysine, arginine, aspartic acid, and phenylalanine have likewise been coupled to alginate (Zhu, 2002). The RGD peptide sequence is used extensively in biomaterials science due to its cell adhesion properties. A number of RGD-containing peptides have been coupled to alginate using carbodiimide chemistry including GRGDY (Rowley et al., 1999), GGGGRGDSP (Drury et al., 2005), and GGGGRGDSY (Connelly, 2007). The coupling efficiency was determined using ^{125}I-labeled peptide and found to be 25% (alginate subunits) without further improvement with increased carboxylic acid activation (Rowley et al., 1999).

Amphiphilic alginate derivatives have been synthesized by reacting the carboxylate groups of alginate, transformed into a tetrabutylammonium salt, in the organic solvent dimethylsulfoxide with alkyl halides (e.g., dodecyl bromide) (Pelletier et al., 2000). In this manner, the long alkyl chains were linked to the alginate biopolymer via ester linkages. An 8% (C_{12} alkyl chain length) and 1.3% (C_{18} alkyl chain length) substitution ratio could be achieved in this way as determined from gas chromatography of the alkyl alcohol obtained from alkaline hydrolysis of the amphiphilic polymer. An alternative method of reacting the carboxylic acid group of alginate, preactivated using 2-chloro-1-methylpyridinium iodide (CMPI), with an alkylamine (e.g., dodecylamine) in dimethyl formamide yielded substitution ratios in the range of 2%—17%, and this could be easily controlled by the amount of CMPI used (Vallee et al., 2009).

14.2.4 Ionically cross-linked alginate hydrogels

Ionically cross-linked alginate hydrogels can be formed in the presence of divalent cations such as Ca^{2+}, Ba^{2+}, Sr^{2+}, Zn^{2+}, Cu^{2+}, Cd^{2+}, and Co^{2+} (Mg^{2+} is an exception) and trivalent cations such as Fe^{3+} and Al^{3+} (Remuñán-López and Bodmeier, 1997; Hermes and Narayani, 2002). These gels shrink during gel formation, leading to loss of water and an increase in the polymer concentration relative to the alginate solution. The association between alginate polymer chain and divalent cations is popularly described using the "egg-box" model. The binding involves coordination of one divalent cation with four oxygen atoms. Braccini and Pérez (2001) proposed that the parallel and antiparallel arrangement of 2_1 helical chains provides a favorable association also providing a compact cavity for a highly cooperative binding with the divalent cation. Li et al. proposed 3_1 helical conformation based on X-ray diffraction of Ca-alginate gels that were formed by a slow gelation method (Li et al., 2007). Divalent cations such as Ca^{2+} and Ba^{2+} bind preferentially to the G blocks, and thus alginate gels with lower G content (higher M content) have lower strength and stability (Mørch et al., 2008; Jørgensen et al., 2007). Grant et al. (1973) demonstrated that although the interactions of alginate with cations are dominated by G blocks, once the G binding sites are saturated, the threshold is passed over to the MG blocks. Thus, a more robust hydrogel can be engineered by using a G-rich alginate or by increasing the amount of G blocks by enzymatic epimerization of the polymer (Mørch et al., 2007).

Ionic cross-linking of chemically modified alginates is often different compared with the unmodified alginate precursor polymer. The gelation ability of ADA in presence of Ca^{2+} ions is maintained only at low oxidation levels (e.g., 10%), and the Ca—ADA hydrogels show reduced mechanical properties. This was related to the reduced molar mass and number of GG blocks participating in binding with divalent cations (Gomez et al., 2007). The amphiphilic alginate biopolymers can form physical gels due to hydrophobic interactions, and these gels can be further strengthened by calcium cross-linking (De Boisseson et al., 2004). Phosphorylated and sulfated alginate can form cross-links with calcium ions, and because of the higher affinity of phosphate and sulfate for calcium compared to carboxylate, these cross-links are more robust (Coleman et al., 2011; Freeman et al., 2008). However, because of the reduced molecular mass of some phosphorylated and sulfated alginate polymers, hydrogels are produced from mixtures of the modified alginate and pure alginate.

Ionically cross-linked alginate microcapsules can be prepared by different methods (Fig. 14.3) including suspending the extruded alginate beads in a cross-linker solution allowing the cross-linking agent to diffuse into the matrix (Skaugrud et al., 1999; Aslani and Kennedy, 1996) or by emulsification followed by addition of a cross-linking solution (Fundueanu et al., 1999). Modifications of the emulsification technique to achieve internal or external gelation have also been reported (Poncelet et al., 1992; Chan et al., 2002). Internal gelation is achieved by adding particulate calcium carbonate and a slow-acting acid to the alginate solution. A combination of two methods can also be used to produce desirable gels or to achieve better control on the gelation process (Tan and Takeuchi, 2007). Ionic cross-linking of alginate membranes can be achieved by both the immersion and internal gelation techniques

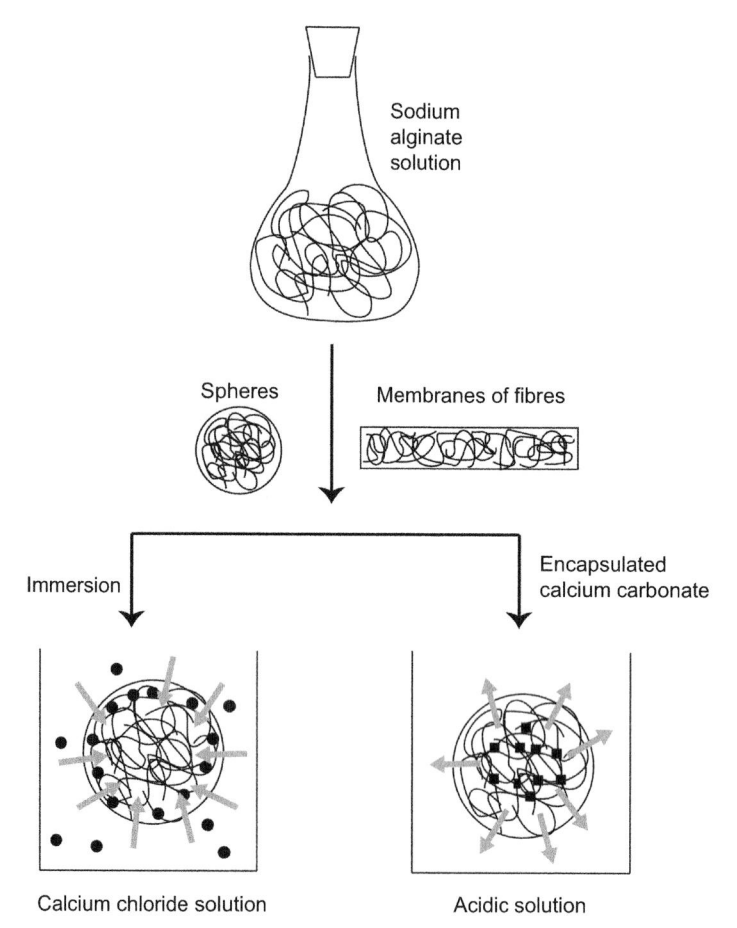

Figure 14.3 Illustration of the common techniques used to produce ionically cross-linked alginate capsules.

(Corkhill et al., 1990; Pavlath et al., 1999; al Musa et al., 1999; Jejurikar et al., 2011; Aston et al., 2016). The internal structure of alginate hydrogels can be investigated by cryogenic scanning electron microscope in combination with correct freezing preparation. This allows evaluation of pore size and distribution of the hydrated gel (Fig. 14.4). These ionically cross-linked hydrogels swell significantly in an aqueous media. Ca-alginate capsules are known to swell more than 90%. The extent of swelling of a cross-linked alginate gel is dependent on temperature, pH, and ion concentration of the swelling media and also on the cross-linking gradient through the gel matrix (Moe et al., 1993). Qin reported that Ca-alginate gels with higher M content swell more than those with higher G content, which reflects the preferential binding of the cations with the G subunits (Qin, 2008).

Degradation of ionically cross-linked alginate gels is initiated by loss of cross-linking cation, followed by loss of high and low molecular weight polymer strands. This affects the gel strength, stability, pore size, and distribution (Shoichet et al., 1995). The rate of

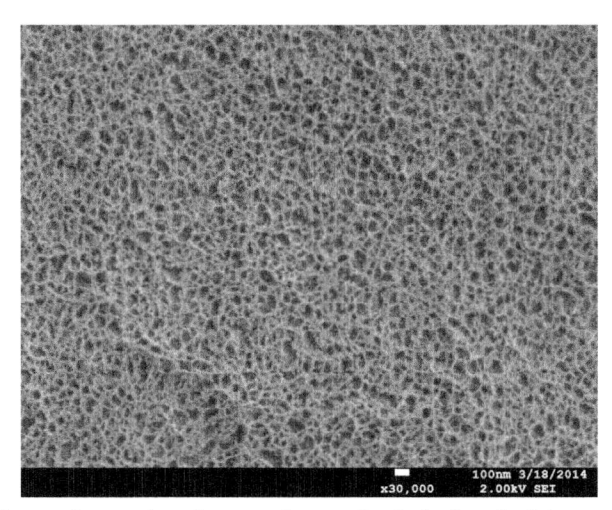

Figure 14.4 Cryogenic scanning electron micrograph of a hydrated calcium cross-linked alginate matrix prepared by internal gelation showing its porous nature. Before imaging, the sample was frozen using high-pressure freezing as this allows retainment of the native hydrogel structure as previously documented (Aston et al., 2016). Magnification 30,000×, scale bar 100 nm.
Image source: image by Robyn Aston and Kim Sewell, The University of Queensland.

degradation is affected by the alginate properties (in particular the M/G ratio) and also by the solution in which the alginate gel is suspended (Tan et al., 2009). The effect of the calcium concentration in the medium was investigated in detail by Kuo and Ma (2008), and it was found that a high concentration (5 mM) resulted in larger retention of the cross-linking density than a lower concentration (e.g., 2—4 mM). It is of course important to realize that the mechanism and rate of degradation in vivo is also dependent on site of implantation and local cellular environment.

14.2.5 Alternative methods for hydrogel fabrication

While ionic cross-linking of alginate hydrogels is very simple and versatile, the low gel stability has led to the development of alternative methods for materials fabrication resulting in greater control over mechanical and swelling properties as well as stability of the gels. As discussed above, the hydroxyl and carboxylic acid functional groups of alginate offer great versatility for chemical modification including cross-linking reactions. Formation of covalently cross-linked hydrogels is thus achieved by reaction of these functional groups with complementary reactive groups (Lee and Mooney, 2012; Pawar and Edgar, 2012). ADA obtained by periodate oxidation of alginate extends the options for covalent cross-linking making use of the highly reactive aldehyde groups. Alternatively, polyelectrolyte complexes (PECs) formed between alginate and a cationic polymer have been utilized in material design.

1-Ethyl-3-(3-dimethylaminopropyl)carbodiimide (EDC) chemistry, when coupled with the co-reactant N-hydroxyl succinimide, facilitates amide bond formation between the carboxylic acid groups of alginate and amine groups of diamines. This strategy has been utilized with methyl ester L-lysine, adipic dihydrazide, and poly(ethylene glycol) diamines (Lee et al., 2000). In addition to the aqueous chemistry route, it has been demonstrated that the TBA-Alg dissolved in DMF can be cross-linked with diamines using CMPI as the activator (Leone et al., 2008). In applications where the presence of the carboxylate groups is important for the function of the hydrogel, covalent cross-linking can be facilitated using the hydroxyl groups of alginate and epichlorohydrin in the presence of NaOH. This approach has been utilized for calcium cross-linked capsules in an aqueous environment (Grasselli et al., 1993). Calcium ions were subsequently extracted leaving covalently cross-linked capsules.

ADA has been used as an aldehyde cross-linker to achieve intramolecular reactions with other parts of the ADA molecule and intermolecular reactions with unmodified alginate (Jejurikar et al., 2012). These gels displayed very high swelling characteristics and good mechanical strength, which could be controlled by the degree of oxidation of ADA as well as the ratio of ADA to alginate. ADA has also been cross-linked with adipic acid hydrazide. The swelling and degradation properties of these hydrogels were controlled by varying the amount of adipic acid hydrazide in the reaction mixture (Lee et al., 2004).

PECs form between cationic and anionic polyelectrolytes through attractive electrostatic interactions. The nature of the PEC is affected by many factors including temperature, ionic strength, and pH as well as polyelectrolyte type, molecular weight, and concentration. By manipulation of these factors, it is possible to form either weakly or strongly associated PECs that possess a range of physicochemical properties. Chemical characterization of both alginate-based and ADA-based PECs has been investigated in detail (Lawrie et al., 2007; Aston et al., 2015). Examples of PECs include coacervates and layer-by-layer (LbL) assemblies, the latter often applied post-material construction to control stability and/or rate of release as discussed below.

14.3 Drug delivery using alginate matrices

Alginate has shown potential as one of the most important biomaterials for delivery of bioactive molecules, cytokines, proteins, growth factors, genes, and other drugs with the ability to deliver either in a localized or in a targeted manner (Lee and Mooney, 2012). As such, some products using alginate as the core material are currently on the market. These include Emdogain gel, which is widely used for oral tissue regeneration and periodontal diseases through the release of amelogenin protein. Emdogain was first introduced in 1996 and is one of the most established and widely studied dental treatments in the United States and European Union. In addition, Flaminal Forte Gel is a commercially available wound dressing containing enzyme systems such as glucose oxidase and lactoperoxidase that kill bacteria. For more complex wounds, products releasing silver aim at eliminating colonization of microorganisms, which may lead to infection and delay healing. One such product is Tromboguard.

The transport of pharmaceuticals to the desired target site and their release in a controlled and sustained way represents the massive research field of drug delivery.

Applications of alginate biopolymer in drug delivery

To simply focus on alginate-based systems and claim that these represent, the state of the art in this field would be naïve. However, a number of strategies in drug delivery can be illustrated through important examples in polysaccharide-based systems. Drug delivery capsules are designed with two key objectives in mind: the safe transport of the active molecule to the active target site and minimal impact on the host biological system. Typically, engineering of the structure of the assembly is either directed toward improved permeability, stability, and retention (passive targeting) or the modification of the functionality of the capsule to improve localized delivery and/or biointegration (active targeting). A number of these strategies are illustrated in Fig. 14.5. The LbL assembly approach has been utilized for a number of different applications with the aim of modulating the release profile of the encapsulated drug through the formation of a diffusional barrier around the alginate capsule. In the work by Matsusaki et al. (2007) encapsulating vascular endothelial growth factor (VEGF) in alginate capsules, it was found that varying the number of layers in the LbL assembly had a significant effect on the rate of VEGF release. Modification of microcapsules to reduce the local inflammatory or tissue response in the host represents an emerging area of interest in biointegration. Strategies include the

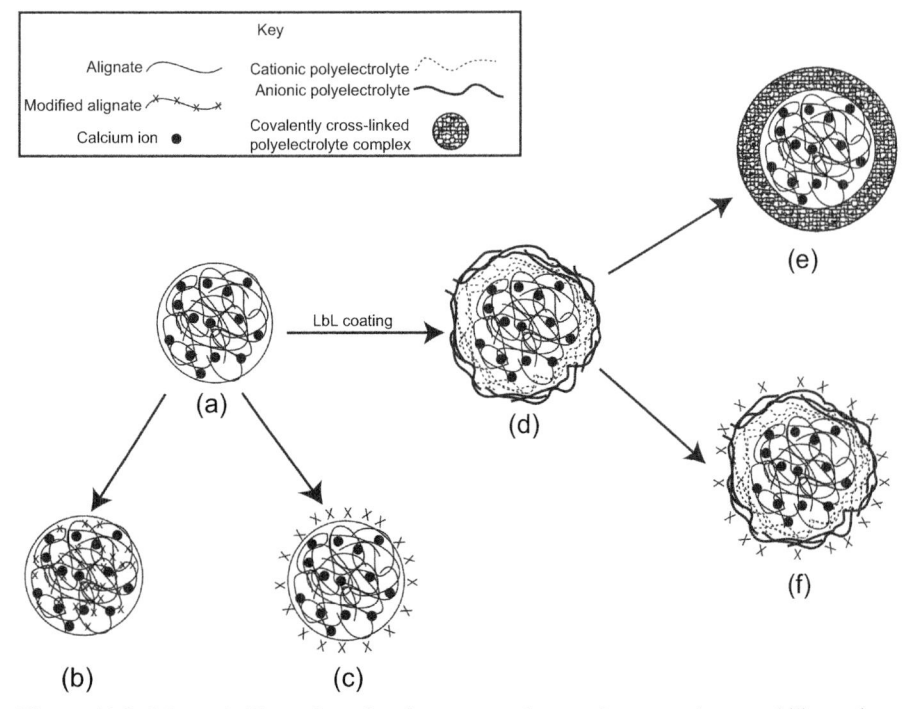

Figure 14.5 Schematic illustration of various approaches used to control permeability and biointegration. (a) Simple cross-linked alginate matrix, (b) cross-linked matrix produced from chemically modified alginate, (c) post-modification of alginate matrix with chemical or biological moieties, (d) layer-by-layer (LbL) assembly around alginate matrix, (e) covalent cross-linking of LbL assembly, and (f) post-modification of LbL assembly.

modification of alginate to incorporate antibodies on the outer surface of the microcapsule (e.g., anti-TNF-α (Leung et al., 2008) and heparin as the outer layer of a multilayered capsule assembled by physical adsorption (Bünger et al., 2003).

The strategies that are implemented depend on the nature of the drug to be delivered, the route of administration, and target environment (Table 14.1). Oral delivery of drugs represents a particular challenge due to the complex gastrointestinal environment, and there are several examples of research directed toward acquiring the desired characteristics for efficacy via this delivery route. These strategies include modification of alginate to introduce binding through disulfide bonds with mucous glycoproteins (Greimel et al., 2007); novel cross-linking strategies to strengthen capsules (Anal and Stevens, 2005); application of a covalently cross-linked chitosan coat (Taqieddin and Amiji, 2004), and the introduction of buoyancy to evade gastric emptying (Shishu et al., 2007). Another challenging route of administration is the intravenous route where the particle size of the drug carrier affects the in vivo fate. While enhanced lifetime within the circulatory system is desirable, nanosized carriers are required for effective uptake in certain applications. Particles with a diameter of 50 nm can cross different barriers (Hoshyar et al., 2016; Tang et al., 2014), and some studies suggested that the particles with a diameter of <12 nm have better tumor permeability (Chauhan et al., 2012). The majority of alginate capsules for drug delivery are above 100 nm in size (Yang et al., 2015; Paques et al., 2014) and can be produced by a variety of methods. Alginate-based particle with sizes of 100–200 nm can be produced using, e.g., ionic gelation (Moradhaseli et al., 2013), emulsification (Yang et al., 2015), a combination of emulsification and calcium ion gelation (Nguyen et al., 2015), or a combination of nanoprecipitation and calcium ion gelation (Kanwar et al., 2012).

14.3.1 Use of chemically modified alginate

The uses of chemically modified alginate in drug delivery applications have explored different aspects of biointegration. The nature of the physicochemical relationship

Table 14.1 Common drug delivery environments.

Route of administration	Environment	Desirable Properties
Oral	Gastrointestinal	Stability in low pH environment. Noncovalent associations with mucous gel layer. Prolonged gastric residence time.
Intravenous	Blood	Enhanced lifetime within circulatory system. Small size to pass through capillaries, cross-different barriers, and exhibits tumor permeability.
Intraperitoneal	Blood	High surface-area-to-volume ratio to maximize contact. In this case also, the diffusion of drug happens through the peritoneal membrane lined within a capillary bed.
Skin	Blood	Efficient transfer to blood or lymphatic system.

between drug molecule and the carrier matrix impacts on the drug release profile, and so the design of capsules represents a balance between the matrix/environment and drug/matrix associations. Examples considered here include the use of thiolated alginate for mucoadhesion, the use of amphiphilic alginate derivatives for enhancing protein retention, the use of oxidized alginate for both formation of a covalently cross-linked matrix and covalent attachment of a drug molecule, and the use of sulfated alginate as a heparin mimetic. While the first example is mainly addressing the interactions of the delivery system with the in vivo environment, the remaining examples explore the drug/matrix interactions.

Thiomers have been proposed as effective mucoadhesive polymers to be used in oral drug delivery applications (Bernkop-Schnürch and Greimel, 2005). They have been proposed to act by thiol/disulfide bond exchange reactions and oxidation reactions with cysteine-rich subdomains of mucus glycoproteins, thus forming covalent linkages to mucin via disulfide bonds. In their study on producing mucoadhesive alginate delivery capsules, Greimel et al. (2007) used mixtures of thiolated alginate and poly(acrylic acid) (PAA) (modified by cysteine attachment) to encapsulate insulin. They found that the introduction of the thiol group offered additional benefit to the delivery system. It increased the encapsulation efficiency from 15% (unmodified alginate) to 65% (thiolated alginate/PAA) and decreased the release rate. In addition, capsule stability in simulated intestinal fluid was significantly enhanced, and this was attributed to the formation of internal disulfide bonds during capsule formation (i.e., 82%−85% of free thiol groups were oxidized). These particles thus show some promising characteristics for use in oral drug delivery.

Hydrophobically modified alginates have been trialed for encapsulation of proteins through the incorporation of long chain alkyl groups, e.g., C_{12} via ester bonds (Leonard et al., 2004). These amphiphilic alginate derivatives were used to encapsulate a series of model proteins with very high encapsulation efficiencies (70%−100%) in which both hydrophobic interactions and calcium ion cross-linking contributed to the bead formation process. In addition, the strong interactions between the proteins and the amphiphilic alginate derivatives were illustrated by retention of the proteins in simple buffers. This was in contrast to rapid release of these proteins from unmodified alginate capsules. Protein release from the amphiphilic capsules was observed only in the presence of the esterase lipase that hydrolyzes the ester bond or in the presence of surfactants that disrupts the hydrophobic interactions in the capsule.

An elegant use of oxidized alginate (ADA) in drug delivery has been demonstrated by Balakrishnan and Jayakrishnan (2005). They produced an injectable scaffold from ADA and gelatin in the presence of small amounts of borax. It was observed that the gelling time decreased with increasing concentration of all reagents, while it increased with the degree of oxidation of ADA. Using primaquine as a model drug, they demonstrated slower drug release rates when using ADA of a high degree of oxidation and attributed this to the ability of the drug to undergo a Schiff's reaction with the aldehyde groups requiring subsequent degradation (rather than simple diffusion) processes to take place to release the drug.

While alginates show low affinity for growth factors, sulfated alginate has been shown to be a heparin mimetic binding to heparin-binding growth factors similar to or stronger than heparin itself for most proteins (Freeman et al., 2008). The higher binding affinity of sulfated alginate has been used to control the release rate from alginate-based drug delivery capsules (Freeman et al., 2008). Furthermore, sequential release of two growth factors could be achieved from injectable alginate biomaterials containing sulfated alginate. Interestingly, the relative rate of release of the two factors could be manipulated by their relative concentrations in the gel, allowing a high affinity growth factor to be released before one with a lower binding affinity (Ruvinov et al., 2011). Evaluation of the chain length, alginate sequence, and degree of oxidation in a series of sulfated alginates has shown a complex interplay between these parameters as well as distinct differences between different growth factors (Arlov et al., 2015). Specifically, a high degree of oxidation (40%) showed the highest affinity for growth factors, approaching that of heparin, and this was, in part, related to the enhanced chain flexibility of these sulfated oxidized alginates. The high binding affinity of sulfated alginate for growth factors has been utilized for fabricating coassembled particles (~200 nm), which can be incorporated into alginate hydrogels (Ruvinov et al., 2016).

14.3.2 Microencapsulation for transplantation

Drug delivery achieved through the transplantation of cells has become a viable therapeutic strategy for organ replacement, tissue engineering, and drug molecule delivery. Cells are typically sourced either from the same species (allograft) or, more commonly, a different species (xenograft) to the host. This strategy enables the sustained release of drug molecules over longer time frames through mechanisms that are not possible using conventional therapeutics. The relatively high biocompatibility and low cytotoxicity of purified alginate have made it the most popular biomaterial for microencapsulation, xenotransplantation, and immunoisolation of cells based on meeting scientific and regulatory standards for human application (Orive et al., 2015). The provenance of this field can be traced to the research of Lim and Sun who, in 1980, reported the assembly of a calcium alginate-poly-L-lysine-alginate (APA) microcapsule surrounding pancreatic islet cells (Lim and Sun, 1980). Efforts have since been directed toward creating the bioartificial pancreas through the microencapsulation and immunoisolation of pancreatic islets. In 2009, human islets were successfully implanted into the peritoneal cavity of diabetic humans that remained viable for 30 months (Tuch et al., 2009), and the field has progressed substantially since. Multiple research reviews consider historical, current, and future achievements along with perspectives in islet encapsulation strategies and technologies to date (Desai and Shea, 2017; Strand et al., 2017; Gurruchaga et al., 2015; Orive et al., 2015; Wilson and Chaikof, 2008; Lacik, 2006; de Vos et al., 2006; Narang and Mahato, 2006). The suite of desirable and undesirable properties of alginate-based microcapsules that contribute to their potential success (Strand et al., 2017; de Vos et al., 2006; Lacik, 2006; Zimmermann et al., 2005) is summarized in Table 14.2.

Despite the demand for the standardization of methodologies surrounding the microencapsulation of pancreatic islets, it is evident that the microcapsule matrix must be tailored for each individual cellular application (Fig. 14.6). A significant

Table 14.2 Desirable attributes in the assembly of alginate-based microcapsules.

Desirable	Undesirable
Alginate of sufficient purity so that it does not induce a cytotoxic response in the hosted cells. Alginates with an endotoxin content >100 EU/g are not suitable for in vivo studies.	Toxins inherent in unpurified alginates include pyrogens, mitogens, polyphenols, and peptides. These substances invoke recognition by macrophages and induce fibrotic overgrowth of the microcapsules.
A robust structure that can withstand localized compression and shear stresses imparted by the target environment. Longevity of the matrix resistant to biodegradation. Tailoring of the capsule through a blend of a high G alginate imparting mechanical strength and a high M alginate promoting elasticity when required.	A low degree of cross-linking results in a structure that swells and degrades in the presence of biological fluids. This also relates to the M/G ratio of the alginate and the viscosity. In contrast, a membrane that is too rigid (high degree of cross-linking) ruptures easily.
A semipermeable membrane that can retain the cells that it surrounds and simultaneously permit diffusion of both the nutrients and the active biomolecules. Oxygen transport to the encapsulated cells is a critical process. The pore dimensions and their interconnectivity in the polymer network control diffusion.	A semipermeable membrane in which the porosity is either too small and prevents diffusion of essential nutrients or oxygen (hypoxia) to the encapsulated cells or is too large enabling cells to escape inducing an immunogenic response, transfection, or for cytotoxic species (e.g., inflammatory cytokines such as TNF α) to enter the microcapsule.
The outer surface of the microcapsule should be immunosilent. In alginate-poly-L-lysine-alginate capsules, high M alginates are preferred as they mask polycations more efficiently through stronger electrostatic interactions.	Exposure of functionalities that cause a host immune response in the form of attack by humoral and T cell—mediated processes or the formation of a collagenous layer to isolate the "foreign" body. In addition, a rough outer surface of the capsule encourages fibrotic overgrowth that inhibits diffusion of nutrients and oxygen to the encapsulated cells.
A surface-to-volume ratio that optimizes implantation efficiency and diffusional transfer processes.	Large microcapsules that restrict diffusion of nutrients and oxygen to the encapsulated cells. The surface-to-volume ratio results in inefficient bioactive molecule release.
The ability to tailor the capsule toward the host environment. Minimal adhesion or fibrotic overgrowth is tolerable for microcapsules implanted in the peritoneal cavity; however, vascularization is required for subcutaneous transplantation.	Physicochemical properties of the outer microcapsule surface, which are detrimental to biointegration.

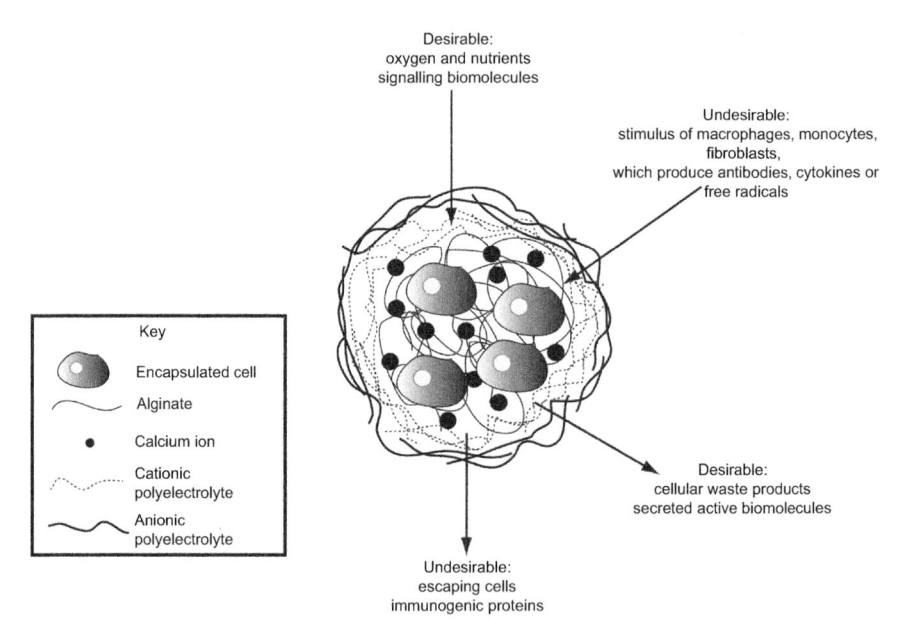

Figure 14.6 Schematic representation of the most common example of a multilayered assembly applied to microencapsulate cells. A summary of the desirable and undesirable processes that can impact on the viability of the cells is shown.

challenge in optimizing the microcapsule for the desired application continues to relate to the strength and permeability of the alginate-based matrix with undesirable outcomes including rupture or the escape of cells during proliferation and differentiation (Li et al., 2008; Lee et al., 2009). The porosity and strength of the matrix can be manipulated through the selection of the type of alginate with the ratio of M/G units dictating the extent of cross-linking (refer to Section 14.2.3). Cellular microencapsulation in alginate-based matrices has been widely adopted in research and clinical practice addressing a broad range of diseases, several examples are provided in Table 14.3, demonstrating the diversity of this therapeutic approach. Several reviews (Gonzalez-Pujana et al., 2017; Gurruchaga et al., 2015; Orive et al., 2014) provide insights into progress in diversifying cellular encapsulation for therapeutic applications including perspectives of the challenges that this research community faces.

Consideration of the strategies summarized in Table 14.3 reinforces the levels of complexity that are required in the structural matrices of the microcapsules to tailor them for encapsulated cells and the target environments in in vivo studies. The complexity in alginate-based structural matrices includes (i) simple cross-linked gels (Fig. 14.5(a)), (ii) multilayered self-assembled PECs (Fig. 14.5(d)), and (iii) postmodification to introduce desirable attributes (Fig. 14.5(e) and (f)). These levels are each described in detail below.

(i) Alginate hydrogels ionically cross-linked by divalent cations are the simplest systems that have been adopted for cell microencapsulation, and, of these, the most widely adopted continues to be Ca-alginate hydrogels. The process of forming hydrogels around cells by this

Table 14.3 Examples of cells encapsulated for treatment of diseases beyond diabetes.

Disease	Cell Line	Secreted biomolecule	Assembly	In vitro or in vivo	References
Fertility and ovarian conditions	Follicles/oocytes	Progesterone, androgen, and estrogen	Ca-Alg microcapsules	In vitro	Vanacker and Amorim, (2017); West et al. (2007)
Nerve regeneration	Schwann	Glial cell line derived neurotrophic factor (Gdnf)	Ba-Alg microcapsules	In vitro	De Guzman et al., 2008
Bone and cartilage tissue repair	Preosteoblastic (MC3T3-E1)	Osteocalcin	RGD-Ca-Alg[a] microcapsules	In vitro	Evangelista et al. (2007)
	Embryonic fibroblasts (C3H10T1/2 line)	Bone morphogenic protein	(Ca-Alg)-Chi-A (Ca-Alg)-PLL-A (Ca removed by chelation)	Peritoneal	Zhang et al. (2008)
	Chrondrocytes	Glycosaminoglycan and collagen	Ca-Alg + chitlac[b]	In vitro	Marsich et al. (2008)
	Bone marrow cells (BMC)	Differentiation and protein release	(Ca-Alg)-PLL-A	In vitro	Abbah et al. (2008)
Stem cell therapy	Crandell Rees feline kidney	Mitochondrial activity	(Tyr-Alg) + (Ca-Alg)	In vitro	Sakai and Kawakami, (2008)
	Fetal myoblasts	Differentiation and therapeutic gene products	(Ca-Alg)-PLL-A	In vitro	Li et al. (2008)

Continued

Table 14.3 Continued

Disease	Cell Line	Secreted biomolecule	Assembly	In vitro or in vivo	References
Cancer immunotherapy	HEK	Recombinant anti-CEA (carcinoembryonic antigen) × anti-CD3 bsAb	(Ca-Alg)-PLL-A	In vitro	Saenz del Burgo et al. (2015)
Neurodegenerative diseases (Parkinson's)	HEK293	Glial cell line–derived neurotrophic factor (GDNF)	Ca-Alg + collagen composite	In vitro	Lee et al. (2009)
	Fischer rat 3T3 fibroblasts	GDNF	(Ca-Alg)-PLL-A	Striatum	Grandoso et al. (2007)
	Hamster kidney (BHK) fibroblasts	Vascular endothelial growth factor	(Ca-Alg)-PLL-A	Intracranial	Antequera et al., 2012
Anemia	C_2C_{12} myoblasts	Erythropoietin	(Ca-A)-PLL-A	Subcutaneously and intraperitoneally	Orive et al. (2005) Ponce et al. (2006)
Hormone replacement Therapy	Sertoli cells	IGF-1	Ba-Alg	Peritoneal	Luca et al. (2013)
Gastrointestinal diseases and disorders	*Escherichia coli* DH5	Urea removal	(Ca-Alg)-Chi-A (Ca-Alg)-PLL-A	Oral delivery to stomach	Lin et al. (2008)
Gene therapy	Recombinant CHO	Endostatin	(Ca-Alg)-PLL-A (Ca removed by chelation)	Peritoneal	Zhang et al. (2007)

[a]Modification of alginate to enhance adhesion, proliferation, and differentiation.
[b]Chitlac = lactose modified chitosan.

approach is very gentle with low shear stresses and minimal exposure to reagents. The characteristic properties of the type of alginate selected for this purpose have a significant impact on the subsequent viability and function of the encapsulated cells. The composition can be altered to vary the mechanical properties and density of the Ca-alginate matrix, which can impact on cell proliferation or differentiation, e.g., in the maturation of encapsulated follicles (West et al., 2007). However, the instability of Ca-alginate capsules in vivo due to exchange of cross-linking ions with native ions has led to the increasing use of Ba^{2+} as the preferred ionic cross-linker. Microcapsules prepared from Ba-alginate hydrogels have been demonstrated to possess improved mechanical properties and increased the capsule lifetime by a factor of two for high G alginates (Mørch et al., 2006), but the ratio of G content needs to be greater than 60%. These microcapsules do not completely prevent the undesirable diffusion of IgG and several cytokines to the cells within the interior of the microcapsule; however, the evidence indicates that the encapsulated islets have sufficient protection from the local environment (Omer et al., 2005).

(ii) As alginate microcapsules are prone to osmotic swelling and rupture, multilayered self-assembled PECs (e.g., the APA system) are typically produced by self-assembling multiple polyelectrolyte layers of opposing charge (e.g., Poly-L-lysine (PLL)) onto the Ca-alginate core through the combination of electrostatic intermolecular forces, hydrogen bonding, and polymer flocculation to strengthen the capsule. This APA assembly is the basis of the majority of cellular encapsulation matrices. The structure of these capsules is often naively represented as discrete layers; however, evidence has been provided for the formation of PECs between alginate and PLL, where PLL penetrates the alginate core up to a depth of 30 μm (de Vos et al., 2006). The most significant challenge is to achieve full screening of the exposed charges on the cationic polyelectrolyte, otherwise these will invoke an immunogenic response. For example, PLL, when inefficiently complexed, results in exposed alginate at the outermost surface of capsule (Tam et al., 2005). Recent studies have explored new cross-linking strategies involving poly-L-ornithine as a substitute for PLL and genipin, resulting in improved biocompatibility (Hillberg et al., 2013). The APA microcapsule offers greater mechanical strength and protection of the encapsulated cells than the Ca-alginate hydrogel capsule; however, it must be considered that the extensive immersion and washing cycles of the coating process may itself be detrimental to the cells.

(iii) Multilayered assemblies may involve the analogues of the APA matrix components modified to introduce desirable attributes such as substitution of divalent ions or introduction of covalent bonds between biopolymers. Ca^{2+} ions are the most commonly adopted cross-linking ion for the innermost core of the standard APA capsule. However, the influence of the size of the cross-linking ion (Ca^{2+}, Ba^{2+}, Sr^{2+}) on the structure of the resulting matrix and encapsulated cell viability has been explored for ARPE-19 cells in vitro (Wikström et al., 2008); Sr^{2+} was found to be unsuitable. The introduction of covalent linkages increases the mechanical strength and reduces the swelling of the capsule matrix in the host environment. The most common modification to the self-assembled LbL matrix is the introduction of covalent bonds between alginate and PLL, for example, those achieved by the introduction of a photo cross-linkable moiety onto the PLL to strengthen the microcapsule (e.g., Dusseault et al., 2008) or the introduction of enzymatically cross-linkable groups (Sakai and Kawakami, 2008).

Methods for preparation of alginate-based microcapsules and cell microencapsulation have become well-established (Somo et al., 2017). Rigorous characterization of physicochemical and physicomechanical properties of microcapsules in parallel with their biocompatibility or biotolerability has become central to evaluating whether a device has potential to be viable (Rokstad et al., 2014). Careful selection of characterization techniques that support biological testing are now recommended as mandatory in the field. The mechanical properties of the capsule become important in the encapsulation of cells that are required to differentiate and proliferate, such as mesenchymal stem cells. An example of a post-modification strategy to alter these properties is the removal of calcium from the internal hydrogel to encourage cell growth and active biomolecule production (Zhang et al., 2008). The modification of alginate through the introduction of peptide moieties (Section 25.2.4) is also a common approach to improving the adhesion of encapsulated cells that are anchorage dependent for survival e.g., osteoblasts (Evangelista et al., 2007), or the structural support required for follicle encapsulation (Vanacker and Amorim, 2017). The addition of short synthetic peptides to the matrix such as arginine-glycine-aspartic acid (RGD) or proteins, for example, collagen or fibronectin, can mimic native physical and biomechanical properties of the target environment. Santos et al. (2014) encapsulated C_2C_{12} myoblasts in biofunctionalized APA capsules with RGD and demonstrated that the peptide density needs to be higher in vivo compared with in vitro for cell viability. An extension of these approaches has been demonstrated in the covalent modification of alginate by YIGSR peptides before encapsulation of neurites followed by the subsequent modification of the capsule via adsorption of laminin (Dhoot et al., 2004) promoting adhesion of the neurites. Post-modification of the encapsulated cell assembly has been attempted to improve integration with their target environment and strategies include PEGylation (Spasojevic et al., 2014; Zhang et al., 2008). Surface functional molecules may also be coupled to the assembly post-encapsulation to improve immunosilence (Leung et al., 2008). Post-implantation vascularization of the matrices is also desirable to optimize oxygen exchange with encapsulated cells (Gandhi et al., 2013); however, fibrotic overgrowth remains as a significant challenge (Wilson and Chaikof, 2008).

The microencapsulation of cells in tandem with active biomolecules is an initiative that triggers a subsequent cellular response in vivo. Bioactive molecules, such as basement membrane extract BD Matrigel (de Guzman et al., 2008), to promote proliferation have been incorporated with the alginate before formation of the Ca-alginate microcapsule. In a separate approach, extracellular matrix proteins such as collagen and laminin have been added during the assembly of the polycationic membrane (Cui et al., 2006).

Injectable alginate hydrogels offer a better opportunity to tailor the matrix encapsulating cells to a specific target environment in tissue engineering compared with preformed matrices (Bidarra et al., 2014). There are a number of studies that demonstrate that researchers are thinking "beyond the sphere" in cell encapsulation by tailoring the scaffolds geometry and matrix properties to suit their specific applications. There are applications where the cells themselves are "delivered" as they migrate from a scaffold that is designed to co-deliver inductive molecules (Hill et al., 2006). An innovative approach is represented in the encapsulation of cells within fibers that may also contain growth factors (Wan et al., 2004). These fibers can then be self-assembled into defined and patterned structures.

14.4 Concluding remarks and future directions

At this point, the influence of the structure of alginate (M/G ratio, chain length) and role of naturally derived impurities have been well characterized. The process of assembling three-dimensional matrices through ionic, covalent, and physical cross-linking of alginate molecules is also well understood and can be applied to control mechanical properties and stability (Lee and Mooney, 2012; Pawar and Edgar, 2012; Simo et al., 2017; Bidarra et al., 2014). The use of chemically modified alginate offers a means to enhance drug matrix interactions and hence control release of active molecules, and this approach is rapidly expanding especially in the use of sulfated alginate as a heparin mimetic (Arlov and Skjåk-Bræk, 2017). The modification of the assemblies is emerging as the popular route for optimizing biointegration of these matrices in their host environment. Research encompasses two approaches: the modification of alginate before capsule assembly and post-modification of assembled matrices. For example, the early work in developing Tyr-modified alginates has translated into viable matrices that mimic soft tissue properties and bind proteins (Schulz et al., 2019). There are now many clinical trials are ongoing for treatment of type 1 diabetes through cell encapsulation (Gurruchaga et al., 2015).

The refinement of the process of assembling alginate-based devices for drug delivery has enabled attention to be turned toward resolving ongoing challenges in translating to clinical practice. One remaining challenge for the use of alginate-based materials in cancer therapy is tuning of the size of the capsules to meet the demand for cellular uptake (Hoshyar et al., 2016; Tang et al., 2014). For cell encapsulation, challenges include developing microcapsule matrices that mimic the microenvironment of the source cells; necrosis due to hypoxia and/or fibrotic growth on capsules after implantation; tailoring the release profiles of the bioactive species; and scaling up the preparation of microencapsulated cells for clinical application. Innovation continues as microcapsules are being designed, which are responsive to external, noninvasive stimuli, which introduce desirable pharmacokinetic properties for both drug and cell delivery (Orive et al., 2015). This builds on earlier research into the introduction of functionalities that can be manipulated through change of environment or time-dependent processes such as pH-responsive systems (Chan et al., 2008). While spherical capsules are the most widespread geometry for devices due to the ease of fabrication, drug delivery is not restricted to matrices of this geometry. There are advantages offered by drug delivery materials as membranes, self-assembled fibers (Wan et al., 2004), and semiinterpenetrating networks (Matricardi et al., 2008).

Acknowledgments

The authors are grateful to the Australian Research Council (Grant No DP0557475) for their support of this work.

References

Abbah, S.A., Lu, W.W., Chan, D., Cheung, K.M.C., Liu, W.G., Zhao, F., Li, Z.Y., Leong, J.C.Y., Luk, K.D.K., 2008. Osteogenic behaviour of alginate encapsulated bone marrow stromal cells: an in vitro study. J. Mater. Sci. Mater. Med. 19, 2113−2119.

al Musa, S., Abu Fara, D., Badwan, A.A., 1999. Evaluation of parameters involved in preparation and release of drug loaded in crosslinked matrices of alginate. J. Control. Release 57, 223−232.

Antequera, D., Portero, A., Bolos, M., Orive, G., Hernández, R.M., Pedraz, J.L., Carro, E., 2012. Encapsulated VEGF-secreting cells enhance proliferation of neuronal progenitors in the hippocampus of AbetaPP/Ps1 mice. J. Alzheimer's Dis. 29, 187−200.

Al-Shamkhani, A., Duncan, R., 1995. Synthesis, controlled-release properties antitumor activity of alginate-cis-aconityl-daunomycin conjugates. Int. J. Pharm. 122 (1−2), 107−119.

Anal, A.K., Stevens, W.F., 2005. Chitosan-alginate multilayer beads for controlled release of ampicillin. Intl. J. Pharm 290, 45−54.

Arlov, Ø., Skjåk-Bræk, G., 2017. Sulfated alginates as heparin analougues: a review of chemical and functional properties. Molecules 22, 778.

Arlov, Ø., Aachmann, F.L., Sundan, A., Espevik, T., Skjåk-Bræk, G., 2014. Heparin-like properties of sulfated alginates with defined sequences and sulfation degrees. Biomacromolecules 15, 2744−2750.

Arlov, Ø., Aachmann, F.L., Feyzi, E., Sundan, A., Skjåk-Bræk, G., 2015. The impact on chain length and flexibility in the interaction between sulfated alginates and HFG and FGF-2. Biomacromolecules 16, 3417−3424.

Aslani, P., Kennedy, R.A., 1996. Studies on diffusion in alginate gels. I. Effect of crosslinking with calcium or zinc ions on diffusion of acetaminophen. J. Control. Release 42, 75−82.

Aston, R., Wimalaratne, M., Brock, A., Lawrie, G., Grøndahl, L., 2015. Interactions between chitosan and alginate dialdehyde biopolymers and their layer-by-layer assemblies. Biomacromolecules 16, 1807−1817.

Aston, R., Sewell, K., Klein, T., Lawrie, G., Grøndahl, L., 2016. Evaluation of the impact of freezing preparation techniques on the characterization of alginate hydrogels by cryo-SEM. Eur. Polym. J. 82, 1−15.

Balakrishnan, B., Jayakrishnan, A., 2005. Self-crosslinking biopolymers as injectable in sity forming biodegradable scaffolds. Biomaterials 26, 3941−3951.

Balakrishnan, B., Lesieur, S., Labarre, D., Jayakrishnan, A., 2005. Periodate oxidation of sodium alginate in water and in ethanol−water mixture: a comparative study. Carbohydr. Res. 340, 1425−1429.

Bernkop-Schnürch, A., Greimel, A., 2005. Thiomers; the next generation of mucoadhesive polymers. Am. J. Drug Deliv. 3 (3), 141−154.

Bidarra, S.J., Barrias, S.S., Granja, P.L., 2014. Injectable alginate hydrogels for cell delivery in tissue engineering. Acta Biomater. 10, 1646−1662.

Bouhadir, K.H., Lee, K.Y., Alsberg, E., Damm, K.L., Anderson, K.W., Mooney, D.J., 2001. Degradation of partially oxidized alginate and its potential application for tissue engineering. Biotechnol. Prog. 17, 945−950.

Braccini, I., Pérez, S., 2001. Molecular basis of Ca^{2+} induced gelation in alginates and pectins: the egg-box model revisited. Biomacromolecules 2, 1089−1096.

Bünger, C.M., Gerlach, C., Freier, T., Schmitz, K.P., Pilz, M., Werner, C., Jonas, L., Scharek, W., Hopt, U.T., deVos, P., 2003. Biocompatability and surface structure of chemically modified immunoisolating alginate-PLL capsules. J. Biomed. Mater. Res. 67A, 1219−1227.

CFR, 2018. https://www.accessdata.fda.gov/scripts/cdrh/cfdocs/cfcfr/CFRSearch.cfm?fr=184.1724.
Chan, L.W., Jin, Y., Heng, P.W.S., 2002. Crosslinking mechanisms of calcium and zinc in production of alginate microspheres. Int. J. Pharm. 242 (1−2), 255−258.
Chan, A.W., Whitney, R.A., Neufeld, R.J., 2008. Kinetic controlled synthesis of ph-responsive network alginate. Biomacromolecules 9 (9), 2536−2545.
Chauhan, V.P., Stylianopoulos, T., Martin, J.D., Popović, Z., Chen, O., Kamoun, W.S., Jain, R.K., 2012. Normalization of tumour blood vessels improves the delivery of nanomedicines in a size-dependent manner. Nat. Nanotechnol. 7, 383−388.
Coleman, R., Lawrie, G., Lambert, L.K., Whittaker, M., Jack, K.S., Grøndahl, L., 2011. Phosphorylation of alginate: synthesis, characterisation and evaluation of in vitro mineralisation capacity. Biomacromolecules 12, 889−897.
Corkhill, P.H., Trevett, A.S., Tighe, B.J., 1990. The potential of hydrogels as synthetic articular cartilage. Proc. Inst. Mech. Eng. H J. Eng. Med. 204 (H3), 147−155.
Connelly, J.T., Garcia, A.J., Levenston, M.E., 2007. Inhibition of in vitro chondrogenesis in RGD-modified three dimentional alginate gels. Biomaterials 28, 1071−1083.
Cui, Y.-X., Shakesheff, K.M., Adams, G., 2006. Encapsulation of RIN-m5F cells within Ba^{2+} crosslinked alginate beads affects proliferation and insulin secretion. J. Micrencaps. 23, 663−676.
de Boisseson, M., Leonard, M., Hubert, P., Marchal, P., Stequert, A., Castel, C., Favre, E., Dellacherie, E., 2004. Physical alginate hydrogels based on hydrophobic or dual hydrophobic/ionic interactions: bead formation, structure, and stability. J. Colloid Interface Sci. 273, 131−139.
de Guzman, R.C., Ereifej, E.S., Broadrick, K.M., Rogers, R.A., Vandevord, P.J., 2008. Alginate-matrigel microencapsulated Schwann cells for inducible secretion of glial cell line derived neurotrophic factor. J. Microencapsul. 25, 487−498.
de Vos, P., de Haan, B.J., Wolters, G.H.J., Van Schilfgaarde, R., 1997. Improved biocompatibility but limited graft survival after purification of alginate for microencapsulation of pancreatic islets. Diabetologia 40, 262−270.
de Vos, P., Faas, M.M., Strand, B., Calafiore, R., 2006. Alginate-based microcapsules for immunoisolation of pancreatic islets. Biomaterials 27, 5603−5617.
Desai, T., Shea, L.D., 2017. Advances in islet encapsulation technologies. Nat. Rev. Drug Discov. 16, 338−350.
Desai, S.J., Simonelli, A.P., Higuchi, W.L., 1965. Investigation of factors influencing release of solid drug dispersed in inert matrices. J. Pharm. Sci. 54, 1459−1463.
Dhoot, N.O., Tobias, C.A., Fischer, I., Wheatley, M.A., 2004. Peptide-modified alginate surfaces as a growth permissive substrate for neurite outgrowth. J. Biomed. Mater. Res. 71A, 191−200.
Donati, I., Holtan, S., Borgogna, M., Dentini, M., Skjak-Braek, G., 2005. New hypothesis on the role of alternating sequences in Calcium−Alginate gels. Biomacromolecules 6 (2), 1031−1040.
Drury, J.L., Boontheekul, T., Mooney, D.J., 2005. Cellular crosslinking of peptide modified hydrogels. J. Biomech. Eng. 127, 220−228.
Dusseault, J., Langlois, G., Meunier, M.-C., Ménard, M., Perreault, C., Hallé, J.-P., 2008. The effect of covalent corss-links between the membrane components of microcapsules on the dissemination of encapsulated malignant cells. Biomaterials 29, 917−924.
Ertestvag, H., Doest, B., Larsen, B., Skjak-Braek, G., Valla, S., 1994. Cloning and expression of an *Azobacter vinelandii* manuunronan C-5 epimerases. J. Bacteriol. 176, 2846−2853.

Evangelista, M.B., Hsiong, S.X., Fernandes, R., Sampiato, P., Kong, H.J., Barrias, C., Salema, R., Barbosa, M.A., Mooney, D.J., Granja, P.L., 2007. Upregulation of bone cell differentiation through immobilization within a synthetic extracellular matrix. Biomaterials 28, 3644−3655.

Folkman, J., Long, D.M., 1964. The use of silicon rubber as a carrier for prolonged drug delivery. J. Surg. Res. 4, 139−142.

Freeman, I., Kedem, A., Cohen, S., 2008. The effect of sulfation of alginate hydrogels on the specific binding and controlled release of heparin-binding proteins. Biomaterials 29, 3260−3268.

Fundueanu, G., Nastruzzi, C., Carpov, A., Desbriesres, J., Rinaudo, M., 1999. Physico-chemical characterization of Ca-alginate microparticles produced with different methods. Biomaterials 20 (15), 1427−1435.

Gacesa, P., 1992. Enzymic degradation of algintes. Int. J. Biochem. 24 (4), 545−552.

Gandhi, J.K., Opara, E.C., Brey, E.M., 2013. Alginate-based strategies for therapeutic vascularization. Ther. Deliv. 4, 327−341.

Gomez, C., Rinaudo, M., Villar, M.A., 2007. Oxidation of sodium alginate and characterization of the oxidized derivatives. Carbohydr. Polym. 67, 296−304.

Gonzalez-Pujana, A., Santos, E., Orive, G., Pedraz, J.L., Hernández, R.M., 2017. Cell microencapsulation technology: current vision of its therapeutic potential through the administration routes. J. Drug Deliv. Sci. Technol. 42, 49−62.

Grandoso, L., Ponce, S., Manuel, I., Arrue, A., Ruiz-Ortega, J.A., Ulibarri, I., Orive, G., Hernandez, R.M., Rodriguez, A., Rodriguez-Puertas, R., Zumarraga, M., Linazasoro, G., Pedra, L., Ugedo, L., 2007. Long-term survival of encapsulated GDNF secreting cells implanted within the striatum of Parkinsonized rats. Int. J. Pharm. 343, 69−78.

Grant, G.T., Morris, E.R., Rees, D.A., Smith, P.J.C., Thom, D., 1973. Biological interactions between polysaccharides and divalent cations: the egg-box model. FEBS Lett. 32 (1), 195−198.

Grasselli, M., Diaz, L.E., Cascone, O., 1993. Beaded matrices from cross-linked alginate for affinity and ion exchange chromatography of proteins. Biotechnol. Tech. 7, 707−712.

Greimel, A., Werle, M., Bernkop-Schnürch, A., 2007. Oral peptide delivery: in-vitro evaluation of thiolated alginate/poly(acrylic acid) microparticles. J. Pharm. Pharmacol. 59, 1191−1198.

Gurruchaga, H., Saenz del Burgo, L., Ciriza, J., Orive, G., Hernández, R.M., Pedraz, J.L., 2015. Advances in cell encapsulation technology and its application in drug delivery. Expert Opin. Drug Deliv. 12, 1251−1267.

Haug, A., Larsen, B., Smidsrød, O., 1963. The degradation of alginates at different pH values. Acta Chem. Scand. 17, 1466−1468.

Haug, A., Larsen, B., Smidsrød, O., 1967. Alkaline degradation of alginate. Acta Chem. Scand. 21, 2859−2870.

Hermes, R.H., Narayani, R., 2002. Polymeric alginate films and alginate beads for the controlled delivery of macromolecules. Trends Biomater. Artif. Organs 15, 54−56.

Hill, E., Boontheekul, T., Mooney, D.J., 2006. Designing scaffolds to enhance transplanted myoblast survival and migration. Tissue Eng. 12, 1295−1304.

Hillberg, A.L., Kathirgamanathan, K., Lam, J.B., Law, L.Y., Garkavenko, O., Elliott, R.B., 2013. Improving alginate-poly-L-ornithine- alginate capsule biocompatibility through genipin crosslinking. J. Biomed. Mater. Res. 101, 258−268.

Holme, H.K., Lindmo, K., Kristiansen, A., Smidsrød, O., 2003. Thermal depolymerisation of alginate in the solid state. Carbohydr. Polym. 54, 431−438.

Hoshyar, N., Gray, S., Han, H., Bao, G., 2016. The effect of nanoparticle size on in vivo pharmacokinetics and cellular interaction. Nanomedicine 11, 673−692.

Jejurikar, A., Lawrie, G., Martin, D., Grøndahl, L., 2011. A novel strategy for preparing mechanically robust ionically cross-linked alginate hydrogels. Biomed. Mater. 6. Article Number: 025010.

Jejurikar, A., Seow, X.T., Lawrie, G., Martin, D., Jayakrishnan, A., Grøndahl, L., 2012. Degradable alginate hydrogels crosslinked by the macromolecular crosslinker alginate dialdehyde. J. Mater. Chem. 22, 9751−9758.

Jørgensen, T.E., Sletmoen, M., Draget, K.I., Stokke, B.T., 2007. Influence of Oligoguluronates on alginate gelation, kinetics, and polymer organization. Biomacromolecules 8 (8), 2388−2397.

Kanwar, J.R., Mahidhara1, G., Kanwar, R., 2012. Novel alginate-enclosed chitosan−calcium phosphate-loaded iron-saturated bovine lactoferrin nanocarriers for oral delivery in colon cancer therapy. Nanomedicine 7, 1521−1550.

Kim, H.-S., Song, M., Lee, E.-J., Ueon, S.S., 2015. Injectable hydrogels derived from phosphorylated alginic acid calcium complexes. Mater. Sci. Eng. C 51, 139−147.

Klock, G., Frank, H., Houben, R., Zekorn, T., Horcher, A., Seibers, U., Wohrle, M., Federlin, K., Zimmermann, U., 1994. Production of purified alginates suitable for use in immunoisolated transplantation. Appl. Microbiol. Biotechnol. 40 (5), 638−643.

Kuo, C.K., Ma, P.X., 2008. Maintaining dimensions and mechanical properties of ionically crosslinked alginate hydrogel scaffolds *in vitro*. J. Biomed. Mater. Res. 84, 899−907.

Lacik, I., 2006. Polymer chemistry in diabetes treatment by encapsulated islets of langerhans: review to 2006. Aust. J. Chem. 59, 508−524.

Lawrie, G., Keen, I., Chandler-Temple, A., Drew, B., Rintoul, L., Fredericks, P., Grøndahl, L., 2007. Interactions between alginate and chitosan biopolymers characterised using FTIR and XPS. Biomacromolecules 8, 2533−2541.

Lee, K.Y., Bouhadir, K.H., Mooney, D.J., 2000. Degradation behavior of covalently crosslinked Poly(aldehyde guluronate) hydrogels. Macromolecules 33, 97−101.

Lee, K.Y., Mooney, D.J., 2012. Alginate: properties and biomedical applications. Prog. Polym. Sci. 37, 106−126.

Lee, K.H., Bouhadir, K.H., Mooney, D.J., 2004. Controlled degradation of hydrogels using multi-functional cross-linking molecules. Biomacromolecules 25, 97−101.

Lee, M., Lo, A.C., Cheung, P.T., Wong, D., Chan, B.P., 2009. Drug carrier systems based on collagen-alginate composite structures for improving the performance of GDNF-secreting HEK293 cells. Biomaterials 30, 1214−1221.

Leonard, M., Rastello De Boisseson, M., Hubert, P., Dalençon, F., Dellacherie, E., 2004. Hydrophobically modified alginate hydrogels as protein carriers with specific controlled release properties. J. Control. Release 98, 395−405.

Leone, G., Torricelli, P., Chiumiento, A., Facchini, A., Barbucci, R., 2008. Amidic alginate hydrogel for nucleus pulposus replacement. J. Biomed. Mater. Res. A 84A, 391−401.

Leung, A., Lawrie, G.A., Nielsen, L.K., Trau, M., 2008. Synthesis and characterization of alginate/poly-L-ornithine/alginate microcapsules for local immunosuppression. J. Microencapsul. 25, 387−398.

Li, L., Fang, Y., Vreeker, R., Mendes, I., 2007. Re-examining the egg-box model in Calcium-Alginate gels by X-ray diffraction. Biomacromolecules 8 (2), 464−468.

Li, A.A., Bourgois, J., Potter, M., Chang, P.L., 2008. Isolation of human foetal myoblasts and its application for microencapsulation. J. Cell Mol. Med. 12, 271−280.

Lim, F., Sun, A.M., 1980. Microencapsulated islets as bioartificial endocrine pancreas. Science 210, 908−910.

Lin, J., Yu, W., Liu, X., Xie, H., Wang, W., Ma, X., 2008. In vitro and in vivo characterization of alginate-chitosan-alginate artificial microcapsules for therapeutic oral delivery of live bacterial cells. J. Biosci. Bioeng. 105, 660−665.

Luca, G., Calvitti, M., Mancuso, F., Falabella, G., Arato, I., Bellucci, C., List, E.O., Bellezza, E., Angeli, G., Lilli, C., Bodo, M., Becchetti, E., Kopchick, J.J., Cameron, D.F., Baroni, T., Calafiore, R., 2013. Reversal of experimental Laron Syndrome by xenotransplantation of microencapsulated porcine Sertoli cells. J. Control. Release 165, 75−81.

Malaprade, L., 1928. Action of polyalcohols on periodic acids. Analytical application. Bull. Soc. Chim. France 43, 683−696.

Marsich, E., Borgogna, M., Donat, I., Mozetic, P., Strand, B.L., Gomez Salvador, S., Vittur, F., Paoletti, S., 2008. Alginate/lactose-modified chitosan hydrogels: a bioactive material for chrondrocyte encapsulation. J. Biomed. Mater. Res. 84A, 364−376.

Matricardi, P., Pontoriero, M., Coviello, T., Casadei, M.A., Alhaique, F., 2008. In situ cross-linkable novel alginate-dextran methacrylate IPN hydrogels for biomedical applications: mechanical characterization and drug delivery properties. Biomacromolecules 9 (7), 2014−2020.

Matsusaki, M., Sakaguchi, H., Serizawa, T., Akashi, M., 2007. Controlled release of vascular endothelial growth factor from alginate hydrogels nano-coated with polyelectrolyte multilayer films. J. Biomater. Sci. Polym. Ed. 18 (6), 775−783.

Moe, S.T., Skjak-Braek, G., Elgsaeter, A., Smidsrød, O., 1993. Swelling of covalently cross-linked alginate gels: influence of ionic solutes and non polar solvents. Macromolecules 26, 3589−3597.

Moradhaseli, S., Mirakabadi, A.Z., Sarzaeem, A., Mohammadpour dounighi, N., Soheily, S., Borumand, M.R., 2013. Preparation and characterization of sodium alginate nanoparticles containing ICD-85 (Venom derived peptides). Int. J. Innov. Appl. Stud. 4, 534−542.

Mørch, Y.A., Donati, I., Strand, B.L., Skjak-Braek, G., 2007. Molecular engineering as an approach to design new functional properties of alginate. Biomacromolecules 8, 2809−2814.

Mørch, Y.A., Donati, I., Strand, B.L., Skjak-Braek, G., 2006. Effect of Ca, Ba, Sr on alginate microbeads. Biomacromolecules 7, 1471−1480.

Mørch, Y.A., Holtan, S., Donati, I., Strand, B.L., Skåk-Bræk, G., 2008. Mechanical properties of C-5 epimerized alginates. Biomacromolecules 9 (9), 2360−2368.

Mucalo, M.R., Yokogawa, Y., Suzuki, T., Kawamoto, Y., Nagata, F., Nishizawa, K., 1995. Further studies of calcium phosphate growth on phosphorylated cotton firbres. J. Mater. Sci. Mater. Med. 6 (11), 658−669.

Narang, A.S., Mahato, R.I., 2006. Biological and biomaterial approaches for improved islet transplantation. Pharmacol. Rev. 58, 194−243.

Nguyen, H.T.P., Munnier, E., Souce, M., Perse, X., David, S., Bonnier, F., Vial, F., Yvergnaux, F., Perrier, T., Cohen-Jonathan, S., Chourpa1, I., 2015. Novel alginate-based nanocarriers as a strategy to include high concentrations of hydrophobic compounds in hydrogels for topical application. Nanotechnology 26, 255101.

Nunamaker, E.A., Purcell, E.K., Kipke, D.R., 2007. In vivo stability and biocompatibility of implanted calcium alginate disks. J. Biomed. Mater. Res. 83A, 1128−1137.

Omer, A., Duvivier-Kali, V., Fernandes, J., Tchipashvili, V., Colton, C.K., Weir, G.C., 2005. Long-term normoglycemia in rats receiving transplants with encapsulated islets. Transplantation 79, 52−58.

Orive, G., de Castro, M., Ponce, S., Hernández, R.M., Gascón, A.R., Boch, M., Alberch, J., Pedraz, J.L., 2005. Long-term expression of erythropoietin from myoblasts immobilized in biocompatible and neovascularised microcapsules. Mol. Ther. 12, 283−289.

Orive, G., Santos, E., Pedraz, R.M., Hernández, R.M., 2014. Application of cell encapsulation for controlled delivery of biological therapeutics. Adv. Drug Deliv. Rev. 67−68, 3−14.

Orive, G., Santos, E., Poncelet, D., Hernández, R.M., Pedraz, J.L., Wahlberg, L.U., De Vos, P., Emerich, D., 2015. Cell encapsulation: technical and clinical advances. Trends Pharmacol. Sci. 36, 537−546.

Paques, J.P., van der Linden, E., van Rijna, C.J.M., Sagis, L.M.C., 2014. Preparation methods of alginate nanoparticles. Adv. Colloid Interface Sci. 209, 163−171.

Pavlath, A.E., Gossett, C., Camirad, W., Robertson, G.H., 1999. Ionomeric films of alginic acid. J. Food Sci. 64 (1), 61−63.

Pawar, S.N., Edgar, K.J., 2012. Alginate derivatization: a review of chemistry, properties and applications. Biomaterials 33, 3279−3305.

Pelletier, A., Hubert, P., Lapicque, F., Payan, E., Dellacherie, E., 2000. Amphillic derivatives of sodium alginate and hyaluronate: synthesis and physico-chemical properties of aqueous dilute solutions. Carbohydr. Polym. 43, 343−349.

Ponce, S., Orive, G., Hernández, R.M., Gascón, Canals, J.M., Muñoz, M.T., Pedraz, J.L., 2006. In vivo evaluation of EPO-secreting cells immobilized in different alginate-PLL microcapsules. J. Control. Release 116, 28−34.

Poncelet, D., Lencki, R., Beaulieu, C., Halle, J.P., Neufeld, R.J., Fournier, A., 1992. Production of alginate beads by emulsification/internal gelation. I. Methodology. Appl. Microbiol. Biotechnol. 38 (1), 39−45.

Qin, Y., 2008. The gel swelling properties of alginate fibers and their applications in wound management. Polym. Adv. Technol. 19 (1), 6−14.

Ranade, V.V., Hollinger, M.A., 2004. Drug Delivery Systems, second ed. CRC Press Inc., Boca Raton, Florida, pp. 63−114.

Reakasame, S., Boccaccini, A.R., 2018. Oxidized alginate-based hydrogels for tissue engineering applications: a review. Biomacromolecules 2018 19, 3−21.

Remuñán-López, C., Bodmeier, R., 1997. Mechanical, water uptake and permeability properties of crosslinked chitosan glutamate and alginate films. J. Control. Release 44, 215−225.

Rokstad, A.M.A., Lacík, I., de Vos, P., Strand, B.L., 2014. Advances in biocompatibility and physico-chemical characterization of microspheres for cell encapsulation. Adv. Drug Deliv. Rev. 67−68, 111−130.

Rowley, J.A., Madlambayan, G., Mooney, D.J., 1999. Alginate hydrogels as synthetic extracellular matrix materials. Biomaterials 20, 45−53.

Ruvinov, E., Leor, J., Cohen, S., 2011. The promotion of myocardial repair by the sequential delivery of IGF- and HGF from an injectable alginate biomaterial in a model of acute myocardial infarction. Biomaterials 32, 565−578.

Ruvinov, E., Freeman, I., Fredo, R., Cohen, S., 2016. Spontaneous coassembly of biologically active nanoparticles via affinity binding of heparin-binding proteins to alginate-sulfate. Nano Lett. 16, 883−888.

Saenz del Burgo, L., Compte, M., Aceves, M., Hernández, R.M., Sanz, L., Álvarez-Vallina, L., Pedraz, J.M., 2015. Microencapsulation of therapeutic bispecific antibodies producing cells: immunotherapeutic organoids for cancer management. J. Drug Target. 23, 170−179.

Sakai, S., Kawakami, K., 2008. Both ionically and enzymatically crosslinkable alginate-tyramine conjugate as materials for cell encapsulation. J. Biomed. Mater. Res. 85A, 345−351.

Santos, E., Garate, A., Pedraz, J.L., Orive, G., Hernández, R.M., 2014. The synergistic effects of the RGD density and the microenvironment on the behavior of encapsulated cells: in vitro and in vivo direct comparative study. J. Biomed. Mater. Res. 102, 3965−3972.

Schulz, A., Gepp, M.M., Stracke, F., von Briesen, H., Neubauer, J.C., Zimmermann, H., January 2019. Tyramine-conjugated alginate hydrogels as a platform for bioactive scaffolds. J. Biomed. Mater. Res. A 107 (1), 114−121.

Shishu, Gupta, N., Aggarwal, N., 2007. Stomach-specific drug delivery of 5-fluoracil using floating alginate beads. AAPS PharmSciTech 8, E1−E7.

Shoichet, M.S., Li, R.H., White, M.L., Winn, S.R., 1995. Stability of hydrogels used in cell encapsulation: an in vitro comparison of alginate and agarose. Biotechnol. Bioeng. 50, 374−381.

Simó, G., Fernández Fernández, E., Vila Crespo, J., Ruipérez, V., Rodríguez Nogales, J.M., 2017. Research progress in coating techniques of alginate gel polymer for cell encapsulation. Carbohydr. Polym. 170, 1−14.

Skaugrud, Ø., Hagen, A., Borgensen, B.M., Dornish, M., 1999. Biomedical and pharmaceutical applications of alginate and chitosan. Biotechnol. Genet. Eng. Rev. 16 (2), 23−40.

Smidsrød, O., Painter, T., 1973. Effect of periodate oxidation upon the stiffness of the alginate molecule in solution. Carbohydr. Res. 26, 125−132.

Smidsrød, O., Skåk-Bræk, G., 1990. Alginates as immobilization matrix for cells. TIBTECH 8, 71−78.

Smisdsrød, O., Haug, A., Larsen, B., 1966. The influence of pH on the rate of hydrolysis of acidic polysaccharides. Acta Chem. Scand. 20, 1026−1034.

Somo, S.I., Khanna, O., Brey, E.M., 2017. Alginate microbeads for cell and protein delivery. In: Opara, E.C., Opara, E.C. (Eds.), Cell Microencapsulation: Methods and Protocols, Methods in Molecular Biology, vol. 1479, pp. 217−224.

Soon-Shiong, P., Otterlie, M., Skjak-Braek, G., Smisdsrød, O., Heintz, R., Lanza, R.P., Espevik, T., 1991. An immunologic basis for the fibrotic reaction to implanted microcapsules. Transplant. Proc. 23 (1 Pt 1), 758−759.

Spasojevic, M., Paredes-Juarez, G.A., Vorenkamp, J., de Haan, B.J., Schouten, A.J., de Vos, P., 2014. Reduction of the inflammatory responses against alginate- Poly-L-lysine microcapsules by antibiofouling surfaces of PEG-b-PLL diblock copolymers. PLoS One 9, e109837.

Strand, B.L., Coron, A.E., Skjak-Braek, G., 2017. Current and future perspectives on alginate encapsulated pancreatic islet. Stem Cells Transl. Med 6, 1053−1058.

Svanem, B.I.G., Skjak-Braek, G., Ertesvag, H., Valla, S., 1999. Cloning and expression of three new *Azobacter vinelandii* genes closely related to a previously described gene family encoding mannuronan C-5 epimerases. J. Bacteriol. 181, 68−77.

Tam, S.K., Dusseault, J., Polizu, S., Ménard, M., Hallé, J.-P., Yahia, L., 2005. Physicochemical model of alginate-poly-L-lysine microcapsules defined at the micrometric/nanometric scale using ATR-FTIR, XPS, and ToF-SIMS. Biomaterials 26, 6950−6961.

Tam, S.K., Dusseault, J., Bilodeau, S., Langlois, G., Hallé, J.-P., Yahia, L.H., 2011. Factors influencing alginate gel biocompatibility. J. Biomed. Mater. Res. 98A (1), 40−52.

Tan, W.-H., Takeuchi, S., 2007. Monodisperse alginate hydrogel microbeads for cell encapsulation. Adv. Mater. 19, 2696−2701.

Tan, C.S., Jejurikar, A., Rai, B., Bostrum, T., Lawrie, G., Grøndahl, L., 2009. Encapsulation of a glucose amino glycan in hydroxyapatite/alginate capsules. J. Biomater. Res. A 91 (3), 866−877.

Tang, L., Yang, X., Yin, Q., Cai, K., Wang, H., Chaudhury, I., Yao, C., Zhou, Q., Kwon, M., Hartman, J.A., Dobrucki, I.T., Dobrucki, L.W., Borst, L.B., Lezmi, S., Helferich, W.G., Ferguson, A.L., Fan, T.M., Cheng, J., 2014. Investigating the optimal size of anticancer nanomedicine. Proc. Natl. Acad. Sci. U. S. A. 111, 15344−15349.

Taqieddin, E., Amiji, M., 2004. Enzyme immobilization in novel alginate-chitosan core-shell microcapsules. Biomaterials 25, 1937−1945.

Tsujino, I., Saito, T., 1961. A new unsaturated uronide isolated from alginase hydrolysate. Nature 192, 970−971.

Tuch, B.E., Keogh, G.W., Williams, L.J., Wu, W., Foster, J.L., Vaithilingam, V., Philips, R., 2009. Safety and viability of microencapsulated human islets transplanted into diabetic humans. Diabetes Care 32, 1887−1889.

Vallée, F., Müller, C., Durand, A., Schimchowitsch, S., Dellacherie, E., Kelche, C., Cassel, J.C., Leonard, M., 2009. Synthesis and rheological properties of hydrogels based on amphiphilic alginate-amide derivatives. Carbohydr. Res. 344, 223−228.

Vanacker, J., Amorim, C.A., 2017. Alginate: a versatile biomaterial to encapsulate isolated ovarian follicles. Ann. Biomed. Eng. 45, 1533−1549.

Vold, I.M.N., Kristiansen, K.A., Christensen, B.E., 2006. A study of the chain stiffness and extension of alginates, in vitro epimerized alginates, and periodate-oxidized alginates using size-exclusion chromatography combined with light scattering and viscosity detectors. Biomacromolecules 7 (7), 2136−2146.

Wan, A.C.A., Yim, E.K.F., Liao, I.-C., Le Visage, C., Leong, K., 2004. Encapsulation of biologics in self-assembled fibres as biostructured units for tissue engineering. J. Biomed. Mater. Res. 71A, 586−595.

West, E.R., Xu, M., Woodruff, T.K., Shea, L.D., 2007. Physical properties of alginate hydrogels and their effects on *in vitro* follicle development. Biomaterials 28, 4439−4448.

Whistler, R.L., Wolfrom, M.L., 1962. In: Methods in Carbohydrate Chemistry, Vol I: Analysis and Preparation of Sugars, second ed. Academic Press Inc, New York and London.

Wikström, J., Elomaa, M., Syväjärvi, H., Kuokkannen, J., Yliperttula, M., Honkakoski, P., Urtti, A., 2008. Alginate-based microencapsulation of retinal pigment epithelial cell line for cell therapy. Biomaterials 29, 869−876.

Wilson, J.T., Chaikof, E.L., 2008. Challenges and emerging technologies in the immunoisolation of cells and tissues. Adv. Drug Deliv. Rev. 60, 124−145.

Yang, J., Han, S., Zheng, H., Dong, H., Liu, J., 2015. Preparation and application of micro/nanoparticles based on natural polysaccharides. Carbohydr. Polym. 123, 53−66.

Zhang, Y., Wang, W., Xie, Y., Yu, W., Lv, G., Guo, X., Xiong, Y., Ma, X., 2007. Optimization of microencapsulated recombinant CHO cell growth, endostatin production and stability of microcapsule *in vivo*. J. Biomed. Mater. Res. B Appl. Biomater. 84B, 79−88.

Zhang, W.-J., Li, B.-G., Zhang, C., Xie, X.-H., Tang, T.T., 2008. Biocompatability and membrane strength of $C_3H_{10}T_{1/2}$ cell-loaded alginate-based microcapsules. Cytotherapy 10, 90−97.

Zhu, H., Ji, J., Lin, R., Gao, C., Feng, L., Shen, J., 2002. Surface engineering of poly(DL-lactic acid) by entrapment of alginate-amino acid derivatives for promotion of chondrogenesis. Biomaterials 23, 3141−3148.

Zimmermann, U., Klock, G., Federlin, K., Hannig, K., Kowalski, M., Bretzel, R.G., Horcher, A., Entenmann, H., Sieber, U., Zekorn, T., 1992. Production of mitogen-contamination free alginates with variable ratios of mannuronic acid to guluronic acid by free flow electrophoresis. Electrophoresis 13 (5), 269−274.

Zimmermann, H., Zimmermann, D., Reuss, R., Feilen, P.J., Manz, B., Katsen, A., Weber, M., Ihmig, F.R., Ehrhart, F., Gebner, P., Behringer, M., Steinbach, A., Wegner, L.H., Sukhorukov, V.L., Vasquez, J.A., Schneider, S., Weber, M.M., Vole, F., Wolf, R., Zimmermann, U., 2005. Towards a medically approved technology for alginate-based microcapsules allowing for long-term immunoisolated transplantation. J. Mater. Sci. Mater. Med. 16, 491−501.

Part Four
Design considerations

Failure mechanisms of medical implants and their effects on outcomes

15

A. Kashi[1], S. Saha[2]
[1]Private Practice, Rochester, NY, United States; [2]University of Washington, WA, Seattle, United States

Chapter outline

- 15.1 Introduction 407
  - 15.1.1 Failure mechanisms of medical implants 409
- 15.2 Manufacturing deficiencies 410
- 15.3 Mechanical factors (e.g., fatigue, overloading, and off-axis loading) 410
- 15.4 Wear 413
  - 15.4.1 Wear and migration 413
- 15.5 Corrosion 416
- 15.6 Clinical factors for implant success and failure 417
  - 15.6.1 Health of patient 417
  - 15.6.2 Surgical errors 417
- 15.7 Failure mechanisms of non−load-bearing implants 418
  - 15.7.1 Soft tissue implants 419
- 15.8 Failure analysis of medical implants 420
- 15.9 Multivariate analysis 422
- 15.10 Ethical issues 423
- 15.11 Conclusion 424
- References 426

15.1 Introduction

Medical implants comprise several types and are indicated in patients to replace lost or damaged tissues. The main goal of implants is to help patients with disabilities to return to normal function for the longest possible duration. Medical implants can be used either to replace missing tissues/organs or to augment existing performance of the body. Because most medical implants are bioengineered for their intended purpose, analyzing their failure mechanism becomes critical not just from a performance perspective but also to help with providing the standard of care that is planned.

Table 15.1 Classification of medical implants.

Soft tissue implants	Metallic	Load-bearing
Hard tissue implants	Nonmetallic	Non—load-bearing
	Hybrid (combination)	

A key challenge with failure analysis of medical implants is the difficulty in translating research/in vitro data to in vivo performance. One of the reasons for this might be the inability to simulate the actual in vivo environment accurately, albeit systems models taking into account certain variables can be designed to closely resemble the in situ milieu. Studying the failure mechanisms of implants is a complex topic and is one of the most important cornerstones of research and lies at the forefront of developing newer and improved generations of prostheses. Understanding the behavior of implants in their failure modes will allow researchers and clinicians to strive for better designs and materials selection for future versions.

A classification of medical implants will be helpful in delineating the various categories under which they can be placed (Table 15.1).

In some instances, medical implants may not strictly fall under the categories mentioned in Table 15.1. For example, the osteo-odonto-keratoprosthesis (OOKP) can be classified as a hybrid organic/inorganic device where the implant is comprised of native tissue (in this case odontogenic tissue or tooth and epithelial tissues harvested from the oral cavity) and artificial optics once it is implanted for restoring vision. Since its introduction by Strampelli in the 1960s, the OOKP has evolved with other clinicians and researchers developing newer designs and utilizing advanced biomaterials with the hopes of making them function better (Falcinelli et al., 2005; Tan et al., 2011, 2012; Weisshuhn et al., 2014; Michael et al., 2008; Zarei-Ghanavati et al., 2017).

For a true understanding of the failure mechanisms of medical implants, the interplay of several in vitro and in vivo factors needs to be studied. Most implants have finite life spans, albeit confounding factors including the wide variations of clinical environments that they are used in, material variations, variations in surgical proficiencies of clinicians, manufacturing processes, and follow-up/after care will influence their success/survival. Some of these aforementioned factors such as manufacturing processes can be controlled in some instances, whereas other influencing factors (e.g., clinical/surgical proficiency) might be difficult to consider while studying their failure modes.

"Medical device error" is the term most commonly used to describe problems with medical devices and their related technologies (Blandford et al., 2014). The main goal of failure analysis is to determine the causative factor for the failure and to work toward minimizing its occurrence in the future. Some of the important information obtained from failure analysis can help in formulating management protocols for patients in general or in specific cohorts as well as help in formulating decisions by regulatory agencies including the need for further testing or withdrawal of devices (McCloy, 2019). For a basic understanding of failure mechanisms of implants, decades of research has led us to adopt standardized testing protocols. Some of these include mechanical and microstructural characterization testing as well as analytical techniques such as computer simulation models. To sustain an aging population of the United States, it is vital that the latest techniques to study medical implant failures adopt multidisciplinary teams of researchers, engineers, clinicians, ethicists, industry/government regulators, and manufacturers. Traditional animal models as well as cutting-edge simulation models need to be encouraged to understand the behavior of implants in different clinical environments.

This chapter will discuss some of the latest updates in the field of medical implant failures. A brief section on ethical issues and statistics as it pertains to medical implant failures is included for the benefit of the readers. Furthermore, a discussion pertaining to the emergence of regenerative organs for tissue replacement is presented. The goal of this chapter will be for researchers and clinicians to better understand the latest advances in this field as well as to stimulate discussions related to this topic.

15.1.1 Failure mechanisms of medical implants

Medical implants can fail by various mechanisms, which can include either single or multiple etiologies. From a biomechanical perspective, it is useful to note that there generally exist two different types of implants—load-bearing and non—load-bearing (Table 15.1). The failure mechanisms associated with these two types can vary, depending on their anatomical location, the loads that they are subjected to during function, and other associated factors including age, sex, and the general systemic health of a patient. As an example, if the same implant is placed in two different patients and if one of the patients is a long-term corticosteroid user, the healing patterns may vary. Typically, in the case of load-bearing implants, the generation of wear debris is a critical factor that can contribute to foreign body reactions and metallosis. This is seen mostly in the case of orthopedic hip and knee implants. However, in the case of dental implant failures, current failures can be attributed to excessive strain on the surrounding bone, bone shielding, failure of surrounding grafted sites, infections, and other comorbid factors. True material failures of dental implants are possible; however, there are several other confounding factors that lead to their failure eventually. Ceramic implants are known to cause lesser wear debris when compared with their metallic counterparts (Higuchi et al., 2016). This is due to better tribological properties of bioceramics (i.e., the possibility to produce a smoother finished surface, thus leading to lesser friction during use) and superior biocompatibility when compared with metals. Apart from metallic implants, authors have studied debris resulting from the wear of failed elbow, wrist, and finger silicone implants (Cook et al., 1999). Another example of a load-bearing implant will be prosthetic heart valves where they are subjected to constant loads from blood flow. Some of the influencing factors in such devices can be the coagulability of the blood, the effect of exercise on the valve dynamics, design of the artificial valve, and the material composition of the valve. In the case of non—load-bearing implants (e.g., maxillofacial implants, including chin, cheek, and cranial prostheses, eye and ear implants, and breast implants), the failure mechanisms are different. For instance, authors have previously reported breast implant failures from folds/flaws, although this is not the most dominant form of failure for these types of prostheses (Brandon et al., 2006).

15.2 Manufacturing deficiencies

The manufacture of an implant typically requires strict adherence to guidelines that have been developed by medical device makers following good manufacturing practices (as mandated by the Food and Drug Administration). Failure to follow these prescribed guidelines can compromise the structural stability of implants. Furthermore, several cases of implant failure have been shown to result from deficiencies in the finished product due to manufacturing defects. These can result as a consequence of improper material selection, poor quality control, or a lack in oversight during handling (e.g., sterilization and packaging). Authors have previously reported that implant failures can be attributed to impurities in the raw materials. These impurities can act as crack initiation sites to allow for further deterioration of the entire implant structure during long-term use. Failures of the past, most notably the Vitek temporomandibular joint (TMJ) implant and the Dow Corning Silicone breast implant failures, have brought to the forefront issues related to appropriate material selection and adequate preclinical testing (Ta et al., 2002; Westermark et al., 2006; Mercuri and Giobbie-Hurder, 2004; Mercuri and Anspach, 2003; Speculand et al., 2000; Kearns et al., 1995). These include in vitro laboratory testing of implants and animal studies. Additionally, the need for liability protection for biomaterial manufacturers and tort reform are important issues that need further discussion. Biomedical device manufacturers are required to follow best manufacturing practices to ensure that the finished products (i.e., implants) conform to industry, physician, and patient requirements. Because an implant might be subjected to contamination and/or deterioration during its manufacture, it is important to characterize the possible failure mechanisms during the manufacturing process.

15.3 Mechanical factors (e.g., fatigue, overloading, and off-axis loading)

Factors that might influence the failure of an implant include the choice of biomaterial, the overall geometry of the prosthesis, the magnitude of forces, and the number of cycles of use that it is subjected to. Typically, implants are designed to withstand millions of cycles of use within the in vivo environment. However, in vitro testing and evaluation might sometimes not be sufficient to predict the lifetime survival rate due to secondary contributing factors including immune suppression, infections, and other secondary illnesses. Furthermore, it is almost impossible to simulate the dynamic physiological changes that occur in situ in in vitro models. For instance, the qualitative properties of bone (as it relates to its mechanical and microstructural properties) are continually changing during an individual's lifetime thus having a potential influence on the survival/success of a prosthesis.

Almost all load-bearing implants (e.g., hips, knees, and dental) are subjected to a combination of axial (i.e., compressive and tensile) bending and torsional forces during function. It is important to note that high stresses would most likely result in single-cycle and low-cycle implant failures, whereas low stresses (i.e., high/multiple cycles that typically last a million or more in number annually) would mostly lead to fatigue types of failure. This behavior is especially important in the context of metallic implants (Geesink et al., 1988). From a biomechanical perspective, the likelihood of failures can be minimized if the extent of the applied loads (stresses) can be reduced.

Another contributing factor for implant failure in orthopedics includes the role of fatigue due to nonhealing of bone. Some of the reasons for this include excessive motion, lack of blood supply, infection, immune suppression, systemic illnesses, and biomechanical factors such as stress shielding of implants. If an implant is subjected to stresses that are beyond its endurance limit, the structure will ultimately fracture/fail. However, if an implant is subjected to low stress levels, it can withstand millions of cycles of use, although these repeated insults can create microscopic fracture zones that can contribute to larger structural failures in the long term due to fatigue. This type of fatigue failure results because of localized stress concentrations on certain points of an implant before other areas of the structure are affected. Furthermore, this can initiate a crack that can eventually propagate to involve the entire implant structure. If a particular material is very stiff, this can lead to stress shielding of the surrounding bone, bone resorption, and therefore leading to implant failure.

Compressive, tensile, and torsional loads are simultaneously acting on a load-bearing implant during function. To fully elucidate the failure criteria of implants, it is essential that biomechanical evaluation incorporates tests that can determine its mechanical properties (i.e., compression, tension, and torsion). We have previously reported the pull-out resistance of dental implants. The objective of this study was to compare the pull-out resistance of small and large diameter (3.25 and 4.5 mm) dental implants and the relationship of these implants to bone density. Two groups of implants, consisting of 18 implants of each diameter, were placed in five embalmed human mandibles. The bone mineral density of the area surrounding the implant site in the coronal cross section was measured by quantitative computed tomography. The initial implant stability was tested with a periodontium diagnostic device, and the pull-out resistance was tested with a mechanical testing system. Results showed the same initial stability for the two implants. However, the maximum pull-out force required for the large diameter implants was 15% greater than that required for the small diameter implants (Kido et al., 1997). Further studies need to be carried out to statistically validate these results.

In some instances, implants may fail if they are subjected to excessive bending moments (Fig. 15.1). In the case of the dental implant (this prosthesis failed after functioning for approximately 10 years) shown in Fig. 15.1, it is evident that the prosthetic portion (i.e., the dental crown) overlying the implant is being subjected to off-axis loading. This region of the crown can act as a cantilever consequently subjecting the implant to excessive tensile loads and bending moments. Additionally, if poor biomechanical considerations are coupled with other underlying factors including dynamic changes in bone quality and comorbidities, they can lead to implant failures.

The fatigue behavior of biomaterials is generally characterized by measuring the applied stress versus number of cycles (S—N curve) for failure (Fig. 15.2). Although in vitro testing and evaluation is helpful in obtaining values for strength, hardness, fracture toughness, and wear, these tests might not be sufficient to predict their lifetime

Figure 15.1 Dental implant crown with an overhanging cantilever leading to bending moments.

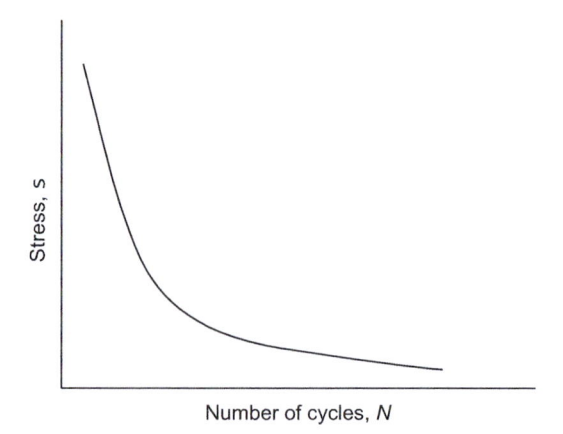

Figure 15.2 S-N curve showing fatigue behavior of biomaterials.

survival rates due to secondary contributing factors, including age, sex, body mass index (in the case of orthopedic hip and knee implants), immune suppression, infections, and other secondary illnesses. In addition to mechanical breakdown due to these previously mentioned factors, it has been reported that mechanical stress leads to corrosion in metals and the degradation in ceramics (Khan et al., 1996; Fehring et al., 2004; Gustafson et al., 1993; Zhang et al., 2004; Santana et al., 2009; Zhang and Lawn, 2009).

The in vivo corrosion of modular hip prosthesis components in mixed and similar metal combinations. The effect of crevice, stress, motion, and alloy coupling (Speculand et al., 2000; Kearns et al., 1995; Geesink et al., 1988)

15.4 Wear

15.4.1 Wear and migration

Wear and migration are problems associated with artificial implants, especially for total or partial joint replacements. In the case of most load-bearing implants, wear is more common compared with migration of the entire implant structure. We will therefore focus in this section mostly to the phenomenon of wear. In the case of metallic implants, particulate disease resulting from wear debris is a frequently encountered clinical problem that can sometimes warrant revision surgery. The problem of wear is often seen in total hip and knee replacements. We have previously reported findings of catastrophic peri-implant bone loss caused by polyethylene and metallic wear in total knees in two cases. The first was a pathologic fracture of the distal femur that was associated with catastrophic polyethylene failure, whereas the second was related to three-body wear of the polyethylene from detached CoCr beads (Gustafson et al., 1993). Many authors have studied wear behavior of total joint replacements and other implants previously (Khan et al., 1996; Fehring et al., 2004; Hirakawa et al., 1996; Griffin et al., 2007; Yamaguchi et al., 1997; Ingham and Fisher, 2000; Sieber et al., 1999; Howie et al., 1988; Al Jabbari et al., 2008; Goodman, 2001; Sumner et al., 1995; Revell et al., 1997; Revell, 2008; Schmalzried et al., 1996; Marshall et al., 2008; Korovessis et al., 2006; Agins et al., 1988) (Posada et al., 2014, 2015; Lehtovirta et al., 2017; Kovochich et al., 2018; Grosse et al., 2015; Wang et al., 2017). Fig. 15.3 is a retrieved polyethylene liner (from a knee implant) showing wear facets. Some of the observations from gross examination of this retrieved prosthesis include delamination, excessive wear, and partial fracture at the edge.

Although the issue of wear is of major concern with load-bearing implants, when compared with metals, ceramic implants are known to have better tribological properties due to their ability to be polished to very smooth surfaces, thus reducing friction and consequently producing less wear during function. However, ceramics are prone to other problems during use. For instance, the formation of cracks following repeated use can act as initiation sites that can progress to catastrophic failure of a ceramic implant. The phenomenon of crack initiation and propagation has been studied extensively with respect to ceramic dental crowns (Zhang et al., 2004, 2008; Zhang and

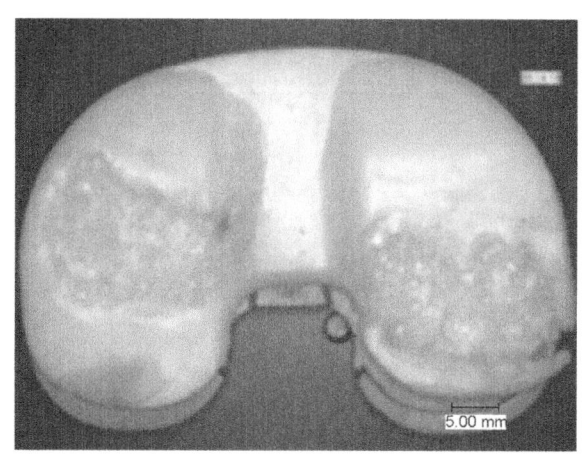

Figure 15.3 Retrieved polyethylene liner from a failed knee implant showing wear facets.

Lawn, 2009; Hermann et al., 2006; Kim et al., 2007). This knowledge might also be useful in understanding flaws and their effects on the structural stability of other orthopedic implants, including artificial hips and knees. Once there is a critical flaw in a brittle material such as a ceramic, the energy needed to drive it further into the structure of the material becomes much lower. Biological fluids in vivo can accelerate crack growth in ceramics by reducing the energy needed for crack propagation. When this is coupled with a lack of conduits (in an implant structure) for the escape of biological fluids, it can increase the hydrostatic pressure within this crack region. If these previously mentioned phenomena are combined with cyclic loading, which is commonly seen in load-bearing implants, it might accelerate crack growth, leading to catastrophic failure. Although ceramics are known to withstand compressive loads, they can easily fail when subjected to tensile loads. The lack of tensile strength makes them less attractive for many load-bearing applications. However, there have also been promising advances in dentistry with newer generation zirconia ceramics with improved mechanical properties. It will be worth investigating whether these materials can be utilized for other load-bearing orthopedic applications in the future.

Another emerging complication with ceramic hip prostheses is the generation of acoustic emissions from stripe wear. Although this has been only anecdotally reported by a few investigators, one of the suggested theories for this occurrence has been attributed to edge loading between the acetabular component and the corresponding ceramic ball (Fig. 15.4). Cases of similar nature have been reported in the medical literature and have resulted in their removal from patients (Restrepo et al., 2008; Taylor et al., 2007; Walter et al., 2006, 2007; Goldhofer et al., 2018; Mai et al., 2010; Tai et al., 2015; Schroder et al., 2011). Fig. 15.4 shows an alumina hip implant that was in situ for 2 years in a 54-year-old patient before it had to be removed due to a persistent squeaking sound during function. On examination, it was observed that the ceramic ball and its corresponding acetabular component showed wear facets. This is seen in squeaking ceramic hip prostheses. The acetabular component and the alumina femoral head (with stripe wear as a result of titanium wear debris in situ) are shown in Figs 15.4

Failure mechanisms of medical implants and their effects on outcomes 415

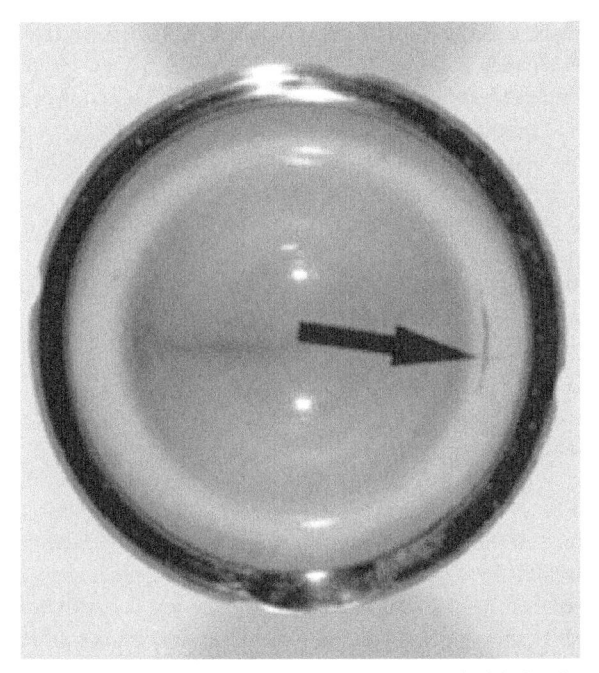

Figure 15.4 Stripe wear on a retrieved acetabular component of a hip implant.

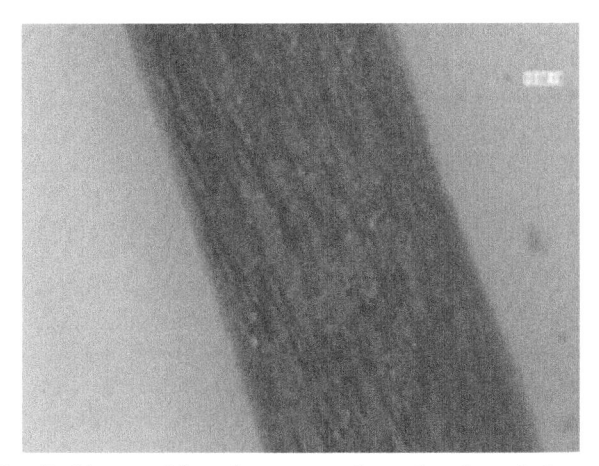

Figure 15.5 Magnified image of the stripe wear on the retrieved acetabular component.

and 15.5, respectively. It is also evident from the acetabular component (area where arrow is shown in Fig. 15.4) that there appears to be some surface damage to an area corresponding to the edge. This might support the theory that edge loading can be a contributing factor to the audible squeak. Because ceramic hip implants have been used for only a relatively short time in the United States, when compared with the more traditional metal-on-metal or metal-on-polyethylene types, clinical complications that have occurred with ceramic implants have been reported only in the past two to three decades.

It will be important to study the failure mechanisms of these newer implant types as they might be used in greater numbers clinically in the future, especially in younger, active adults. This may be attributed to improved material properties and designs, as well as better long-term clinical results. Because younger patients are more active when compared with the elderly, studying the correlation between age and level of physical activity that can produce specific types of implant wear patterns will be useful. Moreover, changes in the surface properties and microstructural features of the ceramic material need to be studied to further delineate the underlying causes for these phenomena.

15.5 Corrosion

The internal milieu of the body is generally comprised of various anions, cations, dissolved oxygen, and organics (Williams and Williams, 2013). Authors have previously mentioned that the anions are mainly chloride, phosphate, and bicarbonate ions and the cations are usually Na^+, K^+, Ca^{2+}, and Mg^{2+} (Williams and Williams, 2013). In addition to these previously mentioned substances, the presence of proteins and dissolved oxygen can influence the onset of corrosion in implants. Corrosion is a complication that can have multiple detrimental effects, both clinically and from a materials standpoint. This is especially important in the context of metallic implants. Examples of some of the types of corrosion include pitting, erosion, galvanic/two-metal corrosion, leaching, stress corrosion, and crevice corrosion. Authors have reported that although the rates of corrosion are less prevalent than in the past, they continue to pose a clinical problem with currently used metal implants (Khan et al., 1996; Korovessis et al., 2006; Brown et al., 1995; Upadhyay et al., 2006; Case et al., 1994; Hallab et al., 2004; Hallab et al., 2001; Rose et al., 1972; Sutow et al., 1985; Urban et al., 1994; Harding et al., 2002; Reclaru and Meyer, 1994; MacDonald et al., 2017; Hallam et al., 2018; Nawabi et al., 2016).

In the case of metallic implants, especially modular types that have one or more components with tight tolerances, there can be higher chances for crevice corrosion to occur. Crevice corrosion can also be witnessed at regions between an implant and adjacent bone. Hard tissue implants are meant to be rigid when they are placed in vivo. However, it is well known that there is micromotion during function. These micromotions can lead to the generation of microscopic particulate debris and other toxic products, thus lowering the pH of the local environment. Implant motion that is greater than 150 μm leads to fixation by formation of mature connective tissue (Pilliar et al., 2006). Animal and human studies have confirmed these findings (Sumner et al., 1995) (Stulberg et al., 1991). Studies have shown that the local pH in corrosion zones of an implant can reach values as low as 2.5 (Hallam et al., 2018). Because of high tolerances between mating components in an implant, biological fluids might not be able to fully enter the crevice regions to flush out these toxic products, leading to corrosion in the long term. X-ray spectrographic microanalysis of failure zones can reveal the elemental composition of the corrosion by-products in an implant.

15.6 Clinical factors for implant success and failure

15.6.1 Health of patient

Implant success or failure depends to a great extent on the systemic health of the individual. For instance, it is well documented that patients with suppressed immune systems (e.g., long-term steroid users) and diabetics are prone to higher rates of infections when compared with nonsteroid users (McCracken et al., 2000; Takeshita et al., 1998; Beikler and Flemmig, 2003; Mombelli and Cionca, 2006; Gerritsen et al., 2000). Studies have shown that there is less osseointegration with titanium implants in diabetic animals (McCracken et al., 2000). However, these results are conflicting as there are other studies that have reported good fixation of percutaneous implants in diabetic animals (Gerritsen et al., 2000). A suppressed immune system is typically more prone to attack by opportunistic infections (e.g., candidiasis), and the presence of a foreign object such as an implant in such cases might lead to higher chances of failure.

Because the average life span of individuals is increasing in the United States and many other countries due to access to improved medical care, the percentage of surviving diabetics in geriatric subgroups of the population has also been increasing. According to the Centers for Disease Control in the United States, in the year 2015, an estimated 30.3 million people or 9.4% of the US population had diabetes. Among this, the percentage of adults with diabetes increased with age with almost 25% of those aged 65 years or older (CDC, 2019). Although it is difficult to predict the exact numbers, it is likely that a significant number of individuals in this age group will require implants. Furthermore, the failure mechanisms of implants in the geriatric population might most likely result from higher chances of infection and other comorbid factors due to poorer general systemic health. This highlights the need for further investigations, including long-term clinical observations and follow-up to study the survival rates of implants in these cases (Morris et al., 2005).

15.6.2 Surgical errors

Surgeon proficiency is an important factor in implant surgery and its success. The placement of even a well-designed implant by an unskilled surgeon will have a high chance of failure. Factors including the surgeons' level of experience with the placement of a particular type of implant, surgical technique, and ease of handling of the armamentarium are critical factors for implant survival outcomes (Shalabi et al., 2006). Failure to follow preoperative, intraoperative, and postoperative protocols can contribute to implant failure. Esposito et al. mention that a combination of surgical trauma and anatomical conditions is probably the most important etiological factor responsible for early oral implant losses. In their review, they relate this to be approximately 3.6% of 16,935 implants (Esposito et al., 1998).

Simple steps, including changes in the orientation of an implant when it is surgically placed, can alter its wear properties. For instance, studies have shown that alignment of acetabular components of total hip prostheses that are greater than 50 degrees can lead to excessive wear of the polyethylene liners (Goodman, 2001).

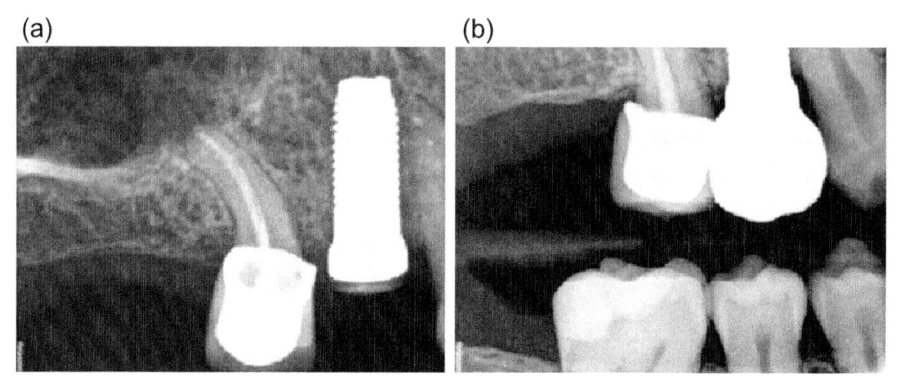

Figure 15.6 (a) Zirconia dental implant; (b) Zirconia dental implant restored with a crown.

Similarly, changes in the orientation of other load-bearing prostheses, including dental implants, can lead to excessive functional loads and accelerated localized bone resorption around them (Isidor, 1997, 2003). Fig. 15.6(a) shows a zirconia dental implant, and Fig. 15.6(b) shows the implant restored with a ceramic−zirconia crown. One of the most critical aspects of choosing this implant to be used clinically will be to also ensure that there are minimal tensile loads when it becomes functional with a crown. This is due to the fact that zirconia/ceramics are weak in tensile strength and more prone to fracture when subjected to off-axis loading.

Implant design and surface finish are critical factors in their long-term clinical success and must take into consideration anatomy and biomaterials (Esposito et al., 1998; Misch, 1999; Sykaras et al., 2000; Beksac et al., 2006; Triplett et al., 2004; Anitua, 2006). Variations in implant designs, materials, and shapes might produce different results, thus making it difficult to assume that similar wear patterns will be seen with all implants (Esposito et al., 1998; Sykaras et al., 2000; Çehreli et al., 2004). Therefore, further biomechanical studies are needed to better understand the different types of implant failures that might most likely be seen.

15.7 Failure mechanisms of non−load-bearing implants

Non−load-bearing implants (e.g., craniofacial prostheses) are generally not subjected to the same mechanical loads as their load-bearing counterparts (e.g., hip, knee, and dental prostheses). Thus, the failure modes that are encountered with these implants are generally different. The issue of wear debris and particulate disease, which is particularly important in the case of load-bearing implants, will not pose a significant threat toward failure in the case of non−load-bearing implants. However, other clinical complications, including secondary infections and bone resorption, are important factors that can lead to failures of non−load-bearing implants.

Poly(methyl methacrylate) (PMMA) is used for cranial defect reconstruction due to its rigidity (unlike ceramics, including hydroxyapatite, which are brittle in nature) (Nassiri et al., 2009; Eppley, 2005; Eppley et al., 2004). However, there still exists scope for developing improved versions of implants; for example, biomaterials that can adapt to physiological growth (e.g., materials that possess the ability to conform to physiological craniofacial skeletal growth), especially in children. For instance, the development of materials that can promote osteoinduction, as well as being capable of resorption so that the defect(s) will be filled in with natural tissue at a desired time, will be useful. Such developments can be particularly helpful in craniofacial reconstruction in the pediatric population, where constant growth of the skeletal tissues is seen. In a biomechanical study by Eppley, different compositions of PMMA cranioplasty materials were evaluated by an impact resistance test (according to ASTM D 3029-78). The results of this study showed that the fracture patterns were different for porous versus solid implants. In the case of porous implants, a radiating stellate pattern originated from the point of impact. However, solid implants showed fracture patterns that were linear and nonstellate (Eppley, 2005). This experimental information might be useful in the design of future versions of biomaterials that take into consideration biomechanical factors including the force needed to fracture the prostheses and the likelihood of fracture in vulnerable populations, including children.

15.7.1 Soft tissue implants

Soft tissue implants are generally used in the body for replacement of a variety of tissues, including ligaments, cartilages, and cardiac valves, and for maxillofacial reconstruction. A few examples of soft tissue implants include silicone gel breast implants, polymeric chin, cheek and lip maxillofacial implants, bovine, porcine and human cadaveric cartilage, ligament and tendon replacements, and bovine or porcine cardiovascular grafts for valve replacements. The biomechanical requirements of soft tissue implants are typically different when compared with implants used for hard tissue replacements. For instance, hip and knee implants are rigid structures, whereas most soft tissue replacements are viscoelastic in nature. Thus, their failure mechanisms (e.g., mechanical) differ when compared with hard tissue implants, although some common factors, including infections, can still lead to failures in any type of implant (Vinh and Embil, 2005; Ehrlich et al., 2005; Widmer, 2002; Bauer et al., 2006; Neut et al., 2003). Additional factors that might influence the survival outcomes of soft tissue implants include the general systemic health and comorbid conditions (e.g., diabetes).

Allogeneic and xenogeneic grafts have been known to cause foreign body reactions and rejections in some individuals. Similarly, silicone gel breast implants and polymeric maxillofacial implants have been associated with infections in some cases. Some rare failure mechanisms, for instance, the occurrence of fold flaw failures in breast implants have been previously reported by clinicians (Brandon et al., 2012). Other authors have studied platinum concentration in body fluids of women who have been exposed to silicone and saline breast implants (Lykissa and Maharaj, 2012). Authors have reported on the use of polyethylene, polypropylene, nylons, carbon fibers, polytetrafluoroethylene, and polyesters as artificial ligament replacement materials. Most of these materials are known to fail clinically, despite their use in other anatomical locations more successfully (Walsh, 2005). This might be related to the complex biomechanics of the knee joint and the effects of wear debris leading to adverse tissue reactions.

15.8 Failure analysis of medical implants

Implant failures can be studied by a variety of analytical/experimental techniques. Some examples include visual inspection, light microscopy, scanning electron microscopy (SEM), fractography, and microhardness measurements. As a first step to study a failed implant, visual inspection can provide clues related to macroscopic fracture zones (Figs. 15.4, 15.5, and 15.7(a–c)). Although it might be difficult to quantitatively assess the failure criteria or to predict what might have occurred on a microscopic scale, visual inspection allows researchers sometimes to gather useful information, including fracture initiation and the directions of crack propagation. The next level of evaluation includes the use of light microscopy. This can highlight the loss of surface coatings on implants and the presence of any pathologic tissue. Figs. 15.7(a) and (b) show close-up images of two retrieved knee implants that have been studied using a handheld digital microscope, while Fig. 15.7(c) shows the surface of a retrieved femoral head from a hip implant.

For a higher resolution and to obtain a detailed microstructural analysis (e.g., grain size and distribution) of fracture zones and normal areas of failed implants, an SEM

Figure 15.7 (a) Retrieved knee implant visualized using a handheld digital microscope; (b) Retrieved knee implant visualized using a handheld digital microscope; (c) Surface of a retrieved femoral head from hip implant showing surface scratches.

study is useful. X-ray spectrographic microanalysis can provide information related to the chemical/elemental composition of retrieved implants. This will allow researchers to study changes in the chemical composition of failed implants, particularly being able to focus on failure and corrosion zones in a structure. In addition, this technique can be useful to analyze oxide layer compositions on implant surfaces, particularly metallic structures.

The surface roughness of an implant is an important factor that can influence wear rate. Surface profilometry can be used to study the surface roughness of an implant. This characterization technique can allow researchers to analyze failed or retrieved implants. We have previously reported on the use of a noncontact surface profilometer to study and compare the surface roughness of a new and a retrieved TMJ implant (Fig. 15.8). The results of our study showed that the roughness parameters of retrieved implants were statistically higher ($P < 0.05$) than those for new specimens. Furthermore, these roughness values were lower than those reported for retrieved metal total knee implants in the literature (Vaderhobli and Saha, 2012).

Some of the other useful techniques for characterizing retrieved implants include microhardness and fractography. Newer characterization techniques that are capable of measuring hardness in the nanoscale range, by employing techniques including nanohardness indenters and of studying surface topographies of implant structures using atomic force and three-dimensional focus variation microscopes, are being used by researchers. They allow the study of materials and implant surfaces at much more detailed resolutions than would have been possible previously. In addition, characterization of implant surfaces at the regions of failure with these newer techniques can provide information related to local events occurring at the failure site. Software programs, including finite element analysis (FEA), can be useful to study or predict the behavior and probable failure mechanisms of implants/prostheses (Zhang et al., 2008). Useful information, such as response to fatigue, stress distribution in implant

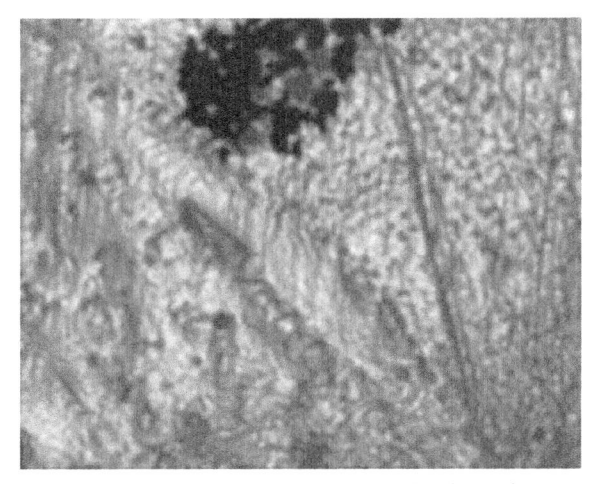

Figure 15.8 Surface roughness analysis of a retrieved TMJ implant using a non-contact surface profilometer.

structures, and design flaws can be investigated using software simulation packages such as FEA. We have previously examined the stress distribution around the screws of a TMJ implant with applied loads using an FEA simulation package (Kashi et al., 2010; Roychowdhury and Pal, 2012; Kashi et al., 2009). The results of our study showed that maximum stresses (von Mises) and strains were seen in the screw hole areas that were closest to the condyle region of the TMJ implant. Other simpler techniques that can be utilized to study contact stresses in joints include the use of pressure-sensitive film (Buechel et al., 1991).

15.9 Multivariate analysis

In addition to the traditional statistical methods employed in engineering and biology such as univariate analysis (where direct correlations of single variables can be plotted on graphs), another important tool that can be used to study the failure modes of medical implants and devices is multivariate statistical methods (Olkin and Sampson, 2004; Hirsch et al., 2011; Katz, 2003). This type of analytical tool is critical to employ because of the multicomponent nature of variables that contribute toward implant failures (Fig. 15.9). For example, if we want to study the strength (i.e., a single variable) of an orthopedic implant when it is subjected to a certain load, the resulting data can be plotted on a simple two-dimensional x—y graph. However, if other multiple variables are introduced including temperature of the environment, viscosity of the biological fluids present, gender/age of the subject, duration of use, and physiologic or pathologic changes in situ, plotting such data in a well-defined manner will not be possible.

When medical devices and implants are tested, they are typically analyzed for their performance in single time points while at best employing lifetime testing (e.g., several millions of cycles of use), whereas the prosthesis is generally meant to function for extended time periods in a constantly changing/dynamic physiologic milieu. During this intended time period, the implant will be subjected to variations in the physiologic milieu not to mention the influence of new comorbid factors that may have not been present when it was placed. Factoring such variables over time periods is very useful

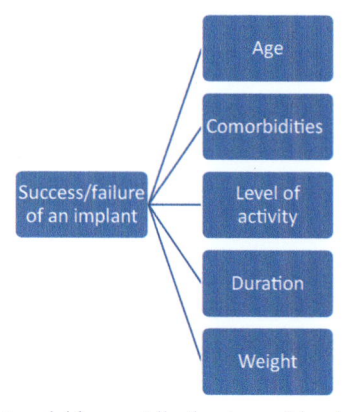

Figure 15.9 Multicomponent variables contributing toward implant failure.

for statistical analysis of such prostheses. However, this can also make statistical analysis complicated. To analyze such complex data, multivariate statistics are key to decipher trends, patterns, and relationships among many variables. Traditional statistical methods will not be able to arrive at meaningful answers in these situations. However, with the introduction of powerful computational devices such as computers, analyzing such complex data has become possible. This will allow researchers to look for trends and patterns among a vast array of variables (or "noise"). Consequently, this can act as a powerful tool for experimental designs in the future and to also develop custom implant solutions for our patients.

15.10 Ethical issues

A detailed overview of the ethical issues related to medical implant failures is beyond the scope of this chapter. However, some of the existing and potentially emerging ethical challenges in this field need a brief discussion. One of these includes the ability of newer generation implants to enhance or augment the physiologic potential of bodily organs. For instance, if an implant or prosthetic device is capable of allowing a person to perform beyond their physiologic limits (e.g., walking, running, or seeing), whether they should be given the right to choose such clinical options needs further debate. Much of the newer age implants and devices are developed by Western nations. However, should there be an obligation for researchers to consider race/gender/ethnic population groups from around the world when they develop such products in the future? There is a dearth of data to demonstrate the short/long-term performance of medical devices and implants specifically in individual population groups. For example, in a preliminary study of osteoporosis, it was found that the strength of skeletal tissues was higher in African-American females when compared with Caucasian females. Not only was the mechanical strength of whole bones higher in the African-American group but the bone cross-sectional area was also found to be larger (Private communication, S. Saha, Feb 22, 2019). In this scenario, if an implant designed for a specific population is used in a different cohort, whether the same failure mechanism criteria can be applied needs further discussion. Whether considerations of racial disparities (i.e., in terms of biological constructs) need to be considered for future versions of customized implants will generate ethical issues that need further discussion.

Another area of debate includes the equitable distribution of implants to areas of the world where the affordability for such technologies may not exist. Whether wealthier nations are obligated to keep the larger worldwide population in mind when they design medical devices and implants needs further discussion. Furthermore, whether retrieved implants from first world countries can be reused in third world countries (who may not have the affordability) provided they are repackaged and verified for their optimal performance needs further debate (Dunn, 2006; Normandin et al., 2008; Amarante et al., 2008). In this situation, if devices/implants fail the responsible parties that can be held accountable is not currently clear. It is important to note that researchers and clinicians strive to follow the Institutional Review Board and Institutional Animal Care and Use Committee guidelines while conducting research and development in the area of medical implants and devices.

Customized implants which are going to become a reality in the near future might pose additional challenges in terms of how their failure mechanisms are evaluated. One of the goals of customization is to benefit patients in terms of either replacing their anatomical defect accurately or with the hope that customized prostheses will perform better than standard/stock prostheses. In such scenarios, we may be lacking in adopting standard procedures to study their failure mechanisms, and newer guidelines might need to be developed in the future.

15.11 Conclusion

Authors have previously reported that biomaterial associated infections in orthopedics range from 1% to 3% (Van de Belt et al., 2001; Jacobs et al., 1998) (Kaufman et al., 2016; Trampuz and Widmer, 2006; Kurtz et al., 2012). Peri-prosthetic infections have been reported to be as high as approximately 10% in the past in the field of orthopedics. However, with the use of modern surgical techniques (e.g., use of laminar air-flow operating rooms), these numbers have been minimized and are currently reported to be as low as 1% (Bauer et al., 2006). Although the majority of elective orthopedic procedures include replacement of hips and knees, it must be noted that other structures, including shoulder, ankle, and craniofacial reconstruction also fall under the category of orthopedic procedures. Artificial ligament and tendon replacements of the knee are also fraught with poor survival outcomes due to the complex biomechanical behavior of the knee. Tissue engineering efforts that might be able to focus on developing newer tendon and ligament replacements will be essential in the future. Developmental disorders (e.g., cleidocranial dysplasia and other congenital head and neck deformities), trauma, and arthritic conditions affecting the maxillofacial region, including the TMJ, warrant the need for developing better versions of implants to rehabilitate these structures. Because the use of TMJ implants is not as common as other implants (such as the hip and knee), their failure analysis has not been studied extensively. Important observations from other total joint replacements, including artificial hips and knees, can help researchers to better understand the behavior of newly developed biomedical implants.

Efforts have been made to improve the survival outcomes of implants by minimizing postoperative infections. For instance, antibiotic-loaded bone cement is used by some surgeons during orthopedic implant placement to minimize infections, although there is no consensus among clinicians to standardize its use (Van de Belt et al., 2001). Furthermore, there is a dearth of evidence-based studies to show if the effect of low dose antibiotics (from bone cements) leaching into the systemic milieu is detrimental, for instance, by promoting long-term antibiotic resistance. Studies have shown that the survival rate of dental implants in Type 2 diabetic patients is improved if chlorhexidine mouthrinse, an antimicrobial, is used at their time of placement (Morris et al., 2005). It might be interesting to study whether similar improvements in the survival rates of other orthopedic implants (e.g., artificial hips and knees) among Type 2 diabetics can be achieved with the use of chlorhexidine during their placement.

The medical device industry has sales of at least US $ 100 billion worldwide and is highly research intensive (Japsen, 2016). For example, over a period of 12 years (from 1990 to 2002), the percentage of revenues reinvested in research and development in the medical device industry almost doubled. Current projections estimate that these medical device sales will exceed US $500 billion by the year 2022 (Japsen, 2016). Consequently, along with this, investments during the past decades have led to encouraging results in terms of reduced incidence of mortality due to heart attacks, strokes, diabetes, and breast cancer (Panescu, 2006). Implant survival can be improved from the combined efforts of clinical and laboratory (e.g., materials development) research. For instance, scientists are working to develop novel ceramic composite materials (e.g., zirconia-toughened alumina) for artificial hips and knees, to counter phase transformation and aging phenomena in zirconia bioceramics (Chevalier, 2006). Researchers have attempted to characterize the microstructure of teeth to elucidate their remarkable resistance to fatigue and their ability to self-heal from microscopic fractures (Chai et al., 2009). Understanding this behavior might potentially be able to help scientists develop new-generation biomaterials with these characteristics.

It is projected that the number of people in the United States who are Medicare beneficiaries will increase to 75 million by the year 2030. This number was approximately 40 million in the year 2000 (Panescu, 2006). Therefore, an aging population will necessitate the need for better implant designs in the future. Studying the failure mechanisms of implants will help researchers design better versions of existing prostheses. The United States has the biggest market share for medical devices, with cardiovascular devices (e.g., stents, defibrillators) representing the major sector and orthopedic devices (e.g., artificial joint replacements and limb replacement devices) being second in sales (Panescu, 2006). These two segments of the medical device market represent a significant portion of implants being used currently. However, research and development of other prostheses, including breast implants, artificial skin, and small joint replacements, need to be carried out for our future patients.

Evidence-based studies in different population groups need to be encouraged. Metaanalyses, systematic reviews of randomized controlled trials, and cohort studies need to be conducted to better delineate implant survival among different types, as well as their treatment outcomes (Lee et al., 2000). Observational, prospective, and retrospective studies of implants and biomaterials can provide useful information, including some factors that might be responsible for their failure (Cook et al., 1999; Buser et al., 1997; Roumanas et al., 2000; Dobbs, 2018; Roos et al., 1997; Brocard et al., 2000; Jones et al., 2012; Liang et al., 2006; Hatanaka et al., 2006). Double-blind controlled studies are essential to provide medical and dental practitioners with Level I evidence as required by Evidenced-Based Medicine, which is presently considered the gold standard in determining the practice guidelines. A major benefit of such endeavors will be to use the results in large population segments worldwide. Consequently, reductions in the number of implant failures and savings in health care costs can be expected.

References

Agins, H.J., Alcock, N.W., Bansal, M., Salvati, E.A., Wilson, P.D., Pellicci, P.M., Bullough, P., 1988. Metallic wear in failed titanium-alloy total hip replacements — a histological and quantitative analysis. J. Bone Joint Surg. — A 70A, 347−356.

Al Jabbari, Y.S., Fournelle, R., Ziebert, G., Toth, J., Iacopino, A.M., 2008. Mechanical behavior and failure analysis of prosthetic retaining screws after long-term use in vivo. Part 1: characterization of adhesive wear and structure of retaining screws. J. Prosthodont. (3).

Amarante, J.M.B., Toscano, C.M., Levin, A.S., Jarvis, W.R., Pearson, M,L., Roth, V., 2008. Reprocessing and reuse of single-use medical devices used during hemodynamic procedures in Brazil: a widespread and largely overlooked problem. Infect. Control Hosp. Epidemiol. (9).

Anitua, E.A., 2006. Enhancement of osseointegration by generating a dynamic implant surface. J. Oral Implantol. (2).

Bauer, T.W., Parvizi, J., Kobayashi, N., Krebs, V., 2006. Diagnosis of periprosthetic infection. J. Bone Joint Surg. −A (4).

Beikler, T., Flemmig, T.F., 2003. Implants in the medically compromised patient. Crit. Rev. Oral Biol. Med. (4).

Beksac, B., Taveras, N.A., Valle, A.G.D., Salvati, E.A., 2006. Surface finish mechanics explain different clinical survivorship of cemented femoral stems for total hip arthroplasty. J. Long Term Eff. Med. Implant. (6).

Blandford, A., Furniss, D., Vincent, C., 2014. Patient safety and interactive medical devices: realigning work as imagined and work as done. Clin. Risk (5).

Brandon, H.J., Taylor, M.L., Powell, T.E., Walker, P.S., 2006. Morphology of breast implant fold flaw failure. J. Long Term Eff. Med. Implants (6).

Brandon, H.J., Taylor, M.L., Powell, T.E., Walker, P.S., 2012. Morphology of breast implant fold flaw failure. J. Long Term Eff. Med. Implant.

Brocard, D., Barthet, D.S.O.P., Baysse, D.S.O.E., Duffort, J.F., Eller, D.S.O.P., Justumus, D.C.D.P., et al., 2000. A multicenter report on 1022 consecutively placed ITI implants: a 7-year longitudinal study. Int. J. Oral Maxillofac. Implant. (5).

Brown, S.A., Flemming, C.A.C., Kawalec, J.S., Placko, H.E., Vassaux, C., Merritt, K., et al., 1995. Fretting corrosion accelerates crevice corrosion of modular hip tapers. J. Appl. Biomater. (1).

Buechel, F.F., Pappas, M.J., Makris, G., 1991. Evaluation of contact stress in metal-backed patellar replacements — a predictor of survivorship. Clin. Orthop. Relat. Res.

Buser, D., Mericske-Stern, R., Bernard, J., Behneke, A., Behneke, N., Hirt, H., et al., 1997. Long term evaluation of non submerged ITI implants. Clin. Oral Implant. Res. (3).

Case, C.P., Langkamer, V.G., James, C., Palmer, M.R., Kemp, A.J., Heap, P.F., et al., 1994. Widespread dissemination of metal debris from implants. J. Bone Joint Surg. (5).

CDC, February 22, 2019. Prevalence of Both Diagnosed and Undiagnosed Diabetes. Internet. Available from: https://www.cdc.gov/diabetes/data/statistics-report/diagnosed-undiagnosed.html.

Çehreli, M., Şahin, S., Akça, K., 2004. Role of mechanical environment and implant design on bone tissue differentiation: current knowledge and future contexts. J. Dent. (2).

Chai, H., Lee, J.J.-W., Constantino, P.J., Lawn, B.R., Lucas, P.W., 2009. Remarkable resilience of teeth. Proc. Natl. Acad. Sci. (18).

Chevalier, J., 2006. What future for zirconia as a biomaterial. Biomaterials (4).

Cook, S.D., Beckenbaugh, R.D., Redondo, J., Popich, L.S., Klawitter, J.J., Linscheid, R.L., 1999. Long-term follow-up of pyrolytic carbon metacarpophalangeal implants. J. Bone Joint Surg. Am. (5).

Dobbs, H., 2018. Survivorship of total hip replacements. J. Bone Joint Surg. Br. (7).

Dunn, D., 2006. Reprocessing single-use devices-the ethical dilemma. AORN J. (5).

Ehrlich, G.D., Stoodley, P., Kathju, S., Zhao, Y., McLeod, B.R., Balaban, N., et al., 2005. Engineering approaches for the detection and control of orthopaedic biofilm infections. Clin. Orthop. Relat. Res. (437).

Eppley, B.L., 2005. Biomechanical testing of alloplastic PMMA cranioplasty materials. J. Craniofac. Surg. (1).

Eppley, B.L., Morales, L., Wood, R., Pensler, J., Goldstein, J., Havlik, R.J., et al., 2004. Resorbable PLLA-PGA plate and screw fixation in pediatric craniofacial surgery: clinical experience in 1883 patients. Plast. Reconstr. Surg. (4).

Esposito, M., Hirsch, J.M., Lekholm, U., Thomsen, P., 1998. Biological factors contributing to failures of osseointegrated oral implants. (I). Success criteria and epidemiology. Eur. J. Oral Sci. (1).

Esposito, M., Hirsch, J.M., Lekholm, U., Thomsen, P., 1998. Biological factors contributing to failures of osseointegrated oral implants: (II). Etiopathogenesis. Eur. J. Oral Sci. (3).

Falcinelli, G., Falsini, B., Taloni, M., Colliardo, P., Falcinelli, G., 2005. Modified osteo-odonto-keratoprosthesis for treatment of corneal blindness: long-term anatomical and functional outcomes in 181 cases. Arch. Ophthalmol. (10).

Fehring, T.K., Murphy, J.A., Hayes, T.D., Roberts, D.W., Pomeroy, D.L., Griffin, W.L., 2004. Factors influencing wear and osteolysis in press-fit condylar modular total knee replacements. Clin. Orthop. Relat. Res. (428).

Geesink, R.G., Groot, K de, Klein, C.P., 1988. Bonding of bone to apatite-coated implants. J. Bone Joint Surg. Br (1).

Gerritsen, M., Lutterman, J.A., Jansen, J.A., 2000. Wound healing around bone-anchored percutaneous devices in experimental diabetes mellitus. J. Biomed. Mater. Res. (6).

Goldhofer, M.I., Munir, S., Levy, Y.D., Walter, W.K., Zicat, B., Walter, W.L., 2018. Increase in benign squeaking rate at five-year follow-up: results of a large diameter ceramic-on-ceramic bearing in total hip arthroplasty. J. Arthroplast. (4).

Goodman, TW. and S., 2001. What surgical related factors contribute to implant wear? In: Goodman, TW. and S. (Ed.), Implant Wear in Total Joint Replacement. American Academy of Orthopedic Surgeons, Rosemont.

Griffin, W.L., Fehring, T.K., Pomeroy, D.L., Gruen, T.A., Murphy, J., 2007. Sterilization and wear related failure in first and second generation press fit condylar total knee arthroplasty. In: Open Scientific Meeting of the Knee-Society. San Diego, CA (464).

Grosse, S., Haugland, H.K., Lilleng, P., Ellison, P., Hallan, G., Høl, P.J., 2015. Wear particles and ions from cemented and uncemented titanium-based hip prostheses — a histological and chemical analysis of retrieval material. J. Biomed. Mater. Res. B Appl. Biomater. (3).

Gustafson, A., Clark, I.C., Saha, S., 1993. Catastrophic peri-implant bone loss caused by polyethylene and metallic wear in total knees. J. Long Term Eff. Med. Implant. (10).

Hallab, N., Merritt, K., Jacobs, J.J., 2001. Metal sensitivity in patients with orthopaedic implants. J. Bone Joint Surg. —A (3).

Hallab, N.J., Messina, C., Skipor, A., Jacobs, J.J., 2004. Differences in the fretting corrosion of metal-metal and ceramic-metal modular junctions of total hip replacements. J. Orthop. Res. (2).

Hallam, P., Haddad, F., Cobb, J., 2018. Pain in the well-fixed, aseptic titanium hip replacement. J. Bone Joint Surg. Br. Hallam.

Harding, I., Bonomo, A., Crawford, R., Psychoyios, V., Delves, T., Murray, D., et al., 2002. Serum levels of cobalt and chromium in a complex modular total hip arthroplasty system. J. Arthroplast. (7).

T Hatanaka, M., Honda, R., Ozawa, Y., Okazaki, Y., Sawaki, M.U., 2006. Long-term results of osseointegrated implant-retained facial prostheses: a 5-year retrospective study. J Jap. Soc. Oral. Implantol. 19 (3-4), 305−311.

Hermann, I., Bhowmick, S., Zhang, Y., Lawn, B.R., 2006. Competing fracture modes in brittle materials subject to concetrated cyclic loading in liquid environments: trilayer structures. J. Mater. Res. (2).

Higuchi, Y., Hasegawa, Y., Seki, T., Komatsu, D., Ishiguro, N., 2016. Significantly lower wear of ceramic-on-ceramic bearings than metal-on-highly cross-linked polyethylene bearings: a 10- to 14-year follow-up study. J. Arthroplast. (6).

Hirakawa, K., Bauer, T.W., Culver, J.E., Wilde, A.H., 1996. Isolation and quantitation of debris particles around failed silicone orthopedic implants. J. Hand Surg. Am. (5).

Hirsch, O., Bösner, S., Hüllermeier, E., Senge, R., Dembczynski, K., Donner-Banzhoff, N., 2011. Multivariate modeling to identify patterns in clinical data: the example of chest pain. BMC Med. Res. Methodol. (155).

Howie, D.W., Vernon-Roberts, B., Oakeshott, R., Manthey, B., 1988. A rat model of resorption of bone at the cement-bone interface in the presence of polyethylene wear particles. J. Bone Joint Surg. −A. (2).

Ingham, E., Fisher, J., 2000. Biological reactions to wear debris in total joint replacement. Proc. Inst. Mech. Eng. H J. Eng. Med. (1).

Isidor, F., 1997. Histological evaluation of peri-implant bone at implants subjected to occlusal overload or plaque accumulation. Clin. Oral Implant. Res. (1).

Isidor, F., 2003. Loss of osseointegration caused by occlusal load of oral implants. A clinical and radiographic study in monkeys. Clin. Oral Implant. Res. (2).

Jacobs, J., Gilbert, J., Urban, R., 1998. Current concepts review. Corrosion of metal orthopaedic implants. J. Bone Joint Surg. Am. (2).

Japsen, B., 2016. Medical Technology Sales to Hit $ 100 B Within Five Years. Internet [cited 2019 Feb 10]. Available from: https://www.forbes.com/sites/brucejapsen/2016/10/17/medical-technology-sales-to-hit-500b-within-five-years/#1d513b9a11be.

Jones, A.P., Sidhom, S., Sefton, G., 2012. Long-term clinical review (10−20 Years) after reconstruction of the anterior cruciate ligament using the leeds-Keio synthetic ligament. J. Long Term Eff. Med. Implant. (1).

Kashi, A., Roy, A., Subrata, S., 2009. ASME 2009 4th frontiers in biomedical devices conference and exposition. In: Finite Element Analysis of TMJ Implant. Irvine.

Kashi, A., Chowdhury, A.R., Saha, S., 2010. Finite element analysis of a TMJ implant. J. Dent. Res. (3).

Katz, M.H., 2003. Multivariable analysis: a primer for readers of medical research. Ann. Intern. Med. (8).

Kaufman, M.G., Meaike, J.D., Izaddoost, S.A., 2016. Orthopedic prosthetic infections: diagnosis and orthopedic salvage. Semin. Plast. Surg. (2).

Kearns, G.J., Perrott, D.H., Kaban, L.B., 1995. A protocol for the management of failed alloplastic temporomandibular joint disc implants. J. Oral Maxillofac. Surg. (11).

Khan, M.A., Williams, R.L., Williams, D.F., 1996. In-vitro corrosion and wear of titanium alloys in the biological environment. Biomaterials (22).

Kido, H., Schulz, E.E., Kumar, A., Lozada, J., Saha, S., 1997. Implant diameter and bone density: effect on initial stability and pull-out resistance. J. Oral Implantol. (4).

Kim, J.W., Kim, J.H., Thompson, V.P., Zhang, Y., 2007. Sliding contact fatigue damage in layered ceramic structures. J. Dent. Res. (11).

Korovessis, P., Petsinis, G., Repanti, M., Repantis, T., 2006. Metallosis after contemporary metal-on-metal total hip arthroplasty: five to nine-year follow-up. J. Bone Joint Surg. −A (6).

Kovochich, M., Fung, E.S., Donovan, E., Unice, K.M., Paustenbach, D.J., Finley, B.L., 2018. Characterization of wear debris from metal-on-metal hip implants during normal wear versus edge-loading conditions. J. Biomed. Mater. Res. B Appl. Biomater. (3).

Kurtz, S.M., Lau, E., Watson, H., Schmier, J.K., Parvizi, J., 2012. Economic burden of periprosthetic joint infection in the United States. J. Arthroplast. (61-5).

Lee, J.J., Rouhfar, L., Beirne, O.R., 2000. Survival of hydroxyapatite-coated implants: a meta-analytic review. J. Oral Maxillofac. Surg. (12).

Lehtovirta, L., Reito, A., Parkkinen, J., Hothi, H., Henckel, J., Hart, A., et al., 2017. Analysis of bearing wear, whole blood and synovial fluid metal ion concentrations and histopathological findings in patients with failed ASR hip resurfacings. BMC Muscoskelet. Disord. (1).

Liang, B., Fujibayashi, S., Fujita, H., Ise, K., Neo, M., Nakamura, T., 2006. Long-term follow-up study of bioactive bone cement in canine total hip arthroplasty. J. Long Term Eff. Med. Implant. (4).

Lykissa, E.D., Maharaj, S.V.M., 2012. Platinum concentration and platinum oxidation states in body fluids, tissue, and explants from women exposed to silicone and saline breast implants. J. Long Term Eff. Med. Implant. (9).

MacDonald, D.W., Chen, A.F., Lee, G.C., Klein, G.R., Mont, M.A., Kurtz, S.M., et al., 2017. Fretting and corrosion damage in taper adapter sleeves for ceramic heads: a retrieval study. J. Arthroplast. (9).

Mai, K., Verioti, C., Ezzet, K.A., Copp, S.N., Walker, R.H., Colwell, C.W., 2010. Incidence of "Squeaking" after ceramic-on-ceramic total hip arthroplasty. Clin. Orthop. Relat. Res. (2).

Marshall, A., Ries, M.D., Paprosky, W., 2008. How prevalent are implant wear and osteolysis, and how has the scope of osteolysis changed since 2000? J. Am. Acad. Orthop. Surg. (1-6).

McCloy, J., February 17, 2019. Failure Analysis 101. Internet. Available from: https://www.meddeviceonline.com/doc/failure-analysis-0003.

McCracken, M., Lemons, J.E., Rahemtulla, F., Prince, C.W., Feldman, D., 2000. Bone response to titanium alloy implants placed in diabetic rats. Int. J. Oral Maxillofac. Implant. (3).

Mercuri, L.G., Anspach, I.E., 2003. Principles for the revision of total alloplastic TMJ prostheses. Int. J. Oral Maxillofac. Surg. (4).

Mercuri, L.G., Giobbie-Hurder, A., 2004. Long-term outcomes after total alloplastic temporomandibular joint reconstruction following exposure to failed materials. J. Oral Maxillofac. Surg. (9).

Michael, R., Charoenrook, V., Paz, M.F., Hitzl, W., Temprano, J., Barraquer, R.I., 2008. Long-term functional and anatomical results of osteo- and osteoodonto-keratoprosthesis. Graefes Arch. Clin. Exp. Ophthalmol. (8).

Misch, C.E., 1999. Implant design considerations for the posterior regions of the mouth. Implant Dent. (4).

Mombelli, A., Cionca, N., 2006. Systemic diseases affecting osseointegration therapy. Clin. Oral Implant. Res. (6).

Morris, H.F., Ochi, S., Winkler, S., 2005. Implant survival in patients with type 2 diabetes: placement to 36 months. Ann. Periodontol. (1).

Nawabi, D.H., Do, H.T., Ruel, A., Lurie, B., Elpers, M.E., Wright, T., et al., 2016. Comprehensive analysis of a recalled modular total hip system and recommendations for management. J. Bone Joint Surg. − Am. (1).

Nassiri, N., Cleary, D.R., Ueeck, B.A., 2009. Is cranial reconstruction with a hard-tissue replacement patient-matched implant as safe as previously reported? A 3-year experience and review of the literature. J. Oral Maxillofac. Surg. (2).

Neut, D., Van Horn, J.R., Van Kooten, T.G., Van Der Mei, H.C., Busscher, H.J., 2003. Detection of biomaterial-associated infections in orthopaedic joint implants. Clin. Orthop. Relat. Res.

Normandin, S., Polisena, J., Moulton, K., Hailey, D., Gardam, M., Jacobs, P., et al., 2008. Reprocessing and reuse of single-use medical devices: a national survey of Canadian acute-care hospitals. Infect. Control Hosp. Epidemiol. (5).

Olkin, I., Sampson, A.R., 2004. Multivariate analysis: overview. In: International Encyclopedia of the Social & Behavioral Sciences.

Panescu, D., 2006. Medical device industry. In: Wiley Encyclopedia of Biomedical Engineering. John Wiley & Sons, New York, pp. 1−10.

Pilliar, R.M., Lee, J.M., Maniatopoulos, C., 2006. Observations on the effect of movement on bone ingrowth into porous-surfaced implants. Clin. Orthop. Relat. Res.

Posada, O.M., Gilmour, D., Tate, R.J., Grant, M.H., 2014. CoCr wear particles generated from CoCr alloy metal-on-metal hip replacements, and cobalt ions stimulate apoptosis and expression of general toxicology-related genes in monocyte-like U937 cells. Toxicol. Appl. Pharmacol. (1).

Posada, O.M., Tate, R.J., Grant, M.H., 2015. Effects of CoCr metal wear debris generated from metal-on-metal hip implants and Co ions on human monocyte-like U937 cells. Toxicol. In Vitro. (2).

Reclaru, L., Meyer, J.M., 1994. Study of corrosion between a titanium implant and dental alloys. J. Dent. (3).

Restrepo, C., Parvizi, J., Kurtz, S.M., Sharkey, P.F., Hozack, W.J., Rothman, R.H., 2008. The noisy ceramic hip: is component malpositioning the cause? J. Arthroplast. (5).

Revell, P.A., 2008. The combined role of wear particles, macrophages and lymphocytes in the loosening of total joint prostheses. J. R. Soc. Interface (28).

Revell, P.A., Al-Saffar, N., Kobayashi, A., 1997. Biological reaction to debris in relation to joint prostheses. Proc. Inst. Mech. Eng. H J. Eng. Med. (2).

Roos, J., Sennerby, L., Lekholm, U., Jemt, T., Grondahl, K., Albrektsson, T., 1997. A qualitative and quantitative method for evaluating implant success: a 5-year retrospective analysis of the Branemark implant. Int. J. Oral Maxillofac. Implant. (4).

Rose, R.M., Schiller, A.L., Radin, E.L., 1972. Corrosion-accelerated mechanical failure of a vitallium nail-plate. J. Bone Joint Surg. Am. (4).

Roumanas, E.D., Freymiller, E.G., Chang, T.L., Aghaloo, T., Beumer, J., 2000. Joint symposium of the American academy of maxillofacial prosthetics/internation congress of maxillofacial prosthetics. In: Implant-retained Prostheses for Facial Defects: An up to 14-year Follow-Up Report on the Survival Rates of Implants at UCLA. Quintessence Publ Co Inc, Kauai, Hawaii.

Roychowdhury, A., Pal, S., 2012. Stress analysis of an artificial temporal mandibular joint. Crit. Rev. Biomed. Eng. (3-4).

Stulberg, B.N., Watson, J.T., Stulberg, S.D., Bauer, T.W., Manley, M.T., 1991. A new model to assess tibial fixation: II. Concurrent histologic and biomechanical observations. Clin. Orthop. Relat. Res.

Santana, T., Zhang, Y., Guess, P., Thompson, V.P., Rekow, E.D., Silva, N.R.F.A., 2009. Off-axis sliding contact reliability and failure modes of veneered alumina and zirconia. Dent. Mater. (7).

Schmalzried, T.P., Peters, P.C., Maurer, B.T., Bragdon, C.R., Harris, W.H., 1996. Long-duration metal-on-metal total hip arthroplasties with low wear of the articulating surfaces. J. Arthroplast. (3).

Schroder, D., Bornstein, L., Bostrom, M.P.G., Nestor, B.J., Padgett, D.E., Westrich, G.H., 2011. Ceramic-on-ceramic total hip arthroplasty: incidence of instability and noise. Clin. Orthop. Relat. Res. (2).

Shalabi, M.M., Gortemaker, A., Van't Hof, M.A., Jansen, J.A., Creugers, N.H.J., 2006. Implant surface roughness and bone healing: a systematic review. J. Dent. Res. (6).

Sieber, H.-P., Rieker, C.B., Köttig, P., 1999. Analysis of 118 second-generation metal-on-metal retrieved hip implants. J Bone Jt Surg (1).

Speculand, B., Hensher, R., Powell, D., 2000. Total prosthetic replacement of the TMJ: experience with two systems 1988−1997. Br. J. Oral Maxillofac. Surg. (4).

Sumner, D.R., Kienapfel, H., Jacobs, J.J., Urban, R.M., Turner, T.M., Galante, J.O., 1995. Bone ingrowth and wear debris in well-fixed cementless porous-coated tibial components removed from patients. J. Arthroplast. (2).

Sutow, E.J., Jones, D.W., Milne, E.L., 1985. In vitro crevice corrosion behavior of implant materials. J. Dent. Res. (5).

Sykaras, N., Iacopino, A.M., Marker, V.A., Triplett, R.G., Woody, R.D., 2000. Implant materials, designs, and surface topographies. Int. J. Oral Maxillofac. Implant. (5).

Ta, L.E., Phero, J.C., Pillemer, S.R., Hale-Donze, H., McCartney-Francis, N., Kingman, A., et al., 2002. Clinical evaluation of patients with temporomandibular joint implants. J. Oral Maxillofac. Surg. (12).

Tai, S.M., Munir, S., Walter, W.L., Pearce, S.J., Walter, W.K., Zicat, B.A., 2015. Squeaking in large diameter ceramic-on-ceramic bearings in total hip arthroplasty. J. Arthroplast. (2).

Takeshita, F., Murai, K., Iyama, S., Ayukawa, Y., Suetsugu, T., 1998. Uncontrolled diabetes hinders bone formation around titanium implants in rat tibiae. A light and fluorescence microscopy, and image processing study. J. Periodontol. (3).

Tan, X.W., Perera, A.P.P., Tan, A., Tan, D., Khor, K.A., Beuerman, R.W., et al., 2011. Comparison of candidate materials for a synthetic osteo-odonto keratoprosthesis device. Investig. Ophthalmol. Vis. Sci. (21-29).

Tan, X.W., Beuerman, R.W., Shi, Z.L., Neoh, K.G., Tan, D., Khor, K.A., et al., 2012. In vivo evaluation of titanium oxide and hydroxyapatite as an artificial cornea skirt. J. Mater. Sci. Mater. Med. (4).

Taylor, S., Manley, M.T., Sutton, K., 2007. The role of stripe wear in causing acoustic emissions from alumina ceramic-on-ceramic bearings. J. Arthroplast. (7).

Trampuz, A., Widmer, A.F., 2006. Infections associated with orthopedic implants. Curr. Opin. Infect. Dis. (4).

Triplett, R.G., Frohberg, U., Sykaras, N., Woody, R.D., 2004. Implant materials, design, and surface topographies: their influence on osseointegration of dental implants. J. Long Term Eff. Med. Implant. (6).

Upadhyay, D., Panchal, M.A., Dubey, R.S., Srivastava, V.K., 2006. Corrosion of alloys used in dentistry: a review. Mater. Sci. Eng. A (1-2).

Urban, R.M., Jacobs, J.J., Gilbert, J.L., Galante, J.O., 1994. Migration of corrosion products from modular hip prostheses. Particle microanalysis and histopathological findings. J. Bone Joint Surg. −A. (9).

Vaderhobli, R., Saha, S., 2012. Profilometric surface roughness analysis of christensen metal temporomandibular joint prostheses. J. Long Term Eff. Med. Implant. (4).

Van de Belt, H., Neut, D., Schenk, W., Van Horn, J.R., Van der Mei, H.C., Busscher, H.J., 2001. Infection of orthopedic implants and the use of antibiotic-loaded bone cements. Acta Orthop. Scand. (6).

Vinh, D.C., Embil, J.M., 2005. Device-related infections: a review. J. Long Term Eff. Med. Implant. (5).

Walsh, W. (Ed.), 2005. Repair and Regeneration of Ligaments, Tendons and Joint Capsule. Humana Press, Totowa.

Walter, W.L., Lusty, P.J., Watson, A., O'Toole, G., Tuke, M.A., Zicat, B., et al., 2006. Stripe wear and squeaking in ceramic total hip bearings. Semin. Arthroplast. (3-4).

Walter, W.L., O'Toole, G.C., Walter, W.K., Ellis, A., Zicat, B.A., 2007. Squeaking in ceramic-on-ceramic hips. The importance of acetabular component orientation. J. Arthroplast. (4).

Wang, Y., Yan, Y., Su, Y., Qiao, L., 2017. Release of metal ions from nano CoCrMo wear debris generated from tribo-corrosion processes in artificial hip implants. J. Mech. Behav. Biomed. Mater. (124-133).

Weisshuhn, K., Berg, I., Tinner, D., Kunz, C., Bornstein, M.M., Steineck, M., et al., 2014. Osteo-odonto-keratoprosthesis (OOKP) and the testing of three different adhesives for bonding bovine teeth with optical poly-(methyl methacrylate) (PMMA) cylinder. Br. J. Ophthalmol. (7).

Westermark, A., Koppel, D., Leiggener, C., 2006. Condylar replacement alone is not sufficient for prosthetic reconstruction of the temporomandibular joint. Int. J. Oral Maxillofac. Surg. (6).

Widmer, A.F., 2002. New developments in diagnosis and treatment of infection in orthopedic implants. Clin. Infect. Dis. (3).

Williams, D.F., Williams, R.L., 2013. Degradative effects of the biological environment on metals and ceramics. In: Biomaterials Science: An Introduction to Materials, third ed.

Yamaguchi, M., Bauer, T.W., Hashimoto, Y., 1997. Three-dimensional analysis of multiple wear vectors in retrieved acetabular cups. J. Bone Joint Surg. Am. (10).

Zarei-Ghanavati, M., Avadhanam, V., Vasquez Perez, A., Liu, C., 2017. The osteo-odonto-keratoprosthesis. Curr. Opin. Ophthalmol. (4).

Zhang, Y., Lawn, B.R., 2009. Competing damage modes in all-ceramic crowns: fatigue and lifetime. Key Eng. Mater.

Zhang, Y., Pajares, A., Lawn, B.R., 2004. Fatigue and damage tolerance of Y-TZP ceramics in layered biomechanical systems. J. Biomed. Mater. Res. B Appl. Biomater. (1).

Zhang, D., Lu, C., Zhang, X., Mao, S., Arola, D., 2008. Contact fracture of full-ceramic crowns subjected to occlusal loads. J. Biomech. (14).

Biointegration of three-dimensional—printed biomaterials and biomedical devices

16

Vamsi Krishna Balla[1,2], Subhadip Bodhak[1], Pradyot Datta[1], Biswanath Kundu[1], Mitun Das[1], Amit Bandyopadhyay[3], Susmita Bose[3]
[1]Bioceramics & Coating Division, CSIR-Central Glass & Ceramic Research Institute, Kolkata, West Bengal, India; [2]Department of Mechanical Engineering, University of Louisville, Louisville, KY, United States; [3]School of Mechanical and Materials Engineering, Washington State University, Pullman, WA, United States

Chapter outline

16.1 Introduction 434
16.2 Metallic implants via three-dimensional printing 438
    16.2.1 Stainless steel (316L) implants 439
    16.2.2 Implants with Ti and its alloys 441
    16.2.3 Porous/cellular implants for orthopedic applications 443
    16.2.4 Additive manufacturing of biodegradable metals 445
16.3 Bioceramic scaffolds using three-dimensional printing 446
    16.3.1 Calcium phosphate ceramics 446
        16.3.1.1 Hydroxyapatite 447
        16.3.1.2 Tricalcium phosphate 448
        16.3.1.3 Other calcium phosphates 449
    16.3.2 Bioactive glass 450
    16.3.3 Ceramic composites 452
    16.3.4 Ceramic—polymer composites 454
16.4 Bioprinting 455
    16.4.1 Bioprinting strategies and classifications 458
        16.4.1.1 Inkjet-based three-dimensional bioprinting 458
        16.4.1.2 Microextrusion-based three-dimensional bioprinting 461
        16.4.1.3 Laser-assisted three-dimensional bioprinting 461
    16.4.2 Bioinks for bioprinting 462
        16.4.2.1 Natural polymers 462
        16.4.2.2 Synthetic polymers 463
    16.4.3 Niche application areas of bioprinting technology 463
        16.4.3.1 Bone tissue 464

        16.4.3.2   Cartilage   464
        16.4.3.3   Skin   466
16.5   **Current challenges and future directions**   467
16.6   **Summary**   467
References   468

16.1   Introduction

Three-dimensional printing (3DP), or additive manufacturing (AM) or solid freeform fabrication (SFF), is a computer-aided manufacturing process where net-shape and complex 3D parts are made by adding materials layer-by-layer (Bandyopadhyay et al., 2011). In this process, commercial 3D computer-aided design (CAD) software is used to design the part to be fabricated, and the 3D CAD model is later sliced into horizontal cross-sections electronically. Depending on the type of 3DP process, each cross-section is built one over the other in 3DP machine. Once all layers are completed, the net-shape functional 3D components are ready for further processing or use. Processing of biomaterials is as important as choosing biomaterials for various biomedical applications. For example, the natural bone is a complex composite consisting of different materials and architecture with functional gradient in composition, macro, and microstructures. The natural gradation of the bone results in large variation in the elastic modulus (20 GPa for cortical bone outside and 0.5 GPa for central cancellous bone, which is highly porous). However, none of the existing bone replacement materials/implants have such characteristics. Therefore, to manufacture artificial implants or tissues-mimicking natural structure and properties, it is important to use state-of-the-art processing technologies such as 3DP.

One important concern with current fully dense metallic implants (hip and knee) is their significantly high elastic modulus, leading to stress shielding followed by implant loosening. The elastic modulus mismatch has been addressed using porous metals (Krishna et al., 2007, 2009; Xue et al., 2007; Bandyopadhyay et al., 2010; Balla et al., 2010a,b) but at the expense of drop in strength of these porous implants. However, implants with designed porosity (size, shape, and placement) found to exhibit desired elastic modulus while significantly improving compressive strength (Balla et al., 2010a,b). Similarly, patient-specific implants having site-specific porosity and properties can be easily manufactured using 3DP (DeVasConCellos et al., 2012). Further, inherently brittle bioceramic and bioactive glass implants have also been fabricated with designed porosity for bone replacement applications (Tarafder et al., 2013a,b). Such design freedom can be easily achieved in functional parts using 3DP (Das and Balla, 2015). 3DP has also demonstrated to have strong potential to fabricate complex patient-specific tissue constructs or whole organ for tissue engineering. In this chapter, we present an overview of different 3DP technologies and materials used in these techniques for printing novel implant structures and scaffolds for tissue replacement and regenerative medicine applications. A brief summary of different 3DP technologies and the materials used to fabricate variety of implants, scaffolds, and tissues is presented in Table 16.1.

Selective laser melting (*SLM*): This is a powder-bed melting process, where high-power laser melts the loose powder and makes a solid part. This process is also known as selective laser sintering (SLS) if the parts are made by sintering of powder bed using laser. The process can be used to fabricate metallic, ceramic, polymeric materials, and

Table 16.1 Summary of three-dimensional printing (3DP) processes and materials used to fabricate implants and scaffolds for biomedical applications.

3DP technique	Materials	Typical applications
Selective laser melting	Ti, Ti6Al4V, 316L stainless steel	Hip implants, acetabular shells, knee implants, custom implants, spinal cages, maxillofacial implants, lattice structures, etc.
Electron beam melting	Ti and Ti6Al4V alloy	
Selective laser sintering	CaPs, bioactive glasses, and their composites	Hard and soft tissue repair, dental and maxillofacial implants
Binder jetting	CaPs, bioactive glasses, biocomposites	Bone defect filling, porous scaffolds with enhanced mechanical and functional performance, custom scaffolds
Fused filament fabrication	CaPs, bioactive glasses and their composites, collagen, poly(ε-caprolactone) (PCL), poly(L-lactic acid) (PLLA), poly(lactide-co-glycolide) (PLGA), polypropylene (PP), Agarose, etc.	Bone defect filling, porous scaffolds with enhanced mechanical and functional performance, custom scaffolds
Material extrusion	Bioceramics and their composites, collagen, PCL, PLLA, PLGA, PP, Agarose, etc.	Bone defect filling, porous scaffolds with enhanced mechanical and functional performance, custom scaffolds
Stereolithography	Biopolymer—ceramic composites, collagen, PCL, PLLA, PLGA, PP, Agarose, etc.	Bone defect filling, porous scaffolds with enhanced mechanical and functional performance, custom scaffolds
3D bioprinting • Microextrusion-based • Inkjet printing—based • Laser-assisted	Wide variety of bioinks comprising mixture of cells, growth factors, therapeutic agents, biomolecules, and biodegradable polymers	3D structures mimicking targeted tissue/organ architecture in structure, dimension, and shape for hard tissue and organ regeneration/building

their composites. In these processes, the laser beam is focused onto a thin layer of loose powder bed (of desired thickness) to melt or sinter the powder, thus forming a solid layer according to the CAD model of the part. The surrounding material is unaffected, remains loose, and forms support for further layers. Then, the build platform is lowered, equivalent to one layer thickness, and fresh layer of loose powder is again spread onto the solidified layer. The process of layer spreading and melting/sintering is repeated to complete the part represented in 3D CAD model. Typical SLS process is schematically shown in Fig. 16.1(a). Often the process is carried out in a protective atmosphere, and at the end, excess power is removed from the part followed by secondary operations, as required. The quality of the parts, in terms of properties and geometrical accuracy, produced by SLS/SLM depends on large number of process parameters such as powder characteristics (size, shape, size distribution, packing density), layer thickness, scan spacing, scan speed, laser power, laser beam size, pulse duration, frequency, etc. Electron beam melting (EBM) is another powder bed–based technology that is very similar to SLM, except that EBM is carried out in high vacuum and used electron beam as energy source. EBM is very popular in processing metallic parts, particularly Ti and its alloys for load-bearing implant applications.

Binder jetting (BJ): This process also starts with spreading a loose powder layer, and a print head prints powder-compatible liquid binder selectively according to CAD model cross section. Depending on the liquid binder used, the printed binder layer is dried using a heater, as shown in Fig. 16.1(b). To obtain high geometrical quality and properties, it is very important to avoid binder spreading deep inside previous layers by proper drying. Once the layer is dried, fresh layer of powder is spread and binder printing process is continued. After completing all layers, the loose powder is removed from the green part followed by thermal treatment to achieve strong part. The BJ is very suitable for processing brittle materials such as bioceramics, bioactive glasses, biopolymers, and their composites. However, build material and binder compatibility are extremely important for biomedical applications. Important BJ parameters that control part properties are binder type, binder amount, binder drop volume, viscosity, printing speed, printing passes, powder characteristics, layer thickness, drying time, temperature, etc.

Fused filament fabrication (FFF): This is an extrusion-based process, popularly known as fused deposition modeling (FDM), and uses feedstock in the form of solid filaments, which are heated in a print head and extruded though a fine nozzle and then deposited in to a desired shape on a build platform, as shown in Fig. 16.1(c). The whole process is typically carried out in a build chamber with appropriate temperature that ensures bonding between layers and deposited roads. Some version of the process uses different material for support structures. The supports can be removed either manually or by melting or dissolving in water. Machine-specific software can be used to create desired tool paths and support structures to minimize the build time and maximize the part properties. Other important process parameters include feed rate, nozzle size, deposition speed, deposition temperature, road width, build chamber temperature, etc. FFF is very popular in fabricating implants using polymers and their composites. For ceramic processing, feedstock filament with maximum solid/ceramic loading in appropriate polymer matrix can be prepared. Then the composite filament

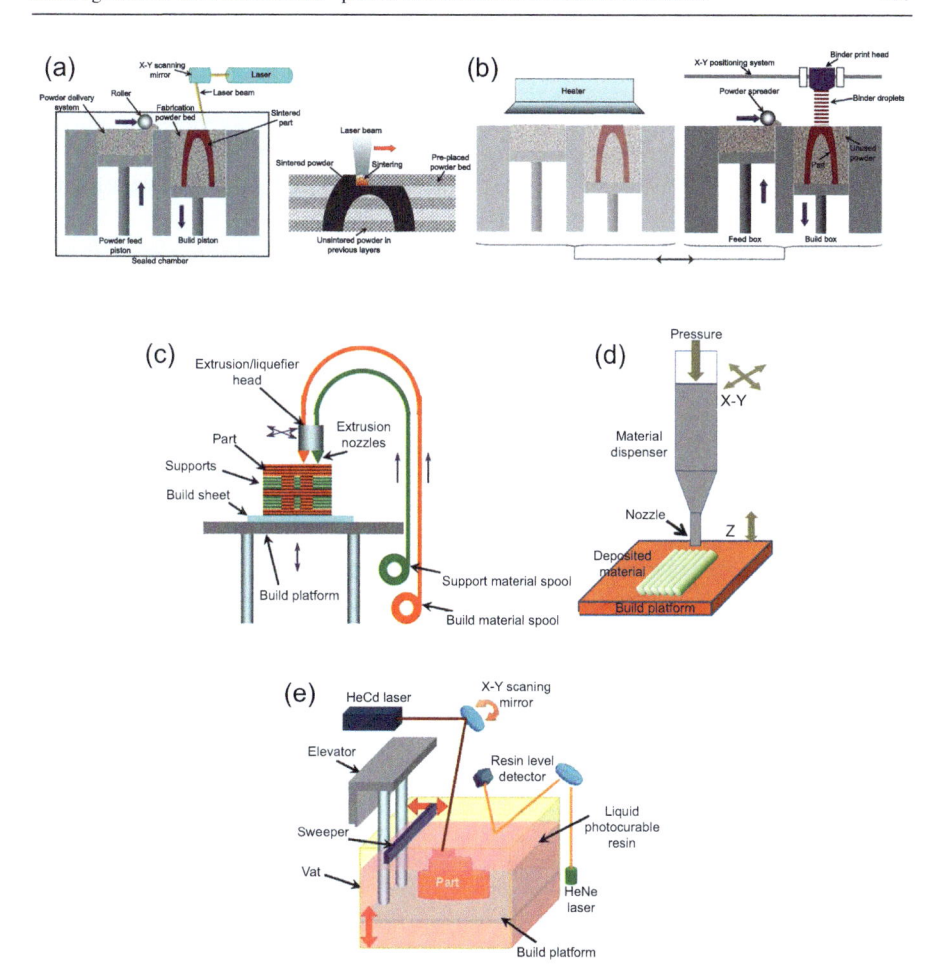

Figure 16.1 Schematic representation of different additive manufacturing (AM) or three-dimensional printing (3DP) technologies used to print unique implant structures and scaffolds for tissue engineering and regenerative medicine applications. (a) Selective laser melting (SLM)/selective laser sintering (SLS) (Bandyopadhyay et al., 2011) (b) binder jetting (BJ) (Bandyopadhyay et al., 2011) (c) Fused filament fabrication (FFF) /fused deposition modeling (FDM) (Bandyopadhyay et al., 2011) (d) RC and (e) stereolithography (SLA) (Bandyopadhyay et al., 2011).

can be used in FFF/FDM machine to print desired implants followed by removal of polymer binder and high-temperature sintering.

Robocasting or direct-write process is another process that is a modified FFF/FDM process. In this process, the feedstock is in the form of viscous solution or paste, which is dispensed through a syringe as shown in Fig. 16.1(d). The feature resolution depends on nozzle size, and the deposited parts are often debinded followed by sintering for ceramic parts and UV hardening for polymer parts. The process is relatively more flexible than standard FFF/FDM as it can operate with variety of feedstock forms (paste, solution, hydrogel, etc.). Further, the process can be used to fabricate parts with compositional gradient. However, overhang structures are difficult to make using this process. Further, the process is highly sensitive to powder-binder combination, and the feedstock must be free of air entrapments, agglomeration, and other defects to achieve high-quality parts. The feedstock must be self-supporting immediately after printing, and therefore, its viscosity is very important.

Stereolithography (SLA): It is the oldest method among different 3DP technologies and depends on solidification of photocurable liquid resin using lasers. As depicted in Fig. 16.1(e), each layer is build by overlapping lines at room temperature. The first layer is formed by solidifying the liquid resin on to a build platform that is lowered just below the liquid resin. The solidification begins by focusing a laser beam onto the liquid and rastering the same to complete a layer as per CAD model. The second layer is made by lowering the build platform (equal to layer thickness), and part building process is completed by repeating these steps. In this process, cure depth and width are critical parameters to be controlled precisely to achieve fully dense parts with high mechanical properties due to strong bonding between layers and interscans. As the process is based on photocurable resin/monomer, the final parts are often cured using heat or UV to achieve high strength by completing the curing (only $\sim$ 80% polymerization takes place during SLA process). The success of the SLA depends on large number of process parameters including type of light source, exposure time and speed, laser power, spot size, layer thickness, etc. For biomedical applications, SLA can be used to make polymers and their composites reinforced with bioceramics and bioactive glasses. However, light reflection from these filler materials can strongly influence curing of resin, but the process provides parts with very high resolution. Finally, the process suffers from limited availability of photocurable resins/monomers that are biocompatible, and the parts are often mechanically weak compared with those fabricated using other 3DP and AM techniques.

The 3DP techniques used for bioprinting are modified versions of techniques originally developed for other materials as discussed above. However, these bioprinting processes have much better resolution and printing precision to achieve cellular micro/nanoarchitectures of natural tissues in the 3D structures. Further, bioprinting uses bioinks that are mixtures of cells, growth factors/therapeutic agents/biomolecules, and biodegradable polymers, which are deposited layer-by-layer while preserving cell functions and viability in the final printed 3D structures/tissue/organ similar in architecture, structure, dimension, and shape of natural tissues. 3D bioprinting techniques used for hard tissue and organ building applications are discussed in "*3D bioprinting and organ manufacturing.*"

16.2 Metallic implants via three-dimensional printing

AM is evolving as the most promising technique to manufacture custom-made implants (Balla et al., 2010a,b; Heinl et al., 2008; Das et al., 2013; Mullen et al., 2009; Yan et al., 2018; Harun et al., 2018). AM of metals has attracted huge attention over the last decade in the field of biomedical implants due to its ability to produce patient-specific, customized implants using data from computed tomography (CT) or magnetic resonance imaging (MRI) (Haleem and Javaid, 2018; Singh and Ramakrishna, 2017). In the past few years, the potential of AM for customized products has been explored in biomedical industries. Few AM technologies, namely powder-bed fusion (PBF), high performance additive manufacturing, ZipDose, pharmacoprinting, bioprinter, and inkjet printing, received Food and Drug Administration (FDA) clearance for biomedical applications (Singh and Ramakrishna, 2017). In 2012, FDA has approved EBM AM technique to fabricate metallic implants such as knee, hip, and dental (Wysocki et al., 2017). A recent report by Wohlers Associates predicted that the market for AM industry would be $21 billion by the year 2020 and also envisioned that 50% of 3D printing will manufacture commercial products by 2020 (Ngo et al., 2018). The biomedical industry presently occupies 11% of total AM market, and it will evolve as one of the major contributors in the near future. Metal AM techniques are predominantly used in defense, automotive, and biomedical industries. Compared with conventional manufacturing, AM of metal provides great freedom for complex geometries with multifunctional components (Harun et al., 2018; Ngo et al., 2018; Herzog et al., 2016). Despite these benefits, metal AM is associated with various challenges such as resolution, surface finish, postprocessing procedures, and anisotropic mechanical behavior (Vaezi et al., 2013). Currently, most common metals used for biomedical implants include austenitic stainless steel (316L), titanium and its alloys, and cobalt-chromium—based alloys, which are readily printable with existing metal AM technologies. The most popular powder-based AM techniques for implant manufacturing (Mota et al., 2015) are SLS, SLM, and EBM. These are categorized under PBF techniques. Among different metal AM techniques, PBF techniques found to be more suitable for implants and scaffolds fabrication for orthopedic and dental applications (Heinl et al., 2008; Wysocki et al., 2017; Sallica-Leva et al., 2013; Tan et al., 2017; Murr et al., 2009; Hao et al., 2016; Sing et al., 2016). SLM and EBM have been successfully used to fabricate variety of bone implants such as knee, hip, mandible replacements, maxillofacial plates, and replacements for zygomatic bone (Jardini et al., 2014; Moiduddin et al., 2017; Rotaru et al., 2015; Gebhardt et al., 2010). A brief comparison of these two technologies is presented in Table 16.2. However, one of the serious drawbacks in metal AM is lack of powder materials availability restricting its widespread industrial use. So far, only a few metallic biomaterials are commercially available in powder form, which includes 316L stainless steel, Ti6Al4V, and CoCrMo alloy.

16.2.1 Stainless steel (316L) implants

Because of high corrosion resistance and relatively low cost, 316L stainless steel has been widely used to manufacture variety of implants. Therefore, significant research has been done on SLM processing of 316L (Casati et al., 2016; Suryawanshi et al., 2017; Liverani et al., 2017; Bartolomeu et al., 2017; Heeling and Wegener, 2018), while the use of EBM has been very low (Olsén et al., 2018; Wang et al., 2018; Rännar et al., 2017; Zhong et al., 2017). Several researchers have studied the effect of powder

Table 16.2 Summary of powder-bed fusion–based metal additive manufacturing (AM) techniques for biomedical metallic implant manufacturing.

Metal AM	Power source	Materials and size range (µm)	Operating environment	Accuracy (mm)	Melting method	Build rate
Selective laser melting	Laser beam (0.5–1 kW)	Metals, polymers, ceramics, composites, etc. ~20–63	- Argon environment - Preheating up to 200°C	High, ±0.04	Contour melting followed by hatch melting	Low, 20–40 cm^3/h
Electron beam melting	Electron beam (3 kW)	Metals and metallic-based composites ~45–105	- High vacuum - Chamber temperature maintained at 600–1000°C	Low, ±0.2	Preheating the bed followed by contour and hatch melting	High, 80 m^3/h

Adapted from X.P. Tan, Y.J. Tan, C.S.L. Chow, S.B. Tor, W.Y. Yeong, 2017. Metallic powder-bed based 3D printing of cellular scaffolds for orthopaedic implants: a state-of-the-art review on manufacturing, topological design, mechanical properties and biocompatibility. Mater. Sci. Eng. C 76, 1328–1343. https://doi.org/10.1016/j.msec.2017.02.094.

size and SLM process parameters on densification and mechanical properties of 316L parts and compared with EBM technique. SLM of 316L has been studied for removable partial denture, fixed bridge customized crowns, and customized orthodontic (Kruth et al., 2005; Bibb et al., 2006; Yang et al., 2012). Yang et al. showed that under optimized SLM parameters, highly dense (>99%) 316L customized brackets can be fabricated with high shape precision (size error < 10 μm) (Yang et al., 2012). Bibb et al. reported case studies of stainless steel surgical guides, used in maxillofacial surgeries and manufactured using SLM (Bibb et al., 2009). It is reported that SLM-produced steel parts obtain fine cellular dendritic microstructure with high tensile strength and reduced ductility than forged counterparts (Sing et al., 2016). In SLM process, melt pool behavior and heat transfer strongly depends on laser process parameters such as the laser power, hatch spacing, layer thickness, beam size, and scanning speed, which control density, surface quality, overall microstructure, and mechanical properties. Cherry et al. (Cherry et al., 2014) found that very high or low laser energy densities create balling effect. They achieved highest density ($\sim$99.62%) in 316L parts fabricated with 104.52 J/mm^3 laser energy density. Yasa et al. (Yasa et al., 2011) studied the effect of laser remelting during SLM production of 316L parts on their surface roughness. It was observed that remelting increased the density and significantly reduced ($\sim$90%) the surface roughness of SLM parts at the expense of longer build time. To reduce the build time, multibeam strategy is one of the advancements in SLM, where a second beam with a larger beam diameter is used to heat the vicinity of the melt pool. Heeling et al. studied multibeam SLM strategies to fabricate 316L parts (Heeling and Wegener, 2018) and found that the scan strategies have strong influence on part's microstructure, surface roughness, and density. Chen et al. (Chen et al., 2018) used fine metal powder ($\sim$16 μm) and achieved high density, UTS (611.9 ± 9.4 MPa), YS (519.1 ± 5.9 MPa), and elongation (14.6 ± 1.9%). Influences of laser power at constant scan speed and different scanning strategies on the microstructure, texture, and tensile properties were examined by Kurzynowski et al (Kurzynowski et al., 2018). Recent study of Kong et al. (Kong et al., 2019) reported anisotropy in microstructure and mechanical properties in SLM processed bulk and porous 316L. On the other hand, very few reports are available for EBM of 316L steel (Sing et al., 2016) (Wang et al., 2018) (Rännar et al., 2017), which could be due to relatively high processing and capital cost of EBM that is not attractive for low-cost materials such as 316L. Rännar et al. (Rännar et al., 2017) studied microstructure, density, and hardness of 316L part as a function of layer thickness during EBM process. Wang et al. studied optimized speed function and focus offset parameters and developed highly dense 316L part and showed high tensile strength (Wang et al., 2018).

16.2.2 Implants with Ti and its alloys

Commercial pure Ti (CP-Ti) and Ti6Al4V alloy have been extensively used for various biomedical implants manufacturing due to their high strength, corrosion resistance, and good biocompatibility (Geetha et al., 2009; Dabrowski et al., 2010). However, majority of these implants have limited life span due to lack of cell adhesion and high elastic modulus that leads to stress shielding (Das et al., 2013; Murr et al., 2011). To improve osseointegration and eliminate stress shielding, porous implants found to be an effective solution (Bandyopadhyay et al., 2010; Mullen et al., 2010; Wang et al., 2016). However, manufacturing porous implants with complex internal architectures, site-specific porosity is not possible with traditional manufacturing techniques. PBF-based AM technologies have been demonstrated to fabricate complex custom-made implants with internal microarchitecture. During the last decade, several research groups have studied AM- of Ti-based implants using SLM and EBM (Wysocki et al., 2017; Sing et al., 2016; Murr et al., 2012a,b), which also resulted in ASTM specification for Ti6Al4V alloy for PBF (ATSM, 2014). SLM and EBM produced Ti6Al4V parts that were found to exhibit different microstructures and hence properties as well. In SLM, typically, the parts experience very high cooling rates compared with EBM process due to powder bed preheating (between 650 and 700°C) in the latter process. Large temperature gradients in SLM generates mismatch in elastic deformation, leading to high residual stresses. Several authors have studied the effective scanning strategy and post-heat treatment to reduce residual stresses in AM metallic parts (Ali et al., 2018; Robinson et al., 2018). In general, EBM-fabricated Ti6Al4V consists of α/β with β phase along the grain boundaries, whereas SLM-fabricated Ti6Al4V parts exhibit martensitic structure with α' acicular grains (Murr et al., 2012a,b; Vilaro et al., 2011) due to high cooling rates associated with SLM. Because of these microstructural variations, microhardness and tensile strength of EBM-fabricated Ti6Al4V parts are lower than SLM-produced Ti6Al4V. Yang et al. (Yang et al., 2016) reported morphology and structure of different types of α' martensite, namely primary, secondary, tertiary, and quartic α' martensites, in SLM-based Ti6Al4V parts, and process parameters were found to change the martensite size. However, almost fully dense parts can be manufactured using EBM, whereas the relative density as high as 99.80% was reported for SLM-fabricated Ti6Al4V (Vandenbroucke and Kruth, 2007). Mechanical properties of Ti6Al4V alloy parts fabricated using SLM and EBM are compared in Table 16.3. Recently researchers also studied novel β-type titanium alloys such as Ti-30Nb-5Ta-3Zr, Ti-35Nb-7Zr-5Ta, and Ti-24Nb-4Zr—8Sn using SLM and EBM technique for biomedical applications (Luo et al., 2018; Liu et al., 2016).

Table 16.3 Mechanical properties of Ti6Al4V alloy processed using selective laser melting (SLM) and electron beam melting (EBM).

	Tensile strength (MPa)	Yield strength (MPa)	Elongation (%)	Microhardness (HV)
SLM	1250—1267	1110—1125	6—7	479—613
EBM	830—1150	915—1200	13—25	358—387

Adapted from S.L. Sing, J. An, W.Y. Yeong, F.E. Wiria, 2016. Laser and electron-beam powder-bed additive manufacturing of metallic implants: a review on processes, materials and designs. J. Orthop. Res 34, 369—385. https://doi.org/10.1002/jor.23075.

16.2.3 Porous/cellular implants for orthopedic applications

Metal AM has been extensively used to produce novel cellular and lattice structures with tailored internal microarchitecture for bone implant applications (Horn et al., 2014; Ahmadi et al., 2014; Wang et al., 2018; Zadpoor, 2018). Cellular solids can be considered as composite of solid and the empty space where solid metallic phase forms a lattice connected by struts (Tan et al., 2017). Typically, lattice structures are grouped in to stochastic and nonstochastic geometries. Stochastic lattice structures are characterized by random cell shape and size. On the other hand, repeated lattice structures form nonstochastic lattice structures, which can be separated based on their shapes and sizes (Hasib et al., 2015). SLM and EBM processes have been successfully used to manufacture implant with such porous architectures with precise control over pore size, orientation, homogeneous pore distribution, and complex 3D shapes of cellular structures for implant applications, which promote bone ingrowth, matching stiffness with natural bone, and reduce total weight of the implant (Tan et al., 2017; Zhang et al., 2018). Mueller et al. (Mueller et al., 2012) proposed another novel approach to reduce stiffness and enhance functionality of femoral stem by incorporating cavities and internal channels using EBM. Fig. 16.2 shows one such hip stem where the internal cavities could be useful for local supply of nutrients for bone growth, other desired materials such as drugs, and filler materials. Such AM-processed implants have strong potential for medical and dental applications (Tan et al., 2017; Wang et al., 2016).

For Ti-based lattice structure fabrication, EBM is found to be more suitable than SLM due to its high vacuum, powder size flexibility (45–105 μm for EBM and 20–45 μm for SLM), and high powder bed temperature resulting no/minimal residual stresses. Further, EBM offers higher production rate of 80 cm^3 h^{-1} than SLM with 20–40 cm^3 h^{-1}. Zhang et al. summarized 23 metallic lattice structures widely investigated to date (Zhang et al., 2018). A large variety of lattice designs, beam-based

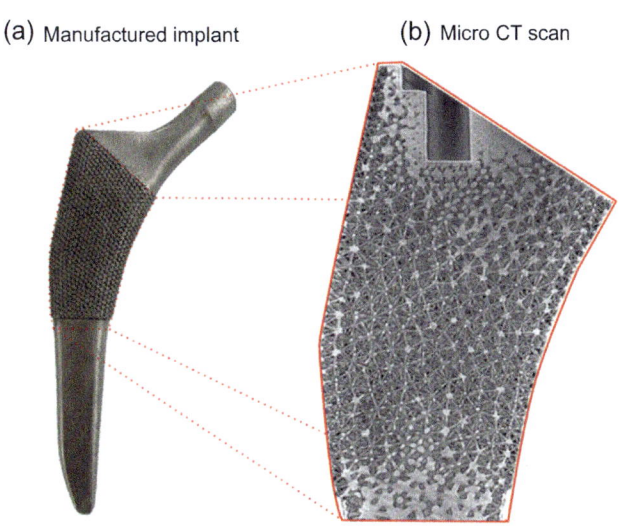

Figure 16.2 Hip stem prototype with site-specific architecture fabricated using powder-bed fusion (PBF)–based additive manufacturing (AM).
Courtesy of www.tctmagazine.com.

(i.e., body-centered cubic) to surface-based (i.e., gyroids), have been fabricated using PBF-based AM techniques. Further, mechanical properties of such cellular solids depend on strut material, lattice structure, strut size, and shape of the pore (Park et al., 2014). Such designs have more flexibility in terms of tailoring mechanical and other functional properties required for different implants that can elicit site-specific properties. Parthasarathy et al. (Parthasarathy et al., 2010) showed that with decrease in strut size of cells, mechanical properties decrease drastically. Similarly, spherical pore shape was found to be stiffer than cylindrical pore in these structures (Hollister, 2005). Some literature also reported that cellular Ti structures with cubic pores fabricated using EBM exhibit mechanical properties (compressive strength and elastic modulus) comparable with those of trabecular and cortical bone (Parthasarathy et al., 2010, 2011; Cheng et al., 2012). The advancement of AM technologies also enabled manufacturing functionally graded lattice structures that closely mimic natural bone. Several researchers reported functional graded porous structure fabricated using SLM and EBM (Choy et al., 2017). For example, EBM can be used to create gradation in porosity mimicking natural bone to enable accelerated vascularization and new bone formation. Typical implants with radial porosity gradient manufactured using EBM are shown in Fig. 16.3. The first lattice structure based implant was developed by Lima-Corporate and Arcam. One such lattice-like structure was developed using EBM on the surface of Ti6Al4V acetabular cup

Figure 16.3 Typical computer-aided design (CAD) models with radial gradation in pore architecture and prototype parts manufactured using electron beam melting (EBM) (Murr et al., 2010).

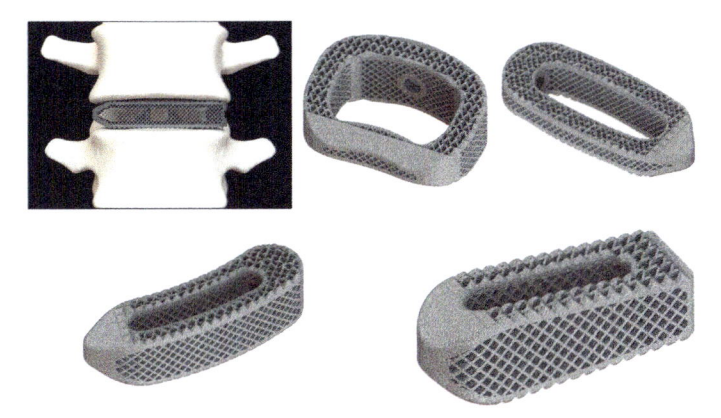

Figure 16.4 Typical Ti6Al4V alloy spinal cage implants manufactured by powder-bed fusion (PBF)−based additive manufacturing (AM) technologies and approved by Food and Drug Administration (FDA) (Zhang et al., 2018).

(Delta-TT cup) and showed great success since last 10 years (Metal A.M., http://www.metal-am.com/celebrating-ten-years-of-metaladditively-manufactured-hip-cups/). Other Ti-based implants approved by FDA are shown in Fig. 16.4. Ti lattice interbody cages developed by Emerging Implant Technologies have already been used in over 10,000 cases in 15 countries (Zhang et al., 2018).

16.2.4 Additive manufacturing of biodegradable metals

Biodegradable metals have recently been studied with great interest due to their attractive degradation behavior coupled with high strength and toughness compared with degradable bioceramics. Typically degradable implants are designed to safely dissolve in physiological fluid after required duration of implantation. Thus, removal of the implant by second surgery can be eliminated. Compared with polymer- and ceramic-based degradable materials, biodegradable metals exhibit high mechanical properties (Li et al., 2014; Zheng et al., 2014). Biodegradable metals based on magnesium, iron, and zinc are mostly processed through conventional manufacturing techniques. There is limited literature available on AM of such biodegradable metals. Among these biodegradable metals, attempts on processing of Mg-based alloys using SLM have been reported (Ng et al., 2011a,b; Wei et al., 2014; Manakari et al., 2016; Li et al., 2018a,b). Prealloyed Mg-5.2Zn-0.5Zr (ZK60) powder has been processed to 94% density using SLM with some loss of Zn. However, fine microstructures of SLM-processed ZK60 alloy resulted in acceptable hardness comparable with that of wrought ZK60. Li et al. (Li et al., 2018a,b) used SLM to fabricate porous magnesium (WE43) scaffolds with diamond unit cell. They observed that Young's modulus of these scaffolds (700−800 MPa) cover the modulus of trabecular bone even after 28 days of biodegradation. Very limited attempts have been reported on AM of biodegradable Fe-alloys and Zn (Montani et al., 2017; Demir et al., 2017; Li et al.,

2018a,b). Montani (Montani et al., 2017) showed higher mechanical properties in laser processed Zn than the as-cast material.

16.3 Bioceramic scaffolds using three-dimensional printing

Among different biomaterials, manufacturing bioceramics using 3DP is relatively more challenging due to their high melting temperature and inherent brittleness. Although process-induced porosity is detrimental to mechanical properties, appropriate porosity characteristics are beneficial for cell adhesion, growth, proliferation, and implant fixation. As a result, judicious balance between biological and mechanical properties originating from porosity of bioceramics is required. Bioceramic, particularly bioactive ceramics based on CaPs, are very popular for dental and orthopedic applications (Bose et al., 2018a,b,c,d). Hence, to capitalize potential of bioceramics, it is natural that manufacturing methods that can overcome processing limitations of brittle ceramics must be developed. 3DP is one such method that not only overcomes many of the inherent limitations of conventional bioceramic processing but also fabricates scaffolds that are reproducible, design-optimized, and tailor-made. There has been huge impetus to use 3DP for ceramic implants, as small, complex, and patient-specific needs can be catered at low cost. However, unlike metals and polymers, 3DP of bioceramics is not widely investigated primarily due to lack of robust 3DP equipment that can handle these ceramics. The following sections describe attempts made in 3DP of common bioceramics, glasses, and their composites for variety of implant applications.

16.3.1 Calcium phosphate ceramics

Calcium phosphate (CaP), owing to its compositional resemblance with bone and teeth, is a very important class of bioceramic materials. It has calcium ions (Ca^{2+}), a multitude of phosphate ions such as metaphosphates (PO^{3-}), orthophosphates (PO_4^{3-}), and pyrophosphates ($P_2O_7^{4-}$), along with occasional hydroxide or carbonate ions. However, depending on Ca/P ratio, various CaP materials with wide range of physical, mechanical, and biological properties are available. In fact, Ca/P ratio is one of the key parameters that affects bioactivity and dissolution of CaPs (Bose and Tarafder, 2012). To underscore the significance of Ca/P ratio, it may be cited that hydroxyapatite ($Ca_{10}(PO_4)_6(OH)_2$) (HA), with Ca/P ratio close to 1.67, shows excellent biocompatibility with natural bone and teeth, but monocalcium phosphate monohydrate ($Ca(H_2PO_4)_2.H_2O$), with 0.5 Ca/P ratio, is not biocompatible due to its high acidic nature. Further, Ca/P ratio between 1 and 2 is found to be useful for osteoblast cell viability and the alkaline phosphatase production (Liu et al., 2008). CaP bioceramics are traditionally produced by conventional ceramic processing techniques such as pressing, hydrothermal exchange, porogen or salt leaching, foaming, freeze drying, and mold impregnation. Although these processes are used to fabricate macroporous CaP scaffolds, they suffer from architectural limitations, internal inhomogeneity, and random pore distribution. As a result, their mechanical properties, biological

16.3.1.1 Hydroxyapatite

Various 3DP technologies have been used to manufacture variety of HA-based parts (Bose et al., 2013, 2018; S Bose et al., 2013; Cox et al., 2015). Vat polymerization or SLA has been used to prepare complex HA parts, where the powder is mixed with photoactive polymer, which is then cured using UV laser followed by thermal treatment to remove the polymeric binder and densification of HA (Leukers et al., 2005; Griffith et al., 1996; Chu et al., 2001; Woesz et al., 2005). Effect of HA concentration on the viscosity and curing has been investigated (Griffith et al., 1996), and a maximum HA loading of 45 vol.% was achieved with inverse relation between HA loading and cure depth. The process has also been used to make resin molds to produce crack-free HA implants (using gel casting) with 26%–52% porosity having pore channels between ⌀ 368 and 968 μm (Chu et al., 2001). Similarly, another study (Woesz et al., 2005) demonstrated that the HA scaffolds with 450 μm pores exhibit good in vitro osteogenesis. An example of porous HA fabricated by impregnation of 3D-printed wax mold is shown in Fig. 16.5.

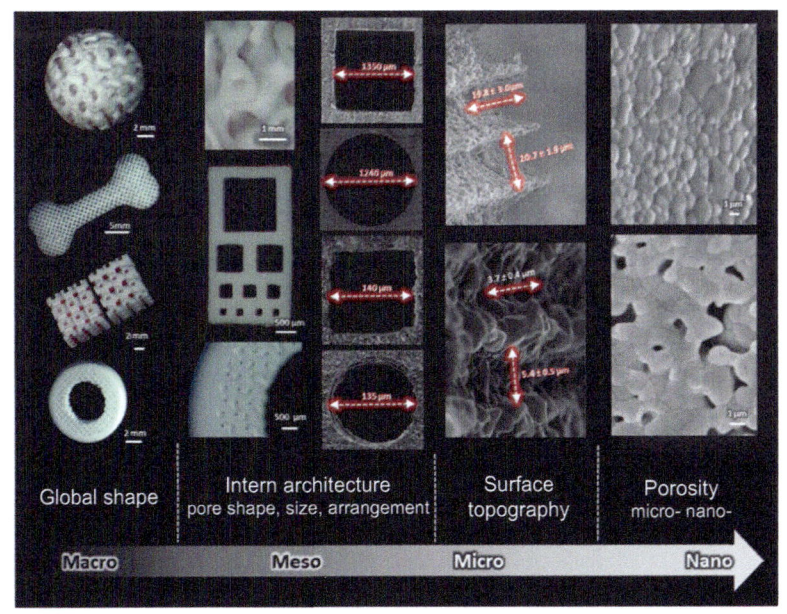

Figure 16.5 Some examples of porous HA fabricated using 3D-printed wax molds and impregnation. In this process, shape, internal architecture, surface microtopography, and micro- and nanopore density of hydroxyapatite (HA) bioceramics can be adjusted (Charbonnier et al., 2016).

Poly(acrylic acid) solution was used as a binder that was spread layer by layer onto HA powders for binding and densified by sintering (Guvendiren et al., 2016). Through meticulous adjustment of ink composition and viscosity, researchers manufactured self-supporting HA scaffolds from high-concentration HA inks (with <1 wt.% organic content) for bone tissue engineering (Michna et al., 2005). Scaffolds with pore size of ~3 mm have also been fabricated using nano-HA powder (Shuai et al., 2011) using laser PBF method, where no decomposition of HA and cracks were observed. Further, the nanoscale structure of HA was retained in the scaffolds demonstrating viability of SLS in processing nano-HA. There are other reports demonstrating fabrication of porous HA scaffolds using binder jet process, and the scaffolds exhibited good biocompatibility during static and dynamic culturing using MC3T3-E1 cells (Leukers et al., 2005). Samples cultured under dynamic culturing conditions were reported to exhibit superior cell growth inside the scaffolds.

Material extrusion—based 3D printing, whose effectiveness depends on rheological properties of HA suspension, has been tried to prepare HA scaffolds. In this process, an optimum level of HA loading enables effective extrusion and sintering. High loading of HA is always preferable to reduce cracking and warpage during sintering but at the cost of increased viscosity, which renders extrusion process difficult. HA-based scaffolds with 70% interconnected porosity, fabricated using this technique, was found to exhibit excellent bone growth (Fierz et al., 2008). Robocasting, another similar method, was also used to fabricate porous HA scaffolds (Dellinger et al., 2006) with a feedstock slurry containing 35 vol. % HA. Compressive strength of sintered scaffolds (1300°C, 2 h) with 39% total porosity was 47 ± 14 MPa and 50 ± 5 MPa in transverse and normal to build directions, respectively. Dellinger et al. also used robocasting process (48 vol.% HA loading) to fabricate scaffolds with bone morphogenetic protein 2 (BMP-2) for onsite delivery (Dellinger et al., 2006). The BMP-2 (10 μg) was added to sintered scaffolds and evaluated in vivo in the goat metacarpal bone (4 and 8 weeks). Significant improvement in the osteogenesis of scaffolds was observed due to synergistic effect of BMP-2 and microporosity.

16.3.1.2 Tricalcium phosphate

Another important class of CaP that is being widely processed using 3D printing is tricalcium phosphate (TCP). α-TCP and β-TCP, as biodegradable materials, have been used in critical size defects in tibiae of mini pigs (Wiltfang et al., 2002) where 70%—80% resorption after 28 weeks was observed in addition to new trabeculae formation. Later, 3DP of TCP has been used to produce patient-specific complex parts. 3D scaffolds with high strength and toughness were printed using a mixture of TCP and alginate powder with phosphoric acid binder (Castilho et al., 2015). When natural polymers such as alginate (~2.5%) was added to TCP, the scaffolds exhibited enhanced strength and MG63 osteoblastic cell compatibility and proliferation, compared to pure TCP scaffolds. Further, natural polymer appears to reduce the possibility of side effect such as deleterious breakdown products and reduced cell attachment, which are common with synthetic polymers (Diogo et al., 2014). Similar results have been reported by Bian et al. where β-TCP/collagen scaffolds were prepared for osteochondral tissue engineering (Bian et al., 2012). Porous β-TCP implants were prepared using vat polymerization, with 45 wt.% β-TCP loading, showed a density of 88% and with a pore size of ~300 μm, and exhibited biaxial strength of 30 MPa (Felzmann et al., 2012).

Porous β-TCP scaffolds were prepared using water as binder with 5 wt.% hydroxypropylmethylcellulose-modified β-TCP as matrix in a binder jetting process (Vorndran et al., 2008). In spite of having low specific surface and low print resolution, the final parts exhibited highest compressive strength of 1.2 ± 0.2 MPa. The effect of microwave sintering and pore size of porous β-TCP scaffolds on their biological and mechanical properties was studied by Tarafder et al (Tarafder et al., 2013a,b). It was observed that microwave sintered (1250°C) scaffolds with designed porosity and pore size of 27% and 500 μm, respectively, have compressive strength of 10.95 ± 1.28 MPa. Further, the scaffolds were found to have excellent in vitro biocompatibility and in vivo bone ingrowth and new bone formation ability. In another study, β-TCP scaffolds were prepared by robocasting with the aim of utilizing them for on-site antituberculosis drug delivery (Yuan et al., 2015). The sintered scaffolds (1100°C for 2 h) resulted in total porosity of 62% (with 43% of 400 μm macroporosity and 19% of 2—8 μm microporosity) and strength of 3.31 ± 0.64 MPa. These scaffolds were loaded with antituberculosis drugs, rifampin, and isoniazid with poly(lactide-co-glycolide) (PLGA) and showed outstanding in vitro biocompatibility, desired degradation, and drug release (Yuan et al., 2015). Indirect AM method has also been used to fabricate porous β-TCP scaffolds, where polymer molds have been prepared by FDM (Bose et al., 2003a,b). These porous β-TCP scaffolds had porosity of around 29% (305 μm pores) and exhibited strength of 1.4 MPa in addition to very good in vitro biocompatibility. Further improvement in the strength of these scaffolds was achieved with addition of strontium and magnesium oxide (Tarafder et al., 2015). Typical morphology and microstructures of 3D-printed pure/doped TCP scaffolds are shown in Fig. 16.6.

16.3.1.3 Other calcium phosphates

Self-setting characteristics of CaP-based cement can be effectively utilized for AM processing. Researchers used cement setting reactions to make samples using CaP powder (process A) and TTCP/DCP composite (process B) with binder jetting (the binder was an acid solution) (Vorndran et al., 2008). They found that printing resolution of method adopted in A was better than the method described in B. Moreover, it was reported that scaffolds prepared by former method had 56% porosity with compressive strength of 7.4 ± 0.7 MPa, while samples prepared by later method showed marginal increase in the porosity (60%) but significant drop in the compressive strength (1.2 ± 0.2 MPa). Interaction between TCP and phosphoric acid was effectively utilized in binder jetting process to make brushite-based implants for maxillofacial and cranial fixations (Klammert et al., 2010). Further, the brushite in these implants was transformed to monetite hydrothermally to achieve high-strength final parts. These parts with 35% porosity exhibited bending strength of 3.9 ± 0.5 MPa. These implants (44% porous brushite implants and 38% moetite implants) were

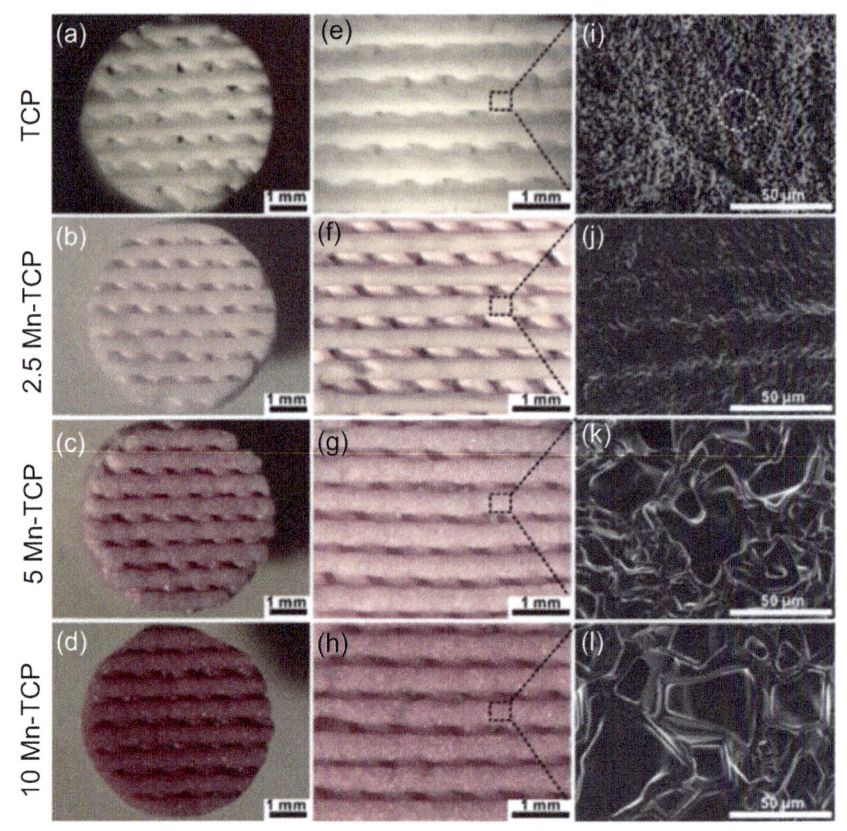

Figure 16.6 Morphology and microstructure of three-dimensional (3D)−printed pure and doped tricalcium phosphate (TCP) scaffolds; light microscope images of pure TCP (a, e), 2.5Mn-TCP (b, f), 5Mn-TCP (c, g), and 10Mn-TCP (d, h). Corresponding scanning electron microscope (SEM) images are shown in (i)−(l), respectively, with micropores shown as white circle for pure TCP (Ma et al., 2018).

evaluated in vivo (Habibovic et al., 2008) and found to exhibit outstanding osteoinductivity supported by histomorphometric data. Post-treatment with poly(L-lactic acid) (PLA)/PGA polymer solution was reported to be beneficial for controlled release of drugs (Gbureck et al., 2007a,b) such as ofloxacin, vancomycin hydrochloride, and tetracycline hydrochloride. Besides, in vivo experiments showed improved angiogenesis in the presence of vascular endothelial growth factor and copper (Gbureck et al., 2007a,b).

16.3.2 Bioactive glass

Bioactive glasses consisting of P_2O_5, SiO_2, CaO, Na_2O, and CaO in different concentrations are being widely used in variety of implant and tissue engineering applications. However, their widespread applications are limited due to their inherently poor mechanical properties. 3DP printing can further the scope of bioactive glasses by incorporating controlled pores and improved mechanical strength (Eqtesadi et al., 2016; Zhang et al., 2016). Tesavibul et al. reported the feasibility of utilizing vat polymerization for making bioactive glass (Tesavibul et al., 2012). Using 43% 45S5 bioactive glass and acrylate-based photoactive polymer, they prepared cellular structures with tailored porosity, which mimic trabecular bone. These cellular structures (50% porous) exhibited biaxial and compressive strengths of 40 and 0.3 MPa, respectively. Further improvements in mechanical strength could be achieved with high solid loadings as high as 70% (Gmeiner et al., 2015), where biaxial bending strength of 124 MPa with characteristic strength of 131 MPa was recorded. These studies opened the path for bioactive glass scaffolds that can be used as load-bearing implants as 3DP can provide scaffolds with best combination of density and mechanical properties. Eqtesadi et al. used a water-based bioactive glass 45S5 suspension, consisting of poly(methyl vinyl ether) and carboxymethyl cellulose, with tailored rheological properties suitable for direct-write process (Eqtesadi et al., 2013). The influence of sintering temperatures on density, microstructure, and compressive strength has been studied (Eqtesadi et al., 2014). It was reported that sintering at 1000°C for 1 h results in scaffolds (60% total porosity and designed macroporosity of 50%) with compressive strength of 13 MPa. It may be noted that melting point of bioactive glass is lower than CaPs, and therefore, PBF processing of bioactive glass is relatively easy than CaPs. Consequently, various laser processing techniques were used to prepare crack-free 3D bioactive glass implants using bioactive glasses such as 45S5 and S520 without altering their bioactivity. Typical 3D printed ion-doped bioactive glass and ceramic composites along with their microstructures are shown in Fig. 16.7.

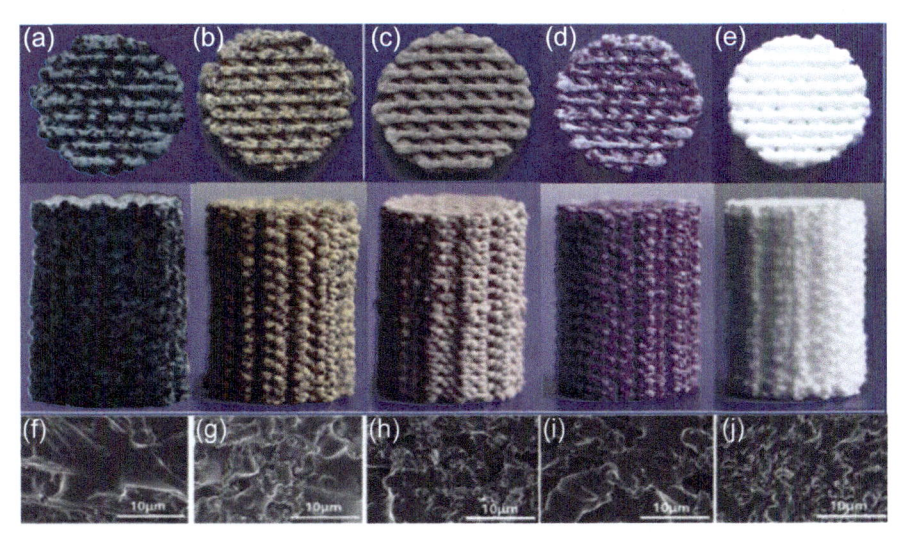

Figure 16.7 Three-dimensional (3D) printed ion-doped bioactive glass and ceramic composite (BGC) with their corresponding microstructures; (a) 5Cu-BGC, (b) 5Fe-BGC, (c) 5Mn-BGC, (d) 5Co-BGC, and (e) BGC. Differing colors and surface microstructures are due to different element doping (Liu et al., 2018).

Mesoporous bioactive glass (MBG) with polyvinyl alcohol as binder was used in binder jet technology to fabricate multifunctional scaffolds for controlled drug release (Wu et al., 2011). In this work, scaffolds with different pore sizes (between 200 and 1300 μm) and pore geometries (square and parallelogram) were designed and fabricated. Maximum compressive strength and modulus of 16.10 ± 1.53 MPa and 155.13 ± 14.89 MPa, respectively, were achieved with scaffolds having 60% porosity and 1001 μm pores. Further, the scaffolds loaded with dexamethasone showed controlled release up to 10 days (Wu et al., 2011). Interestingly, scaffolds reported to have low cell proliferation with high alkaline phosphatase (ALP) expression compared with control samples. Thus, it can be concluded that without second high temperature sintering, controlled porosity scaffolds with low brittleness and high strength can be fabricated using binder jetting technology. New generation bioactive glass such as 13-93 glass shows high modulus, strength, and osteogenesis in vitro and in vivo compared with 45S5 glass due to high silicate content in 13-93 glass (Liu et al., 2013). Therefore, Kolan et al. prepared scaffolds of 13-39 bioactive glass using stearic acid binder in PBF method. The scaffolds with around 50% porosity exhibited compressive strength of 20.4 ± 2.2 MPa (Kolan et al., 2011) with excellent bioactivity. The silicate was substituted with borate in the aforesaid glasses for faster degradation (Kolan et al., 2015). In vitro cell—materials interactions on porous bioactive glass scaffolds with cubic, spherical, diamond, and gyroid pores showed high cell proliferation on gyroid and diamond pore architectures. It is thus concluded that scaffolds with pores having large curvature and surface area can improve cell adhesion and growth (Kolan et al., 2015).

16.3.3 Ceramic composites

Composites with different CaPs were used to prepare scaffolds via 3DP, with improved properties compared with their single constituent materials (Bergmann et al., 2010; Feilden et al., 2017; Inzana et al., 2014; Seidenstuecker et al., 2017; Turnbull et al., 2018). For example, HA that shows good biocompatibility but poor biodegradation rate is normally mixed with TCP to form a composite (biphasic calcium phosphate [BCP]) with tailored biodegradability. Detsch et al. used spherical powder of 60:40 (wt.%) of HA:TCP and fabricated composite scaffolds through binder jetting (Detsch et al., 2011), and the samples were found to have very good surface properties enhancing osteoclastic activation. Similarly, Castilho et al. fabricated BCP scaffolds (Castilho et al., 2014) with different HA/TCP ratios and optimized the printing process. It was found that a ratio of 1.83 would provide compressive strength of 0.42 MPa at 67% porosity and increased to 3.36 MPa with fourfold increase in the roughness when the scaffolds were infiltrated with PLGA. Moreover, these scaffolds exhibited improved osteoblastic cell viability compared with pure TCP. In another study, Khalyfa et al. prepared composite scaffolds and cranial segments of TTCP and β-TCP (Khalyfa et al., 2007). The role of TTCP had been that of a reacting agent (with citric acid binder), whereas β-TCP acted as a biodegradable constituent. To improve osteogenesis, 3DP TCP scaffold has been coated with MBG nanolayer by Zhang et al (Zhang et al., 2015) (Fig. 16.8).

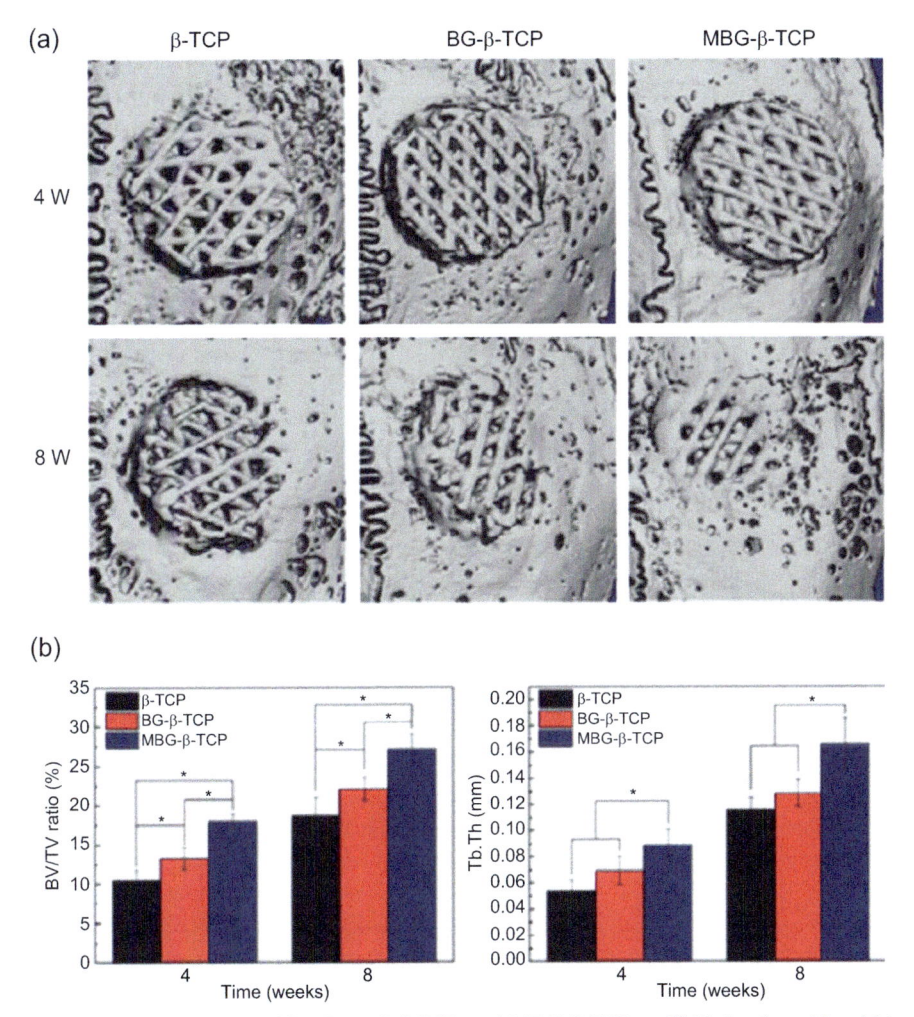

Biointegration of three-dimensional–printed biomaterials and biomedical devices

Figure 16.8 (a) 3D μ-CT of β-TCP, BG-β-TCP, and MBG-β-TCP scaffolds implanted in rabbit calvarial defects at different time intervals to check bone formation ability; (b) kinetics of newly formed bone with respect to volume (by morphometric analysis) and trabecular thickness (after 4 and 8 weeks). MBG-β-TCP expressed fastest bone repair (Zhang et al., 2015). *3D*, three-dimensional; *CT*, computed tomography; *MBG*, mesoporous bioactive glass; *TCP*, tricalcium phosphate.

Brittle composites containing CaPs and bioactive glass are also successfully made using 3D printing. Bergmann et al. fabricated β-TCP:bioactive glass (40:60) composite samples using phosphoric acid solution binder in binder jetting machine (Bergmann et al., 2010). Sintered implants (1000°C) reported to have four-point bending strength of 14.9 ± 3.6 MPa. Same process has also been used to fabricate CaPs:bioactive glass composite comprising HA and apatite–wollastonite glass (Suwanprateeb et al., 2009).

After sintering at 1300°C (3 h), these composite scaffolds exhibited low porosity (3%), maximum flexural strength of 76.82 ± 4.35 MPa, and good bioactivity. Additives such as zinc oxide (ZnO), silica (SiO$_2$), magnesia (MgO), and strontia (SrO) are routinely added to CaPs to improve phase stability, densification, and biological properties of 3DP composite scaffolds (Fielding et al., 2012). Binary doping of 3DP β-TCP scaffolds with SiO$_2$ and ZnO resulted in marked improvement in density and compressive strength. Moreover, doped scaffolds had better in vitro bioactivity and cell—materials interactions than pure β-TCP (Fielding et al., 2012). Similar improvements have been reported with SrO and MgO doping of 3DP scaffolds (Tarafder et al., 2013, 2015). These dopants successfully stopped β to α phase transition, increased compressive strength, and exhibited better osteogenesis in rat femur model (Tarafder et al., 2013a,b). β-TCP scaffolds doped with SiO$_2$ and ZnO also reported to have better osteogenesis and angiogenesis in vivo (Fielding and Bose, 2013).

16.3.4 Ceramic—polymer composites

It is expected that composites with combination of brittle ceramics and flexible polymer can provide desired functional and mechanical performance for implant applications. Such kind of composites has been prepared using many AM processes such as SLS, SLM, FDM, etc. The amount of polymer in the composite plays a decisive role due to its low melting point and directly influence on flowability during material extrusion or 3DP. In SLS and SLM, the laser melts the polymer and glues the ceramic particles together forming a solid part. Among all AM-processing technologies, FDM and 3DP are the most common processes used for production of ceramic—polymer composites. However, the success of these processes primarily depends on making a printable composite filament. Using FDM, porous PP/TCP composite scaffolds with precise dimensions have been prepared for bone replacement applications (Kalita et al., 2003). Scaffolds with 160 μm pores and varying porosity content (36%—52% vol.) were successfully fabricated. Mechanical testing revealed that the scaffolds with 36% porosity have compressive strength and modulus of 12.7 and 264 MPa, respectively. Good in vitro osteoprecusor cells adhesion and proliferation were also reported for these scaffolds (Kalita et al., 2003). FDM was also employed to prepare porous composite scaffolds of poly(ε-caprolactone) (PCL) and 10 wt.% bioactive glass (Korpela et al., 2013). It was reported that addition of bioactive glass increases the compressive modulus of composite from 104 to 147 MPa with excellent in vitro cell—material interactions and proliferation.

Another study reported SLS fabrication of PCL/HA composite scaffolds (Wiria et al., 2007) by optimizing scan speed, laser power, and HA concentration. The authors reported a yield stress (at 2% strain offset) of 11.54 ± 0.80 MPa with a modulus of 102.06 ± 11.26 MPa for PCL-10 wt.% HA composites. SLS has also been used to fabricate composite scaffolds with interconnected porosity using CaP: poly(hydroxybutyrate-co-hydroxyvalerate) and carbonated hydroxyapatite:poly(-L-lactic acid) (Duan et al., 2010). Although these composite scaffolds were mechanically very weak (compressive strength < 1 MPa), they found to have excellent in vitro cell adhesion, and proliferation followed by differentiation. Composite scaffolds of TCP/poly(D,L-lactide) (PDLLA) made by SLM (Lindner et al., 2011) were found to reduce acid environment due to degradation of PDLLA. Composite scaffold with equal amount of PDLLA and TCP using 0.5 W laser power exhibited maximum strength of 23 ± 1 MPa as measured by four-point bending. A novel SLA approach was also used to make composite scaffolds consisting of PDLLA:nano-HA (Ronca et al., 2013). It is known that methacrylate groups of PDLLA can be cross-linked to rigid polymer network leading to formation of PDLLA with high cross-linking density and brittle structure. Therefore, addition of HA to PDLLA is expected to enhance its mechanical and biological behavior. An optimized concentration of 5 wt.% nano-HA appears to provide scaffolds with superior pore network due to its acceptable viscosity and ease of SLA processing. Further, these composites with 5 wt.% nano-HA increased the flexural strength from 3.1 ± 0.4 GPa to 4.1 ± 0.3 GPa.

Polymer composites reinforced with bioactive glass were also made using 3DP. For example, porous composite scaffolds were fabricated using modified PCL, methacrylated poly(ε-caprolactone) reinforced with S53P4 bioactive glass (Elomaa et al., 2013). It was observed that with increase in bioactive glass content, compressive modulus of the scaffolds increased (3.4 MPa) up to maximum bioactive glass content of 20 wt.%. As the bioactive glass was uniformly dispersed in the PCL matrix, the bioactivity of scaffolds was significantly improved with accelerated apatite precipitation in SBF and enhanced cell adhesion and proliferation. Binder jetting is another popular method used to manufacture CaP composite scaffolds with 2 wt.% collagen (Inzana et al., 2014). Incorporation of collagen markedly increased toughness, flexural strength, in vitro cell proliferation and in vivo bone ingrowth after 9 weeks of implantation in murine femoral defect model. Table 16.4 lists some important 3DP technologies used to produce ceramic—polymer composites.

16.4 Bioprinting

Tissue engineering and regenerative medicine is one of the most emerging interdisciplinary fields employing principles from diverse backgrounds including biomaterials science, molecular and developmental biology, pharmaceutical science, bioengineering, and clinical medicines to create functional tissue constructs mimicking native tissue that can repair and/or replace damaged tissues or whole organs either impaired through disease or trauma (Khademhosseini et al., 2009). Contemporary tissue engineering approaches include seeding cells onto biocompatible and/or biodegradable scaffolds along with tissue-specific cytokines/growth factors that can provide appropriate microenvironments (physical, mechanical, and biochemical) to seeded cells to proliferate and differentiate and remodel into three-dimensional (3D) functional tissues. Moreover, a scaffold-free tissue engineering approach has also been recently introduced, which allows production of extracellular matrix (ECM) by cells and self-assembles to build 3D biological structures (Yu et al., 2016). Tissue engineering has been shown some clinical success in fabrication of comparatively less complex and/or avascular tissue constructs such as artificial bone, cartilage, skin, etc.

Table 16.4 Summary of some three-dimensional printing methods for making ceramic–polymer composites.

Ceramic	Polymer	Additive manufacturing method	References
Calcium Phosphate	Collagen	Binder jetting	Inzana et al. (2014)
	poly(ε-caprolactone) (PCL)	Mold by negative material extrusion followed by scaffold making by injection molding	Mondrinos et al. (2006)
Carbonated hydroxyapatite	Poly(L-lactic acid) (PLLA)	Selective laser sintering (SLS)	Duan et al. (2010)
Nano-CaP	Poly(hydroxybutyrate-co-hydroxyvalerate)		
Tricalcium phosphate (TCP)	Poly(lactide-co-glycolide) (PLGA)/PCL	Material extrusion (multihead deposition)	Kim et al. (2010)
	Polypropylene	Fused deposition modeling (FDM)	Kalita et al. (2003)
	PCL	FDM	Lam et al. (2009)
	Poly(D,L-lactide) (PDLLA)	Selective laser melting	Lindner et al. (2011)
	PLLA	Low-temperature material extrusion	Xiong et al. (2002)
Hydroxyapatite (HA)	PCL	Material extrusion (wet spinning)	Puppi et al. (2012)
		SLS	Wiria et al. (2007); Eosoly et al. (2010)
Nano-HA	PDLLA	Stereolithography (SLA)	Ronca et al. (2013)
HA/TCP	Agarose	SLA	Sánchez-Salcedo et al. (2008)
Bioactive glass	PCL	SLA FDM	Korpela et al. (2013); Elomaa et al. (2013)
	Poly(L-lactic acid)/ poly(ethylene glycol)	Material extrusion	Luo et al. (2012); Serra et al. (2013)

(Arai et al., 2018; Bodhak et al., 2018; Kwon et al., 2018). However, fabrication of complex vascular organs such as the heart, kidney, and liver by conventional tissue engineering approaches is not yet feasible due to current challenges with cell source, tissue vascularization, and compatible scaffolds, etc. (Murphy and Atala, 2014). Further, fabrication of anatomically correct patient-specific tissue constructs with well-controlled geometry and pore structure is very challenging in conventional tissue engineering strategies. To this end, 3DP has unfastened as exciting technology and adopted in tissue engineering for fabrication of complex patient-specific tissue constructs or whole organ. 3DP techniques such as SLA, FDM, SLM, SLS, EBM, BJ, and Digital Light Processing have been used to fabricate 3D scaffolds with controlled architecture and tuneable mechanical properties by layer-by-layer deposition of appropriate biomaterials (Ozbolat, 2016). Additional multifunctionality can further be tethered into 3D printed scaffolds by loading cytokines, growth factors, and drugs for targeted tissue regenerations.

Significant amount of research has been invested over past decade to develop tissue-engineered 3D printed blood vessels, heart muscle, nerves, cartilage, bone, liver, and other organs (Jammalamadaka and Tappa, 2018). However, in spite of anatomically correct patient-specific design, these tissue-engineered 3D printed tissue constructs fail to get much success in developing transplantable grade complex tissues and organs (Jammalamadaka and Tappa, 2018). In particular, manual cell seeding, maintaining high cell density, poor spatial cell proliferation, and most importantly lack of vascularization have restricted the 3DP as a stand-alone technique to fabricate complex and vascularized thick tissue constructs and organs. At this juncture, a relatively new strategy called "bioprinting or direct cell printing" has showed remarkable prospects for advancing 3DP or AM-based tissue engineering approach toward developing patient-specific de novo organs. Bioprinting method involves combination of 3DP, tissue engineering, and organ manufacturing techniques that offer controllable AM process, which allows simultaneous placement of cell-laden biologics (i.e., living cells with or without biomaterials, growth factors, proteins, drugs, or other biologically active carriers) in a prescribed layer-by-layer stacking in 3D organization to create tissue-like structures that imitate natural target tissue/organs (Biazar et al., 2018). One of the great advantages of bioprinting is patterning and precisely positioning for spatial placement of multiple cells rather than merely providing scaffold support automatically through a computer-controlled bioprinter into anatomically correct and reproducible patient-specific 3D tissue constructs or organs. In addition, high cell encapsulation density and a precise layout of vascular network can also be realized, which is otherwise very difficult to achieve in 3DP alone. Here, it can be recalled that the need of organ transplants has been steadily growing over the years. According to World Health Organization and the Spanish Transplant Organization, Organización Nacional de Trasplantes joint consortium "Global Observatory on Donation and Transplantation" database, in 2016, along a total of 135,860 organs (such as the kidney, liver, heart, etc.) were transplanted i.e., 7.25% increase compared with 2015, whereas only 34,854 transplants were received from deceased donors (Ozbolat, 2016; Du, 2018). This clearly indicates that organ shortage continues to be a great concern in the coming years, and long-term solution needs biofabrication of artificial organs by 3D bioprinting from patient's own cells for transplantation, thereby saving lives by alleviating organ shortage (Du, 2018). Over the past 10 years, significant research efforts have been put in this area to bioprint tissues, followed by in vivo implantation. However, as of today, none of these attempts have been extended to humans (Ozbolat, 2016).

16.4.1 Bioprinting strategies and classifications

3D bioprinting approaches typically involve three processing steps. (i) Firstly, 3D anatomical printing pattern of the tissue to be engineered is produced using CAD software. The raw imaging data of the tissues, obtained using CT or MRI, are consecutively translated into 3D anatomical representation and subsequently segregated into thin 2D horizontal slices by CAD software. (ii) Secondly, the slice information of tissues/organs is imported into a numerically controlled bioprinting system for manufacturing. Based on the computer-aided instructions, the bioprinter deposits the bioink, i.e., cell-laden biomaterials used as ink, on a suitable substrate in a layer-by-layer assembly via a cartridge or a syringe and patterning them into anatomically correct 3D biological constructs (Murphy and Atala, 2014). (iii) Finally, postprocessing and maturation of 3D biological constructs by suitable in vitro protocols for preparation of transplantable desired tissue/organ constructs is carried out (Zhang et al., 2017). The prerequisites to develop a successful bioprinting process include several basic characteristics such as ability of high-resolution and precise printing to obtain biomimetic cellular micro-/nanoarchitectural environment and 3D structure, multicellular as well as high cell density and cell viability, vascularization along with desirable mechanical strength, biocompatibility, biodegradability, and therapeutic activities, etc. (Murphy and Atala, 2014). Bioprinting technologies differ in four main aspects: cells, growth factors, biomaterials, and bioprinter types. Generally, bioinks, i.e., the mixture of cells, growth factors/therapeutic agents/biomolecules, and biomaterials such as biodegradable polymers are prepared before printing, and bioprinter is being utilized to selectively deposit bioinks in droplets or continuous pattern in a layer-by-layer fashion, where cell viability and functions are preserved into 3D structures mimicking the targeted tissue/organ architecture in structure, dimension, and shape (Chua and Yeong, 2015). In particular, three 3D bioprinting technologies, as shown in Fig. 16.9, were studied extensively in organ and hard tissue building applications, which are (i) inkjet-based, (ii) microextrusion-based, and (iii) laser-assisted.

16.4.1.1 Inkjet-based three-dimensional bioprinting

Inkjet-based bioprinters have been developed by modifying commercially available 2D ink-based printers (Xu et al., 2005). Wilson and Boland first utilized such fully computer-controlled commercial inkjet printers to fabricate cell and protein arrays (Murphy and Atala, 2014). These bioprintings were performed by replacing the printer ink cartridge by cells/proteins and depositing them in a predefined pattern in hydrogel based biopapers (Xu et al., 2005). Later, the biopaper has been replaced by electronic base stage to provide additional z-axis control, and custom-designed cell-printed 3D

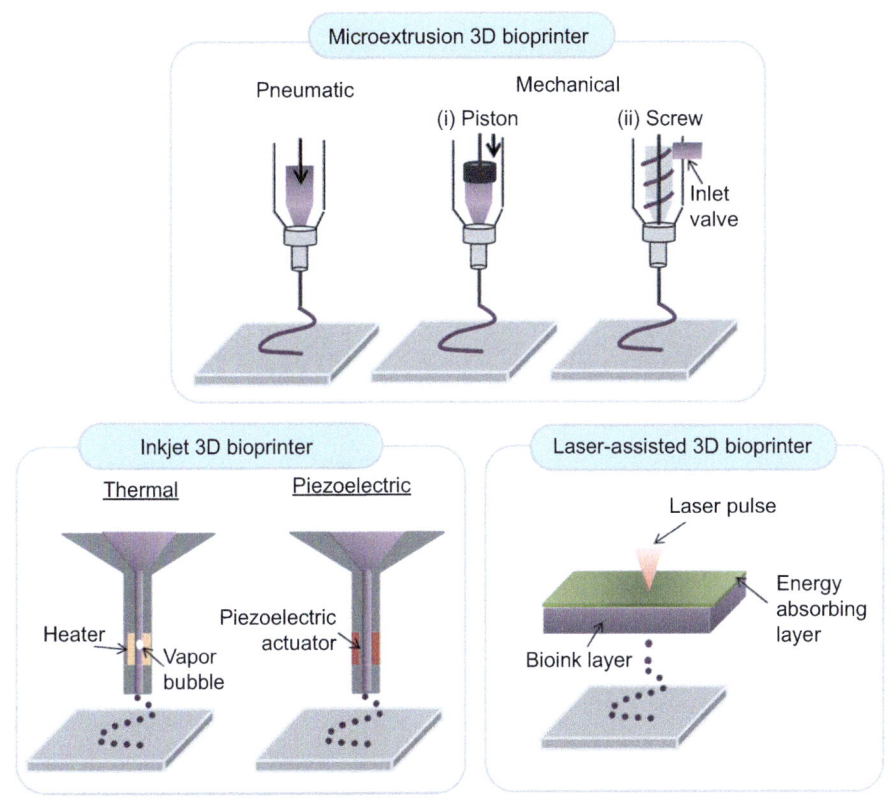

Biointegration of three-dimensional—printed biomaterials and biomedical devices   459

Figure 16.9 Different bioprinting methods utilized for creating three-dimensional (3D) biological constructs in hard tissue and organ-building applications.
Adapted from S. V Murphy, A. Atala, 2014. 3D bioprinting of tissues and organs. Nat. Biotechnol. 32, 773—785. https://doi.org/10.1038/nbt.2958.

biological constructs with tissue vascularization were successfully developed (Xu et al., 2008). Inject printer, also popularly known as drop-on-demand printer, deposits controlled volume of liquid based bioink in a predefined pattern onto a substrate. Based on the working principles, inkjet printers can be classified into two categories: thermal and piezoelectric inkjet bioprinter (Cui et al., 2012a,b; Seol et al., 2014). In thermal inkjet printer, electrical heating element is placed in the print head, which produces pulses of thermal forces or pressure that ejects the bioink droplets from the nozzle onto a substrate (as shown in Fig. 16.9) (Cui et al., 2012a,b). In contrast, piezoelectric inkjet bioprinter uses a piezoelectric crystal in the print head, which undergoes a shape transformation against a voltage pulse that induces an acoustic force inside the print head to break the bioink liquid placed in nozzle into droplets and subsequently deposits onto a substrate in a predefined pattern (Seol et al., 2014). Thermal inkjet printing is relatively cheap, offers high printing speed, and most importantly the exposure of cells to thermal and mechanical stress has shown to have minimal to no effects on cell viability and functionality (Cui et al., 2012a,b). The resolution of inkjet printers is typically between 20 and 100 μm. However, lack of control on droplet volume and uniformity, less directionality, poor cell encapsulations, and clogging at the print head are the major limitations of thermal inkjet printing (Cui et al., 2012a,b). In contrast, piezoelectric inkjet printer offers better control over droplet size, ejection rate, and superior resolution and high directionality of piezoelectric inkjet printer helps in creating concentration gradients of biologics in 3D constructs. However, some concerns have been raised over the frequencies (15−25 kHz) used in piezoelectric inkjet printing, which are found to damage cellular membrane (Cui et al., 2010). Salient advantages and limitations of inkjet bioprinter are highlighted in Table 16.5.

Table 16.5 Salient features of different bioprinting techniques.

Bioprinting Techniques	Advantages	Limitations
Inkjet bioprinter	• High resolution and compatible with many biologics. • High printing speed, low cost, wide availability, and less expensive. • Concentration gradients of biologics in three dimensional (3D) constructs.	• Liquid-based bioink requires postprocessing such as chemical or UV cross-linking to form 3D solid structure. • Low cell concentration. • Not compatible with highly viscous bioinks.
Microextrusion bioprinter	• Ability to print 3D biological constructs with high cell densities at low cost. • Compatible with wide range of bioinks (hydrogel, copolymers, cell spheroids, etc.) with mid to high viscosity. • Offers greater spatial control and beneficial for creating intraorgan branched vascular network.	• Poor cell viability. • Slow print speed and poor resolution.
Laser-assisted bioprinter	• Excellent cell viability and microscale resolution. • Noncontact and nozzle-free printing. • Printability for a wide range of viscous bioinks including peptides and DNA.	• Low flow rate. • Less control over cell positioning. • Moderately expensive.

16.4.1.2 Microextrusion-based three-dimensional bioprinting

Microextrusion-based bioprinter is developed based on the principle of microextrusion of continuous bead-based bioink filament into 3D biological constructs (Khoda et al., 2013). It is the most common method of printing nonbiological materials; however, in recent years, it started gaining popularity in bioprinting as well due to its several advantages. Microextrusion-based bioprinters typically utilizes a temperature-controlled handling and dispensing system (print head) for bioink along with a basement stage, which can be moved along x, y, and z axes controlled by a computer (Khoda et al., 2013). Based on the instructions, the print head deposits continuous bioink beads in 2D pattern (x-y axes) in the form of layer, and then by moving the print head or basement stage in the z axis, the second layer is deposited, thus creating complex 3D tissue constructs. Microextrusion technique has been successfully used to print tissue engineering scaffolds (Zhang and Zhang, 2015). Based on the working principle and dispensing mechanism, microextrusion bioprinters are classified into two categories: pneumatic (Khalil and Sun, 2007) and mechanical microextrusion bioprinters (Visser et al., 2013). Pneumatic extrusion system is based on the force generated by air pressure, whereas the mechanical microextrusion bioprinters can be further divided as piston- and screw-based microextrusion bioprinters, as shown in Fig. 16.9. It can be noted that mechanical dispensing system offers better spatial control over extruded bioink flow rate than pneumatic dispensing system (Khalil and Sun, 2007). Overall, microextrusion printer can be used for a wide range of bioinks with different fluid properties. One of the major advantages of microextrusion printer is its capablity to print 3D biological constructs with high cell densities, which is highly beneficial for tissue regeneration. Further, a wide range of bioinks with moderate to high viscosity (30 mPa/s to > 60 kPa/s) can be printed in microextrusion printers. Several studies have been successfully conducted to create artificial aortic valve (Duan et al., 2013) and vascular tissues (Norotte et al., 2009). However, in spite of successful printing of multicellular tissues/organs, microextrusion-based printing methods still suffer from poor cell viability, slow print speed, and inferior resolution than inkjet-based bioprinters.

16.4.1.3 Laser-assisted three-dimensional bioprinting

Laser-assisted bioprinting (LAB) has recently gained significant interest due to its reliability to print variety of biological materials including peptides and DNA (Colina et al., 2005; Dinca et al., 2008). In particular, LAB is increasingly being utilized in organ/tissue printing for excellent cell viability and microscale resolution (Guillemot et al., 2010). LAB works on the principle of laser-induced forward transfer technology, where a thin layer of laser absorbing layer (dynamic release layer), behind a thick layer of bioink (cells and hydrogel mixture), generated high-pressure bubble from the bioink layer that transfers the bioink onto the substrate on exposing to a focused laser (Fig. 16.9). Printing resolution of LAB is reportedly varied from 10 to 100 μm by changing laser parameters including laser wavelength, pulse duration, pulse energy, focal spot, air gap between coating layers and substrate, as well as on rheological properties of bioink (Koch et al., 2017). Further, chances of bioink clogging are also limited as the LAB is nozzle-free. However, low flow rate and less control over cell positioning are some of the current challenges of this process.

16.4.2 Bioinks for bioprinting

Selection of suitable bioinks with desired functional and mechanical properties is critical to mimic native tissue functionalities. Recently, natural (e.g., collagen, gelatin, alginate, hyaluronic acid, and chitosan) and synthetic polymers (e.g., poly(ethylene glycol) [PEG], PLA, and PCL) have been used as bioinks in 3DP due to their cost-effectiveness, biodegradability, and biocompatibility (Loo et al., 2015). Some of the important bioinks currently in use are briefly described below.

16.4.2.1 Natural polymers

Sodium alginate (alginate): It is a linear anionic polysaccharide (contains β-D mannuronic acid monomers and α-L-guluronic acid blocks domains), which is derived from brown seaweed. Its biomimetic structure, sufficient viscosity, biocompatibility, and temperature-responsive gelation characteristics make this material suitable for 3D printing (Dai et al., 2016). Besides, alginate-based hydrogels may be used in inkjet printing because of their biocompatibility and enough mechanical strength. But it suffers from poor cell adhesion, which can be addressed by modifying with arginylglycylaspartic acid, collagen type I, or oxygenation (Leach et al., 2016).

Collagen: Collagen is an ECM protein and constitutes of various types of amino acids, e.g., hydroxyproline, proline, glycine, etc. Here, hydroxyproline and proline maintain the tertiary structure of the collagen. Superior biocompatibility, easy printability, and excellent cell adhesion properties of collagen make it very popular amongst bioinks for tissue engineering applications (Rhee et al., 2016).

Gelatin: It is a thermal cross-linking, photo-responsive polymer and forms stable 3D structure after UV cross-linking (Billiet et al., 2014). It is biocompatible and has a good cell adhesion that makes cell viability and proliferation easier. But, its highly viscous nature limits its applications in bioprinting.

Chitosan: It is a biocompatible and biodegradable natural polymer composed of β-(1–4)–linked D-glucosamine and N-acetyl-D-glucosamine natural cationic polysaccharide, derived from the alkaline N-deacetylation of chitin. It is biocompatible and biodegradable in nature. It's nontoxic and not expensive, and its easily molded nature makes this material suitable for many applications (Elviri et al., 2017). The structural similarity of chitosan with glycosaminoglycan also makes it a good candidate for chondrogenesis and bone cell colonization. Although, inherent low mechanical properties of chitosan restrict its widespread applications, its properties can be improved by modifying chitosan with various monomers (Zakhem and Bitar, 2015).

Hyaluronic acid: It is biodegradable and biocompatible linear polysaccharide made of (β-1,3) β-1,4-linked D-glucuronic acid and N-acetyl-D-glucosamine disaccharides.

But high hydrophilic nature of hyaluronic acid often limits its printability (Pescosolido et al., 2011). But chemical cross-linking with another hydrophobic material can reduce the hydrophilic nature, but it is still not very much helpful in bioprinting. To this end, blending with photo cross-linkable polymers such as Dex-HEMA as well as physical blends of gelatin-hyaluronic acid have showed improvement in cell viabilities and printability (Skardal et al., 2010a,b).

16.4.2.2 Synthetic polymers

Superior mechanical properties and moderate biocompatibility of synthetic polymers have made them potential candidates for bioprinting applications. Some common synthetic polymers are described below.

Poly(lactide-co-glycolide) (PLGA): PLGA is a copolymer composed of lactide and glycolide. It is synthesized through ring opening polymerization (ROP) technique. Hydrolytic degradation behavior and fast solvent evaporation of PLGA make this polymer a promising bioink for 3D printing (Pirlo et al., 2012).

PEG: It is a biocompatible and hydrophilic polymer. PEG has been widely utilized in various applications such as nanoparticle coating, bioink for printing scaffolds, and encapsulation of cells (Skardal et al., 2010a,b). Very often, PEG is chemically modified into gel with other monomers to improve its cytocompatibility and protein absorption ability (Skardal et al., 2010a,b). It is known that PEG can form physical or chemical cross-linked networks after acrylation, and final PEG-based acrylated polymer may be used as bioink to print vascular grafts (Skardal et al., 2010a,b).

PLA: It is an aliphatic polymer with excellent mechanical strength. Because of its biodegradable, biocompatible, and semicrystalline nature, it is used for various tissue engineering applications. It is also used as bioink because of its low viscosity. Recently, acrylonitrile butadiene styrene-PLA blend was utilized as a bioink material to produce a cartilage graft (Rosenzweig et al., 2015).

PCL: It is a polyester-based thermoplastic with biocompatible and biodegradable nature that can be synthesized through ROP of caprolactone. Because of its low melting point, thermoplastic nature, and excellent mechanical properties, PCL is used as a bioink (Kundu et al., 2015). PCL is a viscous solution and therefore difficult to print. To overcome this drawback, electrohydrodynamic jet technique was used to create 3D construct (Kundu et al., 2015). Because of its high melting point (60°C), PCL itself cannot be used as cell-laden bioink, rather it has been found more suitable as supporting scaffolds in 3DP technology.

16.4.3 Niche application areas of bioprinting technology

In the recent years, several 3D bioprinting attempts have been made to fabricate different vascularized tissues and organs such as the bone, skin, blood vessels, heart muscle, liver, and cartilage, which is a nonvascular structure. Some of the application areas of bioprinting technology are summarized below.

16.4.3.1 Bone tissue

Bioprinting method is combined with bone tissue engineering to fabricate 3D scaffolds mimicking the natural bone tissue functionality to repair critical-sized defects. In one of those first few studies, in 2008, Fedorovich et al. (Fedorovich et al., 2008) has successfully fabricated a cell-laden hydrogel-based 3D constructs using microextrusion bioprinter based on pneumatic dispensing system for vascularized bone grafts. Different types of hydrogels (such as agarose, alginate, methylcellulose, etc.) were loaded with osteoprogenitor cells (such as endothelial progenitors and bone marrow stromal cells) and deposited together in the form of 3D fibers. The viability and osteogenic differentiation potential of the cell-laden 3D constructs were evaluated and compared between printed and as casted (unprinted) samples at different cell culture time periods. Evidently, the embedded cells survived during deposition process and retained their ability to osteogenic differentiation. This study also confirmed that two distinct cell populations can be successfully printed in the same scaffolds by exchanging bioink printing syringes ensuring that bioprinting can be potentially used for bone grafts with different cells. Phillippi et al. showed spatially controlled multilineage differentiation of stem cells (Phillippi et al., 2008). A piezoelectric drop-on-demand system was used to design BMP-2 patterns within a population of primary muscle-derived stem cells. Cooper et al. also patterned immobilized BMP-2 into 3D DermaMatrix human allograft scaffolds using an inkjet bioprinter (Cooper et al., 2010). This study showed that the spatial control of osteogenic lineage of host progenitor cells can be controlled in both in vitro models and in vivo mouse calvarial defect model toward formation of new bone. In another study, Gao et al. have successfully printed human bone marrow–derived mesenchymal stem cells (hBMSCs)–embedded hydrogel (poly(ethylene glycol) dimethacrylate [PEGDMA]) and conjugated with bioactive ceramic nanoparticles (bioglass and HA) by using a thermal inkjet bioprinter (Gao et al., 2014). The study demonstrated that HA is more effective compared with bioglass for hBMSCs osteogenesis in bioprinted bone constructs. In another study, a novel in situ LAB strategy has been adopted to deposit nano-HA particles into the mouse calvaria defect model for repairing of a critical size bone defect (Keriquel et al., 2010).

16.4.3.2 Cartilage

Existing tissue engineering approaches failed to develop functional artificial cartilage due to avascular structure, complex ECM composition, as well as difficult-to-mimic mechanical properties. Researchers are focusing on bioprinting as a viable option to overcome those difficulties in cartilage tissue engineering. For instance, an inkjet bioprinting has been utilized to create human chondrocyte-loaded PEGDMA-based 3D cartilage tissue to repair defects in osteochondral plugs (Cui et al., 2012a,b). Evidently, the cell viability can be increased by utilizing bioprinting methods, and mechanical properties of developed cartilage constructs have been observed to be in similar range of the properties of human articular cartilage. In another work, Hong et al. bioprinted a 3D hybrid stretchable hydrogel from PEG and sodium alginate,

Biointegration of three-dimensional—printed biomaterials and biomedical devices 465

which increased cell encapsulation efficiency along with cell viability and showed greater toughness than natural cartilage (Fig. 16.10) (Hong et al., 2015). LAB is also successfully utilized in printing mesenchymal stem cells, where the progenitor cells could retain their potency during printing and showed both chondrogenic differentiation in the printed MSC graft (Gruene et al., 2011). Recently, Xu et al. fabricated rabbit elastic chondrocyte-laden hydrogel scaffold by using an inkjet bioprinting method by combining 3D bioprinting and electrospinning techniques, where the cell viability was retained and good mechanical properties comparable with native cartilage tissue can be achieved in both in vitro and preclinical studies (Xu et al., 2012).

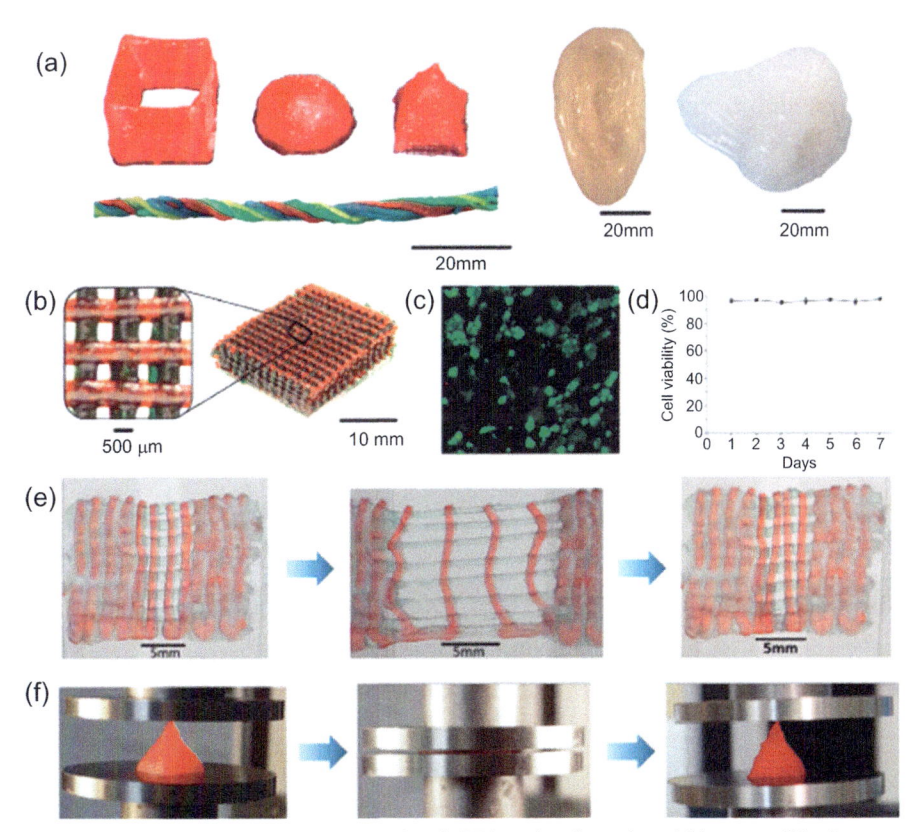

Figure 16.10 (a) Different three-dimensional (3D)—printed tough and biocompatible tissue constructs fabricated from poly(ethyelene glycol) (PEG)-alginate-nanoclay hydrogel. Red food dye was added after printing the constructs for visibility, (b) 3D-printed mesh type scaffolds prepared using hydrogel bioink. HEK cells were infused into the hydrogel scaffolds mesh, (c) fluorescence live/dead staining assay results of the HEK cells embedded hydrogel, (d) HEK cells viability assay results over different culture periods, and (e) stretching behavior of printed bilayer mesh (red is top layer and green is bottom layer) under uniaxial loading. Almost complete recovery of the part after stretching can be seen (f) compression behavior of the printed pyramid-like tissue construct that has undergone almost 95% compressive strain during test but could successfully regain it's origin shape after relaxation (Hong et al., 2015).

16.4.3.3 Skin

3D bioprinting is still a new and evolving technology for artificial skin fabrication by combining collagen-based hydrogel (representing main ECM protein in skin) with fibroblast and/or keratinocyte cells (primary cells found in the skin tissue). For instance, Koch et al. have successfully utilized LAB and successfully deposited 20 layers of fibroblasts (murine NIH-3 T3) and 20 layers of keratinocytes (human HaCaT) encapsulated within collagen matrix and printed in layer-by-layer pattern on a sheet of Matriderm (decellularized dermal matrix) (as shown in Fig. 16.11) to create 3D skin graft (Koch et al., 2012). Fluorescence microscopy results confirmed that the developed bilayered structure is comparable with that of dermis- and epidermis-like structure (Fig. 16.11). Microextrusion bioprinting was used to fabricate skin substitute by depositing layer-by-layer assembly of collagen, dermal fibroblasts, and epidermal keratinocytes, which showed potential to treat full thickness skin damages (Lee et al., 2014).

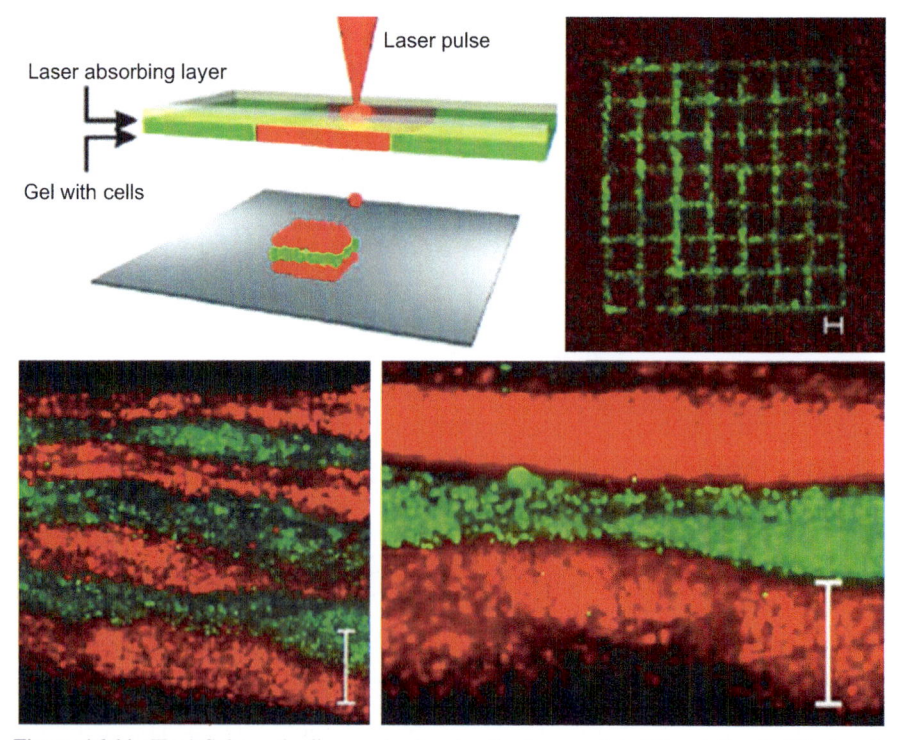

Figure 16.11 (Top) Schematic diagram is showing the laser-assisted bioprinting LaBPset up in which cell-laden hydrogel is printed into a three-dimensional (3D) grid-like pattern via layer-by-layer pattern. The green fluorescence color indicates murine fibroblast cells, and red fluorescence color indicates human keratinocyte cells. (Bottom) A high magnification image reveals that each color layer consists of four printed sublayers (Koch et al., 2012).

16.5 Current challenges and future directions

Although 3DP is an exciting development in the multidisciplinary domain of materials science, biology, manufacturing, and clinical sciences, there are still many challenges that need to be overcome in the coming years. While in the past 20 years processing of tailored porosity ceramics has evolved significantly (Bose et al., 1999, 2003, 2012, 2013; S. Bose et al., 2013; Bandyopadhyay et al., 2015; Darsell et al., 2003), reliable 3DP process that can help to manufacture porous ceramic scaffolds in a reproducible manner still does not exist. While a large number of metallic implants are currently manufactured via 3DP using laser and E-beam−based processes, most of this manufacturing is done based on existing designs. Novel porous implants for load-bearing applications are yet to see commercial success. However, in the field of spinal implants, 3DP of polymers and metals are becoming popular with a variety of novel designs that are only possible via 3DP technology. Multimaterials structures via 3DP is another area that is set to expand in the coming years where manufacturing of devices using multiple materials including sensors can revolutionize the field of biomedical devices (Bandyopadhyay and Heer, 2018). 3DP is also being developed for different surface modification techniques (Bose et al., 2018a,b,c,d; Stenberg et al., 2018; Bose et al., 2018a,b,c,d), and it is expected to make a significant impact in the coming days beyond porous metal coatings. Finally, applications of 3DP in drug delivery devices is also another area that has tremendous growth potential due to site-specific delivery option with unique structural features for the devices (Bose et al., 2018a,b,c,d; Tarafder and Bose, 2014). Although our focus for this article remained 3DP, the world of 4D printing, keeping time as the fourth dimension, is also expected to see some of the same excitements in the coming years.

16.6 Summary

Applications of 3DP in biomaterials and biomedical devices are expanding rapidly. It is estimated that in 2018, over 100,000 biomedical devices were 3D printed, which were actually used commercially. And this trend will continue due to inherent advantages of patient-matched devices, small foot print in manufacturing operation, and flexibility in manufacturing on demand. Although the trend is to transform the world of manufacturing using 3DP or AM, 3DP is allowing massive expansion in innovation toward next generation of biomedical devices. From multimaterials structures to bioprinting or organ tissue engineering are becoming a common research theme across the world due to advancement in 3DP technologies. It is our hope that this chapter helps the reader to gain some fundamental understanding of this exciting field.

References

Ali, H., Ghadbeigi, H., Mumtaz, K., 2018. Effect of scanning strategies on residual stress and mechanical properties of selective laser melted Ti6Al4V. Mater. Sci. Eng. A 712, 175−187. https://doi.org/10.1016/j.msea.2017.11.103.

Ahmadi, S.M., Campoli, G., Amin Yavari, S., Sajadi, B., Wauthle, R., Schrooten, J., et al., 2014. Mechanical behavior of regular open-cell porous biomaterials made of diamond lattice unit cells. J. Mech. Behav. Biomed. Mater. 34, 106−115. https://doi.org/10.1016/j.jmbbm.2014.02.003.

Arai, K., Murata, D., Verissimo, A.R., Mukae, Y., Itoh, M., Nakamura, A., et al., 2018. Fabrication of scaffold-free tubular cardiac constructs using a Bio-3D printer. PLoS One 13, e0209162. https://doi.org/10.1371/journal.pone.0209162.

ATSM, A.S.T.M., 2014. F2924 − 14 Standard Specification for Additive Manufacturing Titanium-6 Aluminum-4 Vanadium With Powder Bed Fusion. https://www.astm.org/Standards/F2924.htm.

Balla, V.K., Bodhak, S., Bose, S., Bandyopadhyay, A., 2010a. Porous tantalum structures for bone implants: fabrication, mechanical and in vitro biological properties. Acta Biomater 6, 3349−3359. https://doi.org/10.1016/j.actbio.2010.01.046.

Balla, V.K., Bose, S., Bandyopadhyay, A., 2010b. Understanding compressive deformation in porous titanium. Philos. Mag. 90, 3081−3094. https://doi.org/10.1080/14786431003800891.

Bandyopadhyay, A., Heer, B., 2018. Additive manufacturing of multi-material structures. Mater. Sci. Eng. R Rep. 129, 1−16.

Bandyopadhyay, A., Espana, F., Balla, V.K., Bose, S., Ohgami, Y., Davies, N.M., 2010. Influence of porosity on mechanical properties and in vivo response of Ti6Al4V implants. Acta Biomater 6, 1640−1648. https://doi.org/10.1016/j.actbio.2009.11.011.

Bandyopadhyay, A., Balla, V.K., Bernard, S.A., Bose, S., 2011. Micro-layered manufacturing. In: Koc, M., Özel, T. (Eds.), Micro-Manufacturing Des. Manuf. Micro-Products. John Wiley & Sons, Inc, pp. 97−158.

Bandyopadhyay, A., Bose, S., Das, S., 2015. 3D printing of biomaterials. MRS Bull. 40 (02), 108−115.

Bartolomeu, F., Buciumeanu, M., Pinto, E., Alves, N., Carvalho, O., Silva, F.S., et al., 2017. 316L stainless steel mechanical and tribological behavior—a comparison between selective laser melting, hot pressing and conventional casting. Addit. Manuf. https://doi.org/10.1016/j.addma.2017.05.007.

Bergmann, C., Lindner, M., Zhang, W., Koczur, K., Kirsten, A., Telle, R., et al., 2010. 3D printing of bone substitute implants using calcium phosphate and bioactive glasses. J. Eur. Ceram. Soc. 30, 2563−2567. https://doi.org/10.1016/J.JEURCERAMSOC.2010.04.037.

Bian, W., Li, D., Lian, Q., Li, X., Zhang, W., Wang, K., et al., 2012. Fabrication of a bio-inspired beta-Tricalcium phosphate/collagen scaffold based on ceramic stereolithography and gel casting for osteochondral tissue engineering. Rapid Prototyp. J. 18, 68−80. https://doi.org/10.1108/13552541211193511.

Biazar, E., Najafi, S.M., Heidari, K.S., Yazdankhah, M., Rafiei, A., Biazar, D., 2018. 3D bioprinting technology for body tissues and organs regeneration. J. Med. Eng. Technol. 42, 187−202. https://doi.org/10.1080/03091902.2018.1457094.

Bibb, R., Eggbeer, D., Williams, R., 2006. Rapid manufacture of removable partial denture frameworks. Rapid Prototyp. J. 12, 95−99. https://doi.org/10.1108/13552540610652438.

Bibb, R., Eggbeer, D., Evans, P., Bocca, A., Sugar, A., 2009. Rapid manufacture of custom-fitting surgical guides. Rapid Prototyp. J. 15, 346−354. https://doi.org/10.1108/13552540910993879.

Billiet, T., Gevaert, E., De Schryver, T., Cornelissen, M., Dubruel, P., 2014. The 3D printing of gelatin methacrylamide cell-laden tissue-engineered constructs with high cell viability. Biomaterials 35, 49−62. https://doi.org/10.1016/j.biomaterials.2013.09.078.

Bodhak, S., de Castro, L.F., Kuznetsov, S.A., Azusa, M., Bonfim, D., Robey, P.G., et al., 2018. Combinatorial cassettes to systematically evaluate tissue-engineered constructs in recipient mice. Biomaterials 186, 31−43. https://doi.org/10.1016/J.BIOMATERIALS.2018.09.035.

Bose, S., Tarafder, S., 2012. Calcium phosphate ceramic systems in growth factor and drug delivery for bone tissue engineering: a review. Acta Biomater 8, 1401−1421. https://doi.org/10.1016/J.ACTBIO.2011.11.017.

Bose, S., Sugiura, S., Bandyopadhyay, A., 1999. Processing of controlled porosity ceramic structures via fused deposition process. Scripta Mater. 41 (9), 1009−1014.

Bose, S., Darsell, J., Kintner, M., Hosick, H., Bandyopadhyay, A., 2003a. Pore size and pore volume effects on alumina and TCP ceramic scaffolds. Mater. Sci. Eng. C 23, 479−486. https://doi.org/10.1016/S0928-4931(02)00129-7.

Bose, S., Darsell, J., Kintner, M., Hosick, H., Bandyopadhyay, A., 2003b. Pore size and pore volume effects on calcium phosphate based ceramics. Mater. Sci. Eng. C 23, 479−486.

Bose, S., Roy, M., Bandyopadhyay, A., 2012. Recent advances in bone tissue engineering scaffolds. Trends Biotechnol. 30 (10), 546−554.

Bose, S., Fielding, G., Tarafder, S., Bandyopadhyay, A., 2013. Trace element doping in calcium phosphate ceramics to understand osteogenesis and angiogenesis. Trends Biotechnol. 31 (10), 594−605.

Bose, S., Vahabzadeh, S., Bandyopadhyay, A., 2013. Bone tissue engineering using 3D printing. Mater. Today 16, 496−504. https://doi.org/10.1016/j.mattod.2013.11.017.

Bose, S., Robertson, S.F., Bandyopadhyay, A., 2018a. Surface modification of biomaterials and biomedical devices using additive manufacturing. Acta Biomater. 66, 6−22.

Bose, S., Banerjee, D., Shivaram, A., Tarafder, S., Bandyopadhyay, A., 2018b. Calcium phosphate coated 3D printed porous titanium with nanoscale surface modification for orthopedic and dental applications. Mater. Des. 151, 102−112.

Bose, S., Emshadi, K., Vu, A.A., Bandyopadhyay, A., 2018c. Effects of polycaprolactone on alendronate drug release from Mg-doped hydroxyapatite coating on titanium. Mater. Sci. Eng. C 88, 166−171.

Bose, S., Ke, D., Sahasrabudhe, H., Bandyopadhyay, A., 2018d. Additive manufacturing of biomaterials. Prog. Mater. Sci. 93, 45−111. https://doi.org/10.1016/J.PMATSCI.2017.08.003.

Casati, R., Lemke, J., Vedani, M., 2016. Microstructure and fracture behavior of 316L austenitic stainless steel produced by selective laser melting. J. Mater. Sci. Technol. 32, 738−744. https://doi.org/10.1016/j.jmst.2016.06.016.

Castilho, M., Moseke, C., Ewald, A., Gbureck, U., Groll, J., Pires, I., et al., 2014. Direct 3D powder printing of biphasic calcium phosphate scaffolds for substitution of complex bone defects. Biofabrication 6, 015006. https://doi.org/10.1088/1758-5082/6/1/015006.

Castilho, M., Rodrigues, J., Pires, I., Gouveia, B., Pereira, M., Moseke, C., et al., 2015. Fabrication of individual alginate-TCP scaffolds for bone tissue engineering by means of powder printing. Biofabrication 7, 015004. https://doi.org/10.1088/1758-5090/7/1/015004.

Charbonnier, B., Laurent, C., Blanc, G., Valfort, O., Marchat, D., 2016. Porous bioceramics produced by impregnation of 3D-printed wax mold: ceramic architectural control and process limitations. Adv. Eng. Mater. 18, 1728−1737. https://doi.org/10.1002/adem.201600308.

Chen, W., Yin, G., Feng, Z., Liao, X., 2018. Effect of powder feedstock on microstructure and mechanical properties of the 316L stainless steel fabricated by selective laser melting. Metals 8, 729. https://doi.org/10.3390/met8090729.

Cheng, X.Y., Li, S.J., Murr, L.E., Zhang, Z.B., Hao, Y.L., Yang, R., et al., 2012. Compression deformation behavior of Ti-6Al-4V alloy with cellular structures fabricated by electron beam melting. J. Mech. Behav. Biomed. Mater. 16, 153−162. https://doi.org/10.1016/j.jmbbm.2012.10.005.

Cherry, J.A., Davies, H.M., Mehmood, S., Lavery, N.P., Brown, S.G.R., Sienz, J., 2014. Investigation into the effect of process parameters on microstructural and physical properties of 316L stainless steel parts by selective laser melting. Int. J. Adv. Manuf. Technol. 76, 869−879. https://doi.org/10.1007/s00170-014-6297-2.

Choy, S.Y., Sun, C.N., Leong, K.F., Wei, J., 2017. Compressive properties of functionally graded lattice structures manufactured by selective laser melting. Mater. Des. 131, 112−120. https://doi.org/10.1016/j.matdes.2017.06.006.

Chu, T.M., Halloran, J.W., Hollister, S.J., Feinberg, S.E., 2001. Hydroxyapatite implants with designed internal architecture. J. Mater. Sci. Mater. Med. 12, 471−478. http://www.ncbi.nlm.nih.gov/pubmed/15348260.

Chua, C.K., Yeong, W.Y., 2015. Bioprinting: Principles and Applications. WORLD SCIENTIFIC. https://doi.org/10.1142/9193.

Colina, M., Serra, P., Fernández-Pradas, J.M., Sevilla, L., Morenza, J.L., 2005. DNA deposition through laser induced forward transfer. Biosens. Bioelectron. 20, 1638−1642. https://doi.org/10.1016/j.bios.2004.08.047.

Cooper, G.M., Miller, E.D., Decesare, G.E., Usas, A., Lensie, E.L., Bykowski, M.R., et al., 2010. Inkjet-based biopatterning of bone morphogenetic protein-2 to spatially control calvarial bone formation. Tissue Eng. A. 16, 1749−1759. https://doi.org/10.1089/ten.TEA.2009.0650.

Cox, S.C., Thornby, J.A., Gibbons, G.J., Williams, M.A., Mallick, K.K., 2015. 3D printing of porous hydroxyapatite scaffolds intended for use in bone tissue engineering applications. Mater. Sci. Eng. C 47, 237−247. https://doi.org/10.1016/j.msec.2014.11.024.

Cui, X., Dean, D., Ruggeri, Z.M., Boland, T., 2010. Cell damage evaluation of thermal inkjet printed Chinese hamster ovary cells. Biotechnol. Bioeng. 106, 963−969. https://doi.org/10.1002/bit.22762.

Cui, X., Boland, T., D'Lima, D.D., Lotz, M.K., 2012a. Thermal inkjet printing in tissue engineering and regenerative medicine. Recent Pat. Drug Deliv. Formulation 6, 149−155. http://www.ncbi.nlm.nih.gov/pubmed/22436025.

Cui, X., Breitenkamp, K., Finn, M.G., Lotz, M., D'Lima, D.D., 2012b. Direct human cartilage repair using three-dimensional bioprinting technology. Tissue Eng. A. 18, 1304−1312. https://doi.org/10.1089/ten.TEA.2011.0543.

Dabrowski, B., Swieszkowski, W., Godlinski, D., Kurzydlowski, K.J., 2010. Highly porous titanium scaffolds for orthopaedic applications. J. Biomed. Mater. Res. B Appl. Biomater. 95, 53−61. https://doi.org/10.1002/jbm.b.31682.

Dai, G., Wan, W., Zhao, Y., Wang, Z., Li, W., Shi, P., et al., 2016. Controllable 3D alginate hydrogel patterning via visible-light induced electrodeposition. Biofabrication 8, 025004. https://doi.org/10.1088/1758-5090/8/2/025004.

Darsell, J., Bose, S., Hosick, H., Bandyopadhyay, A., 2003. From CT scans to ceramic bone grafts. J. Am. Ceram. Soc. 86 (7), 1076—1080.

Das, M., Balla, V.K., 2015. Additive manufacturing and innovation in materials world. In: Bandyopadhyay, A., Bose, S. (Eds.), Addit. Manuf. CRC Press, pp. 295—330. https://doi.org/10.1201/b18893-15.

Das, M., Balla, V.K., Kumar, T.S.S., Manna, I., 2013. Fabrication of biomedical implants using laser engineered net shaping (LENS^{TM}). Trans. Indian Ceram. Soc. 72 https://doi.org/10.1080/0371750X.2013.851619.

Dellinger, J.G., Eurell, J.A.C., Jamison, R.D., Jamison, R.D., 2006. Bone response to 3D periodic hydroxyapatite scaffolds with and without tailored microporosity to deliver bone morphogenetic protein 2. J. Biomed. Mater. Res. A 76A, 366—376. https://doi.org/10.1002/jbm.a.30523.

Demir, A.G., Monguzzi, L., Previtali, B., 2017. Selective laser melting of pure Zn with high density for biodegradable implant manufacturing. Addit. Manuf. 15, 20—28. https://doi.org/10.1016/j.addma.2017.03.004.

Detsch, R., Schaefer, S., Deisinger, U., Ziegler, G., Seitz, H., Leukers, B., 2011. *In vitro* -osteoclastic activity studies on surfaces of 3D printed calcium phosphate scaffolds. J. Biomater. Appl. 26, 359—380. https://doi.org/10.1177/0885328210373285.

DeVasConCellos, P., Balla, V.K., Bose, S., Fugazzi, R., Dernell, W.S., Bandyopadhyay, A., 2012. Patient specific implants for amputation prostheses: design, manufacture and analysis. Vet. Comp. Orthop. Traumatol. 25, 286—296. https://doi.org/10.3415/VCOT-11-03-0043.

Dinca, V., Kasotakis, E., Catherine, J., Mourka, A., Ranella, A., Ovsianikov, A., et al., 2008. Directed three-dimensional patterning of self-assembled peptide fibrils. Nano Lett 8, 538—543. https://doi.org/10.1021/nl072798r.

Diogo, G.S., Gaspar, V.M., Serra, I.R., Fradique, R., Correia, I.J., 2014. Manufacture of β-TCP/alginate scaffolds through a Fab@home model for application in bone tissue engineering. Biofabrication 6, 025001. https://doi.org/10.1088/1758-5082/6/2/025001.

Du, X., 2018. 3D bio-printing review. IOP Conf. Ser. Mater. Sci. Eng. 301, 012023. https://doi.org/10.1088/1757-899X/301/1/012023.

Duan, B., Wang, M., Zhou, W.Y., Cheung, W.L., Li, Z.Y., Lu, W.W., 2010. Three-dimensional nanocomposite scaffolds fabricated via selective laser sintering for bone tissue engineering. Acta Biomater 6, 4495—4505. https://doi.org/10.1016/J.ACTBIO.2010.06.024.

Duan, B., Hockaday, L.A., Kang, K.H., Butcher, J.T., 2013. 3D bioprinting of heterogeneous aortic valve conduits with alginate/gelatin hydrogels. J. Biomed. Mater. Res. A 101A, 1255—1264. https://doi.org/10.1002/jbm.a.34420.

Elomaa, L., Kokkari, A., Närhi, T., Seppälä, J.V., 2013. Porous 3D modeled scaffolds of bioactive glass and photocrosslinkable poly(ε-caprolactone) by stereolithography. Compos. Sci. Technol. 74, 99—106. https://doi.org/10.1016/J.COMPSCITECH.2012.10.014.

Elviri, L., Foresti, R., Bergonzi, C., Zimetti, F., Marchi, C., Bianchera, A., et al., 2017. Highly defined 3D printed chitosan scaffolds featuring improved cell growth. Biomed. Mater. 12, 045009. https://doi.org/10.1088/1748-605X/aa7692.

Eosoly, S., Brabazon, D., Lohfeld, S., Looney, L., 2010. Selective laser sintering of hydroxyapatite/poly-ε-caprolactone scaffolds. Acta Biomater 6, 2511—2517. https://doi.org/10.1016/J.ACTBIO.2009.07.018.

Eqtesadi, S., Motealleh, A., Miranda, P., Lemos, A., Rebelo, A., Ferreira, J.M.F., 2013. A simple recipe for direct writing complex 45S5 Bioglass® 3D scaffolds. Mater. Lett. 93, 68—71. https://doi.org/10.1016/J.MATLET.2012.11.043.

Eqtesadi, S., Motealleh, A., Miranda, P., Pajares, A., Lemos, A., Ferreira, J.M.F., 2014. Robocasting of 45S5 bioactive glass scaffolds for bone tissue engineering. J. Eur. Ceram. Soc. 34, 107−118. https://doi.org/10.1016/J.JEURCERAMSOC.2013.08.003.

Eqtesadi, S., Motealleh, A., Pajares, A., Guiberteau, F., Miranda, P., 2016. Improving mechanical properties of 13−93 bioactive glass robocast scaffold by poly (lactic acid) and poly (ε-caprolactone) melt infiltration. J. Non-Cryst. Solids 432, 111−119. https://doi.org/10.1016/J.JNONCRYSOL.2015.02.025.

Fedorovich, N.E., De Wijn, J.R., Verbout, A.J., Alblas, J., Dhert, W.J.A., 2008. Three-dimensional fiber deposition of cell-laden, viable, patterned constructs for bone tissue printing. Tissue Eng. A. 14, 127−133. https://doi.org/10.1089/ten.a.2007.0158.

Feilden, E., Ferraro, C., Zhang, Q., García-Tuñón, E., D'Elia, E., Giuliani, F., et al., 2017. 3D printing bioinspired ceramic composites. Sci. Rep. 7, 13759. https://doi.org/10.1038/s41598-017-14236-9.

Felzmann, R., Gruber, S., Mitteramskogler, G., Tesavibul, P., Boccaccini, A.R., Liska, R., et al., 2012. Lithography-based additive manufacturing of cellular ceramic structures. Adv. Eng. Mater. 14, 1052−1058. https://doi.org/10.1002/adem.201200010.

Fielding, G., Bose, S., 2013. SiO_2 and ZnO dopants in three-dimensionally printed tricalcium phosphate bone tissue engineering scaffolds enhance osteogenesis and angiogenesis in vivo. Acta Biomater 9, 9137−9148. https://doi.org/10.1016/j.actbio.2013.07.009.

Fielding, G.A., Bandyopadhyay, A., Bose, S., 2012. Effects of silica and zinc oxide doping on mechanical and biological properties of 3D printed tricalcium phosphate tissue engineering scaffolds. Dent. Mater. 28, 113−122. https://doi.org/10.1016/j.dental.2011.09.010.

Fierz, F.C., Beckmann, F., Huser, M., Irsen, S.H., Leukers, B., Witte, F., et al., 2008. The morphology of anisotropic 3D-printed hydroxyapatite scaffolds. Biomaterials 29, 3799−3806. https://doi.org/10.1016/J.BIOMATERIALS.2008.06.012.

Gao, G., Schilling, A.F., Yonezawa, T., Wang, J., Dai, G., Cui, X., 2014. Bioactive nanoparticles stimulate bone tissue formation in bioprinted three-dimensional scaffold and human mesenchymal stem cells. Biotechnol. J. 9, 1304−1311. https://doi.org/10.1002/biot.201400305.

Gbureck, U., Vorndran, E., Müller, F.A., Barralet, J.E., 2007a. Low temperature direct 3D printed bioceramics and biocomposites as drug release matrices. J. Control. Release 122, 173−180. https://doi.org/10.1016/j.jconrel.2007.06.022.

Gbureck, U., Hölzel, T., Doillon, C.J., Müller, F.A., Barralet, J.E., 2007b. Direct printing of bioceramic implants with spatially localized angiogenic factors. Adv. Mater. 19, 795−800. https://doi.org/10.1002/adma.200601370.

Gebhardt, A., Schmidt, F.M., Hötter, J.S., Sokalla, W., Sokalla, P., 2010. Additive manufacturing by selective laser melting: the realizer desktop machine and its application for the dental industry. Phys. Procedia. 5, 543−549. https://doi.org/10.1016/j.phpro.2010.08.082.

Geetha, M., Singh, A.K., Asokamani, R., Gogia, A.K., 2009. Ti based biomaterials, the ultimate choice for orthopaedic implants − a review. Prog. Mater. Sci. 54, 397−425. https://doi.org/10.1016/j.pmatsci.2008.06.004.

Gmeiner, R., Mitteramskogler, G., Stampfl, J., Boccaccini, A.R., 2015. Stereolithographic ceramic manufacturing of high strength bioactive glass. Int. J. Appl. Ceram. Technol. 12, 38−45. https://doi.org/10.1111/ijac.12325.

Griffith, M.L., Chu, T.M., Wagner, W., Halloran, J.W., 1996. Ceramic stereolithography for investment casting and biomedical applications. In: Bourell, J.W.B.D.L., Beaman, J.J., Marcus, H.L., Crawford, R.H. (Eds.), Int. Solid Free. Fabr. Symp., Austin, TX, USA, pp. 31−38.

Gruene, M., Deiwick, A., Koch, L., Schlie, S., Unger, C., Hofmann, N., et al., 2011. Laser printing of stem cells for biofabrication of scaffold-free autologous grafts. Tissue Eng. C Methods 17, 79−87. https://doi.org/10.1089/ten.TEC.2010.0359.

Guillemot, F., Souquet, A., Catros, S., Guillotin, B., 2010. Laser-assisted cell printing: principle, physical parameters versus cell fate and perspectives in tissue engineering. Nanomedicine 5, 507−515. https://doi.org/10.2217/nnm.10.14.

Guvendiren, M., Molde, J., Soares, R.M.D., Kohn, J., 2016. Designing biomaterials for 3D printing. ACS Biomater. Sci. Eng. 2, 1679−1693. https://doi.org/10.1021/acsbiomaterials.6b00121.

Habibovic, P., Gbureck, U., Doillon, C., Bassett, D., Vanblitterswijk, C., Barralet, J., 2008. Osteoconduction and osteoinduction of low-temperature 3D printed bioceramic implants. Biomaterials 29, 944−953. https://doi.org/10.1016/j.biomaterials.2007.10.023.

Haleem, A., Javaid, M., 2018. Role of CT and MRI in the design and development of orthopaedic model using additive manufacturing. J. Clin. Orthop. Trauma. 9, 213−217. https://doi.org/10.1016/j.jcot.2018.07.002.

Hao, Y.L., Li, S.J., Yang, R., 2016. Biomedical titanium alloys and their additive manufacturing. Rare Met 35, 661−671. https://doi.org/10.1007/s12598-016-0793-5.

Harun, W.S.W., Manam, N.S., Kamariah, M.S.I.N., Sharif, S., Zulkifly, A.H., Ahmad, I., et al., 2018. A review of powdered additive manufacturing techniques for Ti-6al-4v biomedical applications. Powder Technol 331, 74−97. https://doi.org/10.1016/j.powtec.2018.03.010.

Hasib, H., Harrysson, O.L.A., West, H.A., 2015. Powder removal from Ti-6Al-4V cellular structures fabricated via electron beam melting. JOM 67, 639−646. https://doi.org/10.1007/s11837-015-1307-x.

Heeling, T., Wegener, K., 2018. The effect of multi-beam strategies on selective laser melting of stainless steel 316L. Addit. Manuf. 22, 334−342. https://doi.org/10.1016/j.addma.2018.05.026.

Heinl, P., Müller, L., Körner, C., Singer, R.F., Müller, F.A., 2008. Cellular Ti-6Al-4V structures with interconnected macro porosity for bone implants fabricated by selective electron beam melting. Acta Biomater 4, 1536−1544. https://doi.org/10.1016/j.actbio.2008.03.013.

Herzog, D., Seyda, V., Wycisk, E., Emmelmann, C., 2016. Additive manufacturing of metals. Acta Mater 117, 371−392. https://doi.org/10.1016/j.actamat.2016.07.019.

Hollister, S.J., 2005. Porous scaffold design for tissue engineering. Nat. Mater. 4, 518−524. https://doi.org/10.1093/jb/mvj031.

Hong, S., Sycks, D., Chan, H.F., Lin, S., Lopez, G.P., Guilak, F., et al., 2015. 3D printing of highly stretchable and tough hydrogels into complex, cellularized structures. Adv. Mater. 27, 4035−4040. https://doi.org/10.1002/adma.201501099.

Horn, T.J., Harrysson, O.L.A., Marcellin-Little, D.J., West, H.A., Lascelles, B.D.X., Aman, R., 2014. Flexural properties of Ti6Al4V rhombic dodecahedron open cellular structures fabricated with electron beam melting. Addit. Manuf. 1−4, 2−11. https://doi.org/10.1016/j.addma.2014.05.001.

Inzana, J.A., Olvera, D., Fuller, S.M., Kelly, J.P., Graeve, O.A., Schwarz, E.M., et al., 2014. 3D printing of composite calcium phosphate and collagen scaffolds for bone regeneration. Biomaterials 35, 4026−4034. https://doi.org/10.1016/j.biomaterials.2014.01.064.

Jammalamadaka, U., Tappa, K., 2018. Recent advances in biomaterials for 3D printing and tissue engineering. J. Funct. Biomater. 9, 22. https://doi.org/10.3390/jfb9010022.

Jardini, A.L., Larosa, M.A., de Carvalho Zavaglia, C.A., Bernardes, L.F., Lambert, C.S., Kharmandayan, P., et al., 2014. Customised titanium implant fabricated in additive manufacturing for craniomaxillofacial surgery. Virtual Phys. Prototyp. 9, 115−125. https://doi.org/10.1080/17452759.2014.900857.

Kalita, S.J., Bose, S., Hosick, H.L., Bandyopadhyay, A., 2003. Development of controlled porosity polymer-ceramic composite scaffolds via fused deposition modeling. Mater. Sci. Eng. C 23, 611−620. https://doi.org/10.1016/S0928-4931(03)00052-3.

Keriquel, V., Guillemot, F., Arnault, I., Guillotin, B., Miraux, S., Amédée, J., et al., 2010. In vivo bioprinting for computer- and robotic-assisted medical intervention: preliminary study in mice. Biofabrication 2, 014101. https://doi.org/10.1088/1758-5082/2/1/014101.

Khademhosseini, A., Vacanti, J.P., Langer, R., 2009. Progress in tissue engineering. Sci. Am. 300, 64−71. http://www.ncbi.nlm.nih.gov/pubmed/19438051.

Khalil, S., Sun, W., 2007. Biopolymer deposition for freeform fabrication of hydrogel tissue constructs. Mater. Sci. Eng. C 27, 469−478. https://doi.org/10.1016/J.MSEC.2006.05.023.

Khalyfa, A., Vogt, S., Weisser, J., Grimm, G., Rechtenbach, A., Meyer, W., et al., 2007. Development of a new calcium phosphate powder-binder system for the 3D printing of patient specific implants. J. Mater. Sci. Mater. Med. 18, 909−916. https://doi.org/10.1007/s10856-006-0073-2.

Khoda, A.K.M., Ozbolat, I.T., Koc, B., 2013. Designing heterogeneous porous tissue scaffolds for additive manufacturing processes. Comput. Des. 45, 1507−1523. https://doi.org/10.1016/J.CAD.2013.07.003.

Kim, J.Y., Jin, G.-Z., Park, I.S., Kim, J.-N., Chun, S.Y., Park, E.K., et al., 2010. Evaluation of solid free-form fabrication-based scaffolds seeded with osteoblasts and human umbilical vein endothelial cells for use in vivo osteogenesis. Tissue Eng. A. 16, 2229−2236. https://doi.org/10.1089/ten.TEA.2009.0644.

Klammert, U., Gbureck, U., Vorndran, E., Rödiger, J., Meyer-Marcotty, P., Kübler, A.C., 2010. 3D powder printed calcium phosphate implants for reconstruction of cranial and maxillofacial defects. J. Cranio-Maxillofacial Surg. 38, 565−570. https://doi.org/10.1016/J.JCMS.2010.01.009.

Koch, L., Deiwick, A., Schlie, S., Michael, S., Gruene, M., Coger, V., et al., 2012. Skin tissue generation by laser cell printing. Biotechnol. Bioeng. 109, 1855−1863. https://doi.org/10.1002/bit.24455.

Koch, L., Brandt, O., Deiwick, A., Chichkov, B., 2017. Laser-assisted bioprinting at different wavelengths and pulse durations with a metal dynamic release layer: a parametric study. Int. J. Bioprint. 3, 42−53.

Kolan, K.C.R., Leu, M.C., Hilmas, G.E., Brown, R.F., Velez, M., 2011. Fabrication of 13-93 bioactive glass scaffolds for bone tissue engineering using indirect selective laser sintering. Biofabrication 3, 025004. https://doi.org/10.1088/1758-5082/3/2/025004.

Kolan, K.C.R., Thomas, A., Leu, M.C., Hilmas, G., 2015. In vitro assessment of laser sintered bioactive glass scaffolds with different pore geometries. Rapid Prototyp. J. 21, 152−158. https://doi.org/10.1108/RPJ-12-2014-0175.

Kong, D., Ni, X., Dong, C., Zhang, L., Man, C., Cheng, X., et al., 2019. Anisotropy in the microstructure and mechanical property for the bulk and porous 316L stainless steel fabricated via selective laser melting. Mater. Lett. 235, 1−5. https://doi.org/10.1016/j.matlet.2018.09.152.

Korpela, J., Kokkari, A., Korhonen, H., Malin, M., Närhi, T., Seppälä, J., 2013. Biodegradable and bioactive porous scaffold structures prepared using fused deposition modeling. J. Biomed. Mater. Res. B Appl. Biomater. 101B, 610−619. https://doi.org/10.1002/jbm.b.32863.

Krishna, B.V., Bose, S., Bandyopadhyay, A., 2007. Low stiffness porous Ti structures for load-bearing implants. Acta Biomater 3, 997−1006. https://doi.org/10.1016/j.actbio.2007.03.008.

Krishna, B.V., Bose, S., Bandyopadhyay, A., 2009. Fabrication of porous NiTi shape memory alloy structures using laser engineered net shaping. J. Biomed. Mater. Res. B Appl. Biomater. 89, 481–490. https://doi.org/10.1002/jbm.b.31238.

Kruth, J.P., Mercelis, P., Van Vaerenbergh, J., Froyen, L., Rombouts, M., 2005. Binding mechanisms in selective laser sintering and selective laser melting. Rapid Prototyp. J. 11, 26–36. https://doi.org/10.1108/13552540510573365.

Kundu, J., Shim, J.-H., Jang, J., Kim, S.-W., Cho, D.-W., 2015. An additive manufacturing-based PCL-alginate-chondrocyte bioprinted scaffold for cartilage tissue engineering. J. Tissue Eng. Regenerat. Med. 9, 1286–1297. https://doi.org/10.1002/term.1682.

Kurzynowski, T., Gruber, K., Stopyra, W., Kuźnicka, B., Chlebus, E., 2018. Correlation between process parameters, microstructure and properties of 316 L stainless steel processed by selective laser melting. Mater. Sci. Eng. A 718, 64–73. https://doi.org/10.1016/j.msea.2018.01.103.

Kwon, S.G., Kwon, Y.W., Lee, T.W., Park, G.T., Kim, J.H., 2018. Recent advances in stem cell therapeutics and tissue engineering strategies. Biomater. Res. 22, 36. https://doi.org/10.1186/s40824-018-0148-4.

Lam, C.X.F., Hutmacher, D.W., Schantz, J.-T., Woodruff, M.A., Teoh, S.H., 2009. Evaluation of polycaprolactone scaffold degradation for 6 months *in vitro* and *in vivo*. J. Biomed. Mater. Res. A 90A, 906–919. https://doi.org/10.1002/jbm.a.32052.

Leach, J., Wang, A., Ye, K., Jin, S., 2016. A RNA-DNA hybrid aptamer for nanoparticle-based prostate tumor targeted drug delivery. Int. J. Mol. Sci. 17, 380. https://doi.org/10.3390/ijms17030380.

Lee, V., Singh, G., Trasatti, J.P., Bjornsson, C., Xu, X., Tran, T.N., et al., 2014. Design and fabrication of human skin by three-dimensional bioprinting. Tissue Eng. C Methods 20, 473–484. https://doi.org/10.1089/ten.TEC.2013.0335.

Leukers, B., Gülkan, H., Irsen, S.H., Milz, S., Tille, C., Schieker, M., et al., 2005. Hydroxyapatite scaffolds for bone tissue engineering made by 3D printing. J. Mater. Sci. Mater. Med. 16, 1121–1124. https://doi.org/10.1007/s10856-005-4716-5.

Li, H., Zheng, Y., Qin, L., 2014. Progress of biodegradable metals. Prog. Nat. Sci. Mater. Int. 24, 414–422. https://doi.org/10.1016/j.pnsc.2014.08.014.

Li, Y., Zhou, J., Pavanram, P., Leeflang, M.A., Fockaert, L.I., Pouran, B., et al., 2018a. Additively manufactured biodegradable porous magnesium. Acta Biomater 67, 378–392. https://doi.org/10.1016/j.actbio.2017.12.008.

Li, Y., Jahr, H., Lietaert, K., Pavanram, P., Yilmaz, A., Fockaert, L.I., et al., 2018b. Additively manufactured biodegradable porous iron. Acta Biomater 77, 380–393. https://doi.org/10.1016/j.actbio.2018.07.011.

Lindner, M., Hoeges, S., Meiners, W., Wissenbach, K., Smeets, R., Telle, R., et al., 2011. Manufacturing of individual biodegradable bone substitute implants using selective laser melting technique. J. Biomed. Mater. Res. A 97A, 466–471. https://doi.org/10.1002/jbm.a.33058.

Liu, H., Yazici, H., Ergun, C., Webster, T.J., Bermek, H., 2008. An in vitro evaluation of the Ca/P ratio for the cytocompatibility of nano-to-micron particulate calcium phosphates for bone regeneration. Acta Biomater 4, 1472–1479. https://doi.org/10.1016/J.ACTBIO.2008.02.025.

Liu, X., Rahaman, M.N., Fu, Q., 2013. Bone regeneration in strong porous bioactive glass (13-93) scaffolds with an oriented microstructure implanted in rat calvarial defects. Acta Biomater 9, 4889–4898. https://doi.org/10.1016/J.ACTBIO.2012.08.029.

Liu, Y., Li, S., Hou, W., Wang, S., Hao, Y., Yang, R., et al., 2016. Electron beam melted beta-type Ti-24Nb-4Zr-8Sn porous structures with high strength-to-modulus ratio. J. Mater. Sci. Technol. 32, 505−508. https://doi.org/10.1016/j.jmst.2016.03.020.

Liu, Y., Li, T., Ma, H., Zhai, D., Deng, C., Wang, J., et al., 2018. 3D-printed scaffolds with bioactive elements-induced photothermal effect for bone tumor therapy. Acta Biomater 73, 531−546. https://doi.org/10.1016/J.ACTBIO.2018.04.014.

Liverani, E., Toschi, S., Ceschini, L., Fortunato, A., 2017. Effect of selective laser melting (SLM) process parameters on microstructure and mechanical properties of 316L austenitic stainless steel. J. Mater. Process. Technol. 249, 255−263. https://doi.org/10.1016/j.jmatprotec.2017.05.042.

Loo, Y., Lakshmanan, A., Ni, M., Toh, L.L., Wang, S., Hauser, C.A.E., 2015. Peptide bioink: self-assembling nanofibrous scaffolds for three-dimensional organotypic cultures. Nano Lett 15, 6919−6925. https://doi.org/10.1021/acs.nanolett.5b02859.

Luo, Y., Wu, C., Lode, A., Gelinsky, M., 2012. Hierarchical mesoporous bioactive glass/alginate composite scaffolds fabricated by three-dimensional plotting for bone tissue engineering. Biofabrication 5, 015005. https://doi.org/10.1088/1758-5082/5/1/015005.

Luo, J.P., Sun, J.F., Huang, Y.J., Zhang, J.H., Zhang, Y.D., Zhao, D.P., et al., 2018. Low-modulus biomedical Ti−30Nb−5Ta−3Zr additively manufactured by selective laser melting and its biocompatibility. Mater. Sci. Eng. C 97, 275−284. https://doi.org/10.1016/j.msec.2018.11.077.

Ma, H., Feng, C., Chang, J., Wu, C., 2018. 3D-printed bioceramic scaffolds: from bone tissue engineering to tumor therapy. Acta Biomater 79, 37−59. https://doi.org/10.1016/J.ACTBIO.2018.08.026.

Manakari, V., Parande, G., Gupta, M., 2016. Selective laser melting of magnesium and magnesium alloy powders: a review. Metals 7, 2. https://doi.org/10.3390/met7010002.

Michna, S., Wu, W., Lewis, J.A., 2005. Concentrated hydroxyapatite inks for direct-write assembly of 3-D periodic scaffolds. Biomaterials 26, 5632−5639. https://doi.org/10.1016/J.BIOMATERIALS.2005.02.040.

Moiduddin, K., Darwish, S., Al-Ahmari, A., ElWatidy, S., Mohammad, A., Ameen, W., 2017. Structural and mechanical characterization of custom design cranial implant created using additive manufacturing. Electron. J. Biotechnol. 29, 22−31. https://doi.org/10.1016/j.ejbt.2017.06.005.

Mondrinos, M.J., Dembzynski, R., Lu, L., Byrapogu, V.K.C., Wootton, D.M., Lelkes, P.I., et al., 2006. Porogen-based solid freeform fabrication of polycaprolactone−calcium phosphate scaffolds for tissue engineering. Biomaterials 27, 4399−4408. https://doi.org/10.1016/J.BIOMATERIALS.2006.03.049.

Montani, M., Demir, A.G., Mostaed, E., Vedani, M., Previtali, B., 2017. Processability of pure Zn and pure Fe by SLM for biodegradable metallic implant manufacturing. Rapid Prototyp. J. 23, 514−523. https://doi.org/10.1108/RPJ-08-2015-0100.

Mota, C., Puppi, D., Chiellini, F., Chiellini, E., 2015. Additive manufacturing techniques for the production of tissue engineering constructs. J. Tissue Eng. Regenerat. Med. 9, 174−190. https://doi.org/10.1002/term.1635.

Mueller, B., Toeppel, T., Gebauer, M., Neugebauer, R., 2012. Innovative features in implants through beam melting − a new approach for additive manufacturing of endoprostheses. In: Innov. Dev. Virtual Phys. Prototyp., pp. 519−523. https://doi.org/10.1201/b11341-84.

Mullen, L., Stamp, R.C., Brooks, W.K., Jones, E., Sutcliffe, C.J., 2009. Selective laser melting: a regular unit cell approach for the manufacture of porous, titanium, bone in-growth constructs, suitable for orthopedic applications. J. Biomed. Mater. Res. B Appl. Biomater. 89, 325−334. https://doi.org/10.1002/jbm.b.31219.

Mullen, L., Stamp, R.C., Fox, P., Jones, E., Ngo, C., Sutcliffe, C.J., 2010. Selective laser melting: a unit cell approach for the manufacture of porous, titanium, bone in-growth constructs, suitable for orthopedic applications. II. Randomized structures. J. Biomed. Mater. Res. B Appl. Biomater. 92, 178−188. https://doi.org/10.1002/jbm.b.31504.

Murr, L.E., Quinones, S.A., Gaytan, S.M., Lopez, M.I., Rodela, A., Martinez, E.Y., et al., 2009. Microstructure and mechanical behavior of Ti-6Al-4V produced by rapid-layer manufacturing, for biomedical applications. J. Mech. Behav. Biomed. Mater. 2, 20−32. https://doi.org/10.1016/j.jmbbm.2008.05.004.

Murr, L.E., Gaytan, S.M., Medina, F., Martinez, E., Martinez, J.L., Hernandez, D.H., et al., 2010. Characterization of Ti-6Al-4V open cellular foams fabricated by additive manufacturing using electron beam melting. Mater. Sci. Eng. A 527, 1861−1868. https://doi.org/10.1016/j.msea.2009.11.015.

Murr, L.E., Amato, K.N., Li, S.J., Tian, Y.X., Cheng, X.Y., Gaytan, S.M., et al., 2011. Microstructure and mechanical properties of open-cellular biomaterials prototypes for total knee replacement implants fabricated by electron beam melting. J. Mech. Behav. Biomed. Mater. 4, 1396−1411. https://doi.org/10.1016/j.jmbbm.2011.05.010.

Murr, L.E., Gaytan, S.M., Ramirez, D.A., Martinez, E., Hernandez, J., Amato, K.N., et al., 2012a. Metal fabrication by additive manufacturing using laser and electron beam melting technologies. J. Mater. Sci. Technol. 28, 1−14. https://doi.org/10.1016/S1005-0302(12)60016-4.

Murr, L.E., Gaytan, S.M., Martinez, E., Medina, F., Wicker, R.B., 2012b. Next generation orthopaedic implants by additive manufacturing using electron beam melting. Int. J. Biom. 2012, 245727. https://doi.org/10.1155/2012/245727.

Ng, C.C., Savalani, M., Man, H.C., 2011a. Fabrication of magnesium using selective laser melting technique. Rapid Prototyp. J. 17, 479−490. https://doi.org/10.1108/13552541111184206.

Ng, C.C., Savalani, M.M., Lau, M.L., Man, H.C., 2011b. Microstructure and mechanical properties of selective laser melted magnesium. Appl. Surf. Sci. 257, 7447−7454. https://doi.org/10.1016/j.apsusc.2011.03.004.

Ngo, T.D., Kashani, A., Imbalzano, G., Nguyen, K.T.Q., Hui, D., 2018. Additive manufacturing (3D printing): a review of materials, methods, applications and challenges. Compos. B Eng. 143, 172−196. https://doi.org/10.1016/j.compositesb.2018.02.012.

Norotte, C., Marga, F.S., Niklason, L.E., Forgacs, G., 2009. Scaffold-free vascular tissue engineering using bioprinting. Biomaterials 30, 5910−5917. https://doi.org/10.1016/j.biomaterials.2009.06.034.

Olsén, J., Shen, Z., Liu, L., Koptyug, A., Rännar, L.E., 2018. Micro- and macro-structural heterogeneities in 316L stainless steel prepared by electron-beam melting. Mater. Char. 141, 1−7. https://doi.org/10.1016/j.matchar.2018.04.026.

Ozbolat, I.T., 2016. 3D Bioprinting : Fundamentals, Principles and Applications. Academic Press is an imprint of Elsevier. First.

Park, S.I., Rosen, D.W., kyum Choi, S., Duty, C.E., 2014. Effective mechanical properties of lattice material fabricated by material extrusion additive manufacturing. Addit. Manuf. 1−4, 12−23. https://doi.org/10.1016/j.addma.2014.07.002.

Parthasarathy, J., Starly, B., Raman, S., Christensen, A., 2010. Mechanical evaluation of porous titanium (Ti6Al4V) structures with electron beam melting (EBM). J. Mech. Behav. Biomed. Mater. 3, 249−259. https://doi.org/10.1016/j.jmbbm.2009.10.006.

Parthasarathy, J., Starly, B., Raman, S., 2011. A design for the additive manufacture of functionally graded porous structures with tailored mechanical properties for biomedical applications. J. Manuf. Process. 13, 160−170. https://doi.org/10.1016/j.jmapro.2011.01.004.

Pescosolido, L., Schuurman, W., Malda, J., Matricardi, P., Alhaique, F., Coviello, T., et al., 2011. Hyaluronic acid and Dextran-based semi-IPN hydrogels as biomaterials for bioprinting. Biomacromolecules 12, 1831−1838. https://doi.org/10.1021/bm200178w.

Phillippi, J.A., Miller, E., Weiss, L., Huard, J., Waggoner, A., Campbell, P., 2008. Microenvironments engineered by inkjet bioprinting spatially direct adult stem cells toward muscle- and bone-like subpopulations. Stem Cell 26, 127−134. https://doi.org/10.1634/stemcells.2007-0520.

Pirlo, R.K., Wu, P., Liu, J., Ringeisen, B., 2012. PLGA/hydrogel biopapers as a stackable substrate for printing HUVEC networks via BioLP^{TM}. Biotechnol. Bioeng. 109, 262−273. https://doi.org/10.1002/bit.23295.

Puppi, D., Mota, C., Gazzarri, M., Dinucci, D., Gloria, A., Myrzabekova, M., et al., 2012. Additive manufacturing of wet-spun polymeric scaffolds for bone tissue engineering. Biomed. Microdevices 14, 1115−1127. https://doi.org/10.1007/s10544-012-9677-0.

Rännar, L.E., Koptyug, A., Olsén, J., Saeidi, K., Shen, Z., 2017. Hierarchical structures of stainless steel 316L manufactured by electron beam melting. Addit. Manuf. 17, 106−112. https://doi.org/10.1016/j.addma.2017.07.003.

Rhee, S., Puetzer, J.L., Mason, B.N., Reinhart-King, C.A., Bonassar, L.J., 2016. 3D bioprinting of spatially heterogeneous collagen constructs for cartilage tissue engineering. ACS Biomater. Sci. Eng. 2, 1800−1805. https://doi.org/10.1021/acsbiomaterials.6b00288.

Robinson, J., Ashton, I., Fox, P., Jones, E., Sutcliffe, C., 2018. Determination of the effect of scan strategy on residual stress in laser powder bed fusion additive manufacturing. Addit. Manuf. 23, 13−24. https://doi.org/10.1016/j.addma.2018.07.001.

Ronca, A., Ambrosio, L., Grijpma, D.W., 2013. Preparation of designed poly(d,l-lactide)/nanosized hydroxyapatite composite structures by stereolithography. Acta Biomater 9, 5989−5996. https://doi.org/10.1016/J.ACTBIO.2012.12.004.

Rosenzweig, D., Carelli, E., Steffen, T., Jarzem, P., Haglund, L., 2015. 3D-Printed ABS and PLA scaffolds for cartilage and nucleus pulposus tissue regeneration. Int. J. Mol. Sci. 16, 15118−15135. https://doi.org/10.3390/ijms160715118.

Rotaru, H., Schumacher, R., Kim, S.-G., Dinu, C., 2015. Selective laser melted titanium implants: a new technique for the reconstruction of extensive zygomatic complex defects. Maxillofac. Plast. Reconstr. Surg. 37, 12. https://doi.org/10.1186/s40902-015-0012-6.

Sallica-Leva, E., Jardini, A.L., Fogagnolo, J.B., 2013. Microstructure and mechanical behavior of porous Ti-6Al-4V parts obtained by selective laser melting. J. Mech. Behav. Biomed. Mater. 26, 98−108. https://doi.org/10.1016/j.jmbbm.2013.05.011.

Sánchez-Salcedo, S., Nieto, A., Vallet-Regí, M., 2008. Hydroxyapatite/β-tricalcium phosphate/agarose macroporous scaffolds for bone tissue engineering. Chem. Eng. J. 137, 62−71. https://doi.org/10.1016/J.CEJ.2007.09.011.

Seidenstuecker, M., Kerr, L., Bernstein, A., Mayr, H., Suedkamp, N., Gadow, R., et al., 2017. 3D powder printed bioglass and β-tricalcium phosphate bone scaffolds. Materials 11, 13. https://doi.org/10.3390/ma11010013.

Seol, Y.-J., Kang, H.-W., Lee, S.J., Atala, A., Yoo, J.J., 2014. Bioprinting technology and its applications. Eur. J. Cardiothorac. Surg. 46, 342−348. https://doi.org/10.1093/ejcts/ezu148.

Serra, T., Planell, J.A., Navarro, M., 2013. High-resolution PLA-based composite scaffolds via 3-D printing technology. Acta Biomater 9, 5521−5530. https://doi.org/10.1016/J.ACTBIO.2012.10.041.

Shuai, C., Gao, C., Nie, Y., Hu, H., Zhou, Y., Peng, S., 2011. Structure and properties of nano-hydroxyapatite scaffolds for bone tissue engineering with a selective laser sintering system. Nanotechnology 22, 285703. https://doi.org/10.1088/0957-4484/22/28/285703.

Sing, S.L., An, J., Yeong, W.Y., Wiria, F.E., 2016. Laser and electron-beam powder-bed additive manufacturing of metallic implants: a review on processes, materials and designs. J. Orthop. Res. 34, 369–385. https://doi.org/10.1002/jor.23075.

Singh, S., Ramakrishna, S., 2017. Biomedical applications of additive manufacturing : present and future. Curr. Opin. Biomed. Eng. 2, 105–115. https://doi.org/10.1016/j.cobme.2017.05.006.

Skardal, A., Zhang, J., McCoard, L., Xu, X., Oottamasathien, S., Prestwich, G.D., 2010a. Photocrosslinkable hyaluronan-gelatin hydrogels for two-step bioprinting. Tissue Eng. A. 16, 2675–2685. https://doi.org/10.1089/ten.TEA.2009.0798.

Skardal, A., Zhang, J., Prestwich, G.D., 2010b. Bioprinting vessel-like constructs using hyaluronan hydrogels crosslinked with tetrahedral polyethylene glycol tetracrylates. Biomaterials 31, 6173–6181. https://doi.org/10.1016/j.biomaterials.2010.04.045.

Stenberg, K., Dittrick, S., Bose, S., Bandyopadhyay, A., 2018. Influence of simultaneous addition of carbon nanotubes and calcium phosphate on wear resistance of 3D printed Ti6Al4V. J. Mater. Res. 33 (14), 2077–2086.

Suryawanshi, J., Prashanth, K.G., Ramamurty, U., 2017. Mechanical behavior of selective laser melted 316L stainless steel. Mater. Sci. Eng. A 696, 113–121. https://doi.org/10.1016/j.msea.2017.04.058.

Suwanprateeb, J., Sanngam, R., Suvannapruk, W., Panyathanmaporn, T., 2009. Mechanical and in vitro performance of apatite–wollastonite glass ceramic reinforced hydroxyapatite composite fabricated by 3D-printing. J. Mater. Sci. Mater. Med. 20, 1281–1289. https://doi.org/10.1007/s10856-009-3697-1.

Tan, X.P., Tan, Y.J., Chow, C.S.L., Tor, S.B., Yeong, W.Y., 2017. Metallic powder-bed based 3D printing of cellular scaffolds for orthopaedic implants: a state-of-the-art review on manufacturing, topological design, mechanical properties and biocompatibility. Mater. Sci. Eng. C 76, 1328–1343. https://doi.org/10.1016/j.msec.2017.02.094.

Tarafder, S., Bose, S., 2014. Polycaprolactone-coated 3D printed tricalcium phosphate scaffolds: in vitro alendronate release behavior and local delivery effect on in vivo osteogenesis. ACS Appl. Mater. Interfaces 6 (13), 9955–9965.

Tarafder, S., Davies, N.M., Bandyopadhyay, A., Bose, S., 2013a. 3D printed tricalcium phosphate scaffolds: effect of SrO and MgO doping on in vivo osteogenesis in a rat distal femoral defect model. Biomater. Sci. 1, 1250–1259. https://doi.org/10.1039/C3BM60132C.

Tarafder, S., Balla, V.K., Davies, N.M., Bandyopadhyay, A., Bose, S., 2013b. Microwave-sintered 3D printed tricalcium phosphate scaffolds for bone tissue engineering. J. Tissue Eng. Regenerat. Med. 7, 631–641. https://doi.org/10.1002/term.555.

Tarafder, S., Dernell, W.S., Bandyopadhyay, A., Bose, S., 2015. SrO- and MgO-doped microwave sintered 3D printed tricalcium phosphate scaffolds: mechanical properties and *in vivo* osteogenesis in a rabbit model. J. Biomed. Mater. Res. B Appl. Biomater. 103, 679–690. https://doi.org/10.1002/jbm.b.33239.

Tesavibul, P., Felzmann, R., Gruber, S., Liska, R., Thompson, I., Boccaccini, A.R., et al., 2012. Processing of 45S5 Bioglass® by lithography-based additive manufacturing. Mater. Lett. 74, 81–84. https://doi.org/10.1016/J.MATLET.2012.01.019.

Trombetta, R., Inzana, J.A., Schwarz, E.M., Kates, S.L., Awad, H.A., 2017. 3D printing of calcium phosphate ceramics for bone tissue engineering and drug delivery. Ann. Biomed. Eng. 45, 23–44. https://doi.org/10.1007/s10439-016-1678-3.

Turnbull, G., Clarke, J., Picard, F., Riches, P., Jia, L., Han, F., et al., 2018. 3D bioactive composite scaffolds for bone tissue engineering. Bioact. Mater. 3, 278–314. https://doi.org/10.1016/J.BIOACTMAT.2017.10.001.

V Murphy, S., Atala, A., 2014. 3D bioprinting of tissues and organs. Nat. Biotechnol. 32, 773−785. https://doi.org/10.1038/nbt.2958.

Vaezi, M., Seitz, H., Yang, S., 2013. A review on 3D micro-additive manufacturing technologies. Int. J. Adv. Manuf. Technol. 67, 1721−1754. https://doi.org/10.1007/s00170-012-4605-2.

Vandenbroucke, B., Kruth, J.P., 2007. Selective laser melting of biocompatible metals for rapid manufacturing of medical parts. Rapid Prototyp. J. 13, 196−203. https://doi.org/10.1108/13552540710776142.

Vilaro, T., Colin, C., Bartout, J.D., 2011. As-fabricated and heat-treated microstructures of the Ti-6Al-4V alloy processed by selective laser melting. Metall. Mater. Trans. A Phys. Metall. Mater. Sci. 42, 3190−3199. https://doi.org/10.4997/JRCPE.2017.305.

Visser, J., Peters, B., Burger, T.J., Boomstra, J., Dhert, W.J.A., Melchels, F.P.W., et al., 2013. Biofabrication of multi-material anatomically shaped tissue constructs. Biofabrication 5, 035007. https://doi.org/10.1088/1758-5082/5/3/035007.

Vorndran, E., Klarner, M., Klammert, U., Grover, L.M., Patel, S., Barralet, J.E., et al., 2008. 3D powder printing of β-tricalcium phosphate ceramics using different strategies. Adv. Eng. Mater. 10, B67−B71. https://doi.org/10.1002/adem.200800179.

Wang, X., Xu, S., Zhou, S., Xu, W., Leary, M., Choong, P., et al., 2016. Topological design and additive manufacturing of porous metals for bone scaffolds and orthopaedic implants: a review. Biomaterials 83, 127−141. https://doi.org/10.1016/j.biomaterials.2016.01.012.

Wang, C., Tan, X., Liu, E., Tor, S.B., 2018. Process parameter optimization and mechanical properties for additively manufactured stainless steel 316L parts by selective electron beam melting. Mater. Des. 147, 157−166. https://doi.org/10.1016/j.matdes.2018.03.035.

Wang, H., Su, K., Su, L., Liang, P., Ji, P., Wang, C., 2018. The effect of 3D-printed Ti6Al4V scaffolds with various macropore structures on osteointegration and osteogenesis: a biomechanical evaluation. J. Mech. Behav. Biomed. Mater. 88, 488−496. https://doi.org/10.1016/j.jmbbm.2018.08.049.

Wei, K., Gao, M., Wang, Z., Zeng, X., 2014. Effect of energy input on formability, microstructure and mechanical properties of selective laser melted AZ91D magnesium alloy. Mater. Sci. Eng. A 611, 212−222. https://doi.org/10.1016/j.msea.2014.05.092.

Wiltfang, J., Merten, H.A., Schlegel, K.A., Schultze-Mosgau, S., Kloss, F.R., Rupprecht, S., et al., 2002. Degradation characteristics of α and β tri-calcium-phosphate (TCP) in minipigs. J. Biomed. Mater. Res. 63, 115−121. https://doi.org/10.1002/jbm.10084.

Wiria, F.E., Leong, K.F., Chua, C.K., Liu, Y., 2007. Poly-ε-caprolactone/hydroxyapatite for tissue engineering scaffold fabrication via selective laser sintering. Acta Biomater 3, 1−12. https://doi.org/10.1016/J.ACTBIO.2006.07.008.

Woesz, A., Rumpler, M., Stampfl, J., Varga, F., Fratzl-Zelman, N., Roschger, P., et al., 2005. Towards bone replacement materials from calcium phosphates via rapid prototyping and ceramic gelcasting. Mater. Sci. Eng. C 25, 181−186. https://doi.org/10.1016/J.MSEC.2005.01.014.

Wu, C., Luo, Y., Cuniberti, G., Xiao, Y., Gelinsky, M., 2011. Three-dimensional printing of hierarchical and tough mesoporous bioactive glass scaffolds with a controllable pore architecture, excellent mechanical strength and mineralization ability. Acta Biomater 7, 2644−2650. https://doi.org/10.1016/j.actbio.2011.03.009.

Wysocki, B., Maj, P., Sitek, R., Buhagiar, J., Kurzydłowski, K., Święszkowski, W., 2017. Laser and electron beam additive manufacturing methods of fabricating titanium bone implants. Appl. Sci. 7, 657. https://doi.org/10.3390/app7070657.

Xiong, Z., Yan, Y., Wang, S., Zhang, R., Zhang, C., 2002. Fabrication of porous scaffolds for bone tissue engineering via low-temperature deposition. Scripta Mater. 46, 771–776. https://doi.org/10.1016/S1359-6462(02)00071-4.

Xu, T., Jin, J., Gregory, C., Hickman, J.J., Boland, T., 2005. Inkjet printing of viable mammalian cells. Biomaterials 26, 93–99. https://doi.org/10.1016/j.biomaterials.2004.04.011.

Xu, T., Kincaid, H., Atala, A., Yoo, J.J., 2008. High-throughput production of single-cell microparticles using an inkjet printing technology. J. Manuf. Sci. Eng. 130, 021017. https://doi.org/10.1115/1.2903064.

Xu, T., Binder, K.W., Albanna, M.Z., Dice, D., Zhao, W., Yoo, J.J., et al., 2012. Hybrid printing of mechanically and biologically improved constructs for cartilage tissue engineering applications. Biofabrication 5, 015001. https://doi.org/10.1088/1758-5082/5/1/015001.

Xue, W., Krishna, B.V., Bandyopadhyay, A., Bose, S., 2007. Processing and biocompatibility evaluation of laser processed porous titanium. Acta Biomater 3, 1007–1018. https://doi.org/10.1016/j.actbio.2007.05.009.

Yan, Q., Dong, H., Su, J., Han, J., Song, B., Wei, Q., et al., 2018. A review of 3D printing technology for medical applications. Engineering 4, 729–742. https://doi.org/10.1016/j.eng.2018.07.021.

Yang, Y., Bin Lu, J., Luo, Z.Y., Wang, D., 2012. Accuracy and density optimization in directly fabricating customized orthodontic production by selective laser melting. Rapid Prototyp. J. 18, 482–489. https://doi.org/10.1108/13552541211272027.

Yang, J., Yu, H., Yin, J., Gao, M., Wang, Z., Zeng, X., 2016. Formation and control of martensite in Ti-6Al-4V alloy produced by selective laser melting. Mater. Des. 108, 308–318. https://doi.org/10.1016/j.matdes.2016.06.117.

Yasa, E., Deckers, J., Kruth, J.P., 2011. The investigation of the influence of laser re-melting on density, surface quality and microstructure of selective laser melting parts. Rapid Prototyp. J. 17, 312–327. https://doi.org/10.1108/13552541111156450.

Yu, Y., Moncal, K.K., Li, J., Peng, W., Rivero, I., Martin, J.A., et al., 2016. Three-dimensional bioprinting using self-assembling scalable scaffold-free "tissue strands" as a new bioink. Sci. Rep. 6, 28714. https://doi.org/10.1038/srep28714.

Yuan, J., Zhen, P., Zhao, H., Chen, K., Li, X., Gao, M., et al., 2015. The preliminary performance study of the 3D printing of a tricalcium phosphate scaffold for the loading of sustained release anti-tuberculosis drugs. J. Mater. Sci. 50, 2138–2147. https://doi.org/10.1007/s10853-014-8776-0.

Zadpoor, A.A., 2018. Mechanical performance of additively manufactured meta-biomaterials. Acta Biomater 85, 41–59. https://doi.org/10.1016/j.actbio.2018.12.038.

Zakhem, E., Bitar, K., 2015. Development of chitosan scaffolds with enhanced mechanical properties for intestinal tissue engineering applications. J. Funct. Biomater. 6, 999–1011. https://doi.org/10.3390/jfb6040999.

Zhang, X., Zhang, Y., 2015. Tissue engineering applications of three-dimensional bioprinting. Cell Biochem. Biophys 72, 777–782. https://doi.org/10.1007/s12013-015-0531-x.

Zhang, Y., Xia, L., Zhai, D., Shi, M., Luo, Y., Feng, C., et al., 2015. Mesoporous bioactive glass nanolayer-functionalized 3D-printed scaffolds for accelerating osteogenesis and angiogenesis. Nanoscale 7, 19207–19221. https://doi.org/10.1039/c5nr05421d.

Zhang, X., Zeng, D., Li, N., Wen, J., Jiang, X., Liu, C., et al., 2016. Functionalized mesoporous bioactive glass scaffolds for enhanced bone tissue regeneration. Sci. Rep. 6, 19361. https://doi.org/10.1038/srep19361.

Zhang, Y.S., Yue, K., Aleman, J., Mollazadeh-Moghaddam, K., Bakht, S.M., Yang, J., et al., 2017. 3D bioprinting for tissue and organ fabrication. Ann. Biomed. Eng. 45, 148–163. https://doi.org/10.1007/s10439-016-1612-8.

Zhang, X.Z., Leary, M., Tang, H.P., Song, T., Qian, M., 2018. Selective electron beam manufactured Ti-6Al-4V lattice structures for orthopedic implant applications: current status and outstanding challenges. Curr. Opin. Solid State Mater. Sci. 22, 75−99. https://doi.org/10.1016/j.cossms.2018.05.002.

Zheng, Y.F., Gu, X.N., Witte, F., 2014. Biodegradable metals. Mater. Sci. Eng. R Rep. 77, 1−34. https://doi.org/10.1016/J.MSER.2014.01.001.

Zhong, Y., Rännar, L.E., Liu, L., Koptyug, A., Wikman, S., Olsen, J., et al., 2017. Additive manufacturing of 316L stainless steel by electron beam melting for nuclear fusion applications. J. Nucl. Mater. 486, 234−245. https://doi.org/10.1016/j.jnucmat.2016.12.042.

Index

Note: 'Page numbers followed by "f" indicates figures, "t" indicates tables'.

A
Abdominal fat, 105
Acellular tissue products, 147–149, 150t–157t
Acid-catalyzed hydrolysis, 378f
Additive manufacturing (AM) techniques, 438–439, 440t
Adhesive peptides, 31–32
Adiponectin, 206
Adipose-derived mesenchymal stem cells (ADSCs), 105, 209
Adipose-derived stromal cells (ADSCs), 317–318
Adipose tissue–derived stem cells (ADMSCs), 28
Adult human MSCs, 106
Adult stem cells (ASCs), 20–21, 21t, 102
Advanced glycation end products (AGEs), 197–198
Agarose, 28
Aging, effects of, 197–198
Alginate-based microcapsules, 389t, 393–394
Alginate biopolymer in drug delivery, 375–376
  alternative methods for hydrogel fabrication, 383–384
  biocompatibility, 377
  chemically modified alginate, 379–380
  degradation, 377–379
  ionically cross-linked alginate hydrogels, 381–383
  using alginate matrices, 384–394
Alginate dialdehyde (ADA), 379
Alginate matrices, 384–394
Alginic acid, 342–343
Alkaline phosphatase (ALP) expression, 452
Allogeneic bone grafts, 247
Allogeneic cells, 210
Allogeneic grafts, 419

Allogeneic limbal stem cell transplant, 271
Allogeneic MSCs, 108
Alloplastic material, 57–58
AlphaCor, 8–9, 9f
Aluminocalcium phosphorous oxide (ALCAP) ceramic capsules, 337
Alzheimer's disease (AD), 111
Amino acids, 380
5-Aminolevulinic acid, 351–352
2-Amino oleic acid, 158–159
AminoPVAlc-SPIONs, 362
Amniotic membrane, 279–280
  transplantation, 270
Amorphous calcium phosphate nanoparticles, 346–347
Amphiphilic alginate biopolymers, 381
Amphiphilic alginate derivatives, 380
Angiogenesis/vascularization, 124
Animal tissue–derived products, 159
Animal tissue products/derivatives, 146
Antiapoptotic signaling, 122
Antibacterial coatings, 302–303
Antibiotic-loaded bone cement, 424
Antibiotic regimen, 176
Antibiotic sterilization, 176
Antiplatelet therapy, 307
Antisense oligonucleotides, 213–214
Antithrombogenicity, 301–302
Aqueous glutaraldehyde sterilization, 176
Atom transfer radical polymerization (ATRP), 282
Attractive therapeutic approach, 107
Auroshell, 353–354
Autoimmune diseases, 108
Autologous conjunctival limbal transplant, 270–271
Autologous stem cells, 127
Autoreactive T cells, 110
Azotobacter vinelandii, 376–377

B

Bacterial metabolites, 238
Base-catalyzed hydrolysis, 378f
bFGF/FGF-2, 205−206
Binder jetting (BJ), 436
Bioabsorbables, 58
Bioactive glass, 450−452
Bioactive knitted silk-collagen sponge, 34−35
Bioactive molecules, 394
Bioactive protein delivery, 307−309
Bioactive self-setting cements, 250−251, 255−256
BioCaS, 255−256
Bioceramics, 337
  scaffolds using three-dimensional printing, 446−455
Biochemical effects, 197−198
Biochemical response, 198−199
Biocompatibility, biostability of, 303−304
Biocompatible coatings for bioactive protein delivery, 307−309
Biodegradables, 58
  metals, 445−446
  polyesters, 312
  synthetic polymers, 312−313
Bioerodibles, 58
Biofilm formation on dental implant, 234
Biofilm ingress, 236
Biofilm-related infection, 236
Bioglass
  and calcium phosphosilicates, 249
  integration of, 252−254
Bioinks for bioprinting, 462−463
Biointegration, 136−138, 245
  AlphaCor, 8−9, 9f
  cell adhesion and vascularization, 4
  definition, 1−2
  dental applications, 7
  designer implant, 11
  future trends, 11−12
  hydroxyapatite (HA), 1−2
  hyperbaric oxygen (HBO) therapy, 10
  Integral Biointegrated Dental Implant System, 7
  orthopedics, 2−7
  osseointegration vs., 3−4, 3f
  percutaneous devices, 10
  physicochemical surface, 6−7
  simulated body fluid (SBF), 3−4
  of synthetic bone graft substitutes, 251−256
  tissue engineering devices, 9−10
Biological materials for corneal regeneration, 278−282
Biomaterials, 26−29
  fatigue behavior of, 411−413
Biomechanical effects, 198−199
Biomedical device manufacturers, 410
Biomerix Rotator Cuff Repair Patch, 212
Biomineralized insulin nanoparticles, 339−340
Bioprinting, 455−457
  bioinks for
    natural polymers, 462−463
    synthetic polymers, 463
  strategies and classifications, 458−462
  technology, niche application areas of, 463−466
Bioresorbable materials, 57−59
Bioresorbables, 58
"Biotin sandwich" approach, 319−320
Biphasic calcium phosphates (BCP), 248
Blood−material interactions, 299−300
Blood supply, 194
Bone, 193
Bone and cartilage diseases, 107
Bone grafts, 246−247
Bone graft substitutes
  biointegration, 245
  of synthetic bone graft substitutes, 251−256
  bone grafts, 246−247
  bone, the hard tissue, 246
  synthetic, 247−251, 257
Bone marrow cells, 121
Bone marrow−derived mesenchymal stem cells (BMSCs), 105, 127, 128t−129t, 208
Bone marrow transplant, 107−108
Bone morphogenetic protein (BMP), 229
  family, 204−205
Bone repair, tissue-engineered scaffolds for, 317−318
Bone tissue, 464
Brown adipose tissue (BAT), 51

C

Ca-alginate hydrogel capsule, 393
Cadaveric keratolimbal allograft (KLAL), 271
Calcified fibrocartilage, 193
Calcium-deficient HA agents, 255
Calcium phosphate (CaP), 449–450
  ceramics, 446–450
  coatings, 64–66
  nanoparticles, 337, 339–340, 344, 346–347, 349
    cancer chemotherapy applications, 346–347
    gene delivery applications, 344–346
    oral insulin delivery applications, 338–340
    theranostic applications, 340–343
    tissue engineering applications, 347–349
  particles, 338–339
Calcium phosphate (CaP)–based alloys, 235–236
Calcium phosphate cements (CPCs), 251
Calcium phosphate-PEG-insulin-casein (CAPIC), 338–339
Calcium sulfate cements (CSCs), 251
Cancellous autografts, 247
Cancer chemotherapy applications, 346–347, 351–354
Cancer drug delivery, nano vs. microparticles in, 309–310
Cancer therapy applications, 356–358
Canonical Wnt signaling, 228–229
Carbonium–oxonium ion, 377–378
Cardiac aging, 124–125
Cardiac cell therapy, 131f
Cardiac progenitor cells, 124–125
Cardiac regeneration, 133
  barriers in stem cell therapy, 127–129
  biointegration, 136–138
  cellular reprogramming, 132
  in children, 133
  controversy, 120–121
  human heart, 120
  hydrogels, 132–133
  mechanisms, 121–125
  stem cell–derived exosomes and small vesicles, 132
  stem cell therapies, 125–127
  tissue engineering, 130–131
  valves, 133–136
Cardiac regenerative therapy, 125–127
Cardiomyocyte mitosis, 124
Cardiomyogenesis, 124
Cardiovascular diseases, 108
Cartilage, 464–465
Cell adhesion, 4
Cell-based approaches—nonlimbal sources, 273
  cultivated oral mucosal epithelial transplantation (COMET), 273–274
  dental pulp stem cells, 276
  hair follicles stem cells (HF-SCs), 275–276
  human embryonic stem cells, 277
  induced pluripotent stem cells (IPSCs), 277–278
  mesenchymal stem cells (MSCs), 274–275
  skin epidermal stem cells, 277
Cell-based therapies, 108–109
Cell–cell communication, 122
Cell cycle, 124
Cell sheet technology, 284
Cellular interactions, 88–94
  cell microintegration, 91–93
  core-shell method, 93
  stem cell studies, 90–91
  in vitro studies, 88–90
  in vivo studies, 93–94
Cellular reprogramming, 132
Cellular solids, 443
Ceramic composites, 452–454
Ceramic–polymer composites, 454–455
Ceramic systems, 254–255
Chemically cross-linked tissue products, 158–159
Chemically modified alginate, 379–380, 386–388
Chemical sterilization method, 174
Chemokine-decorated scaffolds, 34–36
Children, cardiac regeneration in, 133
Chitosan, 462
Chitosan-based hyaluronan hybrid scaffold, 212
Chitosan (CS)–gelatin biopolymers scaffolds, 281
Chloro-1-methylpyridinium iodide (CMPI), 380

Circadian-regulated genes in biointegration, 231
Citrate-stabilized gold nanoparticles, 351
Coatings, integration of, 254—255
Collagen, 278—279, 462
Collagen-based constructs, 210—211
Collagen-based grafts, 211
Collagen-gelatin-hyaluronic acid films, 281
Colloidal gold, 349—352, 354
Combined gene therapy—cell transplantation approach, 317—318
Commercial 3D computer-aided design (CAD) software, 464—465
Commercial pure Ti (CP-Ti), 441—442
Common drug delivery environments, 386t
Complex three-dimensional structure, 190—191
Composites, integration of, 254—255
Computer-aided design (CAD) models, 444f
Confocal microscopy, 120—121
Conjunctival limbal autograft (CLAU), 270—271
Controlled drug delivery, biomaterials for, 304—310
Conventional implants, 347—348
Cornea and corneal layers, 264—265
Corneal blindness and current therapies, 268—269
  allogeneic limbal stem cell transplant, 271
  amniotic membrane transplantation, 270
  autologous conjunctival limbal transplant, 270—271
  corneal stromal stem cells (CSSCs), 272—273
  cultivated limbal epithelial transplantation, 271—272
  simple limbal epithelial transplantation (SLET), 272
Corneal epithelial homeostasis, 266, 268
Corneal epithelium derived IPSC, 277—278
Corneal reconstruction, biomaterials in, 278
  amniotic membrane, 279—280
  collagen, 278—279
  gelatin, 281—282
  silk, 280—281
  synthetic biomaterials, 282—285
    poly(2-hydroxyethyl methacrylate), 282
    poly(lactide-*co*-glycolide), 283

polyethylene glycol diacrylate (PEGDA), 283
poly N-isopropylacrylamide-co-glycidylmethacrylate, 285
polyvinyl alcohol, 282
thermoresponsive polymers, 283—284
Corneal stem cell niche, 267
Corneal stromal stem cells (CSSCs), 272—273
Corneal surface reconstruction, 268—269
Coronary artery bypass graft (CABG) surgery, 131
Corrosion, 416
Craniofacial skeleton, 46—47
Crohn's disease, 108
Cross-linked tissue products, 160t—164t
Cryogenic scanning electron micrograph, 381—382, 383f
Cryopreservation, 175
CTGF/CCN2, 204
Cultivated limbal epithelial transplantation (CLET), 271—272
Cultivated oral mucosal epithelial transplantation (COMET), 273—274

D
3D bioprinting, 138
4D bioprinting, 138
Decellularization, 137
  protocols, 136—137
Decellularized porcine valves, 173
Decellularized tissues, 147—149
Decellularized xenograft, 172—173
Deficiencies, manufacturing, 410
Degradable polymers, 305—306, 306t
Degradation, 377—379
De novo bone formation, 226—227
Dental applications, 7
Dental implants, 221—222
  biointegration of, 222—223
  biological gingival seal, 223—224
  cell signaling and integration of, 228—229
  early inflammatory phase, 224—225
  ECM disorganization, 237
  fixed implant-supported dental prosthesis, 222
  genetic networks in osseointegration, 229—231, 230t

implant failure and enhancement of biointegration, 236
interface biofilms, 234—236
microbial interplay in osseointegration of, 232—234
microbial vs. host cell signaling at interface, 237—239
neovascularization at peri-implant zone, 225
osteoconduction, 225—227
soft tissue healing and biointegration, 227—228
Dental pulp stem cells, 276
Designer implant, 11
Dexamethasone 21-acetate (DXM), 356
Differentiation, 106
Direct-write assembly (DWA), 284
Dopamine-releasing cells, 28
Drug delivery
capsules, 384—386
combination devices with, 321—322
modified polymers for, 306—307
polymers used in, 305—306
Drug-loaded magnetic nanoparticles, 355—356

E

"Egg-box" model, 381
Electromagnetic gamma, 175
Electron beam melting (EBM), 435—436, 442t
Embryonic stem cells (ESCs), 20—21, 102, 125, 209, 321
Emdogain, 384
Endothelial progenitor cells, 124
End-stage organ failure, 19—20
Enhanced neurotrophin delivery, 318
Enhance endothelial cell (EC), 301—302
Epitenon, 190
Ethanol treatment, 175
Ethylene oxide treatment, 174
European Group for Blood and Marrow Transplantation, 108
Exercise, effects of, 198—199
Expanded polytetrafluoroethylene (ePTFE), 61f
Expanded PTFE (ePTFE), 58

Extracellular matrix (ECM), 21—22, 147—149, 188—189, 455—457
proteins, 228
Extracellular matrix (ECM)—based biomaterials, 313
Extraembryonic stem cells (EESCs), 21—22

F

Facial cosmetic surgery market, 56
Facial reconstruction, replacement materials
alloplastic material, 57—58
bioresorbable materials, 57—59
brown adipose tissue (BAT), 51
craniofacial skeleton, 46—47
definition, 48—53
eye/nose and ear, 51—53
facial cosmetic surgery market, 56
facial membranes surface modification, 64—70
calcium phosphate coatings, 64—66
gamma irradiation—induced grafting, 67—70
plasma immersion ion implantation (PIII), 66
future trends, 70
materials, 52f, 53—64, 54t—55t
multilayered bone framework, 46—47
muscles, 50—51
naturally derived materials, 56—57
nonbiodegradable materials, 57—59
plastic surgery, 47
psychological repercussions, 46
repair and regeneration (RR) process, 47—48
skin, 51, 52f
soft-tissue augmentation material, 55—56
special senses, 51—53
subcutaneous injections of fillers, 48—49
tissues at bone interface, 46f, 49—51
white adipose tissue (WAT), 51
Fatigue, 410—413
Fibrils, 189
Fibrinogen, 300
Fibrocartilaginous entheses, 191—192
Fibronectin, 148—149
Fibrous entheses, 191—192
Fibrous sheath, 193
Fine motor movements, 190

Fine-tuning notch signaling, 318—319
Finite element analysis (FEA), 421—422
Flaminal Forte Gel, 384
Foreign body particles, 58—59
Free aldehyde groups, 158—159
Freeze-drying (lyophilization), 175
Fused deposition modeling (FDM), 436—437
Fused filament fabrication (FFF), 436—437

G
Gamma and electron beam irradiation, 175
Gamma irradiation—induced grafting, 67—70
Gamma secretase inhibitor (GSI), 318—319
Gelatin, 281—282, 462
Gelatin-coated iron oxide nanoparticles, 356
Gene delivery applications, 344—346, 354—355, 358—359
General drug delivery and targeting, 361—362
Gene therapy, 213, 344
Gingipains, 238
Glutaraldehyde, 158, 176
Glutaraldehyde-treated bovine pericardium, 173
Glutaraldehyde-treated pericardium, 158—159
Glutaraldehyde-treated xenografts, 158
Glycosaminoglycans, 148—149
Gold-dendrimer nanoparticles, 350—351
Gold nanoparticles, 349—351
　cancer chemotherapy applications, 351—354
　gene delivery applications, 354—355
Graft vs. host disease, 107—108
Gram-negative bacteria, 234—235
Growth hormones, myocardial tissue engineering via, 319—320
Guided bone regeneration (GBR), 61—62

H
Hair follicles stem cells (HF-SCs), 275—276
HA-neodymium nanoplatform, 342—343
Hard tissue implants, 416
Healing, mechanisms of, 201—202
Health of patient, 417
Hemagglutinating virus of Japan envelopes (HVJ-Es), 361—362

Hematopoietic stem cells (HSCs), 106
Hepatocellular carcinoma, 346
Hippo-YAP signaling pathway, 124
Human amniotic membrane (HAM), 279—280
Human corneal epithelial cells, 282
Human embryonic stem cells, 277
Human heart, 120
Human induced pluripotent stem cells (hiPSCs), 125—127
Human leukocyte antigen (HLA) class II antigens, 106
Human serum albumin (HSA), 307—308
Human tissue—derived products, 159
Hyaline cartilage, 109
Hyaluronic acid, 462—463
Hydrogel composite materials for enhanced neurotrophin delivery, 318
Hydrogel fabrication, alternative methods for, 383—384
Hydrogels, 132—133
Hydrolytic degradation, 303—304
Hydroxyapatite (HA), 1—2, 447—448
　woven, 246
Hyperbaric oxygen (HBO) therapy, 10
Hypoxia signaling in peri-implant zone, 228

I
Immobilization, effects of, 199
Immune modulation, 106—107
Immunogenicity, 127
　of tissue products, 172—173
Inadequate cell migration, 136
Induced pluripotent stem cells, 125
Induced pluripotent stem cells (IPSCs), 209—210, 277—278
Inflammation reduction, 122
Inflammatory bowel disease, 108
Inflammatory response, 299—300
Infrapatellar fat pad (IFP), 105
Injectable alginate hydrogels, 394
Injectable device—drug combination, 159
Injectable hydrogels, 127—129
Injection therapy, 130—131
Injury, types of, 200
Inkjet-based three-dimensional bioprinting, 458—460
Innovative therapies, 120
Inorganic nanoparticles, 333—336

Index

Insulin-like growth factor 1, 205
Integral Biointegrated Dental Implant System, 7
Internal architecture, 189–190
Internal basal lamina (IBL), 223–224
Intracoronary injection, 127
Intramyocardial delivery, 127
Ionically cross-linked alginate hydrogels, 381–383
Iron oxide magnetic nanoparticles, 357
Iron oxide nanoparticles, 355–356
   cancer therapy applications, 356–358
   gene delivery applications, 358–359
   general drug delivery and targeting, 361–362
   tissue engineering applications, 359–361

K
Kartogenin, 207

L
Laminin, 148–149
Laser-assisted antimicrobial photodynamic therapy, 236
Laser-assisted three-dimensional bioprinting, 461–462
Limbal deficiency conditions, 268
Limbal epithelial crypts (LECs), 267
Limbal epithelial stem cells and its characteristics, 266
Limbal niche, 267
Limbal stem cell deficiency (LSCD), 268
Limbal stem cells, 267–268
   markers, 267–268
Linalool-conjugated gold nanoparticles, 353
Liver diseases, 108
Living-related conjunctival limbal allograft (Lr-CLAL), 271

M
Machine-specific software, 436–437
Magnetic resonance imaging (MRI), 197–198
Magnetic resonance tomography, 310
Major histocompatibility complex (MHC) class I, 106
Mannan, 361–362
Master switch genes, 229–231
Material extrusion–based 3D printing, 448

Materials applications, biocompatibility of, 298–304
Matrix metalloproteinases (MMPs), 237
Medical applications, biocompatibility of, 298–304
"Medical device error", 408–409
Medical-grade alginates, 377
Medical implants, 407–408
   failure analysis of, 420–422
   failure mechanisms of, 409
Mesenchymal stem cells (MSCs), 125, 272–275
   Alzheimer's disease (AD), 111
   autoimmune diseases, 108
   bone and cartilage diseases, 107
   bone marrow transplant, 107–108
   cardiovascular diseases, 108
   graft *vs.* host disease, 107–108
   liver diseases, 108
   multiple sclerosis, 110
   musculoskeletal diseases, 108–109
   nutshell, 23, 24f
   Parkinson's disease, 110
   properties of, 106–107
   rheumatoid arthritis, 109
   sources of, 104–105
   stem cells, 102
      banking, 111–112
      differentiation, 102–103
      in tissue engineering, 103–104
   systemic lupus erythematosus (SLE), 109
   type 1 diabetes, 110
   types of stem cells, 102
   wound healing, 23–26, 25f
Mesoporous bioactive glass (MBG), 452
Metaanalyses, 425
Metal-based nanoconjugates, 333–336
Metal coating system, 254–255
Metallic implants, via three-dimensional printing, 438–446
Microbial biofilms, 234
Microbial homing, 234–236
Microbial surface components recognizing adhesive matrix molecules (MSCRAMMS), 233–234
Microencapsulation for transplantation, 388–394
Microextrusion-based bioprinter, 461

Microextrusion-based three-dimensional bioprinting, 461
Microporous polyurethane grafts, 302
Migratory capacity, 107
Mildly cross-linked decellularized tissue, 147–148
Mmagnetoliposomes, 361–362
Monoclonal antibodies, 307
Monocrystalline iron oxide nanoparticles, 359–361
Multilayered bone framework, 46–47
Multiple sclerosis, 110
Multivariate analysis, 422–423
Musculoskeletal diseases, 108–109
Myocardial preservation postinjury, 122
Myocardial tissue engineering, via growth hormones, 319–320

N
Nanomedicine, polymeric biomaterials in, 314–315
Nanostructured materials, 333–336
Naturally derived materials, 56–57
Natural polymers, 462–463
N-isopropylacrylamide-co-glycidylmethacrylate (NGMA), 285
Nonbiodegradable materials, 57–59
Noncanonical Wnt signaling, 228–229
Nondetergent-based decellularization, 172
Nonenzymatic cross-linking reaction, 197–198
Non–linear stress–strain curve, 195–196
Non–load-bearing implants, 418–419
Nonviable human tissue, 146
Nonviral vectors, 213–214
Nuclear magnetic resonance (NMR), 379

O
Off-axis loading, 410–413
Oleic acid–coated nanoparticles, 356
Optimal biointegration, 64–70
Oral insulin delivery applications, 338–340
Oral mucosal epithelial cells, 273–274
Orthopedic applications, porous/cellular implants for, 443–444
Orthopedics, 2–7
Osseointegration, 3–4, 3f, 222–223
Osteointegration, 252, 253f
Osteonecrosis, 108–109

Osteo-odonto-keratoprosthesis (OOKP), 408
Osteotransduction, 256f
Ostim, 348
Overuse, 199

P
Paratenon, 194
Parkinson's disease, 110
Pathogen-associated molecular patterns (PAMPs), 239
Pathogenic microbial community, 236
Pediatric cell-based cardiac regeneration, 133, 134t–135t
Peptide drug delivery, 308–309
Percutaneous devices, 10
Peri-implant crevicular fluid, 233
Peri-implant epithelium (PIE), 223–224
Peri-implant invasive diseases (PIIDs), 232
Peri-implant microbiota, 232–233
Periodontal applications, 61–62
Periodontopathogens, 236
PerioGlas, 252–254
Peripheral blood–derived mesenchymal precursor cells, 106
Peripheral blood–derived mesenchymal stem cells, 105
Phosphates, 247–248
Phosphorylation of alginate, 379–380
Physicochemical surface, 6–7
Plasma immersion ion implantation (PIII), 66
Plastic surgery, 47
Platelet-derived growth factor (PDGF), 205–206, 320
Platelet-rich plasma, 207
Pluripotent stem cells (iPSCs), 21–22, 102
Pluronics, 283
Poly(2-hydroxyethyl methacrylate), 282
Poly(acrylic acid) (PAA), 387
Poly(ether urethanes) (PEU), 304
Poly(ethylene glycol) (PEG), 308–309
Poly(ethylene vinyl acetate) (PEVAc), 313–314
Poly(lactide-*co*-glycolide) (PLGA), 283, 449, 463
Poly(methyl methacrylate) (PMMA), 418–419

Poly(N-isopropylacrylamide) (NIPAAm), 283–284
Poly(tetrafluoroethylene) (PTFE), 79–80
Poly(vinyl alcohol) (PVAlc), 357
Polyelectrolyte complexes (PECs), 383
Polyetheretherketone (PEEK), 254–255
Polyethylene (PE), 302–303
Polyethylene glycol diacrylate (PEGDA), 283
Polylactic acid (PLA), 306–307
Polylactide (PLA), 58
Polymer comatrix system, 307
Polymeric biomaterials, 303–304
  in nanomedicine, 314–315
Polymeric materials, biostability of, 303–304
Polymeric nanoparticles, 315
Polymer matrices, 249–250
Poly N-isopropylacrylamide-co-glycidylmethacrylate, 285
Poly(acrylic acid) solution, 448
Polytetrafluoroethylene (PTFE), 59–64, 60f, 302
Polyurethanes, 304
Polyvinyl acetate (PVA)/collagen (PVA-COL) scaffold, 279
Polyvinyl alcohol, 282, 452
Porcine small intestine submucosal grafts, 211
Porous β-TCP scaffolds, 449
Porous/cellular implants for orthopedic applications, 443–444
Porous gelatin hydrogels, 281
Powder-bed fusion (PBF), 438–439
Powder-bed fusion (PBF)—based additive manufacturing, 443f
Processed lipoaspirate (PLA) cells, 106
Proinflammatory cytokines, 200
Proliferation, 124
Pure dense fibrous connective tissue, 192

R
Rabbit corneal endothelial cells, 285
"Race to the surface" theory, 239
Rat subcutaneous implantation model, 148
Reactive oxygen species (ROS), 175
Receptor activator of NF-κB ligand (RANKL), 224–225

Receptor activator of nuclear factor kappa-B ligand (RANKL), 239
Reflection pulleys, 193
"Regenerative engineering.", 311
Regenerative medicine, 311
  biomaterials for, 310–315
Repair and regeneration (RR) process, 47–48
Residual processing chemicals, 171
Retention/engraftment, 127–129
Rheumatoid arthritis, 109
Rhodamine-labeled Pluronic/chitosan nanocapsules, 357–358
RUNX2, 229–231

S
Scaffold decoration, 31–36, 32f
Selective laser melting (SLM), 435–436, 442t
Selective laser sintering (SLS), 435–436
Semisynthetic gelatine methacrylate (GelMA), 282
Sheathed tendons, 194
Silk, 280–281
Silk fibroin, 28
Simple limbal epithelial transplantation (SLET), 272
Simulated body fluid (SBF), 3–4
Sintered calcium phosphate ceramics, 247–248
Sintered ceramics
  integration of, 252
Skin, 466
Skin epidermal stem cells, 277
Small-diameter vascular grafts
  background, 79–80
  cardiovascular diseases (CVDs), 80
  cellular interactions, 88–94
    cell microintegration, 91–93
    core-shell method, 93
    stem cell studies, 90–91
    in vitro studies, 88–90
    in vivo studies, 93–94
  clinical significance, 80
  emerging perspectives, 94–96
  surface modification approach, 85–87
    biological modification, 87
    chemical/physical modification, 86
    click chemistry, 86–87
    cold plasma–based techniques, 86

Small-diameter vascular grafts (*Continued*)
  tissue engineering approach, 81–85
    extracellular matrix, 84–85
    extracellular matrix (ECM), 84
    materials and material properties, 82–83
    native blood vessels, 83
    scaffold fabrication, 81–82
    target architecture, 83
Sodium alginate (alginate), 462
Sodium dodecyl sulfate (SDS), 136–137
Soft-tissue augmentation material, 55–56
Soft tissue implants, 419
Stainless steel (316L) implants, 439–441
Stem cell–based approaches to tendon healing, 207–208
  adipose-derived mesenchymal stem cells (ADSCs), 209
  bone marrow–derived mesenchymal stem cells (BMSCs), 208
  embryonic stem cells (ESCs), 209
  induced pluripotent stem cells (iPSCs), 209–210
  tendon-derived stem cells (TDSCs), 210
Stem cell–based therapeutic approaches, 264
  biomaterials in corneal reconstruction, 278–285
  cell-based approaches—nonlimbal sources, 273
    cultivated oral mucosal epithelial transplantation (COMET), 273–274
    dental pulp stem cells, 276
    hair follicles stem cells (HF-SCs), 275–276
    human embryonic stem cells, 277
    induced pluripotent stem cells (IPSCs), 277–278
    mesenchymal stem cells (MSCs), 274–275
    skin epidermal stem cells, 277
  cornea and corneal layers, 264–265
  corneal blindness and current therapies, 268–269
    allogeneic limbal stem cell transplant, 271
    amniotic membrane transplantation, 270
    autologous conjunctival limbal transplant, 270–271
    corneal stromal stem cells (CSSCs), 272–273
    cultivated limbal epithelial transplantation, 271–272
    simple limbal epithelial transplantation (SLET), 272
  corneal epithelial homeostasis, 266
  corneal stem cell niche, 267
  limbal deficiency conditions, 268
  limbal epithelial stem cells and its characteristics, 266
  limbal stem cell markers, 267–268
  translational and clinical perspective, 286
Stem cell–derived exosomes and small vesicles, 132
Stem cell–driven tissue engineering
  adhesive peptides, 31–32
  adipose tissue–derived stem cells (ADMSCs), 28
  adult stem cells (ASCs), 20–21, 21t
  biomaterials, 26–29
  chemokine-decorated scaffolds, 34–36
  definition, 21–23
  dopamine-releasing cells, 28
  embryonic stem cells (ESCs), 20–21
  end-stage organ failure, 19–20
  extracellular matrix (ECM), 21–22
  extraembryonic stem cells (EESCs), 21–22
  future directions, 36–37
  mesenchymal stem cells (MSCs)
    nutshell, 23, 24f
    wound healing, 23–26, 25f
  pluripotent stem cells (iPSCs), 21–22
  pore size, 31t
  scaffold decoration, 31–36, 32f
  scaffold patterns, 29–37, 30f
  silk fibroin, 28
  surface coating, 33
  systematic evolution of ligands by exponential enrichment (SELEX), 33–34
  tissue engineering (TE), 20
Stem cell factor (SCF), 35
Stem cells, 102
  banking, 111–112
  differentiation, 102–103
  therapies, 125–129, 126f, 208
Stereolithography (SLA), 438
Sterility assurance level (SAL), 174

Index

Stimuli-responsive nanoparticles, 342−343
Stimuli-sensitive systems, 342−343
Storage/shipping stability, 129
Stress relaxation, 195
Subcutaneous injections of fillers, 48−49
Subgingival plaque flora, 232
Superparamagnetic iron oxide nanoparticles (SPIONs), 356−359
Superparamagnetic nanoparticles, 355−356
Surface coating, 33
Surface-engineered biomaterials for tissue engineering, 313−314
Surface modifications
  approach, 85−87
    biological modification, 87
    chemical/physical modification, 86
    click chemistry, 86−87
    cold plasma−based techniques, 86
  biocompatibility of materials, 301−303
Surgeon proficiency, 417
Surgical errors, 417−418
Synovial sheaths, 193
Synthasome X-Repair, 212
Synthetically engineered constructs, 212
Synthetic biomaterials, 282−285
  poly(2-hydroxyethyl methacrylate), 282
  poly(lactide-co-glycolide), 283
  polyethylene glycol diacrylate (PEGDA), 283
  poly N-isopropylacrylamide-co-glycidylmethacrylate, 285
  polyvinyl alcohol, 282
  thermoresponsive polymers, 283−284
Synthetic bone graft substitutes, 247−251, 257
Synthetic calcium phosphates, 337−338
Synthetic OPF-based biomaterials, 212
Synthetic polymers, 283, 314−315, 463
Systematic evolution of ligands by exponential enrichment (SELEX), 33−34
Systemic lupus erythematosus (SLE), 109

T
Targeted drug delivery, 333−336
  calcium phosphate nanoparticles, 337
    cancer chemotherapy applications, 346−347
    gene delivery applications, 344−346
    oral insulin delivery applications, 338−340
    theranostic applications, 340−343
    tissue engineering applications, 347−349
  gold nanoparticles, 349−350
    cancer chemotherapy applications, 351−354
    gene delivery applications, 354−355
  iron oxide nanoparticles, 355−356
    cancer therapy applications, 356−358
    gene delivery applications, 358−359
    general drug delivery and targeting, 361−362
    tissue engineering applications, 359−361
Tendon and tissue regeneration
  biologic and synthetic scaffolds in tendon healing, 210−212
  biomechanical properties, 195−196
  blood supply, 194
  bone, 193
  calcified fibrocartilage, 193
  complex three-dimensional structure, 190−191
  effects of aging, 197−198
  effects of exercise, 198−199
  effects of immobilization, 199
  gene transfer in tendon healing, 213−214
  impacting factors, 196−197
  internal architecture, 189−190
  mechanisms of healing, 201−202
  pure dense fibrous connective tissue, 192
  stem cell−based approaches to tendon healing, 207−208
    adipose-derived mesenchymal stem cells (ADSCs), 209
    bone marrow−derived mesenchymal stem cells (BMSCs), 208
    embryonic stem cells (ESCs), 209
    induced pluripotent stem cells (iPSCs), 209−210
    tendon-derived stem cells (TDSCs), 210
  supporting structures, 193−194
  surgical intervention, 202
  tendon cells and composition, 188−189
  tendon healing, 200−201
  tendon injury, 199−200
  tendon regeneration, 202

Tendon and tissue regeneration (*Continued*)
  tendon to bone insertion, 191–192
  tidemark, 192
  types of injury, 200
  uncalcified fibrocartilage, 192
  utilization of growth factors in tendon healing, 203–207
Tendon bursae, 194
Tendon cells and composition, 188–189
Tendon-derived stem cells (TDSCs), 210
Tendon healing, 200–201
  biologic and synthetic scaffolds in, 210–212
  gene transfer in, 213–214
  stem cell–based approaches to, 207–208
    adipose-derived mesenchymal stem cells (ADSCs), 209
    bone marrow–derived mesenchymal stem cells (BMSCs), 208
    embryonic stem cells (ESCs), 209
    induced pluripotent stem cells (iPSCs), 209–210
    tendon-derived stem cells (TDSCs), 210
  utilization of growth factors in, 203–207
Tendon injury, 199–200
Tendon regeneration, 202
Tendon stem cell activity, 197
Tendon to bone insertion, 191–192
TGF-β signaling, 229
Theranostic applications, 340–343
"Theranostic"-based systems, 340–341
Thermoresponsive polymers, 283–284
Thiomers, 387
Three-dimensional printing (3DP), 434
  bioceramic scaffolds using, 446–455
  metallic implants via, 438–446
  processes, 434–435, 435t
Three-dimensional (3D) scaffolds, 312
Thrombus formation, 299–300
Ti6Al4V alloy, 441–442
Ti and its alloys, implants with, 441–442
Ti-based lattice structure fabrication, 443–444
Tissue-based constructs, 211–212
Tissue-based products, 145
  acellular tissue products, 147–149
  chemically cross-linked tissue products, 158–159
  host response to tissue products, 171–173

  risk management of, 176–177
  sterilization of, 174–176
  tissue-derived products, 159
Tissue culture polystyrene dishes (TCPS), 284
Tissue-derived products, 159, 165t–170t, 177
  sterilization of, 174–176
Tissue-engineered cardiac valve, 133–136
Tissue-engineered heart valves, 136
Tissue-engineered scaffolds for bone repair, 317–318
Tissue-engineered vascular grafts (TEVGs), 359–361
Tissue engineering (TE), 20, 130–131, 455–457
  applications, 347–349, 359–361
  cell sources for, 320–321
  approach, 81–85, 348–349
    extracellular matrix, 84–85
    extracellular matrix (ECM), 84
    materials and material properties, 82–83
    native blood vessels, 83
    scaffold fabrication, 81–82
    target architecture, 83
  biomaterials for, 310–315
  combination devices with, 321–322
  devices, 9–10
  scaffolds, 318–319
  surface-engineered biomaterials for, 313–314
Tissue morphogenesis, 320
Tissue necrosis factor (TNF), 352
Tissue response to materials, 298
Tissue-specific stem cells, 125
Tissue targeting, 129
Traditional animal models, 408–409
Transforming growth factor beta (TGF-β), 203–204
Translational and clinical perspective, 286
Transmission electron microscope (TEM), 340f
Transnational Alliance for Regenerative Therapies in Cardiovascular Syndromes (TACTICS), 121–122
Transparent silk hydrogels, 280–281
Tricalcium phosphate (TCP), 248, 448–449
Tumorigenicity, 127

Type I diabetes, 110, 338–339
Type II collagen (CII), 109
Type V collagen, 188–189

U
Ultrathin AM (UAM), 279–280
Ultraviolet irradiation, 175
Umbilical cord blood (UCB)–derived MSCs, 110
Umbilical cord blood stem cells, 125
Umbilical cord–derived mesenchymal stem cells, 105
Uncalcified fibrocartilage, 192
Un–cross-linked decellularized xenograft, 172
Unsheathed tendons, 194

V
Valves, 133–136
Valvular heart disease (VHD), 133
Valvular tissue engineering, 137
Vascular biology, 137
Vascular endothelial growth factor (VEGF), 206, 320
Vascularization, 4
Vascular tissue engineering, 320
Vitek temporomandibular joint (TMJ) implant, 410

W
Wear, 413–416
Wharton's jelly MSCs (WJMSCs), 275
White adipose tissue (WAT), 51

X
Xenobiotic culture system, 271–272
Xenogeneic grafts, 419
Xenogenic-free culture methods, 286
Xenografts, 211

Y
Young's modulus, 196–198

Z
Zinc calcium phosphate nanoparticles, 339–340

CPI Antony Rowe
Eastbourne, UK
September 26, 2019